Applying Landscape Ecology in Biological Conservation

Springer
New York
Berlin
Heidelberg
Barcelona
Hong Kong
London
Milan
Paris
Singapore
Tokyo

Kevin J. Gutzwiller
Editor

Applying Landscape Ecology in Biological Conservation

With a Foreword by Richard T.T. Forman

With 62 Figures, 2 in Full Color

Kevin J. Gutzwiller
Department of Biology
Baylor University
P.O. Box 97388
Waco, TX 76798-7388
USA
kevin_gutzwiller@baylor.edu

Library of Congress Cataloging-in-Publication Data
Applying landscape ecology in biological conservation/Kevin J. Gutzwiller, editor.
p. cm.
Includes bibliographical references (p.).
ISBN 0-387-98653-7 (hc : alk paper) ISBN 0-387-95322-1 (sc : alk. paper)
1. Landscape ecology. 2. Nature conservation. I. Gutzwiller, Kevin J.
QH541.15.L35 A66 2002
577—dc21 2001031432

Printed on acid-free paper.

Production managed by Timothy Taylor; manufacturing supervised by Jerome Basma.
Composition by Impressions Book and Journal Services, Inc., Madison, WI.
Printed and bound by Maple-Vail Book Manufacturing Group, York, PA.
Printed in the United States of America.

9 8 7 6 5 4 3 2 1

ISBN 0-387-98653-7 SPIN 10696714 (hardcover)
ISBN 0-387-95322-1 SPIN 10841822 (softcover)

Springer-Verlag New York Berlin Heidelberg
A member of BertelsmannSpringer Science+Business Media GmbH

To those who are waging the battle
to conserve native biota

Foreword

I recently visited one of the world's leading ecological research projects. We drove forever through beautiful boring wheat fields. Woods on the horizon occasionally became near misses en route, and shrubby areas with salt-encrusted soil suggested that something was wrong. We looked in vain for roadkilled wallabies and venomous snakes. Fencelines seemingly miles apart tried to give scale to the land, but mainly we faced an endless foreground of dreary verges.

Upon entering some wonderful woods, my ecological juices began to flow. Intriguing plants covered the place, some familiar, some bizarre. Babblers and honeyeaters sang novel songs, and roos had spread scats about. We explored the neighboring area and became landscape detectives. Then the experience sank in. I was in virtually the only long-term, multidimensional, landscape-wide ecological study area on earth. The investigators had measured and knew "everything" about this landscape.

Something jarring then appeared next to a big woods. Several curving rows of young trees, perhaps head high, snaked off toward a distant woodlot. What's that? I queried. The farmer planted those eucalypts a couple of years ago. What for? To connect the small woods to the large woods for the movement of birds. The farmer? For bird movement? What's the farmer know, or care, about birds? Ever since we started the research here we have kept the farmers up to date on our results. We send them the lists of birds, how many Western Greys move along their fencelines, the soil saline levels, and interesting plants discovered. We often hear that as soon as the lists arrive, family members get on the phone and not only take pride in their own species, but also discuss why they don't have a particular species that their neighbor has. Did you tell this farmer where to plant the corridor? No, the farmers talked it over and chose the route. Who decided to make it 80 meters wide? They did. Why does it squiggle across the land? In addition to facilitating species movement, they hope it will reduce soil erosion and salinity problems. While driving away, I thought about how much had been learned from this research, how the researchers had educated the landowners, and how biological conservation had gained from both actions. Landscape ecology had enhanced, not only biological conservation, but also soil conservation and local culture in its caring for the land. Another little epiphany, I noted.

We soon stopped at an old brick building in a town center to find the local Land Care agent. Land care? What's that? The state provides matching funds to support local efforts that channel state-of-the-knowledge information on ecology, soil, farming, water resources, gardening, and hunting to landowners. The Land Care agent then welcomed us, and I could tell from the ready smile, the sense of expertise, and the room bulging with maps, GIS images, bird photos, and more, that she was an effective catalytic agent. What do you really do? She said she could expound on all the children's groups, adult volunteer projects, and packets of information provided to very independent landowners. But first she had to tell me about these landscape ecologists. They regularly stop by with the most amazing useful information, often in a form that is readily translated or passed on. They provide the only broad-scale information that integrates diverse components, and they have the fewest axes to grind. Native species are already benefitting, and both nature and people should gain from a raft of ongoing projects on the ground. I left the brick building feeling that I had witnessed a ray of hope for the future. A concrete basis for optimism.

Since then, the need for a book about applying landscape ecology in biological conservation has accelerated. So, what a pleasure to find the book in your hand by editor Kevin Gutzwiller and 46 perceptive authors! It is a gold mine of insight, providing a remarkable snapshot of the state of our knowledge. The synthesis highlights how far we have come in some 15 years, and yet how much remains unknown and undone.

The genie, landscape ecology, has emerged from Alladin's lamp and has expanded so rapidly that it now permeates fields from forestry to urban and regional planning, wildlife management, landscape architecture, geography, and biological conservation. Road ecology appears to be the sleeping giant of biological conservation, so transportation is next, and any other activity dealing with the land cannot be far behind. That is not surprising, because landscape ecology addresses hydrologic flows, water quality, erosion, sediment deposition, human activities and movements, in addition to animals, plants, and natural communities.

But, as the book strongly indicates, sustaining species and natural communities requires more than landscape ecology. Other useful subjects in biological conservation and conservation biology include genetics, captive breeding, gene flow, physiological ecology of rare species, population dynamics, competition and predation, and mapping of small populations. At the other end of the scale, science, as one slice of human understanding, is but one of the broad keys to success in biological conservation.

The land mosaic paradigm has been especially useful in landscape ecology. Indeed, the patch-corridor-matrix model originally provided conceptual simplicity and a handle to attract scientists to study landscapes. It facilitated hypothesis testing, experimental layout, modeling, comparison of landscapes world-wide, applications of many sorts, and communication among disciplines and the public. It readily linked with function and change and worked equally well from city to wilderness. This handy spatial model can be infinitely subdivided into size-number-type-arrangement categories and should be useful in linking hierarchy and scale to feedback systems in landscapes.

Biological conservation tends to focus on protected areas, yet sustaining most species and natural communities mainly depends on what happens in the unprotected matrix. We can now move beyond the stage of patches-in-an-inhospitable-matrix, source-and-sink, and corridor-connecting-two-patches. The matrix is always heterogeneous, a patchwork with multiple sources, barriers, conduits, attractors, repellers, sinks, avoidance spots, and comfort places. In fact, why couldn't the patch-corridor-matrix model be enriched or even replaced by a functional mosaic model, in which the landscape is composed of such places portraying movements and flows?

In addition, a gaping research frontier exists between patch–corridor differences and landscape-type differences. A simple but universal, spatially focused neighborhood model, for example, could enormously stimulate research, understanding, and applications. Consider a "matrix neighborhood," simply a patch surrounded by matrix, a bog surrounded by spruce forest. A "multipatch neighborhood" has patches in a matrix, a "network neighborhood" has attached corridors in a matrix, and a "patchwork neighborhood" has only adjoining patches. Such a simplified approach could conceptually mimic all local areas in a landscape, would be a basis for studying and portraying internal flows and movements, and would be a useful foundation for planning, design, management, and conservation. The neighborhood models could be expanded in many directions to, for example, have a focal patch in each neighborhood, vary the number-size-type-arrangement of patches, highlight adjacency effects, or link movements of water, sediment, and wildlife with people, vehicle, and money flows in a neighborhood.

Landscapes and regions represent the confluence between natural patterns—processes and where people live. This dynamic confluence represents an area of growing intellectual ferment, which promises exciting discoveries plus solutions to visible ecological and societal problems around us. I suspect that the twenty-aughts will be the decade of landscape ecology, leading to visionary land mosaics in which nature and people both thrive.

This book, *Applying Landscape Ecology in Biological Conservation,* uncovers a cornucopia of new examples and case studies. Several unusual attributes also appear. The editor has instilled a dependable structure in the chapters, which facilitates comparisons among subject areas. Authors present principles as refreshingly brief statements in essentially every chapter. Taken together, these provide a surprisingly valuable body of thought for applying landscape ecology. Some of the principles are expressed graphically or mathematically, but most are in clear verbal-statement form.

Many highly useful tables, plus analogous encapsulations in the text, lucidly summarize world-wide applications in each topic area. This provides a glimpse of both the mean and variance for patterns in a topic area. Each chapter ends with a surprisingly detailed discussion of research needs or voids, a gold mine for students searching for a term paper or thesis topic, as well as for researchers pondering methods and frontiers.

Biological conservation regularly contends with "the bulldozers are running" situation. How can we put a book right to work that bulges with clearly stated principles and clearly presented examples? Will landscape ecology lead to protection of all, or only most, species? Is that good or bad? How can the landscape approach help protect biological diversity through periods of war or poverty or a bad economy? Suppose the focus shifted from protecting species to designing landscapes where both nature thrives and people thrive. Would species suffer or benefit?

My major inspiration and little epiphanies came from, and still come from, nature. But a good book can be a close second. The book in your hand has benchmark attributes, promising a long useful shelf life. I found wisdom in the pages.

RICHARD T.T. FORMAN
Harvard University

Preface

Global declines in the diversity of living organisms are continuing at rates unprecedented in the history of life. Without question, human impacts on natural systems are the primary cause, and with anticipated increases in the world's human population size, additional biotic losses seem inescapable. Can the science of landscape ecology be used to help reduce biological degradation? Yes. Applications of landscape ecology in recent decades have already achieved some success in this regard. These applications have been fruitful because they have capitalized on the fact that an organism's occurrence and persistence at a local site are often influenced by conditions that exist and processes that operate at broader spatial scales. Some conservationists have used knowledge of this ecological relation between local- and broad-scale phenomena to improve the realism of the models and paradigms they use in management decisions.

Can the conservation benefits of applying landscape ecology be expanded significantly? Yes, but not without certain advances. Specifically, more of those who carry out conservation on the ground must understand landscape-ecology concepts and how they can be used. Landscape ecology must be applied more frequently across multiple scales and jurisdictions and among multiple taxa and levels of biological organization. And new knowledge must be developed about how landscape conditions affect organism movement, distribution, and persistence.

The purpose of this book is to spur these advances, by explaining pertinent landscape-ecology concepts; by describing recent applications of landscape ecology as examples of possible management, research, or planning approaches; by distilling principles for applying landscape ecology in conservation settings; by identifying knowledge gaps that prevent applications of landscape ecology; and by describing research approaches to fill those voids.

Chapter authors have written for graduate students who have training in ecology, conservation, and quantitative methods and who will become tomorrow's managers and researchers. This book is directed at managers who want to learn more about landscape ecology, how it has been and can be applied, and what guiding principles to consider in such applications. Also targeted are researchers interested in reducing scientific constraints that limit applications of landscape ecology in biological conservation.

This volume is composed of five sections. The first of three chapters in Section I (Introduction) provides background about landscapes and landscape ecology. This chapter develops landscape-ecology concepts and explains how associated ecological phenomena may influence characteristics of natural systems that are of central significance in conservation, it discusses issues that may influence integration of landscape ecology into conservation efforts, and it briefly considers the future role of landscape ecology in biological conservation. The remaining two chapters in Section I address general concepts and principles that conservationists should be aware of when trying to apply landscape ecology, emerging ideas and issues that are likely to become increasingly important in the years ahead, basic ecological connections between landscape ecology and conservation, and general approaches for advancing applications of landscape ecology in biological conservation.

Building on these general foundations, chapters in Sections II (Multiple Scales, Connectivity, and Organism Movement), III (Landscape Change), and IV (Conservation Planning) provide specific treatment of various topics. In the context of their particular subject matter, these chapters consider landscape-related concepts and principles and how they have been applied (if at all) in conservation. Based on actual applications, management experience, ecological science, and other sources of information, authors of these chapters also derive principles for applying landscape ecology. These principles will undergo refinement as new information about application successes and failures accumulates; the conservation impact of these principles will be maximized by tailoring them to fit a given conservation problem. Thus, these are working principles that will serve as important starting points and guideposts for conservationists.

Chapters in Sections II, III, and IV also reveal major theoretical and empirical knowledge gaps that thwart landscape-ecology applications, and they provide research advice for filling these voids. In the absence of field data and sufficient time to gather them, conservation decisions often have to be based on theoretical considerations. Identifying and filling theoretical voids may improve the effectiveness of decision-making under these circumstances. When one considers the focusing effects that theory (e.g., Theory of Island Biogeography) has had on conservation-related analyses (e.g., estimation of acceptable size and isolation of reserves), developed theory also is invaluable for steering conservation research.

Mounting losses of biological diversity are generating pressure to apply landscape-ecology ideas, many of whose ecological validity remains unsubstantiated. Although crisis situations may justify limited application of incompletely tested ideas, possible consequences of such an approach include misuse of precious management resources, ineffective conservation programs, and the creation of new conservation problems. Identifying and closing major empirical knowledge gaps will improve understanding of organism-landscape relations, reduce dogma, and help conservationists expand science-based applications of landscape ecology.

In some subject areas, theoretical and empirical knowledge gaps are closely intertwined; the research approaches to fill these two types of gaps also may be rather inseparable. This overlap can make it difficult to understand the nature of voids and therefore what research steps should be used to eliminate them. To reduce such confusion, chapters in Sections II, III, and IV distinguish whenever feasible between theoretical and empirical voids and between approaches for theoretical and empirical research.

The single chapter in Section V (Synthesis and Conclusions) identifies principles for applying landscape ecology that have common relevance for a variety of conservation subject areas. This chapter exposes pervasive scientific and nonscientific constraints that limit applications, it discusses general means for reducing these constraints, and it considers the prospects for increasing conservation applications of landscape ecology.

With emphasis on scientific and ecological issues, this volume addresses many current and emerging key interfaces between landscape ecology and conservation. Social, political, and economic factors are clearly relevant to landscape-ecology applications, but they lie outside of the book's ecological focus and are therefore explicitly treated only briefly in a few places. Reflecting the subject matter of conservation research and experience to date, information about vertebrates or higher plants dominates the chapters. Yet, many of the authors' principles for applying landscape ecology and their suggested research approaches for advancing such applications should be valuable for conserving other taxa as well.

By broadening the base of practitioners who are knowledgeable about landscape ecology and its application, and by propelling development of new landscape-ecology knowledge that can be used by managers, my hope is that this book will stimulate additional science-based applications of landscape ecology in biological conservation. Earth's remaining biotic resources deserve no less.

KEVIN J. GUTZWILLER
Baylor University

Acknowledgments

I thank all contributors for their willingness to consider and implement my advice, for their perseverance in the face of personal challenges and busy schedules, and for the wisdom in their writings. Richard J. Hobbs kindly supplied advice about the Foreword. Numerous external reviewers, listed specifically by contributors, provided invaluable suggestions for improving chapter content and clarity. To my wife and son, Pam and Robert Gutzwiller, I express my heartfelt thanks for their perpetual patience and encouragement during the course of my efforts. I am deeply grateful to Jerome and Rita Gutzwiller, my parents, for their life-long support that made my work on this volume possible. And I greatly appreciate assistance received throughout this project from Baylor University.

KEVIN J. GUTZWILLER
Baylor University

Contents

Contributors

H. Resit Akçakaya	Applied Biomathematics 100 North Country Road Setauket, NY 11733, USA E-mail: resit@ramas.com
Peter August	Department of Natural Resources Science University of Rhode Island Kingston, RI 02881, USA E-mail: pete@edc.uri.edu
Richard J. Baker	Natural Heritage and Nongame Research Program Minnesota Department of Natural Resources 500 Lafayette Road, Box 25 St. Paul, MN 55155, USA E-mail: richard.baker@dnr.state.mn.us
Hans Baveco	ALTERRA Green World Research Department of Landscape Ecology P. O. Box 47 NL 6700 AA Wageningen The Netherlands E-mail: j.m.baveco@alterra.wag-ur.nl

DANIEL BERT
Ottawa-Carleton Institute of Biology
Carleton University
1125 Colonel By Drive
Ottawa, Ontario K1S 5B6, Canada
E-mail: dbert@science.uottawa.ca

MICHAEL BEVERS
USDA Forest Service
Rocky Mountain Research Station
2150 Centre Avenue, Building A
Fort Collins, CO 80526, USA
E-mail: mbevers@fs.fed.us

DAVID R. BREININGER
DYN-2
Dynamac Corporation
NASA Biological Sciences Branch
John F. Kennedy Space Center, FL 32899, USA
E-mail: BreinDR@kscems.ksc.nasa.gov

MARY T. BREMIGAN
Department of Fisheries and Wildlife
Michigan State University
East Lansing, MI 48824 ,USA
E-mail: bremigan@msu.edu

HUGH B. BRITTEN
Department of Biology
University of South Dakota
Vermillion, SD 57069, USA
E-mail: hbritten@usd.edu

MARK A. BURGMAN
School of Botany
University of Melbourne
Parkville 3052
Australia
E-mail: m.burgman@botany.unimelb.edu.au

LAURIE W. CARR
Department of Biology
Carleton University
Ottawa, Ontario K1S 5B6, Canada
E-mail: lcarr@ccs.carleton.ca

VIRGINIA H. DALE
Environmental Sciences Division
Oak Ridge National Laboratory
P. O. Box 2008
Oak Ridge, TN 37831, USA
E-mail: vhd@ornl.gov

NANCY DIAZ
Pacific Northwest Research Station
P. O. Box 3890
Portland, OR 97208, USA
E-mail: ndiaz@fs.fed.us

LENORE FAHRIG
Department of Biology
Carleton University
Ottawa, Ontario K1S 5B6
Canada
E-mail: lfahrig@ccs.carleton.ca

CURTIS H. FLATHER
USDA Forest Service
Rocky Mountain Research Station
2150 Centre Avenue, Building A
Fort Collins, CO 80526, USA
E-mail: cflather@fs.fed.us

RICHARD T. T. FORMAN
Graduate School of Design
Harvard University
Cambridge, MA 02138, USA

KATHRYN FREEMARK
Environment Canada
100 Gamelin Boulevard
Hull, Quebec K1A OH3
Canada
E-mail: kathryn.freemark@ec.gc.ca

CARLA J. GRASHOF-BOKDAM
ALTERRA
Green World Research
Department of Landscape Ecology
P. O. Box 47
NL 6700 AA Wageningen
The Netherlands
E-mail: c.j.grashof-bokdam@alterra.wag-ur.nl

ERIC J. GUSTAFSON
North Central Research Station
5985 Highway K
Rhinelander, WI 54501, USA
E-mail: egustafson@fs.fed.us

KEVIN J. GUTZWILLER
Department of Biology
Baylor University
Waco, TX 76798, USA
E-mail: kevin_gutzwiller@baylor.edu

RICHARD J. HOBBS
School of Environmental Science
Murdoch University
Murdoch WA 6150, Australia
E-mail: rhobbs@essun1.murdoch.edu.au

JOHN HOF
USDA Forest Service
Rocky Mountain Research Station
2150 Centre Avenue, Building A
Fort Collins, CO 80526, USA
E-mail: jhof@fs.fed.us

ROBERT M. HUGHES
Dynamac Corporation
200 SW 35th Street
Covallis, OR 97333, USA
E-mail: hughesb@mercury.cor.epa.gov

CAROLYN T. HUNSAKER
USDA Forest Service
2081 East Sierra Avenue
Fresno, CA 93710, USA
E-mail: chunsaker@fs.fed.us

LOUIS IVERSON
USDA Forest Service
359 Main Road
Delaware, OH 43015, USA
E-mail: liverson@fs.fed.us

RICHARD L. KNIGHT
Department of Fishery and Wildlife Biology
Colorado State University
Fort Collins, CO 80523, USA
E-mail: knight@cnr.colostate.edu

ROBERT J. LAMBECK
Greening Australia (WA)
10–12 The Terrace, Fremantle
WA 6160, Australia
E-mail: rlambeck@gawa.comdek.net.au

PETER B. LANDRES
Aldo Leopold Wilderness Research Institute
P. O. Box 8089
Missoula, MT 59807, USA
E-mail: plandres@fs.fed.us

CAROLYN G. MAHAN
114 Eiche Library
Pennsylvania State University
Altoona, PA 16601, USA
E-mail: cgm2@psu.edu

Kees (C. J.) Nagelkerke	Institute for Biodiversity and Ecosystem Dynamics University of Amsterdam Kruislaan 320, 1098 SM Amsterdam The Netherlands E-mail: nagelkerke@bio.uva.nl
Barry R. Noon	Department of Fishery and Wildlife Biology Colorado State University Fort Collins, CO 80523, USA E-mail: brnoon@cnr.colostate.edu
Jarunee Nugranad	Remote Sensing Division National Research Council of Thailand 196 Paholyothin Road Chatuchak, Bangkok 10900, Thailand E-mail: joy@hammerhead.nrct.go.th
Michael A. O'Connell	The Nature Conservancy 1400 Quail Street, Suite 130 Newport Beach, CA 92660, USA E-mail: Moconnell@tnc.org
Paul Opdam	ALTERRA Green World Research P. O. Box 47 NL 6700 AA Wageningen The Netherlands E-mail: p.f.m.opdam@alterra.wag-ur.nl
Shealagh E. Pope	Department of Biology Carleton University Ottawa, Ontario K1S 5B6 Canada E-mail: sepope@consecol.org
D. Graham Roy	Department of Botany Miami University Oxford, OH 45056, USA
Paul W. Seelbach	Institute for Fisheries Research Michigan Department of Natural Resources 212 Museums Annex 1109 North University Ann Arbor, MI 48109, USA E-mail: seelbach@umich.edu

PATRICIA A. SORANNO
Department of Fisheries and Wildlife
Michigan State University
East Lansing, MI 48824, USA
E-mail: soranno@msu.edu

KAREN VAN DE WOLFSHAAR
Group Mathematical and Methods
Dreijenlaan 4
6703 HA Wageningen
The Netherlands
E-mail: Karen.vandewolfshaar@95.student.wau.nl

FRANK VAN DEN BOSCH
Department of Statistics
IACR-Rothamsted
Harpenden, Hertfordshire AL5 2JQ
United Kingdom
E-mail: Frank.vandenbosch@bbsrc.ac.uk

JOHN L. VANKAT
Department of Botany
Miami University
Oxford, OH 45056, USA
E-mail: vankatjl@muohio.edu

JANA VERBOOM
ALTERRA
Green World Research
Department of Landscape Ecology
P. O. Box 47
NL 6700 AA Wageningen
The Netherlands
E-mail: j.verboom-vasiljev@alterra.wag-ur.nl

MARC-ANDRÉ VILLARD
Département de Biologie
Université de Moncton
Moncton, New Brunswick E1A 3E9
Canada
E-mail: villarm@umoncton.ca

CLAIRE C. VOS
ALTERRA
Green World Research
Department of Landscape Ecology
P. O. Box 47
NL 6700 AA Wageningen
The Netherlands
E-mail: c.c.vos@alterra.wag-ur.nl

JOHN A. WIENS	Department of Biology Colorado State University Fort Collins, CO 80523, USA E-mail: jaws@lamar.colostate.edu and National Center for Ecological Analysis and Synthesis 735 State Street, Suite 300 University of California Santa Barbara, CA 93101, USA
MICHAEL J. WILEY	School of Natural Resources and Environment University of Michigan Ann Arbor, MI 48109, USA E-mail: mjwiley@umich.edu
KIMBERLY A. WITH	Division of Biology Kansas State University Manhattan, KS 66506, USA E-mail: kwith@ksu.edu
RICHARD H. YAHNER	Graduate School 114 Kern Building Pennsylvania State University University Park, PA 16802, USA E-mail: rhy@psu.edu

Section I

Introduction

1

Central Concepts and Issues of Landscape Ecology

John A. Wiens

". . . the grand challenge is to forge a conceptual and theoretical synthesis of spatial ecology, embracing in one manner or another individual responses and population dynamics, and explaining patterns in species abundance caused by complex landscapes and patterns driven by complex dynamics" (Hanski 1999:264).

1.1 Introduction

The objective of biological conservation is the long-term maintenance of populations or species or, more broadly, of the Earth's biodiversity. Many of the threats that elicit conservation concern result in one way or another from human land use. Population sizes may become precariously small when suitable habitat is lost or becomes spatially fragmented, increasing the likelihood of extinction. Changes in land cover may affect interactions between predator and prey or parasite and host populations. The spread of invasive or exotic species, disease, or disturbances such as fire may be enhanced by shifts in the distribution of natural, agricultural, or urbanized areas. The infusion of pollutants into aquatic ecosystems from terrestrial sources such as agriculture may be enhanced or reduced by the characteristics of the landscape between source and end point. Virtually all conservation issues are ultimately land-use issues.

Landscape ecology deals with the causes and consequences of the spatial composition and configuration of landscape mosaics. Because changes in land use alter landscape composition and configuration, landscape ecology and biological conservation are obviously closely linked. Both landscape ecology and conservation biology are relatively young disciplines, however, so this conceptual marriage has yet to be fully consummated. My objectives in this chapter are to provide some general background about landscapes and landscape ecology (Section 1.2), to develop the emerging concepts and principles of landscape ecology and show how they may affect the features of ecological systems that are important to conservation efforts or management (Section 1.3), to touch briefly on some issues that may affect the integration of landscape ecology into biological conservation (Section 1.4), and to offer some concluding, philosophical comments about the future role of landscape ecology in conservation (Section 1.5). Many of these points will be developed in more detail in the remainder of this volume.

1.2 General Background

1.2.1 What Are Landscapes and What Is Landscape Ecology?

Although there are many definitions of "landscape" in the geographical and ecological literature (as well as in various dictionaries), all are characterized by two themes: *landscapes* are composed of multiple elements (or "patches"), and the variety of these elements creates heterogeneity within an area. From a conservation perspective, a landscape contains multiple habitats, vegetation types, or land uses. There is more to it than this, however. The elements of a landscape have a particular spatial configuration, which can be portrayed as a map or (more fashionably now) as a geographic information system (GIS) image. It is the spatial relationships among landscape elements as much as their variety that make landscapes important, for these relationships can affect the interactions among the elements in a mosaic as well as what goes on within individual patches.

Agreement about the characteristics of "landscapes" generally stops here, however. It is commonplace, for example, to find references to the "landscape level" in the ecological and conservation literature. Here, the landscape is viewed as a level in an ecological hierarchy: "landscape" is more inclusive than an ecosystem, yet it is nested within a biome—it is a collection of ecosystems (e.g., Forman and Godron 1986; Noss 1991). This view pervades the use of "landscape" in resource management, in which actions at the landscape level are advocated because they encompass more variety than do actions focused on individual habitats, land-cover types, or administrative units such as reserves or parks. Others refer to the "landscape scale," by which they generally mean a spatial scale of resolution that corresponds with human perceptions of their surroundings—a scale of tens of hectares to kilometers (e.g., Forman 1995). One can find "landscape" used in both of these senses throughout this book.

This emphasis on "landscape" as a level or a scale stems in part from the everyday use of the word, which carries with it both human visualizations of landscapes as well as the human desire to order phenomena hierarchically. It also reflects the historical roots of landscape ecology as a discipline. Landscape ecology began in northern and eastern Europe through a merging of holistic ecology with human geography, land-use planning, landscape architecture, sociology, and other disciplines (Naveh and Lieberman 1994; Zonneveld 1995; Wiens 1997). From its birth, then, landscape ecology carried with it a focus on interactions of humans with their environment at broad spatial scales. Although the recent growth of landscape ecology as a discipline has incorporated closer linkages with traditional (i.e., nonhuman) ecology, the utility in management of thinking of landscapes in human terms has perpetuated and reinforced this anthropocentric perspective. After all, decisions about land management or land-use policy are made with reference to the scales of human activities and the hierarchical structure of administrative bodies.

Despite this, there are both logical and operational reasons for arguing that viewing "landscape" as a level or a scale is wrong, or at least unnecessarily restrictive. King (1997, 1999), Allen (1998), and O'Neill and King (1998) have discussed the logical arguments. Briefly, they argue that "level" in an organizational hierarchy must be defined on the basis of similarities in rate processes. Entities that belong to the same level operate at similar rates and therefore can interact with one another, whereas components with different rate structures cannot interact but can only constrain the dynamics of other levels. What is a "level" therefore depends on the scale of observation and the question the investigator asks. In a similar vein, "landscape" represents an arbitrary definition by the observer of a certain kind of object or class, whereas "scale" refers to the physical dimensions of an object or class in space and time. Specifying the class ("landscape") does not necessarily specify a scale, because the way the class is defined may differ among investigators depending on the question or perspective. References to "landscape level" or "landscape scale" therefore mix terms that are logically derived in different ways, and this can lead to imprecision in both meaning and measurement.

Operationally, by restricting consideration of landscape properties and their ecological consequences to certain levels or scales, one essentially denies the relevance of landscape structure to other levels or scales. This can lead to an unintended acceptance of the assumption that neither heterogeneity nor scale is important at those levels or scales. But heterogeneity and scale dependency can affect ecological patterns and processes at the levels of individuals, populations, or communities as well as ecosystems or biomes, and they are expressed on scales covering a few centimeters or meters as well as hectares or kilometers. It is important to realize that what makes landscapes interesting and important to ecology and conservation is not only the emphasis on broad scales or more inclusive levels of organization, but also on how the spatial configuration and dynamics of landscape mosaics influence predation, dispersal, population dynamics, nutrient distribution, or disturbance spread—indeed, virtually all ecological phenomena.

Contained within these varying views about "landscapes" are the elements of three different approaches to landscape ecology, each of which implies something different about how landscape ecology may contribute to biological conservation. One approach derives directly from the European tradition, and it considers landscape ecology as a "holistic, problem-solving approach to resource management" (Barrett and Bohlen 1991). The emphasis is on integrating many aspects of human activities with their environmental consequences—a geographically based resource-management approach.

The second approach emphasizes landscape as a level or scale; in essence, it is ecology writ large. Many of the questions are those that ecologists have traditionally addressed, but they are cast in a broader hierarchical or spatial context. This approach has clear linkages to biogeography and the developing area of macroecology (Brown 1995). It is clearly relevant to regional planning and to geographically defined conservation efforts, such as those dealing with ecoregions (e.g., Ricketts et al. 1999; Poiani et al. 2000) or with the regional distribution of biodiversity "hotspots" (e.g., Reid 1998; Flather et al. 1998).

The third approach deals more explicitly with the causes and consequences of spatial patterns in the environment, with the effects of spatial pattern on ecological processes (Turner 1989; Wiens et al. 1993; Wiens 1995). In this case, the level and scale are determined by characteristics of the organisms or ecological systems of interest and the questions asked (Wiens 1989a; Haila 1999; Mac Nally 1999). The focus of this approach is on the mechanisms by which the spatial structure of the environment influences phenomena of conservation value such as populations or biodiversity. The scales on which these mechanisms are expressed (and thus the "landscape") therefore will differ for different kinds of organisms (e.g., Wiens and Milne 1989). This approach actually embodies two somewhat different perspectives: *spatial ecology,* which considers only how spatial variation in environmental factors affects ecological systems (e.g., Tilman and Kareiva 1997), and *landscape ecology,* which also considers explicit spatial relationships and locational effects.

Without denying the value of the first two approaches to biological conservation, my emphasis in this chapter will be on the third approach, especially the landscape perspective. I emphasize this approach because I believe that it provides the best way to derive insights about how the spatial texture and configuration of landscapes can influence ecological systems and their dynamics. If most conservation issues are indeed ultimately tied to human land use, the importance of such understanding should be obvious.

1.2.2 What Features Characterize Landscapes?

Saying that a landscape approach emphasizes the causes and consequences of heterogeneity or of spatial pattern serves to reinforce the ongoing paradigm shift away from viewing ecological systems as spatially homogeneous (Pickett et al. 1992; Wiens 1995). Words like "heterogeneity" or "spatial pattern," however, are too nebulous to be of much use in characterizing what we need to know about landscapes to gauge their effects. More detail is needed.

Conceptually, the components of a landscape can be partitioned into features of *composition,* the kinds of elements or patches making up a landscape; *structure,* its physical configuration; and *process,* the flows of organisms, materials, or disturbances through the mosaic (Figure 1.1a; Chapter 3). Operationally, we usually express landscapes as maps or images, which incorporate the compositional and structural aspects of landscapes but not process or (except as a time series of maps) dynamics. To derive a map requires drawing boundaries around units so their spatial distribution can be portrayed. Geographers and cartographers have wrestled with the problems of boundary determination and map classification for decades (e.g., Küchler 1974; Bailey 1996; Monmonier 1996); here, I will only note that how these decisions are made affects not only the map, but also all of the analyses and conclusions that follow from it. If vegetation cover is classified in different ways, for example, interpretations of how a wildlife species of concern

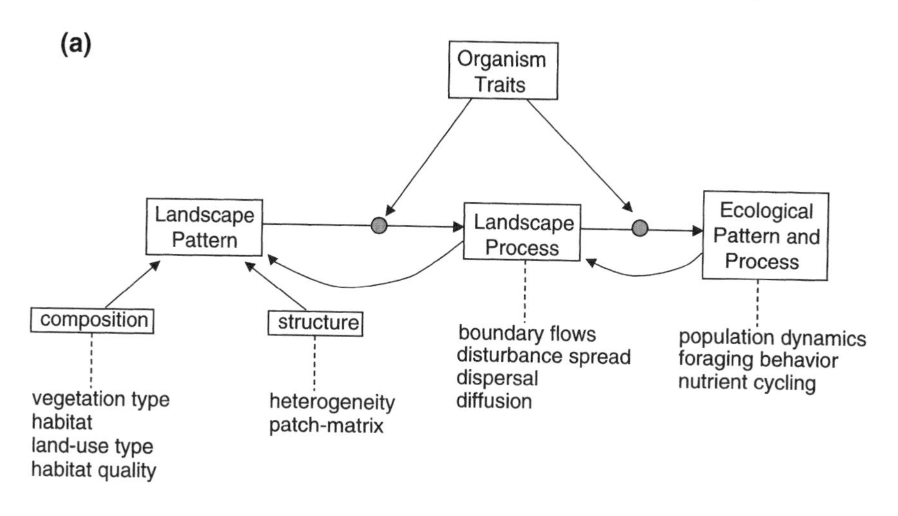

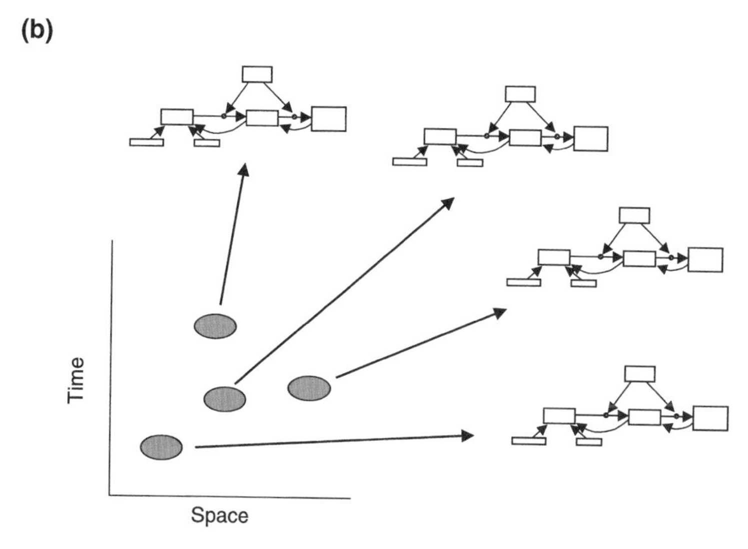

FIGURE 1.1. (a) Elements of a framework for thinking about landscape effects on ecological systems. The spatial pattern of a landscape is derived from its composition (the kinds of elements it contains) and its structure (how they are arranged in space). The spatial pattern of a landscape is translated into spatially dependent landscape processes as a consequence of the interplay between the landscape pattern and the ways in which different kinds of organisms respond to that pattern, which in turn is dictated by ecological, morphological, behavioral, and life-history traits of the organisms. The landscape pattern-process linkage, mediated again by the ecological traits of organisms, produces spatial dependencies in a variety of ecological phenomena. There are feedbacks among all of these relationships, but the most important are those from ecological patterns and processes that influence the nature of landscape processes, which is turn affect the underlying landscape patterns. (b) Because all of the components of the web of spatial interactions shown in (a) may change with changes in scale, the resulting ecological patterns and processes that we study and attempt to manage will probably differ among different space-time scaling domains (shaded ellipses).

responds to "habitat" may differ as well. At a more basic level, a classification approach can easily lead to a neglect of gradients in environmental factors by partitioning or "blocking" the variation that occurs along a gradient. This, in turn, may foster an avoidance of geostatistical analyses that explore the form of spatial correlation in ecological measures (see Fortin 1999). It is easy to regard maps or images as "truth" and to forget that any conclusions are contingent on the classification and boundary determination procedures used to produce the map (Monmonier 1996). Landscape ecologists would be well-advised to heed Austin's (1999) call for closer attention to environmental gradients and the form of species' response functions (what we used to call "niches").

Superficially, landscape pattern can be described in terms of patches, corridors, and the background matrix (Forman 1995). However, this categorization obscures some of the richness of detail that characterizes landscape mosaics, primarily because the "matrix" itself usually contains a variety of patches of different types and properties. In fact, a large array of features and measures can be derived from a map or image of a landscape (Table 1.1). Some of these measures, such as patch size or shape, nearest-neighbor distance, or perimeter:area ratio, portray features of particular patch types independently of their surroundings.

TABLE 1.1. Some measures of landscape structure. For convenience, the measures are separated into those that describe features of individual patches and those that express patterns of the entire landscape mosaic, although the distinction between the two categories is not always sharp. All of the patch-based measures can be characterized by a mean and a variance over the landscape as a whole, reflecting additional aspects of landscape structure. For additional details, see Haines-Young et al. (1993), Forman (1995), McGarigal and Marks (1995), Farina (1998), and Longley et al. (1999).

Patch measures
- Size
- Shape
- Orientation
- Perimeter
- Perimeter:Area ratio
- Context (adjacency, contrast)
- Distance (nearest neighbor, proximity)
- Corridor width, length, shape, linkage (e.g., stream order)

Mosaic measures
- Patch number
- Patch size frequency distribution
- Patch diversity (richness, evenness, dominance, similarity)
- Percent of landscape in a given patch type
- Patch dispersion (contagion)
- Edge density
- Fractal dimension (edge, area)
- Heterogeneity
- Gaps (lacunarity)
- Spatial correlation (semivariance, distance decay, anisotropy)
- Connectedness (network, lattice properties)

Others, such as adjacency or contrast, deal explicitly with what lies across the boundaries of a given patch type. Still other measures—semivariance, lacunarity, fractal dimension, patch diversity, connectedness, or various indices of heterogeneity, for example—characterize features of the mosaic as a whole. Together with spatially referenced records or inventories of ecological variables of interest (e.g., population abundance, species diversity), these measures are the raw materials that are used to assess how landscapes affect ecological phenomena.

1.3 Emerging Concepts of Landscape Ecology

The overarching principle of landscape ecology is that the spatial configuration of landscapes can have important effects on a wide variety of ecological processes. These, in turn, determine the ecological patterns that ecologists are so fond of documenting and theorizing about, and which form the foundation for most thinking in biological conservation. The particular spatial or locational arrangement of the various elements of a landscape produces ecological processes and patterns that are different from those that would emerge in a landscape with a different spatial configuration.

To relate the various quantitative measures of landscape structure and composition to ecological consequences, however, we need something more specific. I suggest that the following five concepts (I hesitate to call them "principles") can serve as a foundation for thinking about landscapes in an ecological or conservation context.

1.3.1 Landscape Elements Differ in Quality

The elements or patches in a landscape or map are distinguished from one another because they differ in some way. Traditionally, patch types are classified by differences in vegetation cover, soils, or geology, or forms of human land use, but other criteria may be used depending on the information available and one's objectives in describing or mapping a landscape. In the context of biological conservation, the focus is often on how organisms or populations are distributed in space, and the underlying premise is that the criteria used to portray the landscape relate in some way to "habitat" and, ultimately, to "habitat quality." Habitat quality, in turn, relates to the probabilities of survival and reproduction of individuals occupying a patch type—the patch-dependent fitness function (Wiens 1997). Differences in reproductive success or survival probability among patches (e.g., vegetation types) are clearly the norm for most (ultimately, all) species, but these components of fitness are difficult to document and indirect measures (especially local population density) are often used as surrogates of fitness-defined patch quality. Substantial problems are involved in using such surrogate measures, however (e.g., Van Horne 1983; Garshelis 2000), so although the notion of patch or habitat quality has clear conceptual appeal, operationally it remains difficult to implement.

Despite these difficulties, it is important to begin to think of landscapes in terms of patches or elements of differing quality rather than remaining content with descriptions of landscape patterns based on arbitrary criteria that have an unknown (but possibly remote) relationship to biological functions or processes. One cannot link spatial patterns to spatial processes using landscape measures that have little to do with process, regardless of how many people have used them in the past or how stunning they appear on a remote-sensing image or a GIS output. Understanding *why* organisms occur where they do or move as they do in a landscape requires a consideration of variations in patch quality. Of course, the quality of landscape elements is not a fixed attribute. Patch quality varies in time as resource levels change, predator or competitor abundances vary, or physiological stresses change. More germane here are the variations in patch quality that may result from the structural configuration of the mosaic in which the patches are embedded. These variations are the focus of the four remaining key concepts of landscape ecology.

1.3.2 Patch Boundaries Influence Ecological Dynamics Both Within and Among Patches

If the underlying premise of landscape ecology is true—if indeed the structural configuration of a landscape can affect both what goes on within as well as between landscape elements—then patch boundaries must play a key role in governing these effects. Boundaries are the "membranes" that enclose patches, and their permeability determines what flows into and out of patches, at what rates, and in what overall directions (Wiens 1992). Part of the concern about the effects of habitat fragmentation, for example, is related to boundary or edge effects. Fragmentation is usually accompanied by a reduction in patch size, increasing the perimeter:area ratio and reducing the proportion of the patch that contains interior or "core" habitat that is immune to edge effects. Many studies have documented reduced nesting success of forest birds close to the boundary of forest fragments, largely due to increased loss to predators crossing the boundary from adjacent patches in the landscape (e.g., Wilcove 1985; Andrén 1992). Apart from changes in predation risk associated with patch boundaries, how species within a patch respond to a boundary may affect their vulnerability to fragmentation. A species that will not cross a boundary into adjacent patch types will be much more likely to suffer reductions in population size and increased extinction probability than will one for which the boundary is more permeable, facilitating dispersal from the patch as well as movement into the patch from elsewhere in the landscape.

Often the boundaries themselves have important properties. Boundaries are often transition zones (ecotones) in microclimatic factors, such as wind speed or radiant energy inputs, and both primary and secondary production may be greater in the boundary zone. This phenomenon underlies the "ecological trap" idea, which hypothesizes that individuals may be attracted to establish breeding territories in boundary situations due to the greater abundance of food there, only to suffer increased predation risk from adjacent areas (Gates and Gysel 1978). The ecological importance of boundaries and ecotones is reflected in an extensive literature on their effects (e.g., Holland et al. 1991; Hansen and di Castri 1992; Gosz 1993).

1.3.3 Patch Context Is Important

Recognition of the importance of patch context is perhaps the essence of landscape ecology. One can assess how differences in patch quality or in movements across patch boundaries affect ecological systems without necessarily considering landscape structure, but it is impossible to address the effects of patch context without a landscape-ecology perspective. What is adjacent to a given patch can have powerful effects on what happens within that patch—its quality, the degree to which the patch boundary filters movements, and the like. For example, the magnitude of edge-related predation on songbird nests in forest patches may be strongly influenced by what lies across the forest boundary (e.g., Wilcove 1985; Andrén 1992). Such patch-context effects may extend to community features as well as to population processes. In desert systems, for example, riparian zones may serve as a source of emigrants for communities occupying adjacent landscape elements, but the extent of this influence (and thus the degree to which communities in the other landscape elements are affected by their adjacency to riparian areas) may differ for different landscape elements (e.g., lowland vs. upland; Szaro and Jakle 1985). Dan Janzen (1983) placed such landscape influences explicitly in a conservation context by observing that "no park is an island," that the surroundings of a park or nature reserve may have important effects on what goes on within the park. The potential of a park or reserve to attain conservation goals, such as preservation of an endangered species, may therefore be compromised by the nature of the surrounding landscape—the complement of predators, parasites, competitors, or disturbances that are available to cross the boundary into the "protected" area. This is why it is so important to know not only *what* a patch is, in terms of its size, boundary length, quality, and so on, but also *where* it is, in terms of its adjacency to different kinds of neighboring patches with different ecological properties.

1.3.4 Connectivity Is a Key Feature of Landscape Structure

Of all the features of landscape structure listed in Table 1.1, corridors and, less often, landscape connectedness have received the greatest attention from conservation biologists. The literature of biological conservation is rife with allusions to the importance of habitat corridors (see Bennett 1999). The usual theme is that *corridors*—more-or-less linear strips of habitat joining patches of similar habitat—may provide essential conduits that enhance movement of individuals between otherwise isolated patches. Corridors facilitate the "rescue effect" (Brown and Kodric-Brown 1977) and lessen the probability of local extinction of small populations in fragmented habitats. Contrary arguments have been raised, having to do primarily with the role that corridors may play in facilitating the spread of diseases or disturbances, or the movements of predators or species of concern (e.g., Rosenberg et al. 1997). Despite their intuitive and logical appeal, evidence for the efficacy of corridors is nowhere near as compelling as the enthusiasm with which corridors have been embraced as a conservation and management tool would seem to suggest (Hobbs 1992; Bennett 1999).

Beyond any debates about the value of corridors in conservation, a focus on corridors tends to perpetuate a simplistic patch-matrix view of landscapes and to obscure the true functional connectivity of landscapes. *Connectivity* (or *connectedness*) is an aggregate property of the structural configuration of elements in a landscape mosaic, their relative viscosities to movements, and the relative permeabilities of their boundaries (Taylor et al. 1993; Wiens 1995; Tischendorf and Fahrig 2000). The probability that an individual will move from one place in a landscape to another (which is what matters in thinking about such things as metapopulation dynamics or fragmentation effects) is therefore determined by the factors underlying the previous three key landscape ecology concepts—patch quality, boundary effects, and patch context—*and* by how different kinds of organisms respond to these features of landscapes. In the Western Australian wheatbelt, for example, Blue-breasted Fairy-wrens (*Malurus pulcherrimus*) are restricted to scattered remnants of native vegetation. Dispersal among such patches is inhibited by gaps in vegetation greater than roughly 60 m (Brooker et al. 1999), so linkages among patches to form dispersal neighborboods are determined largely by the configuration of well-defined vegetated corridors along roadways or fencelines (L. Brooker and M. Brooker, unpublished manuscript). Other species occupying the same habitats, such as Singing Honeyeaters (*Lichenostomus virescens*), are less reluctant to move into and through other patch types, and for them the connectivity of the landscape is much greater (Merriam and Saunders 1993). Dispersal is a key population process, yet the probability that individuals will successfully disperse from some origin (e.g., a birthplace) to some destination (e.g., breeding habitat) involves much more than simple linear diffusion or distance-decay functions. The composition and physical configuration of the landscape can have a profound influence on dispersal pathways (Wiens 2001), with the result that different landscape structures can produce quite different demographics.

1.3.5 Spatial Patterns and Processes Are Scale-Dependent

Perhaps because of the close ties of landscape ecology with geography and cartography, and thus with maps, considerations of scale have been a central focus since its beginnings. Indeed, the emergence of landscape ecology as a discipline has done much to increase the awareness of ecologists of all sorts of the importance of scale. This recognition of scale dependency and scaling relationships runs counter to the reductionist theme that has recently dominated ecology, which has emphasized studies and experiments at fine spatial and temporal scales and simple mathematical models that ignore scale, and often space as well (Wiens 1995). "Scale" has been called a "nonreductionist unifying concept in ecology"

(Peterson and Parker 1998). Despite this, there is yet no formal "theory of scale" in landscape ecology (Meentemeyer and Box 1987; Wiens 1989a), and ecologists use "scale" in many ways (see the long table in Peterson and Parker 1998). I follow O'Neill and King (1998) in insisting that scale can only refer to dimensions in space and time.

Part of the problem is that as these dimensions change, both patterns and processes change, often in complex ways. The physical processes or anthropogenic factors that affect landscape structure, for example, differ at different spatial scales (e.g., Krummel et al. 1987; Ludwig et al. 2000), and different organisms perceive and respond to landscape structure at different scales (Wiens et al. 1993; Haila 1999; Mac Nally 1999). More often than not, the changes in relationships with changes in scale are strongly nonlinear. The thresholds in scale dependencies serve to define *scaling domains,* within which scaling relationships are consistent and extrapolation among scales is possible, but between which the rules change and extrapolation is difficult or impossible (O'Neill 1979; Wiens 1989a). The linkages between landscape pattern, landscape processes, and ecological consequences are therefore likely to be played out in different ways at different scales (Figure 1.1b). As a consequence, virtually all ecological patterns and processes are sensitive to scale.

When we observe these ecological phenomena, we do so through a window whose size is set by the minimum scale of resolution (the *grain*) and the overall scope (the *extent*) of our observations (e.g., the size of individual sampling units and of the area in which they are distributed, respectively). Changing either the grain or the extent changes the observation scale and, thus, the subset of ecological patterns, processes, and relationships that we perceive. It is no wonder that studies of the same phenomena conducted at different scales usually yield different results. In a conservation context, the problem is compounded when an arbitrary scale of management is imposed on ecological systems that are in fact operating at different scales (Wiens et al. In press). As conservation efforts shift from a focus on single species of concern to multiple species, ecosystems, or landscapes (Franklin 1993), the difficulties of dealing with scaling effects will be exacerbated, as both the ways in which landscape structure is affected by human land use or management actions and the ways in which the varied components of the system respond to landscape structure change with changes in scale.

1.4 Integrating Landscape Ecology With Biological Conservation

Changing land use is one of the major forces leading to the changes in population sizes, species distributions, ecosystem functions, or biodiversity that concern conservation biologists. Land use has these effects by altering the features and functions of landscapes that are embodied in the five concepts just described, so there should be little doubt about the relevance of these concepts, and of landscape ecology, to conservation issues. Apart from the details of implementing a landscape approach in biological conservation (e.g., which variables to map and measure, at which scales?), however, several broader issues should be addressed.

1.4.1 Is Landscape Ecology Landlocked?

Conservation, of course, involves more than land. Historically, however, landscape ecology has dealt almost entirely with land—with the components of terrestrial ecosystems and human land use. To be sure, lakes, streams, and rivers have often been included in analyses of landscape patterns, but usually as only one of many elements of the landscape mosaic (e.g., as "water" in remotely sensed images). Ecosystem ecologists have made liberal use of watersheds to integrate land and water dynamics within a defined area, but the spatial patterns within a watershed or exchanges across its boundaries have received much less attention. Stream management has often incorporated consideration of the adjacent riparian vegetation as a buffer zone to maintain stream integrity, but the landscape beyond the bordering riparian strip has often been considered only as a source of water, nutrients, or pollutants, with no explicit spatial structure of its own (but see Malanson 1993; Ward 1998; Wear et al. 1998).

Despite appearances, the central concepts of landscape ecology have been an implicit part of aquatic ecology for some time. Oceanographers, for example, have been dealing with scaling effects for decades (e.g., Steele 1978, 1989), and Hutchinson (1961) explained the paradox of high species diversity among planktonic organisms in terms of the complex, three-dimensional spatial heterogeneity of oceans. Aquatic ecologists have traditionally viewed streams as a mosaic of riffles, pools, and stream segments with high physical connectivity (e.g., Poff and Ward 1990; Robson and Chester 1999). Concepts of patchiness and patch dynamics (e.g., Kling et al. 2000; Palmer et al. 2000; Riera et al. 2000) and of scale (e.g., Poff 1996; Cooper et al. 1998; Lodge et al. 1998) have become central to how aquatic ecologists think about streams, rivers, and lakes. Spatial structure and dynamics may be more difficult to document and measure by remote sensing and GIS in aquatic systems than they are on land, but this does not mean that the interplay of landscape patterns and processes shown in Figure 1.1a is any less important. If landscape ecology is indeed the study of spatial patterns and processes, then it is just as relevant to water as it is to land. The "land" in "landscape ecology" should not be taken too literally.

1.4.2 Does Landscape Ecology Offer More Than Pretty Pictures?

To many people, the power of landscape ecology lies in its maps and images, and in the analyses and modeling that can be done using such pictures. Technological advances have led to rapid increases in the sophistication of such descriptions of landscape patterns. Remote sensing can now supply vast sets of spatial data, and GIS is a magnificent tool for integrating such information and depicting both real and synthetic landscape patterns. Our ability to construct spatially explicit simulation models that track the locations and responses of numerous individuals and their interactions with the landscape is limited more by our skill in structuring logical models and specifying reasonable parameter values than by computational capacity or speed. The value of spatial statistics is now recognized by many ecologists as well as statisticians, and the array of geostatistical tools and the software to enhance their use are expanding rapidly. Both GIS and spatial models are being used in innovative ways to explore the scaling properties of landscape patterns.

Biological conservation requires a rigorous scientific foundation, which landscape ecology should seek to provide. Certainly the quantitative rigor of landscape studies has been greatly enhanced by tools such as remote sensing, GIS, spatial modeling, and spatial statistics. Landscape ecologists also have had some success in approaching landscape problems experimentally, either by designing real experiments using fine-scale experimental model systems (EMS), or by opportunistically studying landscape alterations such as grazing or timber harvesting as quasi-experiments. By and large, however, the approaches we have come to associate with scientific rigor in ecology as a whole—experiments, an emphasis on mechanisms, explicit hypothesis testing, mathematical modeling, and well-developed, predictive theory—are generally not well-suited to dealing with landscapes. The array of possible spatial configurations of landscapes is too great, the range of relevant scales too broad, and the diversity of responses to landscape patterns and processes too large to mesh well with traditional reductionist approaches. It is the classic "middle-number" conundrum of ecology (Allen and Hoekstra 1990; O'Neill and King 1998; Lawton 1999), in which the phenomena studied are not small or simple enough so that one can deal with individual components, nor large enough that one can examine the statistical properties of the systems without worrying about individual details (as in the gas laws of physics). Rather, ecological systems often fall between these extremes: there are too many individual components, with too many complex interactions, to deal with the individuals, yet the individual details affect the dynamics of the system as a whole, so general statistical properties yield incomplete pictures of what is going on. In the case of landscapes, the problem is amplified by spatial variation and interdependencies, scale dependencies, and thresholds.

The difficulties seem especially great when it comes to developing a strong theoretical foundation for landscape ecology (Wiens 1995). Landscape ecologists have developed a lot of verbal theory, which casts ideas in prose rather than in mathematics. Verbal theory is exemplified by the concepts discussed above. However, such theory, being verbal, lacks the rigor and precision we have come to expect of "real" (i.e., mathematical) theory. As a result, the capacity of landscape ecology to provide a theoretical foundation for conservation actions, or even to offer conceptual insights that can be used to generalize among conservation problems, seems limited. Consider an example. Metapopulation theory has become an important element in assessing conservation strategies for threatened or endangered species and in predicting the consequences of habitat fragmentation (see, e.g., the papers in McCullough 1996). It calls explicit attention to the demographic and genetic consequences of the spatial subdivision of populations, and it has been an important contributing factor to the increased focus on corridors in conservation design (Bennett 1999). Yet most metapopulation theory, and virtually all of its application, uses a simple patch-matrix or patch-matrix-corridor characterization of spatial pattern (Hanski 1999). Building more spatial texture and realism into models (or management protocols) diminishes their generality. It is not simply that the core concepts of landscape ecology (its verbal theories) are not recognized, although too often they are not. Rather, the kinds of spatial variance and scaling phenomena they emphasize do not fit well into traditional (i.e., mathematical) ways of theorizing.

1.4.3 Is Landscape Ecology Too Complex?

Landscape ecology has helped to crystalize the ongoing paradigm shift in ecology, from one portraying ecological systems as homogeneous, stable, closed, and scale-insensitive to a view emphasizing their spatial and temporal variability, openness, and scale dependence (Pickett et al. 1992; Wiens 1995). One consequence of this paradigm shift has been a flowering of complexity in time and space. Such complexity, of course, is the nemesis of theory, which thrives on simplification. So, although much of the recent history of ecology has emphasized understanding phenomena by generalization and simplification (i.e., through theory), embracing complexity seems to be the forte of landscape ecology. The devil, as they say, is in the details.

Certainly landscapes *are* complex. By adding spatial effects to the nonspatial way we have traditionally viewed ecological phenomena, contingent effects are increased factorially. Scale, of course, complicates things even more. It is reasonable to ask how much of this detail is really critical, how often we must be spatially explicit, or how much scale really matters. Must the answers to every conservation problem be sought in the idiosyncratic details of each situation? Can some (or perhaps most) of the details of landscape structure be ignored in the interests of generating coarse, but workable, solutions to the problems? An analogy with population dynamics theory and population management may be useful here. Populations, of course, are full of complexity—individual variation in genetics, age, nutritional state, behavior, experience, mating success, and so on. Yet population dynamics models that ignore most of this variation have been the foundation of population management for decades. More often than not, the simplifications have not mattered (or so we think), although there have been some notable failures (e.g., fisheries management; Botkin 1990). Can we somehow simplify our treatment of spatial patterns and processes? Do we need to put much detail into the boxes of Figure 1.1a?

The answer is that we simply don't know. In the history of dealing with population dynamics, simple mathematical theory developed in concert with, or in advance of, empirical studies. As a consequence, our understanding of population processes has largely been channeled in the directions dictated by theory. Because landscape ecology so far lacks such cohesive theory, the empirical findings are mainly responsible for how we view spatial effects and scaling. By and large, these studies consistently show that space and scale *are* important, often in dramatic ways. If our assessment of the habitat associations of a bird species, for example, can change from strongly positive to strongly negative with a change in the scale of analysis (e.g., Wiens 1989b), or if the net reproductive output of a local population changes from positive to negative with a change in patch context (e.g., Pulliam 1988), what does this portend for the success of conservation practices that fail to consider scale or landscape structure? It seems to me that the default position must be that the various landscape effects I have discussed here are likely to be important unless there are good reasons to think otherwise. As Hanski (1999:264) has noted, "the really important issue is whether spatial dynamics are considered at all in landscape management and conservation."

1.5 Concluding Comments

As it has emerged as a discipline in its own right, biological conservation has looked to ecology for general laws to guide conservation actions. If we are to believe John Lawton (1999), such laws will be most likely to emerge at the reductionist (i.e., population) and expansionist (i.e., macroecological) ends of the spectrum. Lawton specifically argues that because it is so plagued by the contingencies of middle-number systems, community ecology "is a mess" and should largely be abandoned. By adding the contingencies produced by spatial patterns, spatial processes, and their interactions (Figure 1.1a), not to mention scale (Figure 1.1b), landscape ecology must be an even greater mess. Perhaps all of this talk about patches, boundaries, connectivity, scale, and spatial processes should be ignored in the interests of getting on with the business of developing general laws that can help us solve conservation problems. Before taking salvation in Lawton's view, however, it might be good to consider the words of E. O. Wilson (2000). Writing explicitly in the context of conservation issues, Wilson suggests that community ecology (and, by inference, landscape ecology) "is about to emerge as one of the most significant intellectual frontiers of the twenty-first century." The contributions to this volume should help you decide who is right.

Acknowledgments

This chapter was written during my tenure as a Sabbatical Fellow at the National Center for Ecological Analysis and Synthesis, with support from the National Science Foundation (DEB-0072909), the University of California, and the University of California-Santa Barbara. The research that formed the foundation for my thinking was supported by the National Science Foundation and, most recently, by the United States Environmental Protection Agency (R 826764-01-0). Bill Reiners and Sarah Gergel helped tune the final version of this chapter.

References

Allen, T.F.H. 1998. The landscape "level" is dead: persuading the family to take it off the respirator. In *Ecological Scale: Theory and Applications,* eds. D.L. Peterson and V.T. Parker, pp. 35–54. New York: Columbia University Press.

Allen, T.F.H., and Hoekstra, T.W. 1990. *Toward a Unified Ecology.* New York: Columbia University Press.

Andrén, H. 1992. Corvid density and nest predation in relation to forest fragmentation: a landscape perspective. *Ecology* 73:794–804.

Austin, M.P. 1999. A silent clash of paradigms: some inconsistencies in community ecology. *Oikos* 86:170–178.

Bailey, R.W. 1996. *Ecosystem Geography.* New York: Springer-Verlag.

Barrett, G.W., and Bohlen, P.J. 1991. Landscape ecology. In *Landscape Linkages and Biodiversity,* ed. W.E. Hudson, pp. 149–161. Washington, DC: Island Press.

Bennett, A.F. 1999. *Linkages in the Landscape: The Role of Corridors and Connectivity in Wildlife Conservation.* Gland, Switzerland: International Union for Conservation of Nature and Natural Resources (IUCN).

Botkin, D.B. 1990. *Discordant Harmonies.* New York: Oxford University Press.

Brooker, L.C., Brooker, M.G., and Cale, P. 1999. Animal dispersal in fragmented habitat: measuring habitat connectivity, corridor use, and dispersal mortality. *Conserv. Ecol.* [online] 3:4 Available from the Internet: www.consecol.org/vol3/iss1/art4.

Brown, J.H. 1995. *Macroecology.* Chicago: University of Chicago Press.

Brown, J.H., and Kodric-Brown, A. 1977. Turnover rates in insular biogeography: effect of immigration on extinction. *Ecology* 58:445–449.

Cooper, S.D., Diehl, S., Kratz, K., and Sarnelle, O. 1998. Implications of scale for patterns and processes in stream ecology. *Australian J. Ecol.* 23:27–40.

Farina, A. 1998. *Principles and Methods in Landscape Ecology.* London: Chapman and Hall.

Flather, C.H., Knowles, M.S., and Kendall, I.A. 1998. Threatened and endangered species geography. *BioScience* 48:365–376.

Forman, R.T.T. 1995. *Land Mosaics: The Ecology of Landscapes and Regions.* Cambridge, United Kingdom: Cambridge University Press.

Forman, R.T.T., and Godron, M. 1986. *Landscape Ecology.* New York: John Wiley and Sons.

Fortin, M.-J. 1999. Spatial statistics in landscape ecology. In *Landscape Ecological Analysis: Issues and Applications,* eds. J.M. Klopatek and R.H. Gardner, pp. 253–279. New York: Springer-Verlag.

Franklin, J.F. 1993. Preserving biodiversity: species, ecosystems, or landscapes? *Ecol. Appl.* 3:202–205.

Garshelis, D.L. 2000. Delusions in habitat evaluation: measuring use, selection, and importance. In *Research Techniques in Animal Ecology: Controversies and Consequences,* eds. L. Boitani and T.K. Fuller, pp. 111–164. New York: Columbia University Press.

Gates, J.E., and Gysel, L.W. 1978. Avian nest dispersion and fledging success in field-forest ecotones. *Ecology* 59:871–883.

Gosz, J.R. 1993. Ecotone hierarchies. *Ecol. Appl.* 3:369–376.

Haila, Y. 1999. Islands and fragments. In *Maintaining Biodiversity in Forest Ecosystems,* ed. M.L. Hunter, Jr., pp. 234–264. Cambridge, United Kingdom: Cambridge University Press.

Haines-Young, R., Green, D.R., and Cousins, S.H., eds. 1993. *Landscape Ecology and GIS*. London: Taylor and Francis.

Hansen, A.J., and di Castri, F., eds. 1992. *Landscape Boundaries: Consequences for Biotic Diversity and Ecological Flows*. New York: Springer-Verlag.

Hanski, I. 1999. *Metapopulation Ecology*. Oxford, United Kingdom: Oxford University Press.

Hobbs, R.J. 1992. The role of corridors in conservation: solution or bandwagon? *Trends Ecol. Evol.* 7:389–392.

Holland, M.M., Risser, P.G., and Naiman, R.J., eds. 1991. *Ecotones: The Role of Landscape Boundaries in the Management and Restoration of Changing Environments*. London: Chapman and Hall.

Hutchinson, G.E. 1961. The paradox of the plankton. *Am. Nat.* 95:137–145.

Janzen, D.H. 1983. No park is an island: increase in interference from outside as park size decreases. *Oikos* 41:402–410.

King, A.W. 1997. Hierarchy theory: a guide to system structure for wildlife biologists. In *Wildlife and Landscape Ecology: Effects of Pattern and Scale,* ed. J.A. Bissonette, pp. 185–212. New York: Springer-Verlag.

King, A.W. 1999. Hierarchy theory and the landscape . . . level? Or: words do matter. In *Issues in Landscape Ecology,* eds. J.A. Wiens and M.R. Moss, pp. 6–9. Guelph, Ontario, Canada: International Association for Landscape Ecology.

Kling, G.W., Kipphut, G.W., Miller, M.M., and O'Brien, W.J. 2000. Integration of lakes and streams in a landscape perspective: the importance of material processing on spatial patterns and temporal coherence. *Freshwater Biol.* 43:477–497.

Krummel, J.R., Gardner, R.H., Sugihara, G., O'Neill, R.V., and Coleman, P.R. 1987. Landscape patterns in a disturbed environment. *Oikos* 48:321–324.

Küchler, A.W. 1974. Boundaries on vegetation maps. In *Tatsachen und Probleme der Grenzen in der Vegetation,* ed. R. Tüxen, pp. 415–427. Lehre, Germany: Verlag von J. Cramer.

Lawton, J.H. 1999. Are there general laws in ecology? *Oikos* 84:177–192.

Lodge, D.M., Stein, R.A., Brown, K.M., Covich, A.P., Brönmark, C., Garvey, J.E., and Klosiewski, S.P. 1998. Predicting impact of freshwater exotic species on native biodiversity: challenges in spatial scaling. *Australian J. Ecol.* 23:53–67.

Longley, P.A., Goodchild, M.F., Maguire, D.J., and Rhind, D.W., eds. 1999. *Geographic Information Systems. Vol. 1: Principles and Technical Issues.* Second Edition. New York: John Wiley and Sons.

Ludwig, J.A., Wiens, J.A., and Tongway, D.J. 2000. A scaling rule for landscape patches and how it applies to conserving soil resources in tropical savannas. *Ecosystems* 3:84–97.

Mac Nally, R. 1999. Dealing with scale in ecology. In *Issues in Landscape Ecology,* eds. J.A. Wiens and M.R. Moss, pp. 10–17. Guelph, Ontario, Canada: International Association for Landscape Ecology.

Malanson, G.P. 1993. *Riparian Landscapes.* Cambridge, United Kingdom: Cambridge University Press.

McCullough, D., ed. 1996. *Metapopulations and Wildlife Conservation Management.* Washington, DC: Island Press.

McGarigal, K., and Marks, B. 1995. *FRAGSTATS: Spatial Analysis Program for Quantifying Landscape Structure.* General Technical Report PNW-GTR-351. Portland, Oregon: USDA Forest Service.

Meentemeyer, V., and Box, E.O. 1987. Scale effects in landscape studies. In *Landscape Heterogeneity and Disturbance,* ed. M.G. Turner, pp. 15–34. New York: Springer-Verlag.

Merriam, G., and Saunders, D.A. 1993. Corridors in restoration of fragmented landscapes. In *Nature Conservation 3: Reconstruction of Fragmented Ecosystems,* eds. D.A. Saunders, R.J. Hobbs, and P.R. Ehrlich, pp. 71–87. Chipping Norton, Australia: Surrey Beatty and Sons.

Monmonier, M. 1996. *How to Lie with Maps.* Chicago: University of Chicago Press.

Naveh, Z., and Lieberman, A.S. 1994. *Landscape Ecology: Theory and Application.* Second Edition. New York: Springer-Verlag.

Noss, R.F. 1991. Landscape connectivity: different functions at different scales. In *Landscape Linkages and Biodiversity,* ed. W.E. Hudson, pp. 27–39. Washington, DC: Island Press.

O'Neill, R.V. 1979. Transmutation across hierarchical levels. In *Systems Analysis of Ecosystems,* eds. G.S. Innis and R.V. O'Neill, pp. 59–78. Fairlands, Maryland: International Cooperative Publishing House.

O'Neill, R.V., and King, A.W. 1998. Homage to St. Michael; or, why are there so many books on scale? In *Ecological Scale: Theory and Applications,* eds. D.L. Peterson and V.T. Parker, pp. 3–15. New York: Columbia University Press.

Palmer, M.A., Swan, C.M., Nelson, K., Silver, P., and Alvestad, R. 2000. Streambed landscapes: evidence that stream invertebrates respond to the type and spatial arrangement of patches. *Landsc. Ecol.* 15:563–576.

Peterson, D.L., and Parker, V.T. 1998. Dimensions of scale in ecology, resource management, and society. In *Ecological Scale: Theory and Applications,* eds. D.L. Peterson and V.T. Parker, pp. 499–522. New York: Columbia University Press.

Pickett, S.T.A., Parker, V.T., and Fiedler, P. 1992. The new paradigm in ecology: implications for conservation biology above the species level. In *Conservation Biology: The Theory and Practice of Nature Conservation, Preservation, and Management,* eds. P.L. Fiedler and S.K. Jain, pp. 65–88. New York: Chapman and Hall.

Poff, N.L. 1996. A hydrogeography of unregulated streams in the United States and an examination of scale-dependence in some hydrological descriptors. *Freshwater Biol.* 36:71–91.

Poff, N.L., and Ward, J.V. 1990. Physical habitat template of lotic systems: recovery in the context of spatiotemporal heterogeneity. *Environ. Manage.* 14:629–645.

Poiani, K.A., Richter, B.D., Anderson, M.G., and Richter, H.E. 2000. Biodiversity conservation at multiple scales: functional sites, landscapes, and networks. *BioScience* 50:133–146.

Pulliam, H.R. 1988. Sources, sinks, and population regulation. *Am. Nat.* 132:652–661.

Reid, W.V. 1998. Biodiversity hotspots. *Trends Ecol. Evol.* 13:275–280.

Ricketts, T.H., Dinerstein, E., Olson, D.M., and Loucks, C. 1999. Who's where in North America? *BioScience* 49:369–381.

Riera, J.L., Magnuson, J.J., Kratz, T.K., and Webster, K.E. 2000. A geomorphic template for the analysis of lake districts applied to the Northern Highland Lake District, Wisconsin, U.S.A. *Freshwater Biol.* 43:301–381.

Robson, B.J., and Chester, E.T. 1999. Spatial patterns of invertebrate species richness in a river: the relationship between riffles and microhabitats. *Australian J. Ecol.* 24:599–607.

Rosenberg, D.K., Noon, B.R., and Meslow, E.C. 1997. Biological corridors: form, function, and efficacy. *BioScience* 47:677–687.

Steele, J.H. 1978. Some comments on plankton patches. In *Spatial Pattern in Plankton Communities,* ed. J.H. Steele, pp. 11–20. New York: Plenum Press.

Steele, J.H. 1989. Scale and coupling in ecological systems. In *Perspectives in Ecological Theory,* eds. J. Roughgarden, R.M. May, and S.A. Levin, pp. 177–180. Princeton: Princeton University Press.

Szaro, R.C., and Jakle, M.D. 1985. Avian use of a desert riparian island and its adjacent scrub habitat. *Condor* 87:511–519.

Taylor, P.D., Fahrig, L., Henein, K., and Merriam, G. 1993. Connectivity is a vital element of landscape structure. *Oikos* 68:571–573.

Tilman, D., and Kareiva, P., eds. 1997. *Spatial Ecology: The Role of Space in Population Dynamics and Interspecific Interactions.* Princeton: Princeton University Press.

Tischendorf, L., and Fahrig, L. 2000. On the usage and measurement of landscape connectivity. *Oikos* 90:7–19.

Turner, M.G. 1989. Landscape ecology: the effect of pattern on process. *Ann. Rev. Ecol. Syst.* 20:171–197.

Van Horne, B. 1983. Density as a misleading indicator of habitat quality. *J. Wildl. Manage.* 47:893–101.

Ward, J.V. 1998. Riverine landscapes: biodiversity patterns, disturbance regimes, and aquatic conservation. *Biol. Conserv.* 83:269–278.

Wear, D.N., Turner, M.G., and Naiman, R.J. 1998. Land cover along an urban gradient: implications for water quality. *Ecol. Appl.* 8:619–630.

Wiens, J.A. 1989a. Spatial scaling in ecology. *Functional Ecol.* 3:385–397.

Wiens, J.A. 1989b. *The Ecology of Bird Communities. Vol. 2: Processes and Variations.* Cambridge, United Kingdom: Cambridge University Press.

Wiens, J.A. 1992. Ecological flows across landscape boundaries: a conceptual overview. In *Landscape Boundaries: Consequences for Biotic Diversity and Ecological Flows,* eds. A.J. Hansen and F. di Castri, pp. 217–235. New York: Springer-Verlag.

Wiens, J.A. 1995. Landscape mosaics and ecological theory. In *Mosaic Landscapes and Ecological Processes,* eds. L. Hansson, L. Fahrig, and G. Merriam, pp. 1–26. London: Chapman and Hall.

Wiens, J.A. 1997. The emerging role of patchiness in conservation biology. In *The Ecological Basis of Conservation: Heterogeneity, Ecosystems, and Biodiversity,* eds. S.T.A. Pickett, R.S. Ostfeld, M. Shachak, and G.E. Likens, pp. 93–107. New York: Chapman and Hall.

Wiens, J.A. 2001. The landscape context of dispersal. In *Dispersal: Individual, Population, and Community,* eds. J. Clobert, E. Danchin, A.A. Dhondt, and J.D. Nichols, pp. 96–109. Oxford, United Kingdom: Oxford University Press.

Wiens, J.A., and B.T. Milne. 1989. Scaling of 'landscapes' in landscape ecology, or, landscape ecology from a beetle's perspective. *Landsc. Ecol.* 3:87–96.

Wiens, J.A., Stenseth, N.C., Van Horne, B., and Ims, R.A. 1993. Ecological mechanisms and landscape ecology. *Oikos* 66:369–380.

Wiens, J.A., Van Horne, B., and Noon, B.R. In press. Integrating landscape structure and scale into natural resource management. In *Integrating Landscape Ecology into Natural Resource Management,* eds. J. Liu and W.W. Taylor. Cambridge, United Kingdom: Cambridge University Press.

Wilcove, D.S. 1985. Nest predation in forest tracts and the decline of migratory songbirds. *Ecology* 66:1211–1214.

Wilson, E.O. 2000. On the future of conservation biology. *Conserv. Biol.* 14:1–3.

Zonneveld, I.S. 1995. *Land Ecology.* Amsterdam, The Netherlands: SPB Academic Publishing.

2

Central Concepts and Issues of Biological Conservation

RICHARD L. KNIGHT AND PETER B. LANDRES

2.1 Introduction

Our chapter introduces concepts and principles of biological conservation that will be useful when applying landscape ecology to a conservation issue. The points raised here expand the general landscape principles of the first chapter to the broad field of biological conservation. Section 2.2 explains concepts and principles that relate directly to the management of biological diversity, beginning with populations, then addressing communities and ecosystems, and concluding with principles at the landscape scale. Section 2.3 focuses on two emerging ideas in biological conservation that have been largely inspired by landscape ecology: the necessity of conserving a landscape and its functions to protect the process of evolution as the ultimate driver of biological diversity, and the complexity of restoring and managing landscapes. Section 2.4 discusses how an understanding of landscape ecology drives the need to consider long time frames and large areas in developing conservation goals and priorities. Section 2.5 considers general approaches for advancing applications of landscape ecology in biological conservation. It stresses the importance of spatial and temporal scales when considering the maintenance of biological diversity, whether the genome of a population or the range of ecosystems within a landscape.

2.2 Concepts and Principles

Biological diversity (often shortened to "biodiversity") has been defined as "the variety of living organisms considered at all levels of organization, including the genetic, species, and higher taxonomic levels, and the variety of habitats and ecosystems, as well as the processes occurring therein" (Meffe and Carroll 1997:675). This definition encompasses the aquatic and terrestrial worlds. Indeed, interfaces between aquatic (marine and freshwater) and terrestrial ecosystems are among the most biologically diverse systems in the world. The study of biodiversity and the means to protect it fall within the domain of an emerging science called conservation biology. The science of conservation biology is ambitious considering the levels of organization that it addresses (genes to biomes). Although the discipline is young, it has been the focus of great intellectual foment, generating ideas and concepts that span the minute to the massive (Knight and George 1995). This chapter addresses concepts of conservation biology central to applying landscape ecology to biological conservation issues. Biological conservation is a more encompassing field than is conservation biology in that it addresses not only the biology, but also the planning, managing, and politics of protecting life's diversity.

2.2.1 Population-Level Considerations

Although the concept of genetic variation can be specific to the level of genetic diversity within an individual, in terms of biodiversity and landscapes, it is best viewed at a population level. In this light, *genetic diversity* refers to the collective level of heterozygosity found within individuals that compose a population. Because within-individual heterozygosity is believed to relate positively with individual fitness, high levels of heterozygosity within a population suggest improved viability in relation to survival and reproduction. *Allelic diversity* refers to the proportion of different alleles at chromosomal loci. Allelic diversity, particularly uncommon alleles, composes the genetic material for adaptive change. Both heterozygosity and allelic diversity may shape a population's persistence over short and long temporal scales.

For a population, some *minimum viable population* exists that has a probability of remaining extant for some time frame, given genetic, demographic, and environmental stochasticity (Figure 2.1). In addition, catastrophic events (anthropogenic and natural occurrences that occur at great spatial and temporal scales) influence a population's viability (Shaffer 1981). A population that consists of spatially discrete subpopulations connected via dispersal is known as a *metapop-*

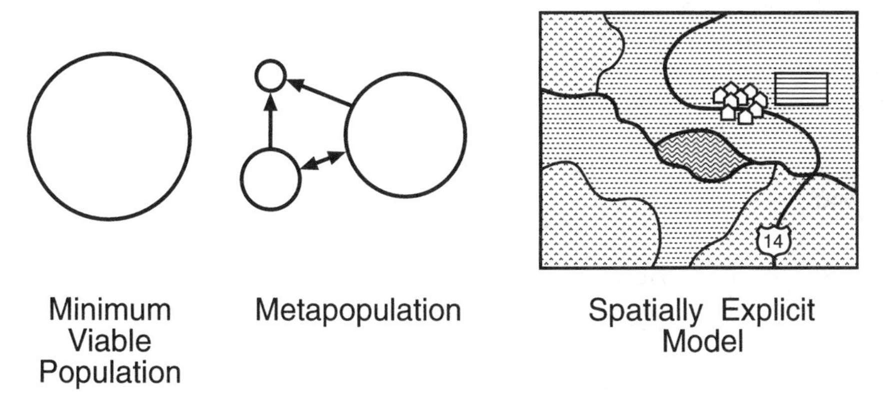

FIGURE 2.1. Three concepts applicable to managing species' populations. A minimum viable population, which is spatially inexplicit; a metapopulation, which has some degree of spatial realism; and a spatially explicit population, which has complete spatial realism.

ulation. Conservation planning for metapopulations can occur at the level of subpopulations as well as for the collective population (Levins 1969) (Figure 2.1). Because subpopulations are subject to various stochastic and catastrophic forces that may vary considerably across subpopulations, demographic factors must often be assessed subpopulation by subpopulation. A subpopulation in which birth rate exceeds death rate ($\lambda > 1$) may be called a *source,* whereas a subpopulation in which death rate exceeds birth rate ($\lambda < 1$) may be referred to as a *sink.* The phenomenon of source populations supplementing sink populations via dispersing individuals has been called the *rescue effect* (Brown and Kodric-Brown 1977).

The concepts of minimum viable populations and metapopulations offer little spatial realism; minimum viable population estimates do not consider dispersal, and metapopulation models do not consider the probability of survival during dispersal between subpopulations. This lack of spatial realism has spurred the development of *spatially explicit models,* which incorporate the geometry of habitat patches and the landscape within which they are embedded (Pulliam et al. 1992) (Figure 2.1). Because species have varying dispersal capacities based on morphological, physiological, or behavioral attributes, spatially explicit models are not only specific to a landscape, they are highly species-specific (Beissinger and Westphal 1998).

2.2.2 Community- and Ecosystem-Level Considerations

Communities are assemblages of species that share an ecosystem. Communities are often characterized by a variety of species-diversity indices that quantify the number of species (*richness*) and the relative abundance of those species (*evenness*) (Whittaker 1975).

Conservationists often use species richness to describe complex spatial patterns of biodiversity (e.g., Knopf and Samson 1994). *Alpha richness* (α) describes the number of species that occur within a single ecosystem; *beta richness* (β) captures the change or turnover in number of species from one ecosystem to another. A high beta richness means that the cumulative number of species recorded increases rapidly as additional ecosystems are censused. *Gamma richness* (γ) is the total number of species observed within all ecosystems of a region. Clearly, these indices are sensitive to the definition of an ecosystem or a patch area and the intensity of a sampling effort.

2.2.3 Landscape-Level Considerations

At the landscape level, the persistence of biodiversity is chiefly affected by human activities, including fragmentation of natural habitat and the degradation of quality of the matrix in which natural areas are embedded.

Habitat fragmentation occurs when a contiguous area of native vegetation is converted to remnant patches of native vegetation surrounded by human uses (Figure 2.2). At the opposite end of this spectrum is *perforation,* in which human

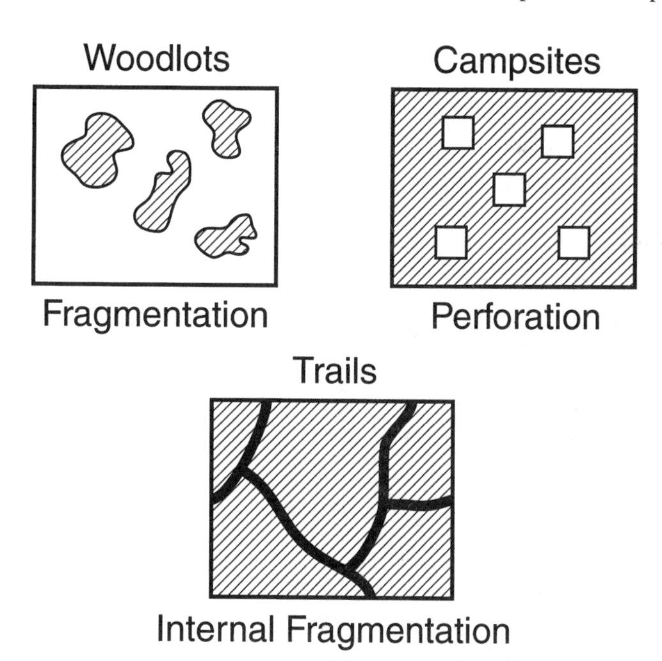

FIGURE 2.2. Three forms of land transformation. Fragmentation occurs when the creation of one land-use type (e.g., developed areas [white]) nearly replaces once-contiguous habitat (e.g., woods [shaded]). Perforation occurs when small openings (e.g., campsites, home sites, forest clear-cuts) are created within otherwise contiguous habitat. Internal fragmentation occurs when rights-of-way (e.g., power lines, roads, trails) dissect contiguous habitat.

uses (e.g., houses, oil wells, campgrounds) alter small areas within an area of natural vegetation (Figure 2.2). *Internal fragmentation* occurs when linear or curvilinear corridors (e.g., roads, power lines, trails) dissect an area (Figure 2.2). Three aspects of fragmentation are a reduction of area, an increase in edge, and increased isolation of the fragmented patches of native vegetation. Reduced area affects *area-sensitive species,* those species that have large area requirements, whether it be due to large body size, specialized needs (e.g., diet), or movement patterns. Increased edge affects *edge-sensitive species,* those species whose fitness is reduced by biotic, abiotic, and human factors associated with edges. The biotic factors include *edge-generalist species,* which may be predators, parasites, or competitors whose fitness is elevated in association with edges. The abiotic factors that may affect edge-sensitive species include sharp gradients in temperature, relative humidity, solar radiation, and moisture. The human factors include all conditions associated with humans, including pets, weapons, their activities, and structures (e.g., Miller et al. 1998). Collectively, these three sets of factors create an *edge effect,* which has a variable depth of impact into habitat fragments. Patch shape plays a critical role in this respect. Compared with patches that are more circular in shape, patches of equal size but more indented and angular in shape have more edge per unit area and therefore have less area immune from edge effects. Increased isolation between patches affects *dispersal-sensitive species,* which have limited capacity to disperse from their native habitat due to morphological, physiological, or behavioral limitations. Importantly, many species are capable of dispersing across human-dominated landscapes, but in so doing they may experience elevated mortality.

The *matrix* is the most extensive, most connected, or most influential landscape element of an area (Forman 1995). In much of the Midwestern USA, the matrix is agricultural; in Orange County, California, the matrix is urban; and in the center of Yellowstone National Park, USA, the matrix is coniferous forest. The matrix is important in that it can often influence ecological processes that may affect biodiversity. For example, nest parasitism by Brown-headed Cowbirds (*Molothus ater*) is common along forest edges where the matrix is agricultural (Figure 2.3a) but uncommon along forest edges where the matrix is forest (Figure 2.3b) (Brittingham and Temple 1983).

At the landscape scale, interactions of native species with non-native species are one of the principle causes of species endangerment. A recent review found that of 877 species listed under the U.S. Endangered Species Act, 305 species were threatened due to invasive species (Czech et al. 2000). Non-native species are typically exotics introduced, intentionally or otherwise, by human activities. Non-native species also include species from North America that have become established in ecosystems outside the limits of their natural range, or those that have become prominent in areas where they were historically less-common.

An equally important landscape-level dilemma facing those committed to biological conservation is the continued degradation of aquatic ecosystems. With a

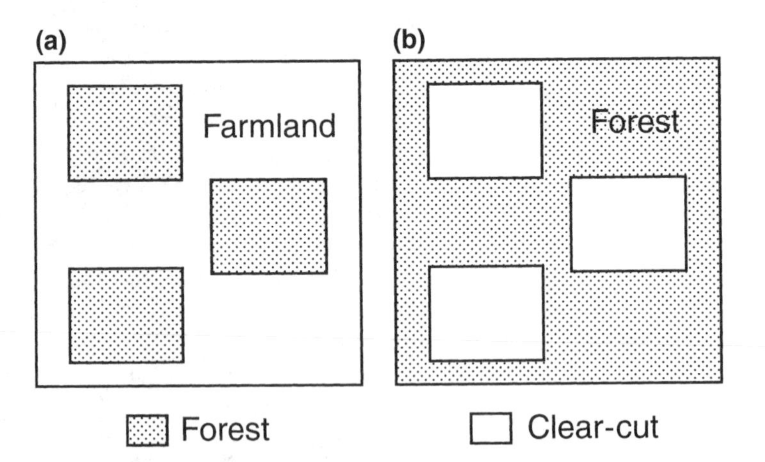

FIGURE 2.3. An example of where the matrix differs even though superficially the landscape may look similar. (a) The matrix is agricultural land with remnant patches of forest embedded within it. (b) The matrix is forest land with openings from forest clear-cuts embedded within it.

burgeoning global population, the issues of water quantity and quality, human demands on this finite resource, and the dependency of many species for aquatic ecosystems are of paramount importance for sustainable societies. In the United States alone, over two million dams support human uses. Water development projects are the greatest cause for endangerment of species on public lands (Losos et al. 1995). Continued changes to aquatic ecosystems are creating a "distress syndrome" resulting in reduced biodiversity, altered primary and secondary productivity, increased prevalence of disease, reduced nutrient cycling efficiency, increased dominance of exotic species, and increased dominance by smaller, shorter-lived opportunistic species (Rapport and Whitford 1999).

2.3 Emerging Ideas and Issues

This section describes how concepts from landscape ecology impel ideas and issues in biological conservation. In a field as diverse as biological conservation, new ideas and issues appear regularly, but there is insufficient space in a single chapter to cover all of these. Instead, this section focuses on two overarching and interconnected themes: protecting evolution unfettered by human desires and actions, and the ecological restoration and management of landscapes.

2.3.1 Evolution Unfettered By Human Desires and Actions

The rich biodiversity of our planet was produced by evolution, and the protection of evolution unaltered and unconstrained by human desires and actions is arguably the ultimate goal of conservation (Landres 1992). A landscape-scale perspective significantly broadens and deepens the traditional range of thinking about the process of evolution, the forces that influence this process, and the actions that conservationists must undertake to protect evolution as the ultimate driver of biological diversity.

Hutchinson (1965) championed "the ecological theater and the evolutionary play" to focus attention on the role of the environment in influencing evolution. Only a landscape, with its full complement of native species, ecological processes, successional types, connectivity, and isolated areas provides the rich, natural "stage" for the unfettered evolutionary "play." For example, a large area has greater opportunities for isolated populations, as well as populations at the margin of their species' range. These populations increase regional genetic diversity and are prime crucibles for evolution. In many cases, the large area of a landscape offers the only opportunity for ecological processes and disturbances to ebb and flow across an area, resulting in more diverse habitats and opportunities for evolution.

Human impacts on evolution also occur at the landscape scale. Habitat fragmentation caused by roads and habitat changes, and increasing simplification and homogeneity of natural ecosystems, alter landscape flows and disturbance regimes, causing further loss of opportunities for evolutionary diversity. Impacts in one area may also reach far into other areas. Fires suppressed in lower-elevation commercial forests, for example, may significantly reduce the natural spread of fire into higher-elevation forests within designated wilderness (Habeck 1985).

Although there is a basic conceptual understanding of how a landscape provides the context for evolution, and how human impacts at the landscape scale affect evolution, most of the details remain unknown. Some of the more urgent questions are: How is unfettered evolution recognized? How big an area and how much time are needed to allow unfettered evolution? What types of habitats, and their abundance and distribution across a landscape are necessary and sufficient for unfettered evolution? How much fragmentation, simplification, homogeneity, and human-modified matrix can occur before evolution becomes human-controlled? Will landscape connectivity that allows dispersal of individuals, but not the flow of ecological processes such as disturbance events, be sufficient to allow unfettered evolution?

2.3.2 Ecological Restoration and Management of Landscapes

If biological diversity is to be sustained, landscapes will need to be restored and managed for their biological values. Landscape-scale restoration and management is necessary because only a landscape offers the area and diversity of environments for isolated populations of species, habitat connectivity, disturbance processes, and a range of successional habitats. With the need to restore and manage landscapes comes a host of issues related to identifying and understanding landscape-scale patterns and processes, and fundamental changes to traditional management concepts.

Restoring and managing a landscape pushes the limits of our understanding about ecological patterns and processes in several areas. For instance, historical climate, vegetation, disturbance processes, and soils have a large influence on the patterns we see today, but we need better understanding about the specific types of historical information needed, and how to use this information to guide restoration and management actions (Landres et al. 1999). Different restoration and management goals require information and action at different spatial and temporal scales. Better understanding and information is needed on how to define the appropriate spatial and temporal scales for a given restoration and management goal (Hobbs 1998). Landscape-scale phenomena interact, yet there is little understanding of how these phenomena affect one another, or the resulting cumulative effects on a landscape. For example, there is virtually no information on how fire and insects affect one another over large areas and long time frames, or the resulting cumulative effects of this interaction on the quality of wildlife habitat. Assessing the success or failure of restoration and management actions requires understanding of and the ability to evaluate ecological integrity of a landscape. However, significant questions remain about how ecological integrity is defined, the use of indicators or indices in assessing ecosystem integrity, the role of ecological resiliency and redundancy in ecological integrity, and the quality of ecological fluxes across seminatural landscapes.

Restoring and managing a landscape requires "thinking like a mountain" (Leopold 1949), a decidedly nontraditional approach to management. First, the area and time frame of planning need to reflect landscape-scale patterns and processes, rather than the currently used sideboards of the closest administrative boundary and a time frame prescribed by national policy. Second, restoring and managing a landscape requires goals and targets that are sufficiently fluid to accommodate natural spatial and temporal variability, and the uncertainty and surprise in outcomes that result from allowing the free play of ecological processes (Christensen 1997). And third, within a landscape, administrative boundaries almost always cause negative impacts to ecological flows and biological diversity (Landres et al. 1998a). To overcome these negative impacts within a landscape, land-use goals need to be coordinated among different landowners.

2.4 Ecological Connections Between Landscape Ecology and Conservation

Landscape ecology directly affects the goals and priorities of conservation. Ideas from landscape ecology have contributed to understanding the importance of historical disturbance regimes, the dispersal of organisms across a landscape, and populations at the margin of their species' range as vital evolutionary crucibles. Most importantly, a landscape scale compels explicit recognition that large areas and long time frames are critical for planning and implementing the conservation of biological diversity. Thinking big, thinking long, and thinking across landscapes have critical implications to successful biological conservation. A large-area perspective significantly broadens the range of conservation goals to include the ecological processes that influence an area, consideration of heterogeneity and fragmentation of landscapes, and the maintenance of natural dispersal. A long-term perspective broadens the range of conservation goals to include unfettered evolution (Landres 1992), planning and managing for the "flux of nature" rather than the "balance of nature" (Pickett and Ostfeld 1995), and identifying factors that cause slow anthropogenic change over large areas that can have just as large an impact as those that cause rapid change in small areas (Hobbs 1998).

A landscape-scale perspective requires an expansion of conservation priorities to include explicit landscape-based goals. The shift from a "balance of nature" to a "flux of nature" paradigm is one priority that has already taken hold. Another priority is to focus on unfettered evolution for two reasons: its primal importance to biological diversity, and "an evolutionary perspective may help to give conservation a permanence which a utilitarian, and even an ecological grounding, fail to provide" (Frankel 1974). Focusing on unfettered evolution casts a different light on traditional conservation priorities. For example, the almost exclusive focus on biodiversity "hot spots" ignores the genetic and evolutionary importance of harsh environments with few species, and populations at the margin of their species' range (Lesica and Allendorf 1995).

A landscape-scale perspective also will place greater attention on the management of areas that are already protected for their conservation value. Environmental advocates and conservationists have discussed, researched, and pushed for allocating new areas to be protected, while almost completely ignoring the management of landscapes that are already protected. It is the ongoing, daily management decisions that determine how well an area protects and sustains biological diversity. For example, within designated wilderness, uncontrolled and unmanipulated landscape-scale patterns and processes are given the highest priority (Hendee et al. 1990), and unfettered evolution has the best chance of being realized. Yet the management of wilderness, our most protected landscape and ecological core for nearly all landscape-scale management plans, is neglected by most conservationists (Landres et al. 1998b).

The effects of landscape-scale goals and priorities are demonstrated in numerous conservation and management projects, reviewed in Gunderson et al. (1995), Yaffee et al. (1996), and Knight and Landres (1998). In all types of ecosystems, from prairie to temperate rainforest, landscape-scale goals and priorities were fundamentally important to the success that was achieved in these projects. In the Pacific Northwest, for example, understanding the long-term, landscape-scale dynamics of the natural fire regime enabled scientists to assess the consequences of alternative management scenarios on biodiversity (Cissel et al. 1999). Similarly, large-area and long-term perspectives were of critical importance in defining targets for the restoration of ponderosa pine (*Pinus ponderosa*) forests in the southwestern United States (Moore et al. 1999).

2.5 General Approaches for Advancing Applications of Landscape Ecology in Biological Conservation

Biological conservation occurs within ecosystems. And because ecosystems compose landscapes, and landscapes compose regions, effective biological conservation requires thinking big, thinking long, and thinking across the full range of spatial and temporal scales. Although simplistic, even the consideration of α, β, and γ richness may promote appropriate management because it will ensure that biodiversity is evaluated at the appropriate scales (e.g., Knopf and Samson 1994).

Landscapes are not closed, self-supporting systems. Instead, they are open, dynamic systems that are strongly affected by administrative lines placed on them (Landres et al. 1998a). More than half of the Earth's surface is under some form of intensive management in which the dominant spatial features are created by human actions (Klopatek et al. 1979). In a very visual way, landscapes show the legacies of our past use. The time for thinking that ecosystems and landscapes are pristine has long passed; effective biological conservation requires thinking and management that acknowledges both ecosystem and administrative boundaries, and it recognizes that the former are more dynamic than are the latter. The first step for effective conservation planning at a landscape level will require being comfortable thinking at a variety of spatial, temporal, and jurisdictional (city, county, state, federal, international) scales (Knight and Landres 1998).

We must remember that landscapes are strongly affected by indirect as well as direct effects. Plant communities can be altered over time when even a single species is removed from a species assemblage, notwithstanding ideas of species redundancy (Jones et al. 1994). Likewise for ecological processes, such as fire (Collins 1992). Managers need to incorporate into their thinking not only single species, but also the critical roles that ecological processes and landscape geometry play in maintaining biodiversity (Knight 1998). Encouragingly, both the ideas and the technology exist for effective means of ensuring that our natural heritage persists (Hansen et al. 1993; D'Erchia 1997; Hansen et al. 1999). Although it will take time to incorporate these ideas, we are at least heading in the right direction.

Acknowledgments

We acknowledge Gary Meffe and Stan Temple for their roles in shaping our thoughts about the conservation of biodiversity. Wendell Gilgert and George Wallace kindly read an earlier draft.

References

Beissinger, S.R., and Westphal, M.I. 1998. On the use of demographic models of population viability in endangered species management. *J. Wildl. Manage.* 62:821–841.

Brittingham, M.C., and Temple, S.A. 1983. Have cowbirds caused forest songbirds to decline? *BioScience* 33:31–35.

Brown, J.H., and Kodric-Brown, A. 1977. Turnover rates in insular biogeography: effect of immigration on extinction. *Ecology* 58:445–449.

Christensen, N.L. 1997. Managing for heterogeneity and complexity on dynamic landscapes. In *The Ecological Basis of Conservation: Heterogeneity, Ecosystems, and Biodiversity,* eds. S.T.A. Pickett, R.S. Ostfeld, M. Shachak, and G.E. Likens, pp. 167–186. New York: Chapman and Hall.

Cissel, J.H., Swanson, F.J., and Weisberg, P.J. 1999. Landscape management using historical fire regimes: Blue River, Oregon. *Ecol. Appl.* 9:1217–1231.

Collins, S.L. 1992. Fire frequency and community heterogeneity in tallgrass prairie vegetation. *Ecology* 73:2001–2006.

Czech, B., Krausman, P.R., and Devers, P.K. 2000. Economic associations among causes of species endangerment in the United States. *BioScience* 50:593–601.

D'Erchia, F. 1997. Geographic information systems and remote sensing applications for ecosystem management. In *Ecosystem Management,* eds. M.S. Boyce and A. Haney, pp. 201–225. New Haven, Connecticut: Yale University Press.

Forman, R.T.T. 1995. *Land Mosaics: The Ecology of Landscapes and Regions.* Cambridge, United Kingdom: Cambridge University Press.

Frankel, O.H. 1974. Genetic conservation: our evolutionary responsibility. *Genetics* 99:53–65.

Gunderson, L.H., Holling, C.S., and Light, S.S., eds. 1995. *Barriers and Bridges to the Renewal of Ecosystems and Institutions.* New York: Columbia University Press.

Habeck, J.R. 1985. Impact of fire suppression on forest succession and fuel accumulations in long-fire-interval wilderness habitat types. In *Proceedings—Symposium and Workshop on Wilderness Fire, General Technical Report INT-182,* technical coordinators J.E. Lotan, B.M. Kilgore, W.C. Fischer, and R.W. Mutch, pp. 110–118. Ogden, Utah: USDA Forest Service Intermountain Research Station.

Hansen, A.J., Garman, S.L., Marks, B., and Urban, D.L. 1993. An approach for managing vertebrate diversity across multiple-use landscapes. *Ecol. Appl.* 3:481–496.

Hansen, A.J., Rotella, J.J., Kraska, M.P.V., and Brown, D. 1999. Dynamic habitat and population analysis: an approach to resolve the biodiversity manager's dilemma. *Ecol. Appl.* 9:1459–1476.

Hendee, J.C., Stankey, G.H., and Lucas, R.C. 1990. *Wilderness Management.* Second Edition. Golden, Colorado: North American Press.

Hobbs, R.J. 1998. Managing ecological systems and processes. In *Ecological Scale: Theory and Applications,* eds. D.L. Peterson and V.T. Parker, pp. 459–484. New York: Columbia University Press.

Hutchinson, G.E. 1965. *The Ecological Theater and the Evolutionary Play.* New Haven, Connecticut: Yale University Press.

Jones, C.G., Lawton, J.H., and Shachak, M. 1994. Organisms as ecosystem engineers. *Oikos* 69:373–386.

Klopatek, J.M., Olson, R.J., Emerson, C.J., and Joness, J.L. 1979. Land-use conflicts with natural vegetation in the United States. *Environ. Conserv.* 6:191–199.

Knight, R.L. 1998. Ecosystem management and conservation biology. *Landsc. Urban Plan.* 40:41–45.

Knight, R.L., and George, T.L. 1995. New approaches, new tools: conservation biology. In *A New Century for Natural Resources Management,* eds. R.L. Knight and S.F. Bates, pp. 279–295. Washington, DC: Island Press.

Knight, R.L., and Landres, P.B., eds. 1998. *Stewardship Across Boundaries.* Washington, DC: Island Press.

Knopf, F.L., and Samson, F.B. 1994. Scale perspectives on avian diversity in western riparian ecosystems. *Conserv. Biol.* 8:669–676.

Landres, P.B. 1992. Temporal scale perspectives in managing biological diversity. *Trans. N. Am. Wildl. Nat. Resour. Conf.* 57:292–307.

Landres, P.B., Knight, R.L., Pickett, S.T.A., and Cadenasso, M.L. 1998a. Ecological effects of administrative boundaries. In *Stewardship Across Boundaries,* eds. R.L. Knight and P.B. Landres, pp. 39–64. Washington, DC: Island Press.

Landres, P.B., Marsh, S., Merigliano, L., Ritter, D., and Normal, A. 1998b. Boundary effects on wilderness and other natural areas. In *Stewardship Across Boundaries,* eds. R.L. Knight and P.B. Landres, pp. 117–139. Washington, DC: Island Press.

Landres, P.B., Morgan, P., and Swanson, F.J. 1999. Overview of the use of natural variability concepts in managing ecological systems. *Ecol. Appl.* 9:1179–1188.

Leopold, A. 1949. *A Sand County Almanac and Sketches Here and There.* New York: Oxford University Press.

Lesica, P., and Allendorf, F.W. 1995. When are peripheral populations valuable for conservation? *Conserv. Biol.* 9:753–760.

Levins, R. 1969. Some demographic and genetic consequences of environmental heterogeneity for biological control. *Bull. Entomol. Soc. Am.* 15:237–240.

Losos, E., Hayes, J., Phillips, A., Wilcove, D., and Alkire, C. 1995. Taxpayer-subsidized resource extraction harms species. *BioScience* 45:446–455.

Meffe, G.K., and Carroll, C.R., eds. 1997. *Principles of Conservation Biology.* Second Edition. Sunderland, Massachusetts: Sinauer Associates.

Miller, S.G., Knight, R.L., and Miller, C.K. 1998. Influence of recreational trails on breeding bird communities. *Ecol. Appl.* 8:162–169.

Moore, M.M., Covington, W.W., and Fule, P.Z. 1999. Reference conditions and ecological restoration: a southwestern ponderosa pine perspective. *Ecol. Appl.* 9:1266–1277.

Pickett, S.T.A., and Ostfeld, R.S. 1995. The shifting paradigm in ecology. In *A New Century for Natural Resources Management,* eds. R.L. Knight and S.F. Bates, pp. 261–278. Washington, DC: Island Press.

Pulliam, H.R., Dunning, J.B., and Liu, J. 1992. Population dynamics in a complex landscape. *Ecol. Appl.* 2:165–177.

Rapport, D.J., and Whitford, W.G. 1999. How ecosystems respond to stress. *BioScience* 49:193–203.

Shaffer, M.L. 1981. Minimum population sizes for species conservation. *BioScience* 31:131–134.

Whittaker, R.H. 1975. *Communities and Ecosystems.* New York: MacMillan.

Yaffee, S.L., Phillips, A.F., Frentz, I.C., Hardy, P.W., Maleki, S.M., and Thorpe, B.E. 1996. *Ecosystem Management in the United States: An Assessment of Current Experience.* Washington, DC: Island Press.

3

Broad-Scale Ecological Science and Its Application

BARRY R. NOON AND VIRGINIA H. DALE

3.1 Introduction

During the last 200 years, the spatial extent of environmental changes and the nature and permanence of these changes have been unprecedented relative to preindustrial times. Concomitant with changes in landscape pattern and composition, there have been dramatic declines in the distribution and abundance of plant and animal species. Given that environmental change is expected to increase in rate and extent over the next several decades, there is an increasing sense of urgency to discover how to ameliorate human impacts on biota and to sustain the ecological services they provide. As a result, new scientific principles and insights are needed to address such broad-scale problems.

One promising new field of science that addresses broad-scale issues is landscape ecology—the study of the physical structure and temporal dynamics of spatial mosaics and their ecological causes and consequences (Wiens 1995, 1999). The focus of this chapter is on the ecological principles and insights provided from the science of landscape ecology that are relevant to human-induced changes in spatial pattern. As a consequence, we emphasize insights relevant to large spatial ($>10^4$ ha) and long temporal (multiple human generations) scales.

In addition, we focus on the components of ecological systems and not on ecosystems per se. We consider ecosystems as just one level in the hierarchy of ecological systems ranging from genetic to global. When we speak of ecosystems, we are referring to underlying processes, such as energy flow and mineral cycling, that organize landscapes and their constituent elements. Although these processes are seldom visible, their expression in the form of landscape patterns and elements (i.e., patches, corridors, and matrix) is visible. Therefore, these elements are the logical choice for the primary units of management and research.

The application of broad-scale science is limited by management constraints. Management actions are temporally and spatially explicit and produce discrete outcomes. Therefore, it is essential to develop an operational framework that expresses management goals in terms of measurable attributes that can be spatially referenced. Such a spatial assessment is possible if based on observable landscape elements and patterns. Because of the ambiguity that surrounds the ecosystem concept, we emphasize the three main components of ecological systems: composition, structure, and process (Franklin and Spies 1991; Committee of Scientists 1999).

The objectives of this chapter are to discuss concepts and principles drawn from ecology that are applicable at broad spatial scales, to describe emerging ideas and issues relevant to the management of ecological systems and to the solution of environmental problems, to explore the ecological connections between landscape ecology and the conservation of biological diversity, and to offer general principles for advancing the application of landscape ecology to nature conservation.

3.2 Concepts and Principles

3.2.1 The Defining Attributes of Ecological Systems

Ecological Composition

Composition refers to the number and variety of biotic and abiotic components of an ecological system. Variability among components of an ecological system can occur at a variety of levels of biological organization, including the gene, species, ecosystem, and landscape. Perhaps the most fundamental level of compositional variation occurs at the level of the gene. The variety of alleles and fitness consequences of genetic variation compose the ultimate currency of natural selection—the capacity for adaptive response to environmental variation. This measure of compositional variation also depends on the level of biological organization; the relative contributions of within- and among-group variation at the individual and population levels are extremely relevant to the overall goal of ecological sustainability.

Compositional diversity also exists at a number of spatial scales. For example, variations in macroscale elements such as the contrast among rivers, ponds, mountains, and plains, represent compositional variation at a landscape scale. Variation in landscape composition also can be described in terms of the variety of plant communities and habitats. Measures of compositional diversity have often defaulted to simple metrics, such as species richness or species diversity. Although these statistics have value, focusing on them without addressing the individual species components can lead to loss of biological diversity at broad spatial scales. Given this concern, some scientists have instead used weighted measures of diversity to account for species that are disproportionately "important" to the ecological system. Scientists agree that species vary in their contributions to ecological processes, but they disagree on the weights that should be assigned to individual species or functional groups of species (e.g., Tilman 1999).

Ecological Structure

Structure refers to the geometry of landscape components, both biogenic and geologic. That is, what are the sizes, shapes, and configurations of the components that define the environment at various spatial scales? At a fine scale, biogenic components can include large live trees, large logs on the forest floor or lake bottom, and fish carcasses. Geologic components can include mountains, canyons, and undammed rivers. They have compositional distinctness, but they also have structural distinctness; for example, large old trees have geometries that are very different from those of young small trees even if they are compositionally identical.

Structural variation also occurs at the scale of the landscape. That is, variation in the size, shape, and spatial relationships of vegetation communities can give rise to very different ecological systems even if composition is held constant. For example, the size, shape, and patterns of interspersion of habitat, and their consequences for connectivity, all influence the types of organisms that can exist in a landscape.

Variations in the physical attributes of ecological systems, especially soil, water, and air, can both constrain and provide opportunities for biological diversity. For example, natural watersheds have many habitats, (e.g., alluvial soils, steep slopes, deep pools, shallow riffles, and waterfalls) that support a diverse biological system. In contrast, a river that has been dammed to create a reservoir or diverted from its natural channel may support fewer habitats and far less diversity.

Ecological Process

Process includes the mechanisms that convert inorganic material and energy into biotic components, as well as physical actions that transfer biotic and abiotic materials among the components of ecological systems. Examples include photosynthesis, energy flow, nutrient cycling, water movement, disturbance, and succession. Processes are the phenomena that link ecosystem components and that initiate renewal as well as transformation.

Each of the three defining attributes of ecological systems can be visualized as an apex of an equilateral triangle (Figure 3.1). Each component is necessary, but by itself insufficient, to characterize the state of an ecological system.

3.2.2 *What Is a Landscape Perspective?*

We use the term *landscape* to refer to "a heterogeneous land area composed of a cluster of interacting ecosystems that is repeated in similar form throughout" (Forman and Godron 1986). At a broad scale, we assume that areas of a landscape have similar geomorphology and a common disturbance regime. A linkage between a broad-scale perspective and the attributes of ecological systems exists because landscape patterns affect all levels of the ecological hierarchy, from genes to ecosystems. This requires the resource manager to assess the effects of landscape change in terms of changes in composition, structure, and process at multiple spatial and temporal scales. Although broad in its spatial scale of assessment, a landscape perspective does not eliminate the need to consider the smaller components of ecological systems with fast dynamics.

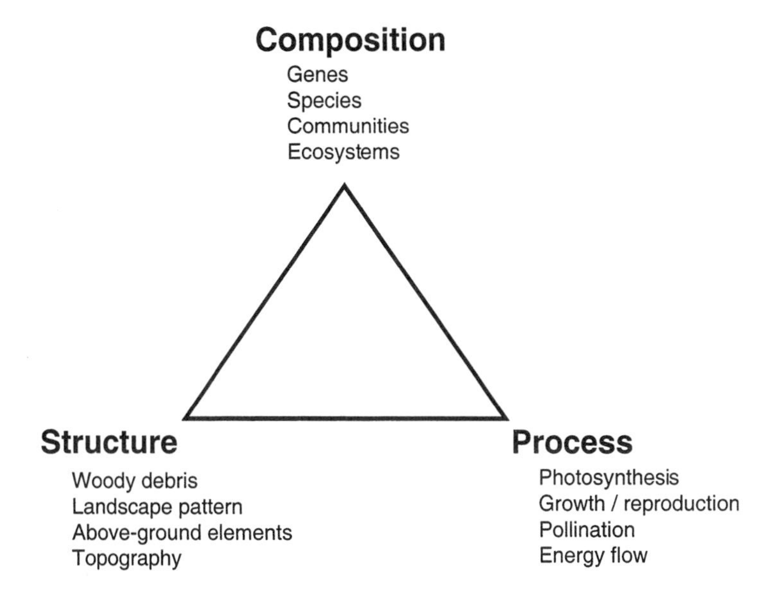

FIGURE 3.1. The key defining elements of all ecological systems, regardless of spatial or temporal scale, shown as an equilateral triangle to emphasize their interdependence. Included are example expressions of composition, structure, and process in ecological systems.

3.2.3 The Nonequilibrium Nature of Ecological Systems

The classic paradigm of ecology has been the stable-state ecological system, sometimes referred to as the "Balance of Nature" or "Nature at Equilibrium" (Pimm 1991). As our understanding of ecological systems has evolved, this view has been replaced by a nonequilibrium paradigm that recognizes the inherent dynamic nature of ecological systems (Pickett et al. 1992). Ecological systems are regularly subject to episodic, natural disturbances that lead to changes in composition, structure, or process.

It is important to note that this new, nonequilibrium paradigm in ecology has the potential for misuse. If nature is often in a state of flux, some may wrongly conclude that whatever changes occur to ecological systems are acceptable. Ecological systems are not infinitely resilient, however, and rates of change are bounded (Chapin et al. 1996). Human impacts must be constrained because ecological systems have adaptational limits that, if surpassed, will lead to undesirable conditions. Sustaining ecological processes so that they operate within their expected bounds of variation is the only way to sustain species diversity for future generations. Because ecological systems are inherently dynamic, change must be taken into account by quantifying standards for components of composition, structure, and process in terms of acceptable ranges and expected distributions rather than fixed values.

Even though we now recognize the nonequilibrium nature of ecological systems, we also recognize that over time frames relevant to human welfare the concept of stability of broad-scale landscapes is well-founded. At scales relevant to human societies, ecological systems have changed sufficiently slowly that there was apparent continuity in landscape processes across human generations.

3.2.4 The Significance of Natural Disturbance Processes

All landscapes, managed or pristine, contain a variety of natural resources that change over time and space. Over long periods, natural catastrophic events are both inevitable and essential in most ecological systems. Chronic but important changes may occur to alter the character of the vegetation and associated resources. These changes include succession, long periods of high or low precipitation, loss of site productivity via compaction or erosion, outbreaks of insects or disease, establishment and spread of non-native species, and loss of native species diversity. Although many natural processes are dynamic and often unpredictable, an appreciation of the expected intensity, frequency, and duration of disturbances must be factored into planning efforts. Alteration of natural disturbance regimes and processes can have profound effects on sustainability, pushing systems to new quasi-stable equilibria that defy attempts to restore them to a previous state (Rapport et al. 1998). An example of this phenomenon is the spread of invasive shrubs across the American Southwest (Cronk 1995).

3.2.5 Expect Ecological Surprises

Management of ecological systems is characterized by uncertainty arising from numerous sources. The primary source of uncertainty is the environment; disturbances that accumulate over time may generate unpredictable changes in ecological systems. This source of uncertainty cannot be controlled by management action. Most important to our discussion here is *scientific uncertainty,* which arises from an incomplete understanding of how ecological systems work or insufficient information to determine relations among processes. To predict the effects of management activities on the composition, structure, and processes of ecological systems requires a model of how the system works. For a model to provide reliable predictions, it must be correctly structured and accurately parameterized. However, most ecological models suffer from both structural and parameter uncertainty (Walters 1986). Often, there is incomplete information about relevant ecological processes, connections among ecosystem components, and the impacts of management. In addition, a system may respond differently to similar disturbances if initial conditions vary. Considering the contingent behavior of ecological systems, as well as additional uncertainties, management alternatives must reflect this uncertainty in model projections. Deterministic projections of a likely ecosystem state are no longer acceptable forms of analysis in natural resources management.

Adaptive management is a structured process of learning how best to manage ecological systems by comparing outcome with expectation and adjusting accordingly. In recent years, adaptive management has been proposed as an explicit way of addressing the uncertainty in management decisions, and a way of gaining understanding to reduce future uncertainties (Walters 1986). The incorporation of uncertainty in the formulation of management strategies distinguishes adaptive management from more traditional approaches to decision making (Williams 1996). Adaptive management has been characterized as embracing uncertainty, because it recognizes uncertainty as an attribute of management and uses management as a tool to accelerate the reduction of uncertainty (Johnson and Williams 1999).

3.2.6 Disturbances Accumulate Over Space and Time

In 1978, the Council on Environmental Quality (CEQ) defined *cumulative effects* as "the impact on the environment resulting from the incremental impact of the action when added to other past, present, and reasonably foreseeable actions regardless of what agency or person undertakes such actions." Implicit in this definition is the assumption that a specific cause-and-effect response to a management action can be identified. Also, the CEQ definition implies a simple additivity of effects, a phenomenon that seldom occurs because of multiple and nonlinear environmental responses to changes in the ecological system.

Natural disturbances and surprise events are a fundamental feature of both managed and unmanaged ecological systems. It is the interaction of these unpredictable events with ongoing management actions or previous land-use changes that must be anticipated in planning. Many types of adverse cumulative effects may not become apparent until disturbance occurs (Paine et al. 1998). Within a large diverse landscape, there is wide variation in site-specific practices and local environmental conditions (e.g., vegetation type, topography, geology, and soils). As a consequence, the collective direct and indirect effects of management practices may not be well-understood or easily predicted. Even when general patterns of cumulative effects become evident at watershed and bioregional scales (e.g., basin-wide and regional patterns of channel incision, reduced abundance or extent of specific plant or animal species, altered water quality), the contribution from individual sites may be difficult to estimate.

Unfortunately, few standard analytical methods are available that effectively address cumulative effects (Committee of Scientists 1999). *Integrated assessments,* conducted by physical, biological, and social scientists working collaboratively, would provide the analytic framework in which to consider feedbacks and cumulative effects; however, this field is just being formalized (Johnson et al. 1999).

3.2.7 Ecological Systems Have Hierarchial Structure

Effective management at broad spatial scales recognizes the hierarchical organization of ecological systems. Our discussion is not restricted to the usual view of the ecological hierarchy (genes, individuals, populations, species, communities, ecosystems, and landscapes) but includes the rates (time derivatives) that characterize specific ecological processes. A hierarchical approach to the assessment of ecological systems recognizes that smaller subsystems change more rapidly than do the larger systems to which they belong (Allen and Hoekstra 1992). At a landscape scale, processes operate slowly and constrain faster and more local processes, affecting structure and composition at smaller spatial scales (e.g., changes in forest canopy affect the distribution of sunlight and therefore rates of photosynthesis at the level of the individual leaf). Salthe (1985) states that entities at a given level in a biological hierarchy are constrained by the fact that they are composed of lower-level entities and constitute the subunits of higher-level entities. The higher-level constraints act as boundary conditions because they constrain the focal-level entities within a distinct subset of possible states.

Given a hierarchial perspective, current scientific understanding suggests that sustaining composition, structure, and process over multiple human generations requires management plans initially focused at a landscape scale. The lack of well-established theories that specify which level of the complex hierarchy of ecological systems is most appropriate for ecological sustainability, however, suggests caution. One danger is to assume that retention of ecological processes at landscape scales eliminates the need to focus on individual species at the local scale. The hierarchial nature of ecological systems suggests a dependency among nested levels within the hierarchy: broad-scale slow dynamics arise from fine-scale faster dynamics.

3.2.8 *Ecological Systems Must Be Assessed at Multiple Spatial Scales*

Patterns in complex ecological systems arise from processes operating at many spatial and temporal scales (Levin 1992). To simplify management practice, it is useful to apply a hierarchical assessment to identify the most relevant spatial and temporal scales for a particular management problem. Regional assessments may be based on bioregional characteristics, specifically those defined by discrete topographic boundaries. The next level down in spatial scale are areas such as watersheds, which follow hydrologic boundaries, or conservation areas that focus on habitats that cut across hydrological boundaries. Because watersheds can range in size from sub-basins to finer scales, watersheds also are represented at the fine scale of resolution. The finest scale of assessment is at the level of the individual project. The importance of multiscale assessments for the conservation of biological diversity is illustrated by considering the dynamics of any metapopulation. Inference about metapopulation stability requires information on the quality of individual patches, the degree of connectivity among neighboring patches, and the spatial extent (and thus degree of spatial independence) of the metapopulation (Wiens 1997).

3.2.9 *Temporal Dynamics of Landscape Change*

Some types and magnitudes of landscape change are accompanied by loss of species and declines in the efficacy of ecological processes necessary for sustained human welfare. This begs the question: What types, rates, and spatial extents of change are acceptable? Here it is useful to distinguish disturbance events that generally do not surpass the adaptational limits of biota (and to which ecological systems are mostly resilient) from human-induced stressors that may lead to novel ecological states.

Guidance for identifying acceptable levels of environmental change is provided by studying characteristics of natural disturbance processes. The assumption here is that the ecological systems we wish to sustain have evolved in the context of such disturbance regimes. Disturbance processes can be described by a number of defining attributes (Table 3.1; Pickett and White 1985; Pickett et al. 1989). Many of these attributes (e.g., return interval, spatial extent, intensity) can be characterized by probability distributions that define the limits of system accommodation without significant ecological change. That is, most natural disturbance processes can be characterized by attributes that can be portrayed as a set of bounded probability distributions. The intersection of such natural-state probability distributions could be used as a first approximation to guide human activities so that they lead to acceptable patterns of ecological change.

A composite abstraction adopted by many resource agencies in the United States is the concept of an *historic range of variability* (HRV). HRV refers to the expected variation in physical and biological conditions caused by natural climatic fluctuations and disturbance regimes. It is derived from an ecological history of a landscape and is estimated from the rate and extent of changes in selected physical and biological variables (Committee of Scientists 1999). HRV depends on the time period evaluated; the longer the time period, the greater the expected range of observed variation.

The concept of HRV of an ecological system is appropriately understood as a set of frequency distributions of physical and biological conditions—distributions with both dynamic shapes and dynamic ranges. It would be inappropriate to consider HRV solely in terms of the upper or lower value of the range of any given distribution. Equally important as a management goal is the shape, as well as the range, of these distributions. Variation in shape simply means that some ecological conditions are more likely than are others.

TABLE 3.1. Defining characteristics of natural disturbance events.

Characteristic	Definition
Type	Classification of the event
Frequency	Return interval
Spatial extent	Area affected
Spatial pattern	Configuration of the effects of the event
Intensity	Physical force
Selectivity	Range of components subject to change
Severity	Biological effect
Synergism	Interaction with other disturbances
Timing and seasonality	When the event occurs
Lag time	Period between the event and its expression

Invoking the HRV concept does not exclude human actions from the landscape. First, actions are often needed to shift altered systems back within the HRV. Second, HRV provides a target distribution of environmental conditions within which human action could operate without significant risk to the integrity of species and ecosystems. Conditions that exceed the HRV serve as warning signals that landscapes are beyond the bounds of historic conditions. Third, it must be recognized that the HRV is clearly no longer possible on some lands that have been permanently transformed by human behavior.

It is important to emphasize the reciprocal relationship between disturbance and landscape pattern. Disturbance events both respond to, and create, landscape pattern. As a consequence, to retain (and attain) desired landscape conditions, the public will have to accept and managers will have to accommodate the occurrence of natural disturbance events within managed ecological systems (Dale et al. 1998).

3.2.10 Methods to Assess Relations Between Landscape Patterns and Ecological Processes

In an *experiment,* events are controlled by the observer and allow for replication of experimental treatments, randomization of treatments across experimental units, and the existence of control units (see Eberhardt and Thomas 1991). Determining causal relations between variation in landscape patterns and variation in the composition, structure, and processes of ecological systems, however, is difficult to accomplish for several reasons. One is that most broad-scale studies cannot meet the strict requirements of experimentation—randomization, replication, and controls—because these are logistically precluded. A second is that even if such experiments were feasible, they often cannot be conducted because of ethical considerations. A final consideration is the complexity and contingencies that characterize most ecological systems. The number of covariates and initial conditions that can affect the state and dynamics of an ecological system is statistically intractable. When studying open systems characterized by multiple variables, it is very difficult to define benchmark conditions or to practice replication.

If the temporal or spatial scales of the landscape of interest are so broad as to preclude experimentation, it may be possible to study a model system that is tractable. Studies of *experimental model systems* (EMS) are done because they are amenable to manipulation and replication. The justification for EMS studies is that observations made at fine spatial scales can be used to obtain insights about pattern-process relations in similar systems operating at broader scales (Wiens et al. 1993). The key assumption required in the application of results from these studies is that the microscopic mechanisms (i.e., movement, connectivity, population dynamics) observed in the experimental systems can be scaled upward to derive macroscopic predictions for the landscape of interest (Ims 1999). Several recent studies have used this approach (e.g., Robinson and Quinn 1988; Wiens et al. 1995) and appear to have observed mechanisms and emergent properties that are transferrable to larger landscapes. However, validating the predictions from EMS to broad landscapes remains difficult and controversial. Despite the difficulty of rejecting the "wrong species, wrong scale" criticisms (Murphy 1989), EMS are valuable for developing the theoretical foundation of landscape ecology.

Quasi-experiments are partial experiments that lack one or more of the components of a true experiment. Often treatments cannot be randomly assigned to landscape units, or replication is not possible. The consequence of a nonrandom assignment of treatments, for example, is that any observed differences may be confounded with initial or subsequent differences among the units. In such cases, insights into cause-and-effect relations and inferences about landscapes beyond those directly studied may not be possible.

Because of increasing interest in the dynamics of complex systems not amenable to traditional experimentation, study of quasi-experimental designs is an active area of research (Manly 1992). Significant advances have been made in addressing designs in which spatial replication is not possible. *Before-after, control-intervention* (BACI) designs have been widely adopted in environmental impact studies (e.g., Underwood 1997) and are appropriate for planned or unplanned experiments occurring at broad spatial scales. The requirements for BACI designs are to have at least one control and treatment landscape that can be measured both before and after an impact has occurred. If multiple time periods are sampled both before and after the impact, the power of statistical inference from such studies can approach that obtained from true experiments.

Another possible solution to design constraints is to conduct *retrospective studies;* that is, use data that have been collected for other purposes, or take advantage of historical accidents in lieu of experiments. Such studies contrast with traditional *prospective studies,* which restrict analysis to data collected as part of the experiment and make no reference to past data or insights. Retrospective studies are particularly valuable for gaining insights about the behavior of complex systems at broad spatial scales. For example, it may not be feasible or ethical to explore experimentally the effects of broad-scale disturbance on ecological processes. However, it would be foolish to ignore "natural experiments" (i.e., changes arising from uncontrolled events) simply because they were not part of a planned experiment. Smith (1998) provides a rich discussion of many types of retrospective studies that improve our understanding of relations between pattern and process in landscape ecology, including, for example, assessment of the effects of landslides on water quality, effects of large fires on endangered species, or effects of other perturbations that would be unethical to conduct as experiments.

Inference in retrospective studies can be limited because of a lack of randomization. For example, it may be that locations where catastrophic fires occur are not representative of the landscape at large. Thus, a retrospective sample may be biased and may not allow reliable inference to the entire landscape. Studies of historical patterns of disturbance events, natural or human-induced, often suffer from a lack of randomization. Nevertheless, they can be very useful for estimating prior probabilities of certain events (e.g., the decline of an endangered species given the occurrence of the disturbance).

Observational studies do not involve a random assignment of treatments or any control exercised by the researcher. Thus, any inferences about causal relations, for example, between measured pattern and observed processes, are necessarily limited. Many observational studies in landscape ecology have attempted to correlate metrics of spatial pattern, or degree of change in these metrics, to one or more putative response variables. Given the difficulty of conducting experiments at broad spatial scales, observational studies will continue to play a key role in landscape studies, generating hypotheses that may only be testable with EMS.

Complex systems not amenable to direct experimentation can often be explored through *simulation modeling,* in which the dynamics of a system are distilled to a small number of linked mathematical equations. The challenge is to build a model that incorporates the key structure of the system, reproduces its essential dynamics in terms of a few state variables, and is not so complex as to be intractable. These models can incorporate the effects of spatial pattern on process or species dynamics by intersecting the model with real or simulated landscapes through a geographic information system interface. Such models have been used to simulate the dynamics of vegetation succession (e.g., Solomon and Bartlein 1992), disturbance regimes (e.g., Romme et al. 1995), or the movement of mobile animals in heterogeneous environments (e.g., Noon and McKelvey 1996). These models allow one to explore the possible effects of changes in landscape pattern when experimentation is not possible.

However, insights provided by simulation models should be viewed skeptically. Complex spatial models are difficult to parameterize and almost impossible to validate. Many theoretical insights have occurred as a result of simulating the dynamics of mobile organisms in simple, binary (suitable, unsuitable habitat) landscapes (e.g., With and King 1997). However, it is unclear to what extent inferences drawn from the analysis of simulated dynamics in such simplified landscapes (e.g., percolation thresholds) apply to movement in real, more complex landscapes.

3.3 Emerging Ideas and Issues

3.3.1 The Dynamics of Spatially Structured Populations and Communities

Future landscapes will be more human-dominated, with less extensive and more fragmented biotic communities, leading to disjunct species distributions. Much recent work, both theoretical and empirical, has demonstrated that the carrying capacity of a landscape for a given species is not a simple function of the amount and quality of its habitat (e.g., Hanski 1999). When habitat has been significantly reduced (e.g., <30% of the landscape within a species' historic range is currently suitable habitat), the spatial arrangement of habitat can be the determinant factor in the persistence or extinction of a species (Fahrig 1997).

Almost all plant and animal species have a period in their life cycle when movement, either passive or active, is the defining behavior. Movement enables organisms to colonize new areas and find essential resources (e.g., food and mates), and it functions to promote and sustain genetic diversity. Successful colonization movements require a source population, a target habitat, and a means of connecting the two. These elements capture the essential spatial components of a species conservation plan. However, it is not clear how we should build on our understanding of single-species dynamics and move toward multispecies conservation planning. Pragmatically, we will need to focus on a small number of species that are somehow representative of the whole. This approach requires some sort of ordination of species along meaningful axes of ecology, life history, behavior, and functional roles. Defining the appropriate axes for ordination and making appropriate choices for surrogate species will challenge researchers for the foreseeable future.

What land allocation strategies will Society implement to conserve biological diversity? If a reserve-design strategy is adopted, then it is important to recognize that a reserve is an open system affected by its internal characteristics and its spatial context. Factors originating external to the boundaries of the reserve may have a greater effect on local population dynamics than may factors intrinsic to the reserve. For a reserve, how small is too small? The ratio of disturbance extent to reserve extent provides some insights (e.g., Turner et al. 1993). If the natural disturbance regime required to sustain the historic heterogeneity of the reserve exceeds the current extent of the reserve, then the reserve is too small and species will be lost.

3.4 Ecological Connections Between Landscape Ecology and Conservation

The collective impression from the observations above is that by judicious control of landscape composition, structure, and processes, it may be possible to affect the expression of biological diversity. The hypothesized chain of causation is as follows: Geomorphic processes → Physical landscape pattern ← → Ecological processes ← → Emergent structure and composition → Observed biological diversity response. Note the existence of reciprocal interactions. Structural and compositional landscape elements, existing in both the physical and biological domains, are measurable expressions of underlying ecological processes. These elements in turn provide the resources and constitute the template for the expression of diverse ecological strategies (i.e., the biological diversity response). This conceptual model suggests a path from theory to application if we focus on the identification, measurement, and use of key landscape elements in biological conservation.

The challenge is to determine what configurations of landscape elements (patches, corridors, matrix) in human-dominated systems lead to conservation of the greatest biological diversity. Landscape ecology has demonstrated theoretically and empirically that the pattern, quality, and context of habitat patches can greatly affect the dynamics of individual species. However, these insights have arisen largely from theoretical studies or from EMS that lack the complexity (social and economic as well as biological) of broad-scale, human systems. As a consequence, only general "rules of thumb" can be offered at this time for the conservation of multispecies communities in real landscapes. If broad-scale ecological science is to better inform conservation biology, then it must more directly investigate the human factors leading to change in landscape pattern and process (see Dale et al. 2000).

3.5 General Approaches for Advancing Applications of Landscape Ecology in Biological Conservation

Approaches to conserving biological diversity focused at the level of the individual species are unlikely to be successful in the long term unless they are balanced by efforts to conserve the larger ecological systems to which species belong. A broad-scale perspective on biological conservation, explicitly incorporating insights from landscape ecology, provides the necessary balance between reductionism and holism needed for long-term conservation.

Incorporating the ecological principles discussed above into planning and management requires a consistent methodology and terminology that is general enough to be applied to different ecological systems and situations. Concepts and working definitions are needed to translate the factors that characterize these systems into measurable attributes. Without such attributes, it will be impossible to assess the degree to which management practices are consistent with an overall goal of sustaining biological diversity. Further, the methods to assess ecological systems should be applicable at multiple spatial scales, and nested, when possible. Nesting allows measurements taken at fine spatial scales to be accumulated so as to allow inference to broad spatial scales.

3.5.1 Essential Concepts

Ecological Integrity

Rapport et al. (1998) suggest devising "fitness tests" for ecosystems similar to those used by doctors. These tests are based on the hypothesis that healthy ecosystems are resistant to stress and recover more quickly, and they use measures based on speed of recovery after perturbation, adjusted for relative stress load, as preliminary assays of ecosystem health. For reasons discussed by Karr (1996), we avoid using the term "health" to characterize the desired state of an ecological system. The concept of *ecological integrity* has been put forth by the scientific community as an appropriate metric, and measurable definitions have been provided (e.g., Angermeier and Karr 1986; Karr 1991, 1996; DeLeo and Levin 1997). A system with integrity has the capacity to support and maintain a balanced, integrated, adaptive community of organisms having the full range of components (genes, species, assemblages) and processes (mutation, demography, biotic interactions, energy dynamics) expected from natural habitats of the region (see Karr 1996).

A fine-scale assessment of integrity stresses the structural and compositional aspects of ecological systems focusing on individual species and their dynamics within specific ecosystems. A more coarse-scale approach focuses on macroscale processes (i.e., primary productivity, nutrient cycling, hydrological regimes) and pays considerably less attention to the structure and composition of the systems from which these processes emerge. Integrity is complex, and its assessment requires a set of indicators measured at different spatial and temporal scales and at different levels in the hierarchy of ecological systems.

A conceptual framework for assessing ecological integrity at broad spatial scales has been proposed by McIntyre and Hobbs (2001). They identify four broad types of landscapes, with intact and completely degraded as extreme conditions, and variegated and fragmented as intermediate states. In variegated landscapes, the desired habitat(s) still forms the matrix, whereas in fragmented landscapes, the desired habitat is found only in the fragments. A landscape's position along this gradient is a function of the percent of desired habitat(s) remaining, the degree of connectivity of the landscape, and the degree and pattern of modification of the remaining habitat.

Desired Future Conditions

To set objectives for, and to assess the success of, management, the desired future states (or probability distribution of states) of ecological systems need to be specified. In general, *desired future conditions* should be those that sustain ecological integrity and native biological diversity over the long term. Desired future conditions can be derived by determining how natural disturbance events influence the distribution of terrestrial and aquatic habitats and the species dependent on them. This distribution then represents the target conditions for landscapes in which management occurs. An assessment procedure that addresses the dynamic aspects of ecological processes in the context of spatial and temporal disturbance history can provide a framework for establishing target ranges for desired future conditions (e.g., Delcourt and Delcourt 1999).

The concept of desired future condition is meaningful at broad spatial scales because it explicitly considers the mix of habitats (type and seral stage) generated by natural processes that are only observable at broad spatial scales. To sustain ecosystems and preserve ecological integrity, management must allow for the dynamic processes that accompany disturbance-recovery cycles, and protect essential energy and material transfers that take place during disturbance events. As a consequence, desired future conditions must include variability as an integral and essential component of habitat and population objectives.

The management challenge is to ensure that human activities do not increase the frequency or severity of disturbance events to such an extent that they exceed the ability of ecosystems to recover, or surpass the adaptational limits of species. To ensure resilience, management practices must not disrupt those energy and material transfers that promote habitat recovery. An appropriate goal for management activities would be to mimic, to the extent possible, natural disturbance events in terms of their severity, spatial extent, type, and frequency.

Focal Species

Because of the impossibility of monitoring the status and assessing the viability of all species, it is necessary to focus on a smaller subset of species. The generic term *focal species* has been proposed by Lambeck (1997) to describe a multispecies indicator or umbrella approach to protecting biological diversity at broad spatial scales. Lambeck (1997) proposed three distinct sets of species that are sensitive to landscape change: area- or habitat-limited species, movement-limited species, and management-limited species. For overall biodiversity protection, the task is to manage the landscape to meet the needs of the most sensitive species (the focal species) in each category.

Recently, a committee of scientists appointed by the U.S. Secretary of Agriculture (Committee of Scientists 1999) proposed that the defining characteristic of a focal species be that its status and temporal trend in abundance and distribution provide insights about the integrity of the larger ecological system to which it belongs. This definition is subtly, but importantly, different from that of Lambeck (1997). The defining characteristic is not necessarily sensitivity to change, but information content in terms of reliable induction from species status to the state of the larger ecological system. Candidates for focal species selection include several existing categories of species used to assess ecological integrity, as follows:

Indicator species—species whose status is believed to be indicative of the status of a larger functional group of species, be reflective of the status of a key habitat type, or act as an early warning to the action of an anticipated stressor to ecological integrity.

Keystone species—species whose effects on one or more critical ecological processes or biological diversity are much greater than would be predicted from their abundance or biomass.

Ecological engineers—species who, by altering the habitat to meet their needs, modify the availability of energy (food, water, or sunlight) and affect the fates and opportunities of other species.

Umbrella species—species whose habitat needs, such as large areas or multiple habitat types, encompass the habitat requirements of many other species.

Link species—species that play critical roles in the transfer of matter and energy among trophic levels or provide a critical link for energy transfer in complex food webs.

Phylogenetically distinct species—species with few or no extant relatives.

Available knowledge of the ecology of species and their functional roles in ecological systems is so limited that it is not always possible, *a priori,* to identify unambiguously the best set of focal species. Therefore, the selection of focal species, based on existing information and criteria for inclusion, should be treated as an interim decision subject to change based on new information. Because of this uncertainty, the assumption that a specific species serves a focal role must be validated by monitoring and research.

3.5.2 Current Research Priorities

The gap between theory and application in the management of ecological systems is extensive. Because of these gaps in our understanding, it is not currently possible to provide firm guidelines for the management of ecological systems at broad spatial scales. At best, general rules of thumb guide the management process. Because the detailed guidance required for on-the-ground implementation remains to be discovered, we end our chapter with a list of some key research priorities required to narrow the gap (Table 3.2).

Acknowledgments

We greatly appreciate the constructive criticism of an earlier version of our chapter by Divya Mudappa and Shankar Raman. V. Dale's work on this paper was funded by a contract from the Conservation Program of the Strategic Environmental

TABLE 3.2. Some key research priorities to enhance the application of broad-scale ecological science in biological conservation.

Theory:
- Develop a general theory explaining the reciprocal relations among the structure, composition, and processes of ecological systems.
- Develop a theoretical framework for predicting the development of landscape pattern following disturbance.

Concepts:
- Develop an assessment framework for managing at broad spatial scales to maximize the likelihood of sustaining biological diversity.
- Clarify the concept of ecological integrity so that it becomes more directly measurable and thus a practical management goal.

Applications:
- Develop methods to estimate ecologically relevant metrics of structure and composition at multiple spatial scales, including:
  - Structure and composition of specific landscape elements,
  - Configuration of elements at local scales,
  - Structure and composition of patches,
  - Configuration of patches at broad spatial scales.
- Develop methods to estimate reliably expected ranges and distributions of natural variation for ecological systems.
- Develop indicators that more accurately assess the status and trends of ecological systems (species, communities, ecosystems, and landscapes) and allow induction to multiple spatial and temporal scales.
- Identify indicators of landscape functionality and integrate these into broad-scale monitoring programs.

Research and Development Program (SERDP) with Oak Ridge National Laboratory (ORNL). ORNL is managed by UT-Battelle, LLC, for the U.S. Department of Energy under Contract DE-AC05-00OR22725.

References

Allen, T.F.H., and Hoekstra, T.W. 1992. *Towards a Unified Ecology.* New York: Columbia University Press.

Angermeier, P.L., and Karr, J.R. 1986. Applying an index of biotic integrity based on stream fish communities: considerations in sampling and interpretation. *N. Am. J. Fisheries Manage.* 6:418–429.

Chapin, F.S. III, Torn, M.S., and Tateno, M. 1996. Principles of ecosystem sustainability. *Am. Nat.* 148:1016–1037.

Committee of Scientists. 1999. *Sustaining the People's Land: Recommendations for Stewardship of the National Forests and Grasslands Into the Next Century.* Washington, DC: USDA Forest Service.

Cronk, Q.C.B. 1995. *Plant Invasions: The Threat to Natural Ecosystems.* London: Chapman and Hall.

Dale, V.H., Brown, S., Haeuber, R.A., Hobbs, N.T., Huntly, N., Naiman, R.J., Riebsame, W.E., Turner, M.G., and Valone, T.J. 2000. Ecological principles and guidelines for managing the use of land. *Ecol. Appl.* 10:639–670.

Dale, V.H., Lugo, A., MacMahon, J., and Pickett, S. 1998. Ecosystem management in the context of large, infrequent disturbances. *Ecosystems* 1:546–557.

Delcourt, H.R., and Delcourt, P.A. 1999. Paleoecological analysis of the legacy of past landscapes. In *Issues in Landscape Ecology,* eds. J.A. Wiens and M.R. Moss, pp. 51–54. Guelph, Ontario, Canada: International Association for Landscape Ecology.

DeLeo, G.A., and S.A. Levin. 1997. The multifaceted aspects of ecosystem integrity. *Conserv. Ecol.* [online] 1: 3. Available from the Internet: www.consecol.org/vol1/iss1/art3.

Eberhardt, L.L., and Thomas, J.M., 1991. Designing environmental field studies. *Ecol. Monogr.* 61:53–73.

Fahrig, L. 1997. Relative effects of habitat loss and fragmentation on population extinction. *J. Wildl. Manage.* 61:603–610.

Forman, R.T.T., and Godron, M. 1986. *Landscape Ecology.* New York: John Wiley and Sons.

Franklin, J.F., and Spies, T.A. 1991. Composition, function, and structure of old-growth Douglas-fir forests. In *Wildlife and Vegetation of Unmanaged Douglas-fir Forests,* eds. L.F. Rugierro, K.B. Aubry, A.B. Carey, and M.H. Huff, pp. 71–82. General Technical Report PNW-285, Portland, Oregon: USDA Forest Service.

Hanski, I. 1999. *Metapopulation Ecology.* New York: Oxford University Press.

Ims, R. 1999. Experimental landscape ecology. In *Issues in Landscape Ecology,* eds. J.A. Wiens and M.R. Moss, pp. 45–50. Guelph, Ontario, Canada: International Association for Landscape Ecology.

Johnson, F., and Williams, B.K. 1999. Protocol and practice in the adaptive management of waterfowl harvests. *Conserv. Ecol.* [online] 3: 8. Available from the Internet: www.consecol.org/vol3/iss1/art8.

Johnson, K.N., Swanson, F., Herring, M., and Greene, S., eds. 1999. *Bioregional Assessments.* Covelo, California: Island Press.

Karr, J.R. 1991. Biological integrity: a long neglected aspect of water resources management. *Ecol. Appl.* 1:66–84.

Karr, J.R. 1996. Ecological integrity and ecological health are not the same. In *Engineering Within Ecological Constraints,* ed. P.C. Schultz, pp. 97–109. Washington, DC: National Academy Press.

Lambeck, R.J. 1997. Focal species: a multi-species umbrella for nature conservation. *Conserv. Biol.* 11:849–856.

Levin, S.A. 1992. The problem of pattern and scale in ecology. *Ecology* 73:1943–1967.

Manly, B.F. 1992. *The Design and Analysis of Research Studies.* London: Cambridge University Press.

McIntyre, S., and Hobbs, R.J. 2001. Human impacts on landscapes: matrix condition and management priorities. In *Nature Conservation 5: Nature Conservation in Production Environments,* eds. J. Craig, D.A. Saunders, and N. Mitchell, pp. 301–307. Chipping Norton, Australia: Surrey Beatty and Sons.

Murphy, D.D. 1989. Conservation and confusion: wrong species, wrong scale, wrong conclusions. *Conserv. Biol.* 3:82–84.

Noon, B.R., and McKelvey, K.S. 1996. Management of the Spotted Owl: a case history in conservation biology. *Ann. Rev. Ecol. Syst.* 27:135–162.

Paine, R.T., Tegner, M.J., and Johnson, E.A. 1998. Compounded perturbations yield ecological surprises. *Ecosystems* 1:535–545.

Pickett, S.T.A., Kolasa, J., Armesto, J., and Collins, S.L. 1989. The ecological concept of disturbance and its expression at various hierarchial levels. *Oikos* 54:129–136.

Pickett, S.T.A., Parker, V.T., and Fiedler, P.L. 1992. The new paradigm in ecology: implications for conservation biology above the species level. In *Conservation Biology: The Theory and Practice of Nature Conservation, Preservation, and Management,* eds. P.L. Fiedler and S.K. Jain, pp. 65–88. New York: Chapman and Hall.

Pickett, S.T.A., and White, P.S., eds. 1985. *The Ecology of Natural Disturbance and Patch Dynamics.* New York: Academic Press.

Pimm, S.L. 1991. *The Balance of Nature? Ecological Issues in the Conservation of Species and Communities.* Chicago: University of Chicago Press.

Rapport, D.J., Whitford, W.G., and Hilden, M. 1998. Common patterns of ecosystem breakdown under stress. *Environ. Monitor. Assess.* 51:171–178.

Robinson, G.R., and Quinn, J.R. 1988. Extinction, turnover, and species diversity in an experimentally fragmented California annual grassland. *Oecologia* 76:71–82.

Romme, W.H., Turner, M.G., Wallace, L.L., and Walker, J.S. 1995. Aspen, elk, and fire in northern Yellowstone National Park. *Ecology* 76:2097–2106.

Salthe, S.N. 1985. *Evolving Hierarchial Systems: Their Structure and Representation.* New York: Columbia University Press.

Smith, G.J. 1998. Retrospective studies. In *Statistical Methods for Adaptive Management Studies,* eds. V. Sit and B. Taylor, pp. 41–53. Victoria, British Columbia: British Columbia Ministry of Forests.

Solomon, A.M., and Bartlein, P.J. 1992. Past and future climate change: response by mixed deciduous-coniferous ecosystems in northern Michigan. *Can. J. For. Res.* 22:1727–1738.

Tilman, D. 1999. The ecological consequences of changes in biodiversity: a search for general principles. *Ecology* 80:1455–1474.

Turner, M.G., Romme, W.H., Gardner, R.H., O'Neill, R.V., and Kratz, T.K. 1993. A revised concept of landscape equilibrium: disturbance and stability on scaled landscapes. *Landsc. Ecol.* 8:213–227.

Underwood, A.J. 1997. *Experiments in Ecology.* London: Cambridge University Press.

Walters, C.J. 1986. *Adaptive Management of Renewable Resources.* New York: MacMillan.

Wiens, J.A. 1995. Landscape mosaics and ecological theory. In *Mosaic Landscapes and Ecological Processes,* eds. L. Hansson, L. Fahrig, and G. Merriam, pp. 1–26. London: Chapman and Hall.

Wiens, J.A. 1997. Metapopulation dynamics and landscape ecology. In *Metapopulation Biology: Ecology, Genetics, and Evolution,* eds. I.A. Hanski and M.E. Gilpin, pp. 43–62. New York: Academic Press.

Wiens, J.A. 1999. The science and practice of landscape ecology. In *Landscape Ecological Analysis: Issues and Applications,* eds. J.M. Klopatek and R.H. Gardner, pp. 371–383. New York: Springer-Verlag.

Wiens, J.A., Crist, T.O., With, K., and Milne, B.T. 1995. Fractal patterns of insect movement in microlandscape mosaics. *Ecology* 76:663–666.

Wiens, J.A., Stenseth, N.C., Van Horne, B., and Ims, P.A. 1993. Ecological mechanisms and landscape change. *Oikos* 66:369–380.

Williams, B.K. 1996. Adaptive optimization and the harvest of biological populations. *Math. Biosci.* 136:1–20.

With, K.A., and King, A.W. 1997. The use and misuse of neutral landscape models in ecology. *Oikos* 79:219–229.

Section II

Multiple Scales, Connectivity, and Organism Movement

4

Spatial Factors Affecting Organism Occurrence, Movement, and Conservation: Introduction to Section II

KEVIN J. GUTZWILLER

Conservation effectiveness frequently depends on knowledge about multiscale factors that affect organism distributions, and about environmental conditions that enable organisms to move across landscapes. Such information is valuable for accomplishing conservation objectives as diverse as maintaining natural distributions and dispersal processes, ensuring genetic diversity, establishing corridors and networks, and controlling invasion by non-native species. Section II addresses these interrelated topics.

4.1 Multiscale Effects

Environmental conditions at different spatial scales can affect an organism's occurrence at a given location (e.g., Wiens 1989; Knick and Rotenberry 1995), and patterns of habitat use by a species at one scale are not necessarily predictable from those at another scale (Gutzwiller and Anderson 1987). The relative importance of fine-, intermediate-, and broad-scale influences on organism distributions seems to vary widely among species, perhaps because of differences among species' habitat requirements, biological traits (e.g., body size, vagility, migratory status), and abilities to perceive stimuli from different scales (see Bissonette et al. 1997 for a related discussion). If a given scale's influences are ecologically significant, it is conceivable that a lack of information about those influences or failure to apply such knowledge could seriously compromise the success of a management program. Multiscale effects, discussed in Chapter 5, should be addressed for a full realization of benefits from conservation efforts.

4.2 Movement Across Landscapes

For most organisms of conservation concern, movement across landscapes is a common and necessary process that enables them to meet their life-history needs in heterogeneous environments (Forman 1995). Organisms disperse from natal areas to establish their territories and home ranges. They travel to different parts of landscapes to secure resources such as food, water, shelter, and breeding habitat. Movement across landscapes is especially common for large and vagile species, as well as for those that require diverse habitats during their annual cycle. Not all organisms require contiguous conditions (e.g., soil, water, vegetation) for these movements; indeed, stepping-stone and other configurations of suitable patches may suffice for some species (Forman 1995).

To maintain natural movement patterns and their important biological consequences, it is crucial for conservationists to understand translandscape movement and the landscape conditions (especially the degree of spatial continuity, or *connectivity*) that make it possible. Species dispersal and landscape corridors, discussed in Chapter 6, are becoming increasingly important conservation issues as degradation of many natural habitats continues. Chapter 7 considers percolation theory and how it can be used to determine landscape connectivity and effects of habitat fragmentation.

4.3 Consequences of Landscape Connectivity and Species Movements

Landscape connectivity and the associated movement of organisms have important ramifications for biological conservation. Landscapes that exhibit connectivity of natural land-cover types often support more native species than do those in which such connections have been severed (Soulé and Terborgh 1999). This is not surprising. Many organisms evolved in highly connected systems (Dobson et al. 1999) and have had little time to adapt to human-caused landscape changes that retard or preclude movement. Patterns of organism distribution and abundance result from movements that are controlled largely by landscape linkages of useable habitat (Saunders and Hobbs 1991), and such organism patterns are often the crux of important conservation concerns (e.g., population viability, gene flow, evolutionary potential).

Landscape connectivity and organism movement are key underlying factors affecting the distribution and maintenance of genetic diversity, the establishment and management of functional habitat networks, and the control of invasion by non-native species. Chapter 8 examines relations between landscape connections and genetic diversity. Habitat networks and how they can be used in conservation are the topics of Chapter 9. And Chapter 10 addresses landscape invasibility by exotic species. As we develop a better understanding of these issues, and as associated concepts of landscape ecology are applied more widely, success in conserving biological resources should improve.

References

Bissonette, J.A., Harrison, D.J., Hargis, C.D., and Chapin, T.G. 1997. The influence of spatial scale and scale-sensitive properties on habitat selection by American marten. In

Wildlife and Landscape Ecology: Effects of Pattern and Scale, ed. J.A. Bissonette, pp. 368–385. New York: Springer-Verlag.

Dobson, A., Ralls, K., Foster, M., Soulé, M.E., Simberloff, D., Doak, D., Estes, J.A., Mills, L.S., Mattson, D., Dirzo, R., Arita, H., Ryan, S., Norse, E.A., Noss, R.F., and Johns, D. 1999. Corridors: reconnecting fragmented landscapes. In *Continental Conservation: Scientific Foundations of Regional Reserve Networks,* eds. M.E. Soulé and J. Terborgh, pp. 129–170. Washington, DC: Island Press.

Forman, R.T.T. 1995. *Land Mosaics: The Ecology of Landscapes and Regions.* Cambridge, United Kingdom: Cambridge University Press.

Gutzwiller, K.J., and Anderson, S.H. 1987. Multiscale associations between cavity-nesting birds and features of Wyoming streamside woodlands. *Condor* 89:534–548.

Knick, S.T., and Rotenberry, J.T. 1995. Landscape characteristics of fragmented shrub-steppe habitats and breeding passerine birds. *Conserv. Biol.* 9:1059–1071.

Saunders, D.A., and Hobbs, R.J., eds. 1991. *Nature Conservation 2: The Role of Corridors.* Chipping Norton, Australia: Surrey Beatty and Sons.

Soulé, M.E., and Terborgh, J., eds. 1999. *Continental Conservation: Scientific Foundations of Regional Reserve Networks.* Washington, DC: Island Press.

Wiens, J.A. 1989. *The Ecology of Bird Communities. Volume 2. Processes and Variations.* Cambridge, United Kingdom: Cambridge University Press.

5

Patch-, Landscape-, and Regional-Scale Effects on Biota

KATHRYN FREEMARK, DANIEL BERT, AND MARC-ANDRÉ VILLARD

5.1 Introduction

Modification of landscape mosaics by human activities is a critical issue in the conservation of biodiversity (Heywood and Watson 1995). Although the current focus on recovery of endangered species is necessary to prevent their extinction, a single species approach is impractical and impossible over the longer term (Noss et al. 1997). Conserving viable ecosystems and ecological integrity by maintaining both their structure and function (including appropriate change and disturbance regimes) across a hierarchy of spatial scales over time (Figure 5.1) is a more proactive and efficient approach than is creating habitat conditions specific to a single endangered species. In this context, "fine-filter" approaches for particular populations or species are mainly useful to monitor the efficiency of "coarse-filter" approaches for habitats and ecosystems.

Our biological focus in this chapter is at the species level, particularly nonfish vertebrates (mostly birds) and, to a lesser extent, plants and insects. A multiscale approach to conserving biodiversity requires interfacing with human social systems that influence allocation of resources and land use across a variety of scales (Brunckhorst and Rollings 1999; Rollings and Brunckhorst 1999; White et al. 1999), but we do not consider this interaction further because it is beyond the scope of this book.

In Section 5.2, we define key concepts that relate to conserving biota across multiple scales. Next (Section 5.3), we review examples of recent applications of those concepts at patch, landscape, and regional scales. In Section 5.4, we extract a set of principles useful for applying landscape ecology to biological conservation at each scale. Then we highlight a few key theoretical and empirical knowledge gaps (Section 5.5) and briefly describe research approaches (Section 5.6) to address those gaps.

5.2 Concepts, Principles, and Emerging Ideas

The following definitions were derived in consultation with a variety of sources, including Forman (1995) and chapters in Soulé and Terborgh (1999a).

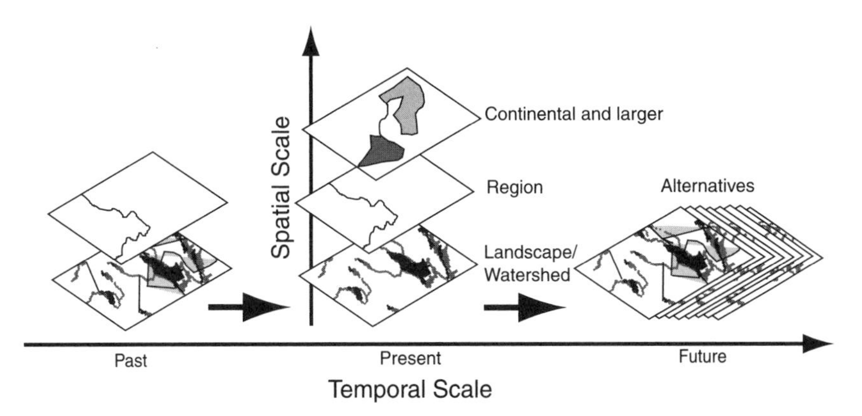

FIGURE 5.1. A multiscale, spatiotemporal perspective for conserving biodiversity.

Connectivity or *permeability:* degree to which patches or landscapes are linked by the flow of organisms through intervening patch types possibly via habitat corridors or stepping stones.

Core or *interior:* the inner portion of a patch where the environment differs significantly from the edge of the patch.

Corridor: a physical linkage between habitat patches within a landscape that may serve as a pathway by which organisms move or interchange, or as a habitat in which organisms can feed or breed en route from one patch to another.

Disturbance regime: a set of natural or anthropogenic events that significantly alters the pattern of variation in the structure or function of an ecological system.

Edge: the portion of a patch near the perimeter where the environment differs significantly from the core or interior of the patch. Edge width differs around a patch (e.g., wider on sides facing the predominant wind direction and solar exposure). The term also is used in reference to the periphery of a species' range.

Edge contrast: degree of difference between a patch and the patch types surrounding it.

Habitat buffer zone: an area around a patch of interest that retains some degree of naturalness but allows sustainable economic uses that are compatible with the goals for the patch they surround; a transition zone that surrounds and protects natural core areas and primary corridors; an area that permits a greater range of human uses than do core reserves but that is managed with native biodiversity as the preeminent concern.

Habitat location: the spatial position of a habitat patch within a landscape.

Habitat target: amount of suitable habitat required for the long-term persistence of a species within a landscape.

Interspersion: degree to which a given patch or landscape type is scattered rather than aggregated or clumped.

Juxtaposition: adjacency of different patch or landscape types.

Landscape: a mosaic of patches. The relevant scale depends on the organism (or ecological process) being studied and the questions being asked.

Landscape linkage or *corridor:* a large connection that is meant to facilitate animal movements and other essential flows across and among regions; primarily facilitates the dispersal of animals from their place of birth to their adult home range where they breed.

Minimum area: the least amount of suitable habitat required for the long-term persistence of a species within a patch or landscape.

Mosaic: collection of patches at the landscape scale, or a collection of landscapes at the regional scale.

Native range: The historic range of a species.

Nested subsets and *nestedness:* when species composition for sites with low species richness (e.g., small patches) are largely subsets of species compositions for sites with higher species richness (e.g., larger patches).

Orientation: the geographical orientation of a patch.

Patch: a relatively homogeneous area that differs from its surroundings. For the purposes of this chapter, it is synonymous with "stand" in a forestry context. Patches are described by their size distribution, including minimum, mean, and variance.

Proximity: distance among similar habitat patches or landscapes within a region.

Region: an area composed of landscapes with the same macroclimate and suite of human activity. For the purposes of this chapter, a region is intermediate in scale between landscape and continental.

Shape: the geometry of a patch. A more convoluted patch will have a higher proportion of edge habitat and therefore a higher edge:interior ratio.

Sink: an area where input of immigrants exceeds output of emigrants; opposite of source.

SLOSS: acronym for "single large or several small" in relation to conservation of patches that support the most species in a landscape or region.

Source: an area where output of emigrants exceeds input of immigrants; opposite of sink.

Spatially hierarchical approaches: methods that simultaneously consider multiple spatial scales nested within each other.

Stepping stone: a patch of suitable habitat between larger patches of suitable habitat at landscape, regional, and broader scales.

Wildlands: Vast wilderness-like areas without roads, dams, motorized vehicles, and so on, that are strictly protected and managed for conservation of nature.

5.3 Recent Applications

Table 5.1 summarizes the 25 documents reviewed in Table 5.2 for the key concepts related to conserving biodiversity (e.g., endangered species; landbirds; wildlife in riparian areas, wetlands, grasslands, farmlands, forest lands, wildlands, or mosaic landscapes) across scale. All sources are planning documents. We reviewed case studies in Arnold (1995), Dramstad et al. (1996), Williams et al. (1997), and Peck (1998). We also reviewed the North American Shorebird

TABLE 5.1. Documents reviewed for consideration of spatial scale in conserving biodiversity.

Document	Focus	Reference
1	Endangered species	Noss et al. 1997
2	Biodiversity	Peck 1998
3	USA watersheds	Williams et al. 1997
4a	Great Lakes basin wetlands	Environment Canada, Ontario Ministry of Natural Resources, and Ontario Ministry of Environment 1998
4b	Great Lakes basin riparian areas	Environment Canada, Ontario Ministry of Natural Resources, and Ontario Ministry of Environment 1998
4c	Great Lakes basin forests	Environment Canada, Ontario Ministry of Natural Resources, and Ontario Ministry of Environment 1998
5	Grasslands in Massachusetts	Massachusetts Audubon Society 1998
6	Grasslands in Wisconsin	Sample and Mossman 1997
7	Wilderness areas around Chicago	Chicago Wilderness Project, unpublished report
8	Greenways in Maryland	Weber and Wolf 2000
9	Landbirds in the USA	Bonney et al. 2000; Pashley et al. 2000
10	Forests in Ontario	Strobl 1998
11	Forests in Canada	Canadian Council of Forest Ministers 1995
12	Forests in Canada	Canadian National Strategy Coalition 1998
13	European farmland	LANDECONET Research Consortium 1997
14	Natural areas in Ontario	Ontario Ministry of Natural Resources 1998
15	Australian farmland	Lambeck 1999
16a	Landscape mosaics	Harrison and Fahrig 1995; Kozakiewicz 1995; Wiens 1995
16b	Conservation programs	Arnold 1995
17	Forest	Hansen et al. 1999
18	Wildlands	Soulé and Noss 1998; Soulé and Terborgh 1999a; Soulé and Terborgh 1999b
19	Land-use planning projects	Dramstad et al. 1996

Conservation Plan, the North American Waterfowl Management Plan, and the Partners in Flight program as examples of conservation initiatives spanning local to intercontinental scales; we were only able to source reference documents from Partners in Flight with sufficient detail for inclusion in this chapter (Bonney et al. 2000; Pashley et al. 2000). We did not include recent project developments to aid wildlife in crossing barriers (e.g., roads) because of their local and specific nature and because they are covered elsewhere in this book (Chapter 13).

5.4 Principles for Applying Landscape Ecology

The following principles for applying landscape ecology to biological conservation were derived from our knowledge of the literature and our review of the documents identified in Table 5.1 and summarized in Table 5.2. Our coverage is extensive in its

TABLE 5.2. Concepts, principles, and emerging ideas across scale included in documents identified in Table 5.1. Documents are listed in decreasing order of the number of concepts, principles, and emerging ideas identified. Within each scale, concepts, principles, and emerging ideas are listed in decreasing order of inclusion in the documents with slight deviation to provide for grouping into principles (indicated by different symbols).

Chapter section	Concepts, principles, emerging ideas	Documents																					
		16b	18	3	15	6	16a	19	1	14	2	8	13	4a	5	4c	9	7	11	17	4b	10	12
5.4.1	**Patch scale**																						
	η more contiguous area	■	■	■	■	■	■	■	■	■	■	■	■	■	■	■					■		
	η larger, more compact area		■	■	■		■			■	■	■	■				■	■	■			■	
	η more core and lower edge:interior				■	■	■	■	■	■		■	■	■	■	■						■	
	η minimum area		■		■	■	■	■		■			■		■	■	■			■	■		
	B maintain suitable habitat composition and quality	■	■	■	■	■	■			■				■	■	■		■		■			
	B maintain suitable vegetation composition, structure, and age class			■	■					■							■		■	■			
	B less adverse interspecific interactions					■									■			■	■				
	B less human access	■		■					■		■												
	B maintain or simulate natural change and disturbance regime					■									■						■		
	B less edge contrast	■				■									■								
	B less adverse management practices	■				■									■								
	> optimize geographic orientation	■			■		■	■				■											

5.4.2 Landscape scale

\# higher connectivity and permeability

\# provide habitat buffer zones

\# less interpatch distance

\# more corridors, linkages, and stepping stones

\# consider habitat location

\# consider habitat juxtaposition/interspersion

< enhance mosaic composition

< maintain minimum habitat targets

< higher suitable habitat proportion

< maintain or simulate natural change and disturbance regime

< fewer roads and less human access

∀ provide variance in patch sizes

∀ minimum large patch size

∀ larger mean patch size

∀ consider patch size distribution

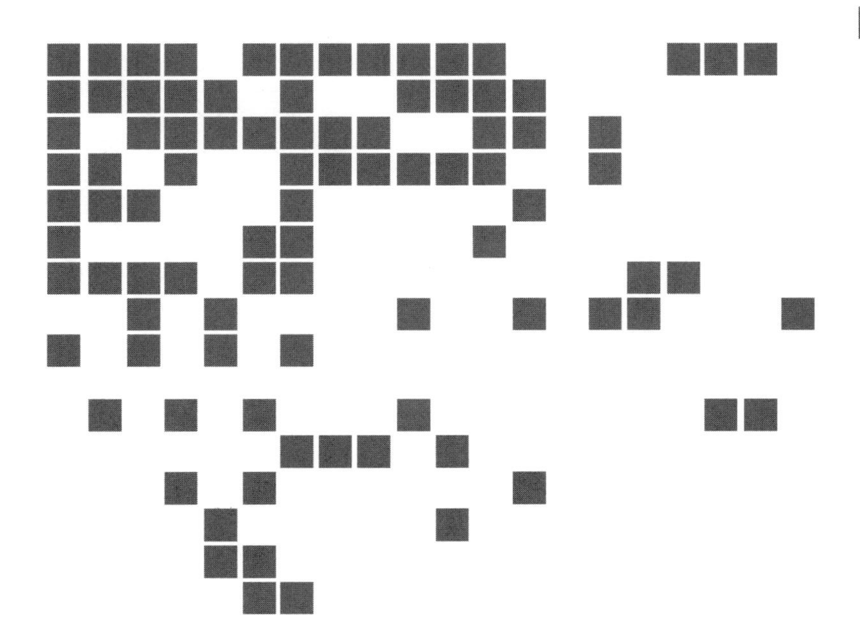

TABLE 5.2. *Continued.*

Chapter section	Concepts, principles, emerging ideas	Documents																					
		16b	18	3	15	6	16a	19	1	14	2	8	13	4a	5	4c	9	7	11	17	4b	10	12
5.4.3	**Regional and larger scales**																						
	# higher interconnectivity	■	■	■					■		■												
	# more landscape linkages, corridors, and stepping stones	■	■	■					■														
	# closer proximity to source landscapes or regions	■	■						■														
	< fewer roads and less human access	■	■	■					■														
	< maintain natural change and disturbance regime		■						■														■
	< maintain habitat across species native range		■		■				■														
5.4.4	**Spatially hierarchical approaches**	■				■		■	■								■			■			

breadth but not in its depth of any given subject area (particularly at the patch and landscape scales) because of the burgeoning literature. For brevity, we have concentrated on reviews as much as possible. The principles are not intended as recipes, but to provide a background for the professional to use in planning and implementing strategies for biological conservation in a variety of contexts (particularly agricultural, but also forestry and, to a lesser extent, urban).

5.4.1 Patch Scale

Principle: Maintain Suitable Habitat Composition and Quality Within Patches.

This principle is well-recognized in the existing literature and was noted in most of the documents included in Table 5.2, either explicitly or in relation to other descriptors. In harvested landscapes (e.g., farmland, forestry), this may require managing at the landscape scale to provide continuing availability of suitable habitat.

The occurrence and abundance of species is strongly influenced by patch or stand conditions (cf., Freemark et al. 1995; Drapeau et al. 2000). Each species requires an array of critical resources (e.g., vegetation, food types) to ensure its survival and reproduction. For Neotropical migrant birds, species richness, abundance, and reproductive success within a patch also can be adversely affected by elevated densities of deer (DeCalesta 1994; McShea et al. 1995), and nest predators and brood parasites associated with high ratios of habitat edge to interior (Freemark et al. 1995; Donovan et al. 1997). To provide suitable habitat over time, it is essential to maintain (and in some situations even simulate) natural change (e.g., succession) or disturbance regimes (Pickett and White 1985; Allen and Hoekstra 1992), although this has not typically been an explicit goal of either reserve design or management (Arnold 1995). This may require managing the degree of human disturbance (Knight and Gutzwiller 1995). For example, urban forests experience significantly higher levels of disturbance than do their rural counterparts, with the probability of disturbance in at least some cases depending more on the distance to the nearest dwelling than on local human population density (Matlack 1994; Friesen et al. 1995; Friesen 1998). Effects of management practices on wildlife and their habitats continues to be an active area of research, particularly for Neotropical migrant landbirds (Martin and Finch 1995) and in relation to forestry (Chapter 14), grazing (Dobkin et al. 1998), and cropland agriculture (Freemark 1995; Freemark and Boutin 1995; Rodenhouse et al. 1995; Pain and Pienkowski 1997). Practices that have been found to have adverse effects should be avoided.

Principle: Maintain Larger, More Compact Areas.

This principle has become well-recognized in the existing literature and was noted either explicitly or in relation to other descriptors in all but two of the documents included in Table 5.2.

In a review of 61 studies considering patch effects on a wide variety of taxa (insects, amphibians, birds, mammals), Mazerolle and Villard (1999) found that in addition to habitat characteristics, patch area, and to a lesser degree patch shape, significantly influenced species presence and abundance of most taxa examined. Effects of patch area have received a great deal of attention in the island biogeography literature (reviewed by Shafer 1990). Area effects on species richness have been the object of considerable debate because increases in species richness with area would be expected simply because larger patches contain more individuals and, therefore, would be expected to contain more species. Deterministic effects of patch area on the composition of species assemblages have been suggested based on the observation of nested subsets, whereby species assemblages of depauperate patches are subsets of assemblages of species-rich patches (Wright et al. 1998). The degree of nestedness is then related back to patch characteristics such as patch area or habitat diversity. Blake (1991) found a significant degree of nestedness in the composition of forest bird assemblages in the midwestern USA, especially for forest-interior species and for Neotropical migrants. Calmé and Desrochers (1999) also found significant nestedness in bird assemblages of peatlands of southern Québec, Canada. However, the influence of peatland area on the degree of nestedness was mainly attributable to microhabitat diversity. Honnay et al. (1999) also found a stronger effect from habitat diversity than from fragment area on the degree of nestedness of forest-interior plant species in Belgium.

The nature of the edge of a patch affects habitat suitability and movement (Forman 1995; Wiens 1995). Single-brooded, low-nesting, Neotropical migratory species are believed to be particularly vulnerable to the higher rates of predation and nest parasitism associated with nonforest edges (Whitcomb et al. 1981). As a result, mated individuals often breed less successfully in small forests or larger forests with high ratios of nonforest edge to interior (Donovan et al. 1997). However, as Villard (1998) points out, differential density responses to patch area among patch-interior, patch-edge, or generalist species strongly depend on the definition of these ecological categories. Current classifications for many species are not adequately supported by empirical data on distributions relative to edges. For example, few so-called forest-interior species have been shown to avoid edges and to reproduce mainly in the interior of patches.

Burke and Nol (1998a) found in their study area in central Ontario that forest size had little influence on forest-interior conditions because edge effects penetrated less than 20 m. In the eastern USA, Matlack (1994) recommended that 92 m from an edge was required to protect interior habitat from vegetational changes.

Sensitivity to smaller patch area has been observed in the occurrence or density of forest and grassland birds (Freemark et al. 1995); species richness of birds, mammals, herptiles, and plants in wetlands in Ontario (Findlay and Houlahan 1997); forest birds in Europe (Hinsley et al. 1995); and in the density of animals in general (Bender et al. 1998). Quantifying area thresholds, particularly at broad spatial scales, is currently an active topic of research (Trzcinski et al. 1999; Villard et al. 1999). In northeastern and central North America, forest fragments smaller than 10 ha are unsuitable for many bird species (Robbins et al. 1989; Freemark and Collins 1992). Forests as large as 3,000 ha may be required to retain all species of the forest-interior avifauna (Robbins et al. 1989). When habitat is severely limited, smaller forests can provide important habitat for bird species during migration (Blake 1986) and for some breeding species (Blake and Karr 1987). Based on nest monitoring, Trine (1998) suggests that forests as large as 2,500 ha may be required in the midwestern USA to maintain source populations for some Neotropical migrant bird species.

Population demography within fragments can vary with the quantity and quality of different habitats within a fragment, fragment size, and shape. Smaller populations of habitat-specialist plants in smaller patches can have higher probabilities of extinction (Quintana-Ascencio and Menges 1996). Villard et al. (1992) and Villard et al. (1995) observed year-to-year changes in the distribution of four species of Neotropical migrant birds among forest fragments in agricultural landscapes of eastern Ontario. Patches that exhibited population turnovers between years were smaller and more isolated from other occupied patches.

Principle: Orient Patch to Intercept Wildlife Movement and Enhance Habitat Suitability.

Bird species abundances were significantly different between west-slope and east-slope canyons in Nevada (Dobkin and Wilcox 1986). Larger, oblong patches of riparian woodland oriented perpendicular to migration paths intercepted more migrating birds and thereby influenced species richness and abundance during the breeding season (Gutzwiller and Anderson 1992). In their review, Mazerolle and Villard (1999) found that when patch orientation was considered, it significantly influenced species presence and abundance of most taxa, although less so than did habitat characteristics and patch area. Orientation also may influence vegetation composition (Burke and Nol 1998a) and microclimate (e.g., litter desiccation; Burke and Nol 1998b).

5.4.2 *Landscape Scale*

The scale of a landscape is an emergent property of the perspective of the species or process of interest, resulting from an interaction between the scale at which the species or process operates and the scale of the landscape pattern (Freemark et al. 1995; Gustafson 1998). Therefore, the choice of an appropriate spatial scale and resolution (grain) for managing for biological conservation at a landscape scale needs to be made in relation to the organisms and processes of interest over a particular time frame. In general, landscape boundaries defined on the basis of ecological criteria related to the management problems at hand are more useful for conservation planning (especially the scientific aspects of planning) than are those defined by conventional political or administrative jurisdictions (Noss et al. 1997). Watersheds are becoming an increasingly popular spatial unit for natural resource management. However, because many physical and ecological processes do not respect watershed boundaries, it is critical to recognize that many natural resource management decisions made inside a given watershed will affect resources and people in other geographic areas (Williams et al. 1997).

In the review by Mazerolle and Villard (1999), only a small proportion of studies on birds, mammals, reptiles, and amphibians failed to detect significant effects of landscape context on the species richness or presence and abundance of species, whereas most studies conducted on invertebrates did not detect such effects. Therefore, landscape context should be considered in wildlife management and planning at least for vertebrates. Figure 5.2 illustrates key concepts of spatial (and temporal) effects on the probability that species will persist in a landscape or region.

Principle: Maintain Landscape Mosaics That Are More Permeable.

Considering a landscape as a mosaic of sites (with their associated successional change and natural and anthropogenic disturbance regime) with a complexity of spatial interactions is emerging as a more instructive concept than is considering the landscape as a set of "patches" and "corridors" in a "matrix" (Wiens 1995; Dale et al. 2000; Drapeau et al. 2000). Higher connectivity of the landscape, which in turn is related to the permeability of intervening habitats, enhances the suitability of a landscape mosaic to biota. However, connectivity of a landscape has to be defined from the perspective of the species or process of interest. Species operating at the same spatial scale may have different perceptions of whether a given landscape is connected depending on their particular habitat preferences and their ability or willingness to cross gaps of unsuitable habitat. Dispersal or gap-crossing abilities dictate the scale at which organisms interact with landscape pattern, and the gap or patch structure of a landscape is a function of the scale of disturbance or habitat destruction, whether natural or anthropogenic. Habitat corridors and stepping stones can enhance connectivity among patches and permeability of the mosaic for at least some species (Dobson et al. 1999).

A number of recent studies have documented increasingly adverse effects on forest birds from changes in landscape composition along a forest–agriculture–urban gradient (Blair 1996; Bolger et al. 1997). In Saskatchewan, Bayne and Hobson (1997) found that the percentage of artificial ground nests (but not shrub nests) destroyed at the edge or interior of forest patches in logged and contiguous forest landscapes was significantly lower than that at the edge or interior of patches in an agricultural landscape. Donovan et al. (1997) found that predation of artificial ground nests in forests in the midwestern USA was significantly higher in edge habitats than in core habitats, and in highly fragmented landscapes compared with unfragmented landscapes. For birds in western USA riparian habitats, large woodland patches surrounded by natural landscape should be maintained and acquired before conserving large patches surrounded by agriculture, if maintaining high species richness of native birds is a management objective (Saab 1999). Friesen et al. (1995) found that Neotropical migrant bird species breeding in forest fragments declined or disappeared altogether as the level of residential development increased (from 0 to >25 houses within 100 m of the forest edge). Because of these types of effects, some authors (Noss et al. 1997; Groom et al. 1999) have argued for a concentric zoning model, with protection increasing inward and

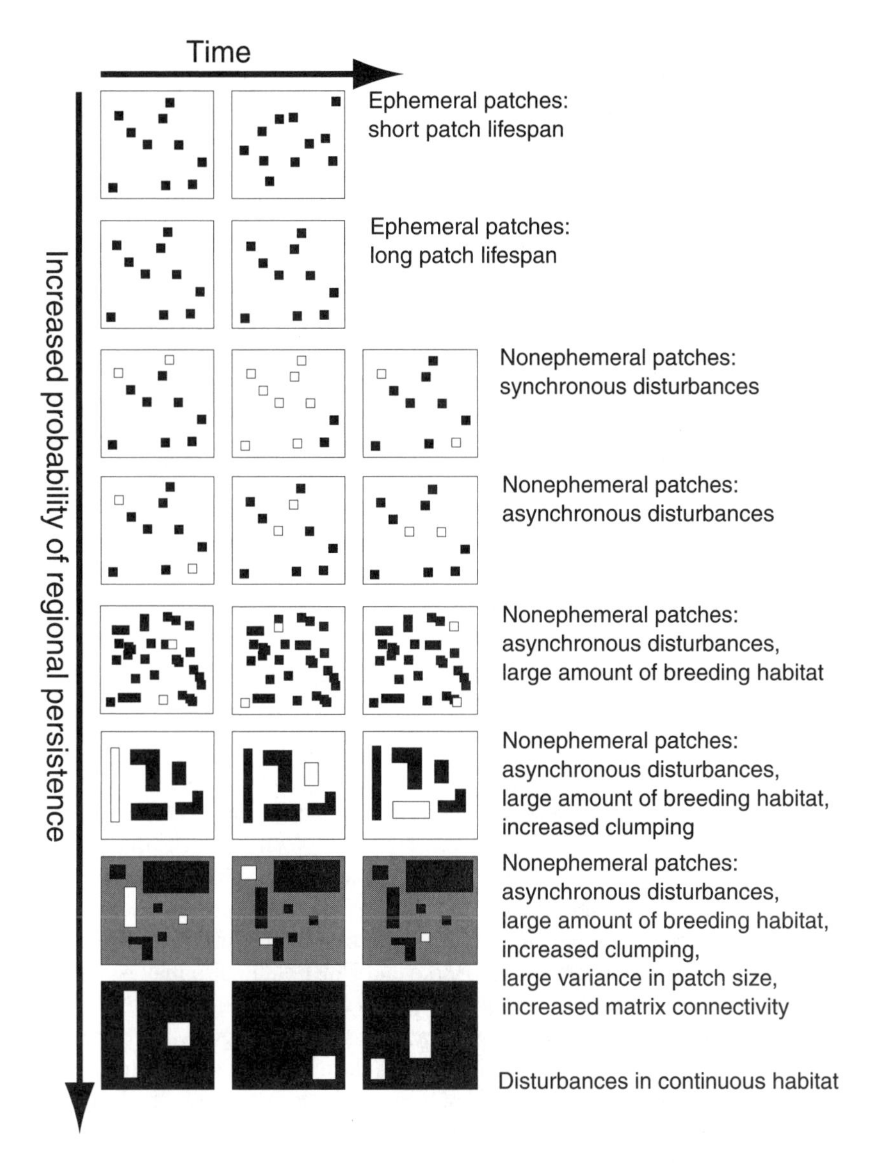

FIGURE 5.2. Key spatial and temporal effects derived from modeling studies on the probability that species will persist in a set of patches in a landscape or region (adapted from Harrison and Fahrig 1995). Ephemeral patches (i.e., the dark boxes in the first two panels) change spatial location over time. Nonephemeral patches (i.e., the dark and light boxes in the lower panels) tend to remain over time, although their populations may (dark boxes) or may not (white boxes). Increased matrix connectivity is shown by gray shading.

intensity of human use increasing outward from sites of high conservation value or sensitivity, at least in situations in which buffer zones are not likely to be population sinks for sensitive species or sources for invasions of exotics or other opportunistic species into reserves.

More corridors, habitat linkages, or stepping stones can facilitate the movement of some animals (Haas 1995). In Europe, Hinsley et al. (1995) found that the length of hedgerow in the landscape surrounding woodland fragments was positively related to the presence of a number of breeding bird species in particular woodland fragments. Because narrow corridors are subject to disruption by natural disturbances, more than one linkage between isolated natural areas is needed to maintain the benefits of connectivity through time. Some authors argue that broader corridors that display landscape-scale processes (such as patch dynamics) are required if plants and less-mobile animals are to maintain contact with other populations of their species (Csuti 1991). Small habitat patches that function as stepping stones can be important for some species, such as birds on migration (Blake 1986).

Principle: Maintain Landscape Mosaics With Sufficient Proportions of Suitable Habitat.

It is not surprising that species benefit from an increase in the proportion of suitable habitat at the landscape scale. For birds in native habitats, this occurs in part because landscape composition influences relative rates of nest parasitism (Robinson et al. 1995; Donovan et al. 1997) and nest predation (Hartley and Hunter 1998; Bergin et al. 2000), as well as abundance of predators (Oehler and Litvaitis 1996; Pedlar et al. 1997). Heavily forested landscapes (which tend to be found in areas managed for timber) have lower rates of predation and nest parasitism than do agricultural or suburban landscapes (Bayne and Hobson 1997; Hartley and Hunter 1998). Furthermore, fragmentation effects on microclimate within patches, such as the desiccation of litter and its effect on soil invertebrates (Burke and Nol 1998b), also may be less significant when the proportion of woodland exceeds a certain threshold value.

Based on simulation modeling, Venier and Fahrig (1996) argued that by increasing the success rate of dispersers, an increase in the amount of suitable habitat will also increase the occurrence and local abundance of a species across the landscape. This, in turn, would be expected to increase species persistence. From an empirical study, Gibbs (1998) suggested that the tolerance of amphibians to woodland fragmentation is inversely related to their dispersal ability, probably because with less woodland, most dispersers fail to locate suitable habitat and to recruit into breeding populations. Trzcinski et al. (1999) found that the proportion of woodland was a better predictor of the presence of 31 forest bird species in 10 × 10-km landscapes than was woodland configuration. Villard et al. (1999) found that configuration as well as proportion of woodland were significant predictors of landscape occupancy for 15 forest bird species in fragments (>5 ha) surveyed in smaller landscapes (2.5 × 2.5 km) over two successive years. The area of surrounding habitat was positively related to the breeding presence in particular woodland fragments for a number of bird species in Europe (Hinsley et al. 1995; Jansson and Angelstam 1999) and for herptile and mammal (but not bird or plant) species richness in wetlands in southeastern Ontario, Canada (Findlay and Houlahan 1997).

From a conservation perspective, it would be valuable to know whether thresholds exist in the proportion of habitat above which species persistence in a landscape is virtually ensured, and if so, whether these thresholds are species-specific or relatively general among ecologically similar species. As noted by Noss et al. (1997), what may be a substantial amount of land in one region may be far too little in another. The amount of protected land required to meet conservation goals will depend on many factors, including the specific objectives of the plan, the physical and biotic heterogeneity of the planning region (regions that are more heterogeneous or that have higher endemism will require more protected area), the area requirements of target taxa, and the area necessary for natural disturbances and other ecological processes to function normally. In highly fragmented and degraded landscapes, conservation goals may be best achieved by setting aside less land and concentrating more on habitat restoration and active management to enhance and sustain habitat quality. In some of these landscapes, the land available for conservation, even if protected in its entirely, may be too little to attain many conservation goals.

Using simulation modeling, Fahrig (1997) found that above 20% of suitable habitat, the probability of persistence of a hypothetical population was virtually ensured, irrespective of the spatial arrangement of the habitat. For birds and mammals, effects of habitat fragmentation, beyond simple habitat loss, are not as pronounced when landscapes have at least 30% suitable habitat cover (Robbins et al. 1989; Freemark and Collins 1992; Andrén 1994). Thresholds also have been found in the reproductive success of some songbird species in relation to the proportion of woodland in the landscape (Robinson et al. 1995).

Specifying area thresholds can be problematic because they can be highly species-specific (Villard et al. 1999). Furthermore, population demography within fragments can vary not only with habitat quantity and quality, fragment size, shape, and spatial configuration, but also with the regional context of the study area and the location within the geographic range of the species of interest. Ideally, managers should use regionally specific thresholds exhibited by the most sensitive species rather than average values across species and regions (Monkkonen and Reunanen 1999). Specification of habitat targets is beginning to show up in land-use planning guidelines (Table 5.2).

Harrison and Fahrig (1995) contend that when landscape pattern changes over time, the rate of change is far more important than is the spatial pattern in affecting population survival (Figure 5.2). Important aspects of spatiotemporal pattern for disturbances are rate, size, and temporal correlation (synchronicity). For ephemeral patches, rates of patch formation and patch lifespan also influence a species' persistence probability.

Historically, conservation efforts have focused on protection of resources. Inclusion of restoration is becoming increasingly significant to management agencies (Williams et al. 1997). This requires creating a system similar but not equivalent to the natural one to accommodate human uses. Human intervention may be needed to simulate natural change and disturbance regimes that have been disrupted by anthropogenic disturbances (Allen and Hoekstra 1992). The impacts of landscape transformation by roads and subsequent degradation in habitat suitability is extensively reviewed by Forman and Alexander (1998) and discussed in Chapter 13.

Principle: With Sufficient Suitable Habitat,
Patch Size Distribution Is of Secondary Importance.

Conservation approaches, particularly for birds, should focus on the overall proportion of habitat in the landscape; they should consider the size and shape of individual fragments and the variance among them only in the most fragmented landscapes or in agricultural landscapes in general, where negative edge effects have been shown to be more intense than in landscapes managed for timber. The degree to which species in smaller patches constitute nested subsets of species in larger patches will influence whether maintaining single large or several small (SLOSS) patches will be necessary for meeting conservation objectives.

5.4.3 Regional Scale

Current consensus among biologists is that protected areas are necessary but not sufficient for conserving biodiversity (Noss et al. 1997). Interconnected reserves enveloped in well-managed, multiple-use buffer zones or surrounding landscapes may offer the best hope of maintaining biodiversity and meeting human needs over the long term, particularly in the face of climate change. This strategy is generally what regional conservation plans attempt to achieve, except in urban landscapes where the role of buffer zones is often questionable. In regions where expansive reserve networks cannot be developed, smaller protected areas potentially play the role of safe-guarding some of the species and habitats most sensitive to human activities, in addition to supplying educational and aesthetic value (Shafer 1990). If reserves are too small in the context of the surrounding land use—or if adequate buffers are lacking—these values will be compromised by edge effects and the inability of the reserves to support even temporary populations of species with large home ranges or naturally low densities.

Principle: Maintain Closer Proximity to, and Higher Connectivity With,
Source Landscapes and Regions.

Broad-scale regional analyses coupled with studies of reproductive success are beginning to show that source-sink population dynamics may be operating across landscapes and regions for top carnivores (Mladenoff et al. 1995) and forest songbirds (Donovan et al. 1995a,b; Robinson et al. 1995), particularly Neotropical migrants (Anders et al. 1997; Bollinger et al. 1997; Trine 1998). For some species, smaller and lower-quality patches within a landscape may only be occupied in years when regional populations are high (Probst and Weinrich 1993).

Csuti (1991) and Soulé and Terborgh (1999a,b) argue that the best way to maintain connectivity is to have a network of large natural areas connected by landscape linkages. If future changes in land use were planned to maintain the connectivity of wildlands, whole regions or even continents could act as functional biosphere reserves—a planned mosaic of exploited, seminatural, and protected areas that function both economically and biologically.

Principle: Maintain Sufficient Suitable Habitat Across Species' Native Ranges.

Researchers in the midwestern USA (Donovan et al. 1995a,b; Robinson et al. 1995; Donovan et al. 1997) argue that until the spatial scale at which source and sink populations interact can be determined, large tracts of forest should be maintained throughout the breeding range of at least some migratory bird species to ensure their long-term viability. A similar concern has been expressed for the eastern USA (Hoover et al. 1995). Interregional or continental-scale ecology is still in its infancy, and more research will be necessary to document and understand phenomena taking place across this spatial scale. Implementation of conservation strategies at these broad scales is further complicated by the number of owners and jurisdictions that need to be involved.

As at the patch and landscape scales, it is important to maintain natural change and disturbance regimes at the regional scale to provide suitable habitat for species. For many species, this may include maintaining large expanses of roadless areas (Soulé and Terborgh 1999a,b) and areas with little or no human intrusion (Knight and Gutzwiller 1995).

5.4.4 Spatially Hierarchical Approaches

A number of researchers (Freemark 1995; Freemark et al. 1995; Probst and Thompson 1996; Noss et al. 1997; White et al. 1999; Donovan et al. 2000) have argued that a comprehensive conservation strategy needs to consider multiple geographic scales (spanning, for example, local to continental scales) and multiple temporal scales (Figure 5.1). Some of the documents reviewed in Table 5.2 noted the need to use spatially hierarchical approaches.

5.5 Knowledge Gaps

Based on our review of the existing literature and planning documents, the following emerged as some of the most important gaps to address to further our understanding of how to incorporate a multiscale perspective into biodiversity conservation.

5.5.1 Theoretical Voids

Cross-Scale Extrapolation

Integration and extrapolation among different scales is becoming a central problem in landscape ecology and other disciplines. There is great concern about how well predictions based on knowledge from one scale (e.g., patch) can be extrapolated to a broader scale (e.g., region). Loss of data and error propagation are major problems with changes in grain size and study extent (Gustafson 1998; Withers and Meentemeyer 1999). Withers and Meentemeyer (1999) note that the explosive growth in digital data collected from satellites, and widespread distribution of global and regional data sets on the World Wide Web, will create great demand for development of methods for cross-scale extrapolation and error analysis.

Multiscale Population Modeling

Spatially explicit population models (Dunning et al. 1995) are increasingly being used in landscape ecology and other disciplines to predict spatial patterns of species, populations, and vegetation types; to identify critical habitat for species conservation; or to simulate effects of landscape change scenarios on individual species. Typically, models are constructed at a single spatial scale. To understand population dynamics more fully, modeling needs to be linked across different scales.

5.5.2 Empirical Voids

Habitat Targets and Area Thresholds

Predicting threshold effects from habitat loss and fragmentation is a major challenge facing conservation biologists. However, most researchers continue to treat habitat loss and habitat fragmentation as one process because the two rarely occur independently in the natural world (Fahrig 1997). Consequently, the effects of fragmentation per se, without confounding effects from habitat loss, are only beginning to be understood (Fahrig 1997; Trzcinski et al. 1999). More empirical studies are needed to answer questions such as those posed by Harrison and Fahrig (1995): To what extent can alteration of landscape pattern compensate for loss of habitat? Can a high probability of population persistence be maintained while reducing the amount of habitat available, by carefully selecting the sizes and spatial locations of the remaining habitat fragments? Under what circumstances (and for what kinds of species in what kinds of landscapes) is spatially explicit modeling necessary for predicting the effects of habitat fragmentation on population persistence? How does the nature and quality of the intervening habitat affect area thresholds observed empirically? To what extent do area sensitivity and area thresholds for a species vary among regions? How can the effects of alteration of habitat pattern on species diversity in the landscape be predicted?

Empirical Data to Parameterize Spatially Explicit Population Models

Simulation modeling based on solid empirical data on a variety of taxa and spatial scales should enable us to understand better how species respond to the loss and fragmentation of natural habitats, and to identify threshold levels or ranges in this response. Regional conservation strategies could then be designed using the thresholds exhibited by those species most sensitive to loss and fragmentation (Monkkonen and Reunanen 1999). However, the trend in the biological sciences is toward an increasing number of professionals with expertise in computer sciences and modeling and a declining number of field naturalists with expertise and experience to collect reliable, basic data on species that can inform population models (Noss et al. 1997). Basic field data also are needed for reliable assessments of the status and trends of target species, habitat suitability (e.g., density vs. productivity), adult and juvenile dispersal (when, how, how far), and other factors for target species and their interactions, particularly for taxa other than birds. For birds, consistently designed field studies are needed across species' ranges to collect the data needed to assess spatiotemporal variability in species distributions, productivity, and survival, and to assess persistence and source-sink metapopulation dynamics across scales.

5.6 Research Approaches

Research at landscape and regional scales has been problematic to date because of the difficulty of replicating studies and conducting experiments at such broad scales (Wiens 1995). A couple of solutions have been to use small-scale experimental model systems and computer modeling, both of which are most useful when validated with field studies.

5.6.1 Approaches for Theoretical Research

Cross-Scale Extrapolation

Freemark et al. (1995), Probst and Thompson (1996), White et al. (1999), and Donovan et al. (2000) illustrate explicit hierarchical frameworks that attempt to include goals, objectives, and specific actions at different scales in space and time for conserving landbirds; these approaches also can be applied to biodiversity more generally. A hypothetical example for a landscape in Florida is presented in Box 4.3 in Noss et al. (1997). Withers and Meentemeyer (1999) recommend Ehleringer and Field (1993) as a particularly useful volume to consult in relation to the development of methods for cross-scale extrapolation and error analysis.

Multiscale Population Modeling

Spatially explicit models have been linked across scales (quadrat, community, and landscape) to allocate human activities in management areas (Childress et al. 1999). Holt et al. (1995) linked different spatially explicit models to demonstrate scale-sensitive problems. Withers and Meetenmeyer (1999) cite Cherrill et al. (1995) as an example of a spatially hierarchical approach to predict plant species distributions across a region in northern England.

5.6.2 Approaches for Empirical Research

Habitat Targets and Area Thresholds

Brennan et al. (In press) discuss alternative statistical designs for conducting landscape-scale studies to investigate issues such as effects of habitat loss versus fragmentation, or habitat-area thresholds. They view a focal-patch approach, with nonoverlapping landscapes as the study unit, as the most feasible design (i.e., in each study unit, only a single patch at the center of the landscape is sampled). It represents a reasonable compromise between sample size and sampling intensity. We have generalized their design to a "focal point" or "focal station" approach (Figure 5.3) to accommodate landscapes in which patches are not easily defined (e.g., forested landscapes such as those studied by Drapeau et al. 2000).

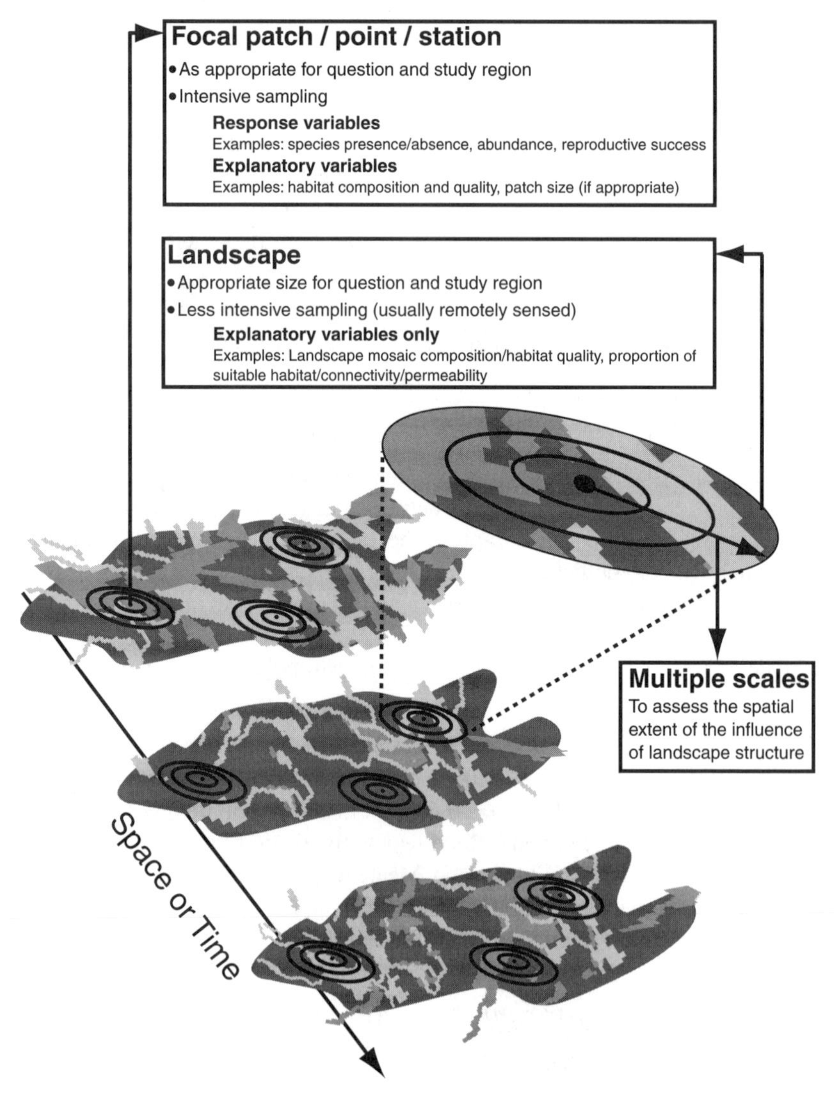

FIGURE 5.3. A focal patch, point, or station approach to conducting landscape-scale studies (adapted from Brennan et al. In press).

Empirical Data to Parameterize Spatially Explicit Population Models

It is beyond the scope of this chapter to describe how to collect basic natural-history information needed for simulation modeling for the range of species relevant to understanding multiscale effects on biota. Although some techniques are readily available and easily used, others are labor-intensive (Sutherland et al. 2000) or still need to be developed to solve particular challenges in working with certain taxonomic groups and life-history traits (e.g., tracking dispersal of songbirds). A variety of broad-scale survey techniques, some of which include measurement of fitness parameters, have recently been designed (Bowman et al. 2000; Gunn et al. 2000).

Acknowledgments

We thank Kevin Gutzwiller and the editorial staff at Springer-Verlag for assistance in production of this chapter. Stéphane Menu, Jeff Bowman, and David Kirk provided reviews that were timely and helpful. Vanessa Lyon and Laura Kenney provided invaluable technical assistance in the closing hours. Funding for D. Bert was provided by a NSERC research operating grant to K. Freemark; M.-A. Villard was funded by his NSERC research operating grant.

References

Allen, T.F.H., and Hoekstra, T.W. 1992. *Toward a Unified Ecology.* New York: Columbia University Press.

Anders, A.D., Dearborn, D.C., Faaborg, J., and Thompson, F.R., III. 1997. Juvenile survival in a population of Neotropical migrant birds. *Conserv. Biol.* 11:698–707.

Andrén, H. 1994. Effects of habitat fragmentation on birds and mammals in landscapes with different proportions of suitable habitat: a review. *Oikos* 71:355–366.

Arnold, G.W. 1995. Incorporating landscape pattern into conservation programs. In *Mosaic Landscapes and Ecological Processes,* eds. L. Hannson, L. Fahrig, and G. Merriam, pp. 309–337. London: Chapman and Hall.

Bayne, E.M., and Hobson, K.A. 1997. Comparing the effects of landscape fragmentation by forestry and agriculture on predation of artificial nests. *Conserv. Biol.* 11:1418–1429.

Bender, D.J., Contreras, T.A., and Fahrig, L. 1998. Habitat loss and population decline: a meta-analysis of the patch size effect. *Ecology* 79:517–533.

Bergin, T.M., Best, L.B., Freemark, K.E., and Koehler, K.J. 2000. Effects of landscape structure on nest predation in roadsides of a midwestern agroecosystem: a multiscale analysis. *Landsc. Ecol.* 15:131–143.

Blair, R.B. 1996. Land use and avian species diversity along an urban gradient. *Ecol. Appl.* 6:506–519.

Blake, J.G. 1986. Species-area relationship of migrants in isolated woodlots in east-central Illinois. *Wilson Bull.* 98:291–296.

Blake, J.G. 1991. Nested subsets and the distribution of birds on isolated woodlots. *Conserv. Biol.* 5:58–66.

Blake, J.G., and Karr, J.R. 1987. Breeding birds of isolated woodlots: area and habitat relationships. *Ecology* 68:1724–1734.

Bolger, D.T., Scott, T.A., and Rotenberry, J.T. 1997. Breeding bird abundance in an urbanizing landscape in southern California. *Conserv. Biol.* 11:406–421.

Bollinger, E.K., Peer, B.D., and Jansen, R.W. 1997. Status of Neotropical migrants in three forest fragments in Illinois. *Wilson Bull.* 109:521–526.

Bonney, R., Pashley, D.N., Cooper, R.J., and Niles, L., eds. 2000. *Strategies for Bird Conservation: The Partners in Flight Planning Process.* Proceedings RMRS-P-16. Ogden, Utah: USDA Forest Service, Rocky Mountain Research Station.

Bowman, J., Forbes, G., and Dilworth, T. 2000. The spatial scale of variability in small mammal populations. *Ecography* 23:328–334.

Brennan, J.M., Bender, D.J., Contreras, T.A., and Fahrig, L. In press. Focal patch landscape studies for wildlife management: optimizing sampling effort across scales. In *Integrating Landscape Ecology into Natural Resource Management,* eds. J. Liu and W.W. Taylor. London: Cambridge University Press.

Brunckhorst, D.J., and Rollings, N.M. 1999. Linking ecological and social functions of landscapes: I. Influencing resource governance. *Nat. Areas J.* 19:57–64.

Burke, D.M., and Nol, E. 1998a. Edge and fragment size effects on vegetation of deciduous forests in Ontario, Canada. *Nat. Areas J.* 18:45–53.

Burke, D.M., and Nol, E. 1998b. Influence of food abundance, nest-site habitat, and forest fragmentation on breeding Ovenbirds. *Auk* 115:96–104.

Calmé, S., and Desrochers, A. 1999. Nested bird and micro-habitat assemblages in a peatland archipelago. *Oecologia* 118:361–370.

Canadian Council of Forest Ministers. 1995. *Defining Sustainable Forest Management: A Canadian Approach to Criteria and Indicators.* Ottawa: Natural Resources Canada, Canadian Forest Service.

Canadian National Strategy Coalition. 1998. *National Forest Strategy '98.* Ottawa: Government of Canada.

Cherrill, A.J., McClean, C., Watson, P., Tucker, K., Rushton, S.P., and Sanderson, R. 1995. Predicting the distributions of plant species at the regional scale—a hierarchical matrix model. *Landsc. Ecol.* 10:197–207.

Childress, W.M., McLendon, T., and Price, D.L. 1999. A multiscale ecological model for allocation of training activities on U.S. Army installations. In *Landscape Ecological Analysis: Issues and Applications,* eds. J.M. Klopatek and R.H. Gardner, pp. 80–108. New York: Springer-Verlag.

Csuti, B. 1991. Conservation corridors: countering habitat fragmentation. In *Landscape Linkages and Biodiversity,* ed. W.E. Hudson, pp. 81–90. Covelo, California: Island Press.

Dale, V.H., Brown, S., Haeuber, R.A., Hobbs, N.T., Huntly, N., Naiman, R.J., Riebssame, W.E., Turner, M.G., and Valone, T. 2000. Ecological principles and guidelines for managing the use of land. *Ecol. Appl.* 10:639–670.

DeCalesta, D.S. 1994. Effect of white-tailed deer on songbirds within managed forests in Pennsylvania. *J. Wildl. Manage.* 58:711–718.

Dobkin, D.S., Rich, A.C., and Pyle, W.H. 1998. Habitat and avifaunal recovery from livestock grazing in a riparian meadow system of the northwestern Great Basin. *Conserv. Biol.* 12:209–221.

Dobkin, D.S., and Wilcox, B.A. 1986. Analysis of natural forest fragments: riparian birds in the Toiyabe Mountains, Nevada. In *Wildlife 2000: Modeling Habitat Relationships of Terrestrial Vertebrates,* eds. J. Verner, M.L. Morrison, and C.J. Ralph, pp. 293–299. Madison, Wisconsin: University of Wisconsin Press.

Dobson, A., Ralls, K., Foster, M., Soulé, M.E., Simberloff, D., Doak, D., Estes, J.A., Mills, L.S., Mattson, D., Dirzo, R., Arita, H., Ryan, S., Norse, E.A., Noss, R.F., and Johns, D.

1999. Corridors: reconnecting fragmented landscapes. In *Continental Conservation: Scientific Foundations of Regional Reserve Networks,* eds. M.E. Soulé and J. Terborgh, pp. 129–170. Covelo, California: Island Press.

Donovan, T.M., Freemark, K.E., Maurer, B.A., Petit, L., Robinson, S.K., and Saab, V.A. 2000. Setting local and regional objectives for the persistence of bird populations. In *Strategies for Bird Conservation: The Partners in Flight Planning Process,* eds. R. Bonney, D.N. Pashley, R.J. Cooper, and L. Niles, pp. 53–59. Proceedings RMRS-P-16. Ogden, Utah: USDA Forest Service, Rocky Mountain Research Station.

Donovan, T.M., Jones, P.W., Annand, E.M., and Thompson, F.R., III. 1997. Variation in local-scale edge effects: mechanisms and landscape context. *Ecology* 78:2064–2075.

Donovan, T.M., Lamberson, R.H., Kimber, A., Thompson, F.R., III, and Faaborg, J. 1995a. Modeling the effects of habitat fragmentation on source and sink demography of Neotropical migrant birds. *Conserv. Biol.* 9:1396–1407.

Donovan, T.M., Thompson, F.R., III, Faaborg, J., and Probst, J.R. 1995b. Reproductive success of migratory birds in habitat sources and sinks. *Conserv. Biol.* 9:1380–1395.

Dramstad, W.E., Olson, J.D., and Forman, R.T.T. 1996. *Landscape Ecology Principles in Landscape Architecture and Land-Use Planning.* Washington, DC: Island Press.

Drapeau, P., Leduc, A., Giroux, J.-F., Savard, J.-P.L., Bergeron, Y., and Vickery, W.L. 2000. Landscape-scale disturbances and changes in bird communities of boreal mixed-wood forests. *Ecol. Monogr.* 70:423–444.

Dunning, J.B., Stewart, D.J., Danielson, B.J., Noon, B.R., Root, T.L., Lamberson, R.H., and Stevens, E.E. 1995. Spatially explicit population models: current forms and future uses. *Ecol. Appl.* 5:3–11.

Environment Canada, Ontario Ministry of Natural Resources, and Ontario Ministry of Environment. 1998. *A Framework for Guiding Habitat Rehabilitation in Great Lakes Areas of Concern.* Downsview, Ontario: Environment Canada, Environmental Conservation Branch, Conservation Strategies Division.

Ehleringer, J.R., and Field, C.B., eds. 1993. *Scaling Physiological Processes: Leaf to Globe.* San Diego, California: Academic Press.

Fahrig, L. 1997. Relative effects of habitat loss and fragmentation on population extinction. *J. Wildl. Manage.* 61:603–610.

Findlay, C.S., and Houlahan, J. 1997. Anthropogenic correlates of species richness in southeastern Ontario wetlands. *Conserv. Biol.* 11:1000–1009.

Forman, R.T.T. 1995. *Land Mosaics: The Ecology of Landscape and Regions.* Cambridge, United Kingdom: Cambridge University Press.

Forman, R.T.T., and Alexander, L.E. 1998. Roads and their major ecological effects. *Ann. Rev. Ecol. Syst.* 29:207–231.

Freemark, K. 1995. Assessing effects of agriculture on terrestrial wildlife: developing a hierarchical approach for the US EPA. *Landsc. Urban Plan.* 31:99–115.

Freemark, K., and Boutin, C. 1995. Impacts of agricultural herbicide use on terrestrial wildlife in temperate landscapes: a review with special reference to North America. *Agric. Ecosys. Environ.* 52:67–91.

Freemark, K.E., and Collins, B. 1992. Landscape ecology of birds breeding in temperate forest fragments. In *Ecology and Conservation of Neotropical Migrant Landbirds,* eds. J.M. Hagan and D.W. Johnston, pp. 443–454. Washington, DC: Smithsonian Institution Press.

Freemark, K.E., Dunning, J.B., Hejl, S.J., and Probst, J.R. 1995. A landscape ecology perspective for research, conservation and management. In *Ecology and Management of Neotropical Migratory Birds: A Synthesis and Review of Critical Issues,* eds. T.E. Martin and D.M. Finch, pp. 381–427. New York: Oxford University Press.

Friesen, L. 1998. Impacts of urbanization on plant and bird communities in forest ecosystems. *For. Chron.* 74:855–860.

Friesen, L.E., Eagles, P.F.J., and MacKay, R.J. 1995. Effects of residential development on forest-dwelling Neotropical migrant songbirds. *Conserv. Biol.* 9:1408–1414.

Gibbs, J.P. 1998. Distribution of woodland amphibians along a forest fragmentation gradient. *Landsc. Ecol.* 13:263–268.

Groom, M., Jensen, D.B., Knight, R.L., Gatewood, S., Mills, L., Boyd-Heger, D., Mills, L.S., and Soulé, M.E. 1999. Buffer zones: benefits and dangers of compatible stewardship. In *Continental Conservation: Scientific Foundations of Regional Reserve Networks,* eds. M.E. Soulé and J. Terborgh, pp. 171–197. Covelo, California: Island Press.

Gunn, J.S., Desrochers, A., Villard, M.-A., Bourque, J., and Ibarzabal, J. 2000. Playbacks of mobbing calls of Black-capped Chickadees as a method to estimate reproductive activity of forest birds. *J. Field. Ornith.* 71:472–483.

Gustafson, E.J. 1998. Quantifying landscape spatial pattern: what is the state of the art? *Ecosystems* 1:143–156.

Gutzwiller, K.J., and Anderson, S.H. 1992. Interception of moving organisms: influences of patch shape, size, and orientation on community structure. *Landsc. Ecol.* 6:293–303.

Haas, C.A. 1995. Dispersal and use of corridors by birds in wooded patches on an agricultural landscape. *Conserv. Biol.* 9:845–854.

Hansen, A.J., Rotella, J.J., Kraska, M.P.V., and Brown, D. 1999. Dynamic habitat and population analysis: an approach to resolve the biodiversity manager's dilemma. *Ecol. Appl.* 9:1459–1476.

Harrison, S., and Fahrig, L. 1995. Landscape pattern and population conservation. In *Mosaic Landscapes and Ecological Processes,* eds. L. Hannson, L. Fahrig, and G. Merriam, pp. 293–337. London: Chapman and Hall.

Hartley, M.J., and Hunter, M.L., Jr. 1998. A meta-analysis of forest cover, edge effects and artificial nest predation rates. *Conserv. Biol.* 12:465–469.

Heywood, V.H., and Watson, R.T. 1995. *Global Biodiversity Assessment.* New York: Cambridge University Press.

Hinsley, S.A., Bellamy, P.E., Newton, I., and Sparks, T.H. 1995. Habitat and landscape factors influencing the presence of individual breeding bird species in woodland fragments. *J. Avian Biol.* 26:94–104.

Holt, R.D., Pacala, S.W., Smith, T.W., and Liu, J.G. 1995. Linking contemporary vegetation models with spatially explicit animal population models. *Ecol. Appl.* 5:20–27.

Honnay, O., Hermy, M., and Coppin, P. 1999. Nested plant communities in deciduous forest fragments: species relaxation or nested habitats? *Oikos* 84:119–129.

Hoover, J.P., Brittingham, M.C., and Goodrich, L.J. 1995. Effects of forest patch size on nesting success of Wood Thrushes. *Auk* 112:146–155.

Jansson, G., and Angelstam, P. 1999. Threshold levels of habitat composition for the presence of Long-tailed Tit (*Aegithalos caudatus*) in a boreal landscape. *Landsc. Ecol.* 14:283–290.

Knight, R.L., and Gutzwiller, K.J., eds. 1995. *Wildlife and Recreationists: Coexistence Through Management and Research.* Washington, DC: Island Press.

Kozakiewicz, M. 1995. Resource tracking in space and time. In *Mosaic Landscapes and Ecological Processes,* eds. L. Hannson, L. Fahrig, and G. Merriam, pp. 136–148. London: Chapman and Hall.

Lambeck, R.J. 1999. *Landscape Planning for Biodiversity Conservation in Agricultural Regions: A Case Study from the Wheatbelt of Western Australia.* Biodiversity Technical Paper, Number 2. Canberra, Australia: Department of the Environment and Heritage.

LANDECONET Research Consortium. 1997. *Farm Landscapes for Biodiversity: Guide to Using Landscape Ecology to Assess and Improve the Quality of Northern European Farmed Landscapes for Biodiversity.* Grange-over-Sands, Cumbria, United Kingdom: Institute of Terrestrial Ecology, Merlewood Research Station.

Martin, T.E., and Finch, D.M., eds. 1995. *Ecology and Management of Neotropical Migratory Birds: A Synthesis and Review of Critical Issues.* New York: Oxford University Press.

Massachusetts Audubon Society. 1998. *Conserving Grassland Birds.* Lincoln, Massachusetts: Massachusetts Audubon Society, Center for Biological Conservation.

Matlack, G.R. 1994. Vegetation dynamics of the forest edge—trends in space and successional time. *J. Ecol.* 82:113–123.

Mazerolle, M.J., and Villard, M.-A. 1999. Patch characteristics and landscape context as predictors of species presence and abundance: a review. *Ecoscience* 6:117–124.

McShea, W.J., McDonald, M.V., Morton, E.S., Meier, R., and Rappole, J.H. 1995. Long-term trends in habitat selection by Kentucky Warblers. *Auk* 112:375–381.

Mladenoff, D.J., Sickley, T.A., Haight, R.G., and Wydeven, A.P. 1995. A regional landscape analysis and prediction of favorable gray wolf habitat in the northern Great Lakes Region. *Conserv. Biol.* 9:279–294.

Monkkonen, M., and Reunanen, P. 1999. On critical thresholds in landscape connectivity: a management perspective. *Oikos* 84:302–305.

Noss, R.F., O'Connell, M.A., and Murphy, D.D. 1997. *The Science of Conservation Planning: Habitat Conservation Under the Endangered Species Act.* Covelo, California: Island Press.

Oehler, J.D., and Litvaitis, J.A. 1996. The role of spatial scale in understanding responses of medium-sized carnivores to forest fragmentation. *Can. J. Zool.* 74:2070–2079.

Ontario Ministry of Natural Resources. 1998. *Natural Heritage Reference Manual for Policy 2.3 of the Provincial Policy Statement.* Version Number 3. Kemptville: Government of Ontario.

Pain, D., and Pienkowski, M., eds. 1997. *Farming and Birds in Europe: The Common Agricultural Policy and Its Implications for Bird Conservation.* San Diego, California: Academic Press.

Pashley, D.N., Beardmore, C.J., Fitzgerald, J.A., Ford, R.P., Hunter, W.C., Morrison, M.S. and Rosenberg, K.V. 2000. *Partners in Flight: Conservation of the Land Birds of the United States.* The Plains, Virginia: American Bird Conservancy.

Peck, S., ed. 1998. *Planning for Biodiversity: Issues and Examples.* Covelo, California: Island Press.

Pedlar, J.H., Fahrig, L., and Merriam, H.G. 1997. Raccoon habitat use at 2 spatial scales. *J. Wildl. Manage.* 61:102–112.

Pickett, S.T.A., and White, P.S., eds. 1985. *The Ecology of Natural Disturbance and Patch Dynamics.* New York: Academic Press.

Probst, J.R., and Thompson, F.R., III. 1996. A multi-scale assessment of the geographic and ecological distribution of midwestern Neotropical migratory birds. In *Management of Midwestern Landscapes for the Conservation of Migrant Landbirds,* ed. F.R. Thompson, III., pp. 1–19. General Technical Report NC-187. St. Paul, Minnesota: USDA Forest Service, North Central Forest Experiment Station.

Probst, J.R., and Weinrich, J. 1993. Relating Kirtland's Warbler population to changing landscape composition and structure. *Landsc. Ecol.* 8:257–271.

Quintana-Ascencio, P.F., and Menges, E.S. 1996. Inferring metapopulation dynamics from patch-level incidence of Florida scrub plants. *Conserv. Biol.* 10:1210–1219.

Robbins, C., Dawson, D., and Dowell, B. 1989. Habitat area requirements of breeding forest birds on the Middle Atlantic States. *Wildl. Monogr.* 103:1–34.

Robinson, S.K., Thompson, F.R., III, Donovan, T.M., Whitehead, D.R., and Faaborg, J. 1995. Regional forest fragmentation and the nesting success of migratory birds. *Science* 267:1987–1990.

Rodenhouse, N.L., Best, L.B., O'Connor, R.J., and Bollinger, E.K. 1995. Effects of agricultural practices and farmland structures. In *Ecology and Management of Neotropical Migratory Birds: A Synthesis and Review of Critical Issues,* eds. T.E. Martin and D.M. Finch, pp. 269–293. New York: Oxford University Press.

Rollings, N.M, and Brunckhorst, D.J. 1999. Linking ecological and social functions of landscapes: II. Scale and modeling of spatial influence. *Nat. Areas J.* 19:65–72.

Saab, V. 1999. Importance of spatial scale to habitat use by breeding birds in riparian forests: a hierarchical analysis. *Ecol. Appl.* 9:135–151.

Sample, D.W., and Mossman, M.J. 1997. *Managing Habitat for Grassland Birds: A Guide for Wisconsin.* Madison, Wisconsin: Wisconsin Department of Natural Resources, Bureau of Integrated Science Services.

Shafer, C.L. 1990. Values and shortcomings of small reserves. *BioScience* 45:80–88.

Soulé, M.E., and Noss, R.F. 1998. Rewilding and biodiversity conservation as complementary goals for continental conservation. *Wild Earth* 8:18–28.

Soulé, M.E., and Terborgh, J., eds. 1999a. *Continental Conservation: Scientific Foundations of Regional Reserve Networks.* Covelo, California: Island Press.

Soulé, M.E., and Terborgh, J. 1999b. Conserving nature at regional and continental scales—a scientific program for North America. *BioScience* 49:809–817.

Strobl, S. 1998. Towards a list of science priorities for the conservation and management of southern Ontario forests—results of a workshop. *For. Chron.* 74:838–849.

Sutherland, G.D., Harestad, A.S., Price, K. and Lertzman, K.P. 2000. Scaling of natal dispersal distances in terrestrial birds and mammals. *Conserv. Ecol.* [online] 4:16. Available from the Internet: www.consecol.org/vol4/iss1/art16.

Trine, C.L. 1998. Wood Thrush population sinks and implications for the scale of regional conservation strategies. *Conserv. Biol.* 12:576–585.

Trzcinski, M.K., Fahrig, L., and Merriam, G. 1999. Independent effects of forest cover and fragmentation on the distribution of forest breeding birds. *Ecol. Appl.* 9:586–593.

Venier, L., and Fahrig, L. 1996. Habitat availability causes the species abundance-distribution relationship. *Oikos* 76:564–570.

Villard, M.-A. 1998. On forest-interior species, edge avoidance, area sensitivity, and dogmas in avian conservation. *Auk* 115:801–805.

Villard, M.-A., Freemark, K.E., and Merriam, H.G. 1992. Metapopulation theory and Neotropical migrant birds in temperate forests: an empirical investigation. In *Ecology and Conservation of Neotropical Migrant Landbirds,* eds. J.M. Hagan and D.W. Johnston, pp. 474–482. Washington, DC: Smithsonian Institution Press.

Villard, M.-A., Merriam, G., and Maurer, B.A. 1995. Dynamics in subdivided populations of Neotropical migratory birds in a fragmented temperate forest. *Ecology* 76:27–40.

Villard, M.-A., Trzcinski, M.K., and Merriam, G. 1999. Fragmentation effects on forest birds: relative influence of woodland cover and configuration on landscape occupancy. *Conserv. Biol.* 13:774–783.

Weber, T., and Wolf, J. 2000. Maryland's green infrastructure—using landscape assessment tools to identify a regional conservation strategy. *Environ. Monit. Assess.* 63:265–277.

Whitcomb, R.F., Robbins, C.S., Lynch, J.F., Whitcomb, B.L., Klimkiewicz, K., and Bystrak, D. 1981. Effects of forest fragmentation on avifauna of the eastern deciduous forest. In *Forest Islands Dynamics in Man-dominated Landscapes,* eds. R.L. Burgess and D.M. Sharpe, pp. 125–205. New York: Springer-Verlag.

White, D., Preston, E.M., Freemark, K.E., and Kiester, A.R. 1999. A hierarchical framework for conserving biodiversity. In *Landscape Ecological Analysis: Issues and Applications,* eds. J.M. Klopatek and R.H. Gardner, pp. 127–153. New York: Springer-Verlag.

Wiens, J.A. 1995. Landscape mosaics and ecological theory. In *Mosaic Landscapes and Ecological Processes,* eds. L. Hannson, L. Fahrig, and G. Merriam, pp. 1–26. London: Chapman and Hall.

Williams, J.E., Wood, C.A., and Dombeck, M.P., eds. 1997. *Watershed Restoration: Principles and Practices.* Bethesda, Maryland: American Fisheries Society.

Withers, M.A., and Meentemeyer, V. 1999. Concepts of scale in landscape ecology. In *Landscape Ecological Analysis: Issues and Applications,* eds. J.M. Klopatek and R.H. Gardner, pp. 205–252. New York: Springer-Verlag.

Wright, D.H., Patterson, B.D., Mikkelson, G.M., Cutler, A., and Atmar, W. 1998. A comparative analysis of nested subset patterns of species composition. *Oecologia* 113:1–20.

6

Corridors and Species Dispersal

CLAIRE C. VOS, HANS BAVECO, AND CARLA J. GRASHOF-BOKDAM

6.1 Introduction

After introducing corridor concepts, we explore how those concepts have been applied and whether the applications were effective. Based on empirical data, simulation models, and on-the-ground applications, general principles for developing effective corridors will be presented. In the last two sections, major knowledge gaps and research approaches for filling them are discussed.

6.2 Concepts, Principles, and Emerging Ideas

Corridors are one of the strategies to reduce the negative effects of habitat fragmentation in human-dominated landscapes. *Corridors* can be defined as landscape structures that enhance the dispersal of organisms between suitable habitat patches in fragmented landscapes where isolates of suitable habitat are surrounded by a matrix of inhospitable habitat types. This definition involves structural as well as functional aspects. The structural aspect lies in the fact that the corridor, which is not necessarily linear or continuous, differs from its surroundings. The functional aspect entails that the dispersal from source patch to target patch is indeed enhanced. Corridors also can have an important function as routes for daily or seasonal movements. In this chapter, we focus on corridors that facilitate *dispersal,* the one-way movement of an individual away from the natal area to a new area. Corridors contribute to the *functional connectivity* of a landscape—how connected an area is for a process such as an animal moving through different types of landscape elements (Forman 1995), and more particularly the degree to which different parts of the landscape facilitate or impede movement among resource patches (Taylor et al. 1993).

The number of corridor studies has increased in recent years (Beier and Noss 1998; Bennett 1999). This growing interest is triggered by the realization that connectivity, which can be translated into successful dispersal and recolonization, is a key prerequisite for long-term survival in increasingly fragmented landscapes (Opdam 1991; Forman 1995; Wiens 1997). The corridor concept has its roots in theoretical population modeling, in which a higher conservation value was predicted for fragments that are linked by corridors than for isolated fragments (Diamond 1975; Wilson and Willis 1975). The corridor concept also has been incorporated into metapopulation theory, in which increased survival is predicted when more patches in a metapopulation are connected by corridors (Fahrig and Merriam 1985).

Corridor requirements vary widely in dimension and habitat type, depending on the combination of species-specific characteristics (e.g., habitat choice, dispersal capacity, and individual area requirements) and landscape characteristics (e.g., degree of habitat fragmentation, scale of habitat heterogeneity, and the intensity of human land use). Some species require corridors that consist of a continuous linear strip of a particular habitat type to effectively connect habitat patches. For species that operate on broader spatial scales, corridors are not necessarily linear habitat strips but may consist of a heterogeneous zone of habitat types or landscape elements that enhance dispersal and that differ from the surrounding matrix. To emphasize that connectivity can be increased not only by continuous corridors, but also by a range of habitat configurations, Bennett (1999) proposed the term *linkage,* an arrangement of habitat that enhances the movement of animals or the continuity of ecological processes through the landscape.

Species differ widely in their dispersal behavior, scale of movements, and sensitivity to human disturbance. Habitat-specific movement velocity, mortality during dispersal, habitat preferences, and boundary behavior in complex heterogeneous landscapes will all influence the dispersal stream between habitat patches, and thus connectivity. The large group of ground-dwelling species (e.g., mammals, amphibians, reptiles, nonflying insects) as well as some flying species (e.g., butterflies, forest birds, bats) will all depend to some extent on landscape structure in their movement patterns. Plant species are dependent on passive dispersal that may or may not be influenced by landscape structure. For wind-dispersed plants, the dominant wind direction and the occurrence of storms are the most important factors that affect dispersal distance and direction. Water-dispersed seeds are directed by the stream; they can cover long distances, not only downstream but, during floods, also perpendicular to the river. Seeds dispersed by animals also are directed by the landscape and indirectly depend on the foraging behavior of the animal species (Riffell and Gutzwiller 1996).

6.3 Recent Applications

The corridor concept has appealed to landscape planners and managers and has turned up in many regional plans for the conservation of biodiversity. However, as lines on a map between natural areas are easily drawn, the pitfall is that the concept may be applied too easily, without thorough study of its potential effectiveness. Data on recent on-the-ground applications of corridor concepts for nature conservation purposes were collected by the following methods. We searched recent literature and the Internet for mention of corridors in conservation plans. In addition, we contacted colleagues for information on recent applications in their country. We present a selection that gives a world-wide overview of different types of corridor application. We extracted the following information from each case: the conservation problem and how corridors should help solve the problem; the target species or communities for which corridors were intended; details about the corridor design such as habitat types and corridor length and width; and evaluation of corridor functioning after implementation.

An overview of the inventory is given in Table 6.1. Many studies mentioned loss of habitat, natural areas that have become too small, or habitat fragmentation and isolation as the conservation problems that should be solved by corridors. In addition, broad problems, such as climate change, loss of biodiversity, and loss of viability of species, were mentioned. Rarely, the studies contained specific information, such as the distribution of species, experiments, or modeling studies, to underpin the conservation problems. There were only two examples (cases 13 and 15, Table 6.1) in which corridors were planned to solve fragmentation problems for specific target species. In many cases, species were mentioned, but only as examples or as indicator species. Often, it was expected that both animal and plant species would benefit from the corridor. For instance, in case 2, it was assumed that propagules of plant species would be dispersed through the corridor by large animals. In most cases, two or several core areas were being connected, sometimes incorporating other nature areas in between. Also, some plans in the USA and Europe were designed to create a complete network of core areas, buffer zones, and corridors (cases 2, 5, 7, and 10, Table 6.1).

In most cases, the habitat type of the corridor was the same as the core area. If core areas were heterogeneous natural parks (cases 2, 3, and 13), habitat in the corridor was supposed to support dispersal of the target species, or corridor habitat was supposed to be heterogeneous to enable dispersal of all flora and fauna. In cases 2 and 3, target species and corridor type were described roughly. In those cases in which corridors were intended to function as migration routes for specific species (cases 13 and 15), a specific description of the corridor habitat type was given. For instance, the corridor type for the African elephant (case 13) consisted of forest and grass.

The size of the corridors mentioned in the cases varied greatly because corridors were applied from local to supraregional scales. Corridor length varied from 80 m to 60 km, and corridor width ranged from 10 m to 15 km. In almost all applications, corridor width was at least 10% of its total length; thus, corridor width increased with corridor length. But in cases 12 and 17, corridor width was 2% and 3% of the corridor length, respectively. It is questionable whether narrow corridors over long distances will be effective for the broad range of species that is expected to use them. In case 5, planned corridors were up to 40 km long. It was expected that due to limited available space, the implemented corridors would become too narrow. In the recent Nature Policy Plan (Ministry of Agriculture, Nature Conservation and Fisheries 2000), the National Ecological Network was extended with robust supraregional corridors that are more than 1 km wide, in which additional reproduction habitat patches (nodes) are incorporated.

TABLE 6.1. Examples of applications of corridors throughout the world.

Reference	Country/region	Conservation problems	Target species	Habitat types	Size	Evaluation
USA / CANADA						
1. Bonner 1994	USA	Rapid global warming may out pace species' abilities to adapt to changing conditions	Ecosystem	Not mentioned	Length: not mentioned Width: 300 m	Not mentioned
2. The Nature Conservancy, unpublished data	USA	Santa Rosa Plateau is a "biological island," connection with Cleveland National Forest needed	14 target species: mammals, reptiles, and amphibians	Riparian corridors with three nodes (grassland and forest)	Length: 500–6000 m Width: 250 m Node: 160 ha	Movement of plant propagules regarded as sufficient if movement of large wildlife is ensured
3. Smith 1993	USA	Osceola National Forest too small for endangered species; connect with Okefenokee National Wildlife Refuge	Red-cockaded Woodpecker (*Dendrocopus borealis*), black bear (*Ursus americanus*), and other endangered species	Swamp habitat	Length: 16 km Width: minimum ~ 8 km	Not mentioned
4. Machtans et al. 1996	Canada	Fragmentation of forest by clear cut logging or by agricultural development	Passerine birds	Buffer strips of mature trees along lakes and creeks	Length: 200–600 m Width: 100 m	Not mentioned

TABLE 6.1. *Continued.*

Reference	Country/region	Conservation problems	Target species	Habitat types	Size	Evaluation
EUROPE						
5. Ministry of Agriculture, Nature Conservation and Fisheries 1990, 2000; Reijnen and Koolstra 1998	The Netherlands	Fragmentation of habitat for target species	Originally based on six species of large mammals and fish, subsequently extended with small mammals, amphibians, reptiles, birds, and insects	Diverse	Length: 100 m to >10 km Width: ~10 m to ~1 km Nodes: additional reproduction areas	Evaluation before implementation in Province of Gelderland: most corridors will be effective in combination with enlargement of habitat patches. Road barriers must be solved
6. The World Conservation Union 1995	Hungary	Loss of biodiversity and human welfare; network of fragmented wetlands needed	Not mentioned	Riverine areas	Length: not mentioned Width: not mentioned	Prerestoration experiments, environmental risk assessment, and mapping of recolonization sources
7. Kubeš 1996	Czech Republic	Loss of ecological function in a human-dominated environment	Not mentioned	Diverse habitat types; Biocenters (Bc) are connected by corridors	Length: not mentioned Bc: 0.5–5 ha Width: 10–20 m Bc: 10–50 ha Width: 40 m Bc > 1000 ha Width: valleys, rivers	Not mentioned

8. Ferris-Kaan 1991	United Kingdom	Scarcity of open areas in Bernwood Forest; network of open rides and glades needed	Butterflies and moths	Grassy vegetation	Major ride length: not mentioned Major ride width: >15 m	Study of Ringlet Butterfly (*Aphantopus hyperantus*) by Sutcliffe and Thomas (1996)
9. Baudry and Burel 1984	France	Destruction of hedgerow networks	Birds, insects, small mammals, and reptiles	Hedgerows	Length: not mentioned Width: not mentioned	Study of ecological function of hedgerows
10. Foppen et al. 2000	Europe	Restoration of connectivity between large nature reserves in Europe	Many terrestrial and freshwater vertebrates	Diverse	Length: not mentioned Width: not mentioned	Will be carried out after implementation
AFRICA						
11. A. Tye, unpublished report	Tanzania	Low densities of bird populations in five forest areas; connection needed	Flora and fauna	Forest habitat	Length: ~1–2 km Width: ~1 km	Not mentioned
12. Mwalyosi 1991; Bennett 1999	Tanzania	Loss of traditional migration routes from Tarangire National Park to Lake Manyara National Park	Large mammals, e.g., wildebeest (*Connochaetes* spp.) and zebra (*Equus* spp.)	Not mentioned	Length: 60 km Width: >1 km	Evaluation before implementation
13. Baranga 1991	Uganda	Loss of corridor function in Kibale Game Corridor between Kibale Forest Reserve and Queen Elizabeth National Park	Large mammals, especially African elephant (*Loxodonta africana*) and buffalo (*Syncerus caffer*)	Medium-altitude forest and grass	Length: ~35 km Width: ~15 km	Not mentioned

TABLE 6.1. *Continued.*

Reference	Country/region	Conservation problems	Target species	Habitat types	Size	Evaluation
ASIA						
14. Harris and Scheck 1991	Sri Lanka	Floodplain National Park connecting expanded Wasgomuwa and Somawathiel National Parks to compensate for consequences of dam building	Endemic flora and fauna, large mammals like Indian elephant (*Elephas maximus bengalensis*)	Jungle forest	Length: ~20 km Width: minimum ~5 km	Not mentioned
15. McKinnon and De Wulf 1994	China	Connecting fragmented habitat in Minshan area	Giant panda (*Ailuropoda melanoleuca*)	Montane forests with arrow bamboo (*Bambusa*) stands	Length: not mentioned Width: minimum 1 km	Evaluation before implementation
AUSTRALIA / NEW ZEALAND						
16. Land Conservation Council 1994	Australia	Fragmentation of three forest areas in Ash Ranges National Park; connection needed	Leadbeater's possum (*Gymnobelideus leadbeateri*), species of older-aged forest	Mature wet forest, cool temperate rainforest	Length: ~5 km Width: minimum 1 km	Not mentioned
17. Deptartment of Conservation and Natural Resources 1995	Australia	Reduce impact of logging, connect Coopracambra National Parks by network of protection zones (stepping stones)	Forest-dependent wildlife and Croajinolong	Eucalypt (*Eucalyptus* spp.) forest	Length: >6 km Width: ~200 m	Not mentioned
18. Bennett 1999	Australia	Protection and management of roadside vegetation for conservation values	Flora and fauna	Roadside verges	Length: not mentioned Width: not mentioned	Assessment by volunteer observers
19. O'Donnell 1991	New Zealand	Harvesting of beech (*Nothofagus*) forest; connection of two larger forest areas (Paparoa Range and Southern Alps) needed	Forest birds	Temperate rainforest	Length: ~10 km Width: >2 km	Not mentioned

In most of the cases, no evaluation of corridor functioning was mentioned. Most applications were still in a planning phase. Only in some cases was an evaluation conducted with simulation models, field experiments, or by gathering field data (cases 5, 6, 9, 15, and 18, Table 6.1). This is surprising, as a preevaluation is needed to define the dimensions and character of the planned corridor(s), and monitoring is required to evaluate corridor effectiveness after implementation. Quantitative studies of dispersal between habitat patches before and after corridor implementation were not found. We could not find any cases in which the question "Do corridors sufficiently solve fragmentation problems on a regional scale?" was answered.

However, many empirical and simulation studies have provided some evidence of the usefulness of corridors for dispersal. In the next three sections, we describe applications of the corridor concept in a research context. Specifically, we discuss corridor function at the landscape scale (Section 6.3.1) and corridor scale (Section 6.3.2), and we present some modeling approaches that help to predict corridor effectiveness (Section 6.3.3).

6.3.1 Studies at the Landscape Scale

Studies at the landscape scale are primarily focused on the landscape as a whole and not on single corridors. In many descriptive or observational studies at the landscape scale, a positive relation is found between the probability of occupation or colonization of a suitable habitat patch and the density of landscape elements considered to function as corridors (e.g., Pahl et al. 1988; Bright et al. 1994; Vos and Stumpel 1996). For instance, in a study of the red squirrel (*Sciurus vulgaris*) (Verboom and Van Apeldoorn 1990), the occupation probability of a suitable habitat patch depended on the number of hedgerows surrounding a woodlot within 200–600 m. In another study, the probability of occurrence of holly (*Ilex aquifolium*) appeared to be higher with an increasing number of hedgerows within a range of 1,000 m in an agricultural landscape (Grashof-Bokdam 1997). In both examples, other fragmentation factors and habitat quality were accounted for.

In a second type of analysis at the landscape scale, habitat patches that are actually connected by corridors are compared with unconnected habitat patches. In several studies, a higher colonization probability or a higher frequency of visiting individuals was found for habitat patches connected by corridors than was found for unconnected patches (e.g., Dmowski and Kozakiewicz 1990; Dunning et al. 1995). Haas (1995) found for the American Robin (*Turdus migratorius*) that the average number of dispersal events between pairs of patches connected by corridors was 2.50, but it was only 0.17 between unconnected patches. In this study, other factors that may influence dispersal, such as patch size, distance to nearest wooded habitat, and habitat quality, were similar for connected and unconnected patches.

A few experimental studies at the landscape scale have demonstrated positive effects of corridors (e.g., Mansergh and Scotts 1989). Machtans et al. (1996) and Desrochers and Hannon (1997) studied the use of linear forest strips between forest patches by forest birds before and after harvesting of adjacent forest. The use of strips by juveniles increased, and movement rates through the forest clear-cuts were significantly lower, indicating the use of these strips as dispersal corridors. Haddad (1999a) demonstrated in an experimental study with two butterfly species that corridors increased interpatch movement rates.

6.3.2 Studies at the Corridor Scale

In many studies at the corridor scale, an inventory is made of species present in the corridor. However, species presence in corridors does not provide evidence that corridors actually function as dispersal routes between habitat patches. An alternative and perhaps more accurate explanation is that the corridor provides extra habitat for these species. Still, for very immobile species such as many plants, a corridor that consists of reproduction habitat may be the only effective way to increase connectivity between populations. Most studies on the use of corridors are based on measurement of animal movements by radiotelemetry, mark-recapture studies, and behavioral observations. The advantages of radiotelemetry and direct observations over mark-recapture techniques are that behavior and actual routes taken in heterogeneous landscapes can be recorded. An indirect indication for the corridor function of particular landscape types can be derived from the relative preference for certain habitat types by moving individuals. For instance, in several studies conducted in agricultural landscapes, preference for hedgerows and avoidance of open fields were documented (e.g., Bright and Morris 1991).

A method to mimic behavior of dispersers in unfamiliar landscapes, which may differ from movement patterns within the home range, is to translocate individuals into various landscape configurations, (e.g., Ruefenacht and Knight 1995; Mauritzen et al. 1999; Vos 1999). Merriam and Lanoue (1990) compared the behavior of three groups of the white-footed mouse (*Peromyscus leucopus*) in farmland. All groups (residents, translocated individuals from woodland, and translocated individuals from comparable farmland) preferred fencerows to other habitat types in the landscape. These observations of habitat preference do not enable one to distinguish whether the fencerows functioned as suitable habitat, dispersal routes, or both.

More convincing evidence of the use of corridors for dispersal comes from observation of movements through corridors from one habitat patch to another, in combination with avoidance of surrounding landscape types (e.g., Wauters et al. 1994; Beier 1995; for an extensive review, see Bennett 1999). Sutcliffe and Thomas (1996) showed with mark-recapture methods and behavioral observations in both habitat patches and connecting open tracks that the Ringlet Butterfly (*Aphantopus hyperantus*) used the tracks as dispersal corridors. Dispersal between patches via open tracks explained exchange rate better than did direct distance. Also, the invasion of weeds along corridors proves that corridors indeed facilitate dispersal, but not necessarily for the target species (Benninger-Truax et al. 1992; Pyšek and Prach 1993).

6.3.3 Modeling Approaches to Predict Corridor Effectiveness

Dispersal models help to predict the effectiveness of corridors, as they extrapolate individual movement patterns to differences in connectivity between populations on the landscape or regional scale. In general, a model-based approach to investigate the effectiveness of planned or realized corridors requires spatially realistic models (Wiens 1997). When focusing on movement, such models take into account landscape composition and configuration in combination with the species' landscape-specific movement parameters (Wiens et al. 1997).

Spatially realistic movement models are based on spatial information provided by geographic information systems. Two formats are in common use to represent spatial information: a grid (raster) and a vector format. In the representation of movement, we observe a similar distinction between models with cell-to-cell movement across a grid, and models in which movement is represented as a sequence of moves (vectors), characterized by move length, angle, and duration (Turchin 1998).

In grid-based movement models (Johnson et al. 1992a,b; Schippers et al. 1996), dispersers select a destination cell from the neighbors of the current cell through a set of rules. These models can easily accommodate species-specific details, like preferences for cells with certain habitats. However, boundary behavior displayed by dispersers encountering an edge does not fit well within the approach, as the grid format defies a faithful rendering of edges and linear elements in the landscape. Most grid-based models (percolation-, diffusion-, and random-walk models) applied to date quantify the general connectivity of the landscape without pretending to give an accurate representation of actual movement through the landscape (Gustafson and Gardner 1996; With et al. 1997), but see Johnson et al. (1992b) for simulation models used in combination with empirical observations. Brooker et al. (1999) applied a grid-based movement model to quantify the importance of corridor continuity and to estimate distance-dependent dispersal mortality for two forest bird species. This was accomplished through a comparison of model-derived and observed colonization events of forest habitat patches. The individual-based approach they used enabled them to incorporate fairly complex local decision making by dispersers. They concluded that one of the species relied much more on corridor continuity than did the other species. Van Dorp et al. (1997) used a grid-based (cellular automaton) model to investigate the efficacy of linear landscape elements as corridors for perennial grassland species with short-range seed dispersal. Local population dynamics were represented by a stage-classified transition matrix. They concluded that linear elements were not effective because estimated dispersal rates were generally too low. With increasing corridor width (up to 20 m), however, the rate of dispersal approximated that in continuous habitat.

In vector-based movement models (Vermeulen 1995; Tischendorf and Wissel 1997; Vos 1999), a moves-generating algorithm is applied within homogeneous areas, and transition rules are triggered at the edges of these areas. Combined with a vector representation of the landscape (homogeneous areas as polygons), these models can accommodate details such as habitat preferences and boundary behavior, including edge-tracking displacements. Landscape heterogeneity is necessarily reduced to a limited number of element (polygon) types; otherwise, these models become unwieldy. A vector-based movement model is easily parameterized from empirical data on tracked individuals (e.g., Tischendorf et al. 1998; Turchin 1998), especially when movement can be represented as a correlated random walk—an assumption that clearly needs to be tested first. A correlated random walk (CRW) is a random-walk variant in which the direction of each move correlates to some degree with the direction of the preceding move (Kareiva and Shigesada 1983).

Tischendorf and Wissel (1997) used a CRW-movement model to investigate the impact of corridor width and movement attributes (correlation strength and boundary behavior) on the probability of successful corridor passage. They found that with stronger movement autocorrelation, individuals covered longer distances within corridors. Corridor width, determining the frequency of boundary encounters, had a large impact on passage probability, but an optimal corridor width was not found. Instead, the likelihood of arriving increased asymptotically with corridor width (as was also found by Soulé and Gilpin 1991). With a similar model, Haddad (1999b) predicted that corridors would increase interpatch movement rates of bufferfly species (as was observed experimentally). Local behavior (e.g., at habitat edges) appeared to be a key factor in predicting the conservation potential of the corridors.

Grashof-Bokdam and Verboom (in Grashof-Bokdam 1997) developed a vector-based colonization model for forest plants in which suitable habitat could be colonized from source patches by bird-transported seeds, either along forest corridors or independently of landscape structure. They used the model to evaluate effects of alternative landscape configurations on colonization success of secondary forest habitat. Although colonization improved by adding corridors to the landscape, colonization of habitat adjacent to old forest habitat (source patches) was most successful.

6.4 Principles for Applying Landscape Ecology

We have distilled some general principles for applying corridor concepts from both the on-the-ground and research applications described in the previous section. Probably the dominant principle for effectively applying these concepts is to consider as a starting point the differences in scale at which various species function. A species' individual-area requirements and dispersal capacity determine the level of habitat fragmentation that causes survival problems (Vos et al. 2001). These same characteristics also determine the geographic scale at which corridors may be effective. For the protection of biodiversity, it may be necessary to implement corridors on different scales in the same landscape. Figure 6.1 represents a landscape with corridors at three hierarchical scales, zooming in from a supraregional

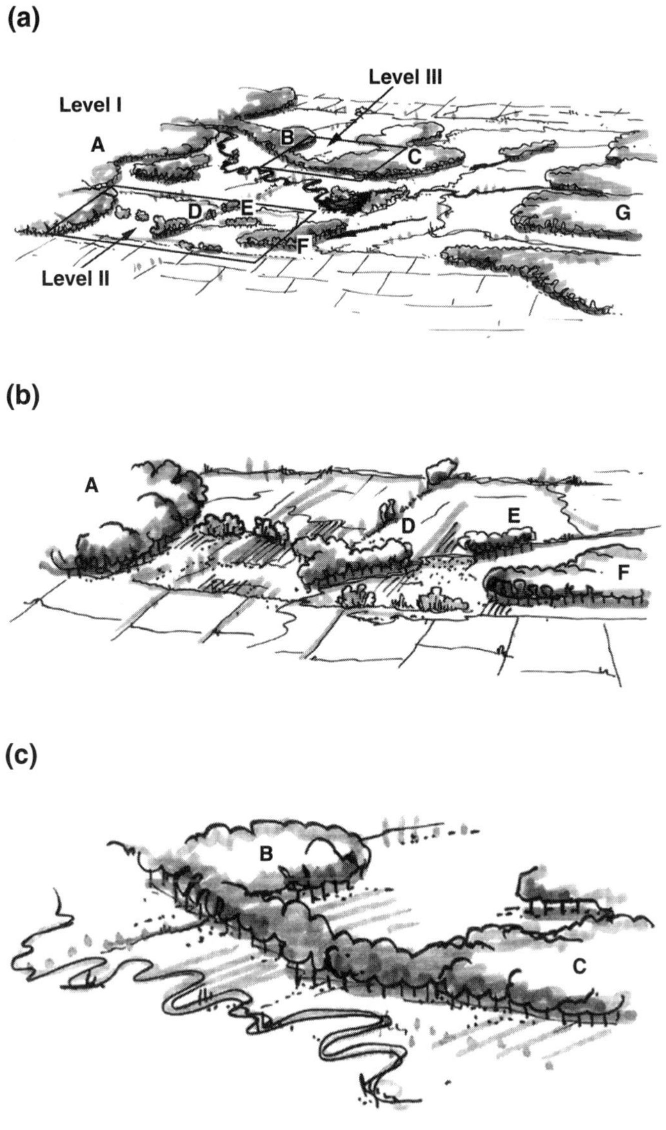

(a)

Level III
Level I
A
B
C
D
E
G
F
Level II

(b)

A
D
E
F

(c)

B
C

FIGURE 6.1. Examples of corridors at three different hierarchical scales. (a) Level I, supraregional scale. Corridor length: >10 km. Corridor width: >1 km. Corridor connects two large areas (A and G), is a mosaic of different habitat types, and has only a dispersal function (no reproduction). (b) Level II, regional scale. Corridor length: 1 km to ≤10 km. Corridor width: 10 m to ≤1 km. Corridor connects two habitat patches (A and F), it consists of specific habitat types, and only small gaps in the corridor are allowed. If the corridor becomes longer than the dispersal distance of the target species, additional stepping stones (nodes D and E) are necessary. (c) Level III, local scale. Corridor length: <1 km. Corridor width: generally <10 m but depends on whether a buffer zone to maintain habitat quality is required. The corridor consists of continuous reproduction habitat without gaps.

scale, to a regional and a local scale. Case 5 (Table 6.1) is an example in which this distinction between corridors on three hierarchical scales was made.

6.4.1 Species' Spatial Scales

The first species to disappear when habitat becomes fragmented and human land use becomes dominant are ground-dwelling species with large individual-area requirements. For large mammals, especially predators, even extensive natural areas are too small for long-term survival. Because these species often have considerable dispersal capacity, it may be effective to connect several areas over long distances (supraregional corridor, Figure 6.1a). These corridors consist of a mosaic of several habitat types, are relatively broad (>1 km) to keep the mobile species inside of the corridor, and can be long (>10 km). Factors that will increase the effective limit on corridor length are presence of forage and shelter and a low mortality risk (e.g., when crossing roads). Examples in which supraregional corridors were applied are cases 3, 5, 12, 13, and 14 (Table 6.1).

Many species, such as small mammals and amphibians, are able to form viable metapopulations on a regional scale. They have moderate individual-area requirements and are able to disperse over several kilometers. If the landscape becomes too fragmented in combination with intensive land use in the matrix, local population extinctions will occur and corridors are necessary for successful recolonization from the surrounding populations. These species require corridors that are 1 km to ≤10 km long, 10 m to ≤1 km wide, and consist of specific habitat types that provide shelter and food (regional corridor, Figure 6.1b). Only small gaps (10–100 m) within the corridor can be crossed. If the corridor has to link habitat patches over a distance that exceeds the dispersal distance of the target species, additional reproduction patches or "nodes" (Bennett 1999) should be incorporated to maintain a sufficient stream of dispersers. Examples in which regional corridors were applied are cases 2 and 5 (Table 6.1).

When habitat fragmentation is extensive, it affects even species with very small individual-area requirements and low dispersal capacity (<1 km), such as nonflying insects or plant species. Within a single nature reserve, natural fragmentation due to succession also may affect species at this scale. To facilitate dispersal between patches, these species need corridors that are <1 km long and <10 m wide to pass unsuitable habitat types (local corridor, Figure 6.1c). For these relatively immobile species, only corridors that consist of suitable reproduction habitat that harbors a resident population will be effective. Local corridors consist of continuous reproduction habitat without gaps. Because long and narrow corridors become ineffective due to "leaking" of dispersers into the matrix (Vermeulen 1995), the enlargement of existing patches may be more effective than using corridors. Local corridors were applied in case 5 (Table 6.1).

The quantitative dimensions of corridor scale listed above should be regarded as rules of thumb, generally indicating that corridor width should increase with corridor length. However, there will be exceptions to these dimensions that may be caused by species-specific requirements or conditions outside of the corridor (see Section 6.4.2).

6.4.2 Buffer Zones and Ecosystem Corridors

When habitat quality of the corridor is very important, the corridor must be sufficiently buffered against edge effects to maintain that particular habitat quality. For instance, to maintain forest interior conditions in the corridor core, an additional buffer zone of 400–600 m may be necessary (Bennett 1999). Examples in which these relatively wide corridors were applied are cases 11, 16, and 19 (Table 6.1). When the intensity of the land use in the matrix is high (e.g., intense farming), the required width of the buffer zone will increase.

If corridors are intended for whole ecosystems, as in the case of global warming (case 1, Table 6.1), the requirements of the most critical species should be met. In general, this implies that the corridor should include all ecosystem habitat types. Especially for ecosystem corridors, the development time of a corridor is a critical factor. For forests that take hundreds of years to develop, ecosystem corridors will only be an option if remnants of the former ecosystem can be incorporated into the corridor (cases 4, 14, 15, 16, 17, and 19, Table 6.1).

6.4.3 Using Existing Corridors

An effective strategy to maintain connectivity in a landscape is to optimize natural corridors, such as rivers and streams, provided that they actually connect suitable habitat patches (cases 2, 5, 6, and 7, Table 6.1). These linear elements can support species of several habitat types (Pyšek and Prach 1993) and can function as corridors at supraregional and regional scales. In addition, the potential of human-made corridors such as strips of natural habitat along roads and railroads should be considered because such strips can compose a considerable part of the landscape. The use of road verges as effective corridors has been demonstrated for relatively immobile species, such as ground beetles and plants (Vermeulen 1995). Verges will generally function as corridors on a local scale. For good habitat quality and avoidance of too much "leaking" of individuals to the surroundings, road verges should be at least 20 m wide (Van Dorp et al. 1997). Because roads and railroads form barriers that can cause high mortality, the benefits of verges should be balanced against these negative impacts, or measures should be taken to minimize barrier effects (see Chapter 13 for more information about roads).

6.5 Knowledge Gaps

Lack of model predictions and empirical data that underpin effective corridor design are important general voids that should be filled with priority. The application of corridors that insufficiently enhance species dispersal will not contribute to the long-term survival of species in fragmented habitat.

6.5.1 *Theoretical Voids*

Because it is impossible to study all species separately, there is a need to translate knowledge into a general framework for corridor requirements. More effort should be put into determining key characteristics of species that predict corridor effectiveness in landscapes with different degrees of habitat fragmentation, and into determining species groups with comparable corridor requirements. We still have a very incomplete understanding of how factors such as habitat-specific movement velocity, dispersal mortality, habitat preferences, and boundary behavior in complex heterogeneous landscapes collectively affect corridor effectiveness.

Another important void that blocks effective corridor application is the lack of techniques for multispecies optimization of the spatial distribution of suitable habitat and the placement of corridors in a landscape. In this respect, the construction of corridors is only one of the options to solve fragmentation problems. Its potential benefits should be weighed against or combined with other options to improve the connectivity within a population network (e.g., the creation of new habitat patches, enlargement of existing patches, or improvement of habitat quality).

6.5.2 *Empirical Voids*

For many species, basic knowledge of dispersal behavior, dispersal distances, and the influence of landscape heterogeneity on dispersal direction is incomplete or lacking totally. We need to know which species are sensitive to barriers and what kind of corridors may be effective for a broader range of species.

To determine the effectiveness of corridors and optimal corridor dimensions, quantitative studies on dispersal behavior in relation to landscape heterogeneity are needed. Empirical evidence of the use of corridors is often incomplete, and knowledge about the relation between dispersal behavior and landscape structure is developing slowly. Studies based on the individual level and often on a very local scale are necessary. Extrapolation with dispersal models from individual movement patterns to differences in connectivity between populations on the landscape or regional scale has been rare.

There is a need for well-underpinned, spatially explicit movement models for umbrella species. However, sufficient empirical data to test and calibrate spatially explicit movement models are lacking. The scarcity of empirical data on dispersal attributes has been considered the most urgent problem in the application of complex, spatially realistic models (Wennergren et al. 1995).

6.6 Research Approaches

It should be stressed that studies to improve our knowledge of corridors should not postpone the implementation of corridors, if this implementation is based on existing knowledge from which threatened species are likely to benefit. As was pointed out by Merriam (1991) and Soulé and Gilpin (1991), general principles, even if imprecise, are of relevance to conservation right now.

6.6.1 Approaches for Theoretical Research

To develop a framework for determining key characteristics of species that affect corridor effectiveness, simulation studies with model species in computer-generated landscapes are a useful approach. These studies need to especially consider interactions among species characteristics such as habitat-specific movement velocity, dispersal mortality, habitat preferences, and boundary behavior in different landscape configurations. The modeling of the impact of corridor width and movement attributes by Tischendorf and Wissel (1997) is a good example of this approach. This approach could form the basis for an integration of corridor requirements for an array of species. The study by Vos et al. (2001) is an example of how a general framework based on key characteristics of species can be developed. Vos et al. (2001) used model species, artificial landscapes, and empirical data to group species into "ecological profiles for fragmentation sensitivity" and to link these profiles to requirements for habitat configuration in the landscape.

Application of spatial optimization techniques (Hof and Bevers 1998) requires a quantitative measure of corridor effectiveness. A useful measure would probably need to be related to the flow of individuals through and from the corridor. These values can be compared for different spatial configurations of corridors. Using optimization techniques, from the set of possible configurations, the most promising one can be selected. A complication is to deal with conflicting spatial corridor demands posed by different species. Either priorities would need to be set for the species that have to benefit from a corridor, or a constraint-solving approach could be taken to design a corridor as a compromise that meets the minimum demands of all species.

6.6.2 Approaches for Empirical Research

For details about how to evaluate the effectiveness of corridors as conduits for movement, and useful checklists for optimal research design, see Dawson (1994), Beier and Noss (1998), and Turchin (1998). For an adequate understanding of the effectiveness of corridors in species conservation, a combination of different research approaches is needed. These approaches also can provide basic knowledge of dispersal behavior, dispersal distances, and the influence of landscape heterogeneity. For plant species, it is difficult to observe dispersal events of seeds, but the use of genetic techniques can give some answers on the effectiveness of corridors to facilitate gene flow between populations (Meagher and Thompson 1987; Grashof-Bokdam et al. 1998).

The landscape approach, in which a correlation is sought between corridor density or presence and the recolonization or presence of local populations, can be used to determine whether a species may benefit from corridors. In observational studies at the landscape scale, an important pitfall is the presence of confounding factors. To avoid confounding factors, other variables that may influence dispersal should be kept constant (e.g., Haas 1995) or be incorporated as covariables (Verboom and Van Apeldoorn 1990).

Studies at the corridor scale and of the actual routes taken by individuals should provide empirical evidence of the effectiveness of corridors and information for optimal corridor dimensions. In many corridor studies, all attention is focused on the potential corridor, whereas movements in the matrix outside of the corridor are neglected. To determine whether particular habitat types serve as conduits for dispersing organisms, their use must be quantified in comparison to dispersal frequency through other habitat types. It also has been pointed out (Dawson 1994; Beier and Noss 1998) that a distinction should be made between presence in a corridor that can be explained by a habitat function and presence due to a dispersal function. However, for very immobile species, extreme habitat specialists, and corridors over long distances, a corridor that also provides reproduction habitat may very well be the only effective method to facilitate exchange between habitat patches.

Experiments at the landscape scale can provide strong empirical evidence about corridor effectiveness (Haddad 1999a,b). But it is difficult to carry out landscape-scale experiments that provide unambiguous evidence for the use of corridors for dispersal because requirements of replication and control sites are almost impossible to fulfill. However, the problem of replication can be overcome to some extent by studying evidence from numerous studies via meta-analysis. Closer cooperation among researchers, conservation agencies, and other landowners would greatly improve opportunities for landscape-scale experiments. The experiments on bird dispersal in forests mentioned earlier (Machtans et al. 1996; Desrochers and Hannon 1997) are good examples of this broad-scale experimental approach.

The development of well-underpinned, spatially explicit movement models requires a closer link between the collection of empirical data in the field and the development of predictive models. Too often, theoretical and empirical lines of research operate in isolation. One should carefully balance the complexity and realism of model formulations against the empirical data available, and incorporate only those aspects and relationships that can be supported with data. Vector-based movement models (e.g., Turchin 1998; Haddad 1999b) are especially suitable for parameterization with empirical data for tracked individuals. Empirical data on observed dispersal events or colonization are required to test these models (e.g., Brooker et al. 1999).

Acknowledgments

We thank Andrew Bennett for his useful comments about this chapter.

References

Baranga, J. 1991. Kibale Forest Game Corridor: man or wildlife? In *Nature Conservation 2: The Role of Corridors,* eds. D.A. Saunders and R.J. Hobbs, pp. 371–375. Chipping Norton, Australia: Surrey Beatty and Sons.

Baudry, J., and Burel, F. 1984. Landscape project 'remembrement': landscape consolidation in France. *Landsc. Plan.* 11:235–241.

Beier, P. 1995. Dispersal of juvenile cougars in fragmented habitat. *J. Wildl. Manage.* 59:228–237.

Beier, P., and Noss, R.F. 1998. Do habitat corridors provide connectivity? *Conserv. Biol.* 12:1241–1252.

Bennett, A.F. 1999. *Linkages in the Landscape: The Role of Corridors and Connectivity in Wildlife Conservation.* Gland, Switzerland and Cambridge, United Kingdom: The World Conservation Union (IUCN) Forest Conservation Programme.

Benninger-Truax, M., Vankat, J.L., and Schaefer, R.L. 1992. Trail corridors as habitat and conduits for movement of plant species in Rocky Mountain National Park, Colorado, USA. *Landsc. Ecol.* 6:269–278.

Bonner, J. 1994. Wildlife's roads to nowhere? *New Scientist* 143(1939):30–34.

Bright, P.W., Mitchell, P., and Morris, P.A. 1994. Dormouse distribution: survey techniques, insular ecology and selection of sites for conservation. *J. App. Ecol.* 31:329–339.

Bright, P.W., and Morris, P.A. 1991. Ranging and nesting behavior of the dormouse, *Muscardinus avellanarius* in diverse low growing woodland. *J. Zool.* 224:177–190.

Brooker, L., Brooker, M., and Cale, P. 1999. Animal dispersal in fragmented habitat: measuring habitat connectivity, corridor use, and dispersal mortality. *Conserv. Ecol.* [online] 3:4. Available from the Internet: www.consecol.org/vol3/iss1/art4.

Dawson, D. 1994. *Are Habitat Corridors Conduits for Animals and Plants in a Fragmented Landscape? A Review of the Scientific Evidence.* English Nature Research Report 9. Peterborough, United Kingdom: English Nature.

Department of Conservation and Natural Resources. 1995. *Forest Management Plan for the East Gippsland Forest Management Area.* Melbourne, Australia: Department of Conservation and Natural Resources.

Desrochers, A., and Hannon, S.J. 1997. Gap crossing decisions by forest songbirds during the post-fledging period. *Conserv. Biol.* 11:1204–1210.

Diamond, J.M. 1975. The island dilemma: lessons of modern biogeographic studies for the design of natural reserves. *Biol. Conserv.* 7:129–146.

Dmowski, K., and Kozakiewicz, M. 1990. Influence of a shrub corridor on movements of passerine birds to a lake littoral zone. *Landsc. Ecol.* 4:99–108.

Dunning, J.B., Stewart, D.J., Danielson, B.J., Noon, B.R., Root, T.L., Lamberson, R.H., and Stevens, E.E. 1995. Spatially explicit population models: current forms and future uses. *Ecol. Appl.* 5:3–11.

Fahrig, L., and Merriam, G. 1985. Habitat patch connectivity and population survival. *Ecology* 66:1762–1768.

Ferris-Kaan, R. 1991. *Edge Management in Woodlands.* Occasional Paper 28. Edinburgh, United Kingdom: Forestry Commission.

Foppen, R.P.B., Bouwma, I.M., Kalkhoven, J.T.R., Dirksen, J., and Van Opstal, S. 2000. *Corridors of the Pan-European Ecological Network: Concepts and Examples for Terrestrial and Freshwater Vertebrates.* ECNC Technical Report Series. Tilburg, The Netherlands: European Centre for Nature Conservation.

Forman, R.T.T. 1995. *Land Mosaics: The Ecology of Landscapes and Regions.* Cambridge, United Kingdom: Cambridge University Press.

Grashof-Bokdam, C.J. 1997. *Colonization of Forest Plants: The Role of Fragmentation.* IBN Scientific Contributions 5. Wageningen, The Netherlands: Institute for Forestry and Nature Research.

Grashof-Bokdam, C.J., Jansen, J., and Smulders, M.J.M. 1998. Dispersal patterns of *Lonicera periclymenum* determined by genetic analysis. *Molec. Ecol.* 7:165–174.

Gustafson, E.J., and Gardner, R.H. 1996. The effect of landscape heterogeneity on the probability of patch colonization. *Ecology* 77:94–107.

Haas, C.M. 1995. Dispersal and use of corridors by birds in wooded patches on an agricultural landscape. *Conserv. Biol.* 9:845–854.

Haddad, N.M. 1999a. Corridor and distance effects on interpatch movements: a landscape experiment with butterflies. *Ecol. Appl.* 9:612–622.

Haddad, N.M. 1999b. Corridor use predicted from behaviors at habitat boundaries. *Am. Nat.* 153:215–227.

Harris, L.D., and Scheck, J. 1991. From implications to applications: the dispersal corridor principle applied to the conservation of biological diversity. In *Nature Conservation 2: The Role of Corridors,* eds. D.A. Saunders and R.J. Hobbs, pp. 189–220. Chipping Norton, Australia: Surrey Beatty and Sons.

Hof, J., and Bevers, M. 1998. *Spatial Optimization for Managed Ecosystems.* New York: Columbia University Press.

Johnson, A.R., Milne, B.T., and Wiens, J.A. 1992a. Diffusion in fractal landscapes: simulations and experimental studies of tenebrionid beetle movements. *Ecology* 73:1968–1983.

Johnson, A.R., Wiens, J.A., Milne, B.T., and Crist, T.O. 1992b. Animal movements and population dynamics in heterogeneous landscapes. *Landsc. Ecol.* 7:63–75.

Kareiva, P.M., and Shigesada, N. 1983. Analyzing insect movement as a correlated random walk. *Oecologia* 56:234–238.

Kubeš, J. 1996. Biocentres and corridors in a cultural landscape. A critical assessment of the 'territorial system of ecological stability.' *Landsc. Urban Plan.* 35:231–240.

Land Conservation Council. 1994. *Melbourne Area District 2 Reviews. Final Recommendations.* Melbourne, Australia: Land Conservation Council.

Machtans, C.S., Villard, M.-A., and Hannon, S.J. 1996. Use of riparian buffer strips as movement corridors by forest birds. *Conserv. Biol.* 10:1366–1379.

Mansergh, I.M., and Scotts, D.J. 1989. Habitat continuity and social organization of the mountain pygmy-possum restored by tunnel. *J. Wildl. Manage.* 53:701–707.

Mauritzen, M., Bergers, P.J.M., Andreassen, H.P., Bussink, H., and Barendse, R. 1999. Root vole movement patterns: do ditches function as habitat corridors? *J. Appl. Ecol.* 36:409–421.

McKinnon, J., and De Wulf, R. 1994. Designing protected areas for giant pandas in China. In *Mapping the Diversity of Nature,* ed. R.I. Miller, pp. 128–142. London: Chapman and Hall.

Meagher, T.R., and Thompson, E. 1987. Analysis of parentage for naturally established seedlings of *Chamaelirium luteum. Ecology* 68:803–812.

Merriam, G. 1991. Corridors and connectivity: animal populations in heterogeneous environments. In *Nature Conservation 2: The Role of Corridors,* eds. D.A. Saunders and R.J. Hobbs, pp. 133–142. Chipping Norton, Australia: Surrey Beatty and Sons.

Merriam, G., and Lanoue, A. 1990. Corridor use by small mammals: field measurements for three experimental types of *Peromyscus leucopus. Landsc. Ecol.* 4:123–131.

Ministry of Agriculture, Nature Conservation and Fisheries 1990. *Nature Policy Plan* (in Dutch). The Hague, The Netherlands: Ministry of Agriculture, Nature Conservation and Fisheries.

Ministry of Agriculture, Nature Conservation and Fisheries 2000. *Nature for People and People for Nature* (in Dutch). The Hague, The Netherlands: Ministry of Agriculture, Nature Conservation and Fisheries.

Mwalyosi, R.B.B. 1991. Ecological evaluation for wildlife corridors and buffer zones for Lake Manyara National Park, Tanzania, and its immediate environment. *Biol. Conserv.* 57:171–186.

O'Donnell, C.F.J. 1991. Application of the wildlife corridors concept to temperate rainforest sites, North Westland, New Zealand. In *Nature Conservation 2: The Role of Corridors,* eds. D.A. Saunders and R.J. Hobbs, pp. 85–98. Chipping Norton, Australia: Surrey Beatty and Sons.

Opdam, P. 1991. Metapopulation theory and habitat fragmentation: a review of Holarctic breeding bird studies. *Landsc. Ecol.* 5:93–106.

Pahl, L.I., Winter, J.W., and Heinsohn, G. 1988. Variation in responses of arboreal marsupials to fragmentation of tropical rainforest in northeastern Australia. *Biol. Conserv.* 46:71–82.

Pyšek, P., and Prach, K. 1993. Plant invasions and the role of riparian habitats: a comparison of four species alien to central Europe. *J. Biogeog.* 20:413–420.

Reijnen, R., and Koolstra, B. 1998. *Evaluation of Ecological Corridors in the Province of Gelderland* (in Dutch). Report 372. Wageningen, The Netherlands: Institute for Forestry and Nature Research.

Riffell, S.K., and Gutzwiller, K.J. 1996. Plant-species richness in corridor intersections: is intersection shape influental? *Landsc. Ecol.* 11:157–168.

Ruefenacht, B., and Knight, R.L. 1995. Influences of corridor continuity and width on survival and movement of deermice *Peromyscus maniculatus. Biol. Conserv.* 71:269–274.

Schippers, P., Verboom, J. Knaapen, J.P., and Van Apeldoorn, R.C. 1996. Dispersal and habitat connectivity in complex heterogeneous landscapes: an analysis with a GIS-based random walk model. *Ecography* 19:97–106.

Smith, D.S. 1993. Greenway case studies. In *Ecology of Greenways: Design and Function of Linear Conservation Areas,* eds. D.S. Smith and P.C. Hellmund, pp. 161–208. Minneapolis: University of Minnesota Press.

Soulé, M.E., and Gilpin, M. 1991. The theory of wildlife corridor capability. In *Nature Conservation 2: The Role of Corridors,* eds. D.A. Saunders and R.J. Hobbs, pp. 3–8. Chipping Norton, Australia: Surrey Beatty and Sons.

Sutcliffe, O.L., and Thomas, C.D. 1996. Open corridors appear to facilitate dispersal by Ringlet Butterflies (*Aphantopus hyperantus*) between woodland clearings. *Conserv. Biol.* 10:1359–1365.

Taylor, P.D., Fahrig, L., Henein, K., and Merriam, G. 1993. Connectivity is a vital element of landscape structure. *Oikos* 68:571–573.

The World Conservation Union. 1995. *River Corridors in Hungary: A Strategy for the Conservation of the Nanube and Its Tributaries (1993–94).* Gland, Switzerland and Budapest, Hungary: The World Conservation Union.

Tischendorf, L., Irmler, U., and Hingst, R. 1998. A simulation experiment on the potential of hedgerows as movement corridors for forest carabids. *Ecol. Model.* 106:107–118.

Tischendorf, L., and Wissel, C. 1997. Corridors as conduits for small animals: attainable distances depending on movement pattern, boundary reaction and corridor width. *Oikos* 79:603–611.

Turchin, P. 1998. *Quantitative Analysis of Movement: Measuring and Modeling Population Redistribution in Animals and Plants.* Sunderland, Massachusetts: Sinauer Associates.

Van Dorp, D., Schippers, P., and Van Groenendael, J.M. 1997. Migration rates of grassland plants along corridors in fragmented landscapes assessed with a cellular automaton model. *Landsc. Ecol.* 12:39–50.

Verboom, B., and Van Apeldoorn, R. 1990. Effects of habitat fragmentation on the red squirrel, *Sciurus vulgaris* L. *Landsc. Ecol.* 4:171–176.

Vermeulen, H.J.W. 1995. *Road-Side Verges: Habitat and Corridor for Carabid Beetles of Poor Sandy and Open Areas.* Ph.D. dissertation. Wageningen, The Netherlands: Wageningen Agricultural University.

Vos, C.C. 1999. *A Frog's-Eye View of the Landscape.* Ph.D. dissertation. Wageningen, The Netherlands: Wageningen University and DLO Institute for Forestry and Nature Research.

Vos, C.C., and Stumpel, A.H.P. 1996. Comparison of habitat-isolation parameters in relation to fragmented distribution patterns in the tree frog (*Hyla arborea*). *Landsc. Ecol.* 11:203–214.

Vos, C.C., Verboom, J., Opdam, P.F.M., and Ter Braak, C.J.F. 2001. Towards ecologically scaled landscape indices. *Am. Nat.*157:24–41.

Wauters, L., Casale, P., and Dhondt, A. 1994. Space use and dispersal of red squirrels in fragmented habitats. *Oikos* 69:140–146.

Wennergren, U., Ruckelshaus, M., and Kareiva, P. 1995. The promise and limitations of spatial models in conservation biology. *Oikos* 74:349–356.

Wiens, J.A. 1997. Metapopulation dynamics and landscape ecology. In *Metapopulation Biology: Ecology, Genetics, and Evolution,* eds. I.A. Hanski and M.E. Gilpin, pp. 43–68. San Diego: Academic Press.

Wiens, J.A., Schooley, R.L., and Weeks, R.D., Jr. 1997. Patchy landscapes and animal movements: do beetles percolate? *Oikos* 78:257–264.

Wilson, E.O., and Willis, E.O. 1975. Applied biogeography. In *Ecology and Evolution of Communities,* eds. M.L. Cody and J.M. Diamond, pp. 522–534. Cambridge, Massachusetts: Harvard University Press.

With, K.A., Gardner, R.H., and Turner, M.G. 1997. Landscape connectivity and population distribution in heterogeneous environments. *Oikos* 78:151–169.

7

Using Percolation Theory to Assess Landscape Connectivity and Effects of Habitat Fragmentation

Kimberly A. With

7.1 Introduction

Maintaining landscape connectivity has become a management imperative for many agencies (Salwasser 1991; Petit et al. 1995). It is therefore essential that landscapes in danger of becoming disconnected can be identified before they become too fragmented, after which management actions are less likely to be successful and cost-effective. Landscape connectivity is far more complex than is implied by the notion of habitat corridors linking fragments (*structural connectivity*). For example, if an organism is able to move through the intervening matrix, then isolated habitat patches may be functionally connected, if not structurally connected, by dispersal. Landscape connectivity, therefore, must ultimately be defined by the extent to which different habitat types and other elements of landscape structure facilitate movement among patches (*functional connectivity;* Taylor et al. 1993; With et al. 1997). The theoretical basis for understanding ecological flows, such as dispersal, gene flow, or the propagation of disturbances (e.g., spread of nonindigenous invasive species) across landscapes has emerged within the discipline of landscape ecology primarily as applications of percolation theory.

In this chapter, I focus on applications of percolation theory, such as neutral landscape models, which provide a theoretical framework for quantifying landscape connectivity, modeling movement or dispersal in spatially complex landscapes, and predicting the ecological consequences of habitat loss and fragmentation. My objectives are to demonstrate how percolation theory and its neutral landscape derivatives provide a useful framework for predicting when landscapes become disconnected and for modeling the ecological consequences of habitat loss and fragmentation; identify recent planning and research applications of percolation theory in biological conservation; distill general principles for applying percolation theory in conservation biology; identify the theoretical and empirical voids that still need to be addressed if the principles distilled from percolation theory are to be widely adopted in conservation; and suggest research approaches to address these theoretical and empirical voids.

7.2 Concepts, Principles, and Emerging Ideas

7.2.1 What Is Percolation Theory?

Percolation theory examines how connectivity is disrupted within spatially structured systems (Zallen 1983; Stauffer and Aharony 1991). *Connectivity,* in the parlance of percolation theory, refers to linkages among sites that facilitate flows through the system. The system is usually a square grid (*lattice*) in which some fraction (p) of the grid cells (sites) are filled or occupied; for example, consider a raster landscape map in which cells are either habitat (filled cells) or matrix (empty cells). Life is more than just a square lattice, however (Stauffer and Aharony 1991). Other lattice geometries can be created using triangles or hexagons (honeycomb lattice), and lattices also can extend into three or more dimensions (e.g., simple cubic lattice). For example, honeycomb lattices have been used to characterize landscapes in some spatially explicit models of bird populations because hexagons permit better packing of territories in space (e.g., Pulliam et al. 1992). Sites are connected if they are adjacent (neighboring cells) or can be linked by some process that permits flows across empty cells (e.g., dispersal across habitat gaps). Connected sites form *clusters,* and the properties of these clusters are of particular interest to students of percolation theory, especially when a single cluster spans the entire system. This spanning cluster confers overall connectivity because flows can *percolate* across the entire system, and the spanning cluster is therefore called the *percolation cluster* (Figure 7.1).

Percolation on a lattice is referred to as *site percolation,* because the focus is on the flows through filled sites. Alternatively, one could assume that the entire lattice is filled (all cells are potentially accessible) and that conduits exist between sites that are either open with probability, p, or closed ($1 - p$). A cluster is then defined as a group of sites linked via open conduits, through which flows can occur; this is *bond percolation.* Note that a lattice is not strictly necessary for bond percolation. For example, a landscape can be represented as discrete habitat patches in which one could assess connectivity simply by quantifying the fraction connected by corridors, or by determining what fraction of sites fall within the dispersal range of a given organism (patches are still linked by dispersal even in the absence of a physical corridor). Site percolation models are more prevalent in ecological applications and thus will dominate the discussion in this chapter, but an analysis of habitat connectivity employing bond percolation will be presented in an analysis of critical habitat for the Mexican Spotted Owl (*Strix occidentalis lucida;* Section 7.3.1).

7.2.2 Quantifying Thresholds in Connectivity

Percolation theory is thus the quantitative analysis of connectivity in spatially structured systems. The degree of connectivity is usually assessed as either the probability of percolation (i.e., probability that the system contains a percolation cluster, $P_{(p)}$) or by deriving an *order parameter* (Ω), which defines the transition

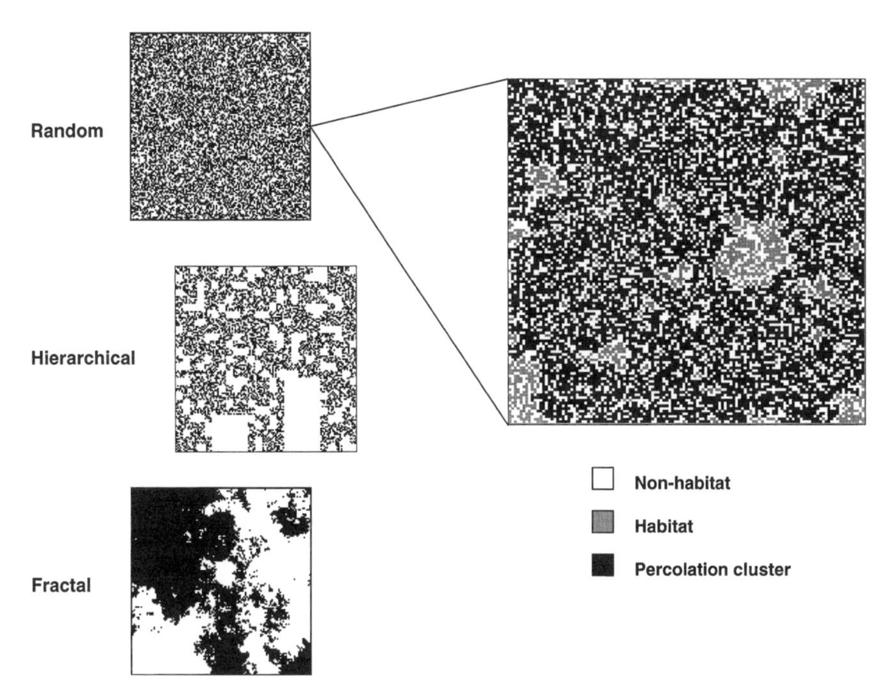

FIGURE 7.1. Left: Neutral landscape models. Right: The percolation cluster is highlighted in a random neutral landscape ($p = 0.6$).

from a connected system to a disconnected one. For example, the relative size of the largest cluster (LC_{rel}) is an order parameter given by LC/LC_{max}, where LC is the number of cells in the largest cluster, and $LC_{max} = pm^2$, where p is the fraction of filled cells (e.g., habitat) and m is the number of sites along one side of the lattice (Bascompte and Solé 1996; Pearson et al. 1996). The transition from a connected to a disconnected system is predicted to occur abruptly. Stochastic disturbances that disrupt linkages among sites (e.g., the sites themselves are destroyed) reduce the size of the percolation cluster until it fragments and the system no longer percolates. The level of disturbance (e.g., fraction of sites destroyed or bonds disrupted) at which this transition from a connected to a disconnected system occurs is the *critical* or *percolation threshold* (p_c). Above the critical threshold ($p > p_c$), the probability of having a percolation cluster is high ($P_{(p)} \rightarrow 1$) and the largest cluster *is* the percolation cluster, which dominates the system ($\Omega \rightarrow 1$). Below the critical threshold ($p < p_c$), no percolation cluster occurs ($P_{(p)} = 0$) and the system consists of numerous small clusters (i.e., the system is disconnected, $\Omega \rightarrow 0$). Because the threshold is often a sharp transition, it is more convenient to refer to systems simply as being either "connected" or "disconnected" depending on where they lie with respect to the critical threshold. The small, finite size of lattices used in ecological applications (and the neighborhood rule used to define connectivity, to be discussed later) may produce a more gradual transition, such that the percolation threshold is defined as the critical amount of habitat at which the probability of having a percolation cluster exceeds some threshold level (usually ≥50%). In practice, the probability of percolation ($P_{(p)}$) is determined as the fraction of replicate landscapes that percolate (i.e., neutral landscape models; Section 7.2.3).

Another measure commonly employed to quantify connectivity is the *correlation length* (C), which is defined as the size-weighted average radius of clusters (Plotnick and Gardner 1993; Keitt et al. 1997). Clusters generally have irregular shapes, and thus, cluster size is defined as the "radius of gyration" (R) calculated as

$$R = \frac{\sum_{i=1}^{n} \sqrt{(x_i - \bar{x})^2 + (y_i - \bar{y})^2}}{n}, \tag{7.1}$$

where x_i and y_i denote the coordinates of individual sites within a cluster, and n is the total number of sites within an individual cluster. The correlation length, C, is therefore obtained as

$$C = \frac{\sum_{s=1}^{m} (n_s R_s)}{\sum_{s=1}^{m} n_s}, \tag{7.2}$$

where R_s is the radius of gyration for cluster s, m is the number of clusters, and n_s is the number of sites in cluster s. Greater correlation lengths reflect an overall increase in the sizes of clusters and, hence, landscape connectivity.

7.2.3 From Percolation Models to Neutral Landscape Models

Percolation theory thus has some attractive features that can facilitate the definition and analysis of habitat connectivity in ecological landscapes. The definition of landscape connectivity is explicitly process-based and is determined by how individual movement behavior or dispersal ability interacts with the patch structure of the landscape. In percolation theory, connectivity also is defined by the extent to which flows can occur through a spatially structured medium. In lattice percolation models, grid cells are randomly assigned or connected at some density (p). Lattice percolation models bear a striking resemblance to raster data sets, such as land coverages within a geographic information system (GIS). Robert Gardner and his colleagues thus used lattice percolation models as *neutral landscape models* (Gardner et al. 1987; Gardner and O'Neill 1991). Neutral landscape models (NLMs) are theoretical habitat distributions and are thus neutral to the biophysical processes, such as topography and disturbance history, that shape real landscapes. They thus provide a statistical baseline or null model for exploring the relationship between spatial pattern and ecological processes (With and King 1997).

Although random connectance is the simplest lattice percolation model (and hence NLM), a diverse array of neutral landscape patterns can be generated (Figure 7.1). Hierarchical random landscape maps (O'Neill et al. 1992; Lavorel et al. 1993) generate geometric landscape patterns resembling those created by human land-use activities (e.g., timber harvest, agriculture; Forman and Godron 1986). Such maps are "hierarchical" because habitat is randomly assigned at desired levels of p at each of several nested levels (i.e., the distribution of habitat at the broadest scale, p_1, constrains the distribution of habitat at the next finer scale, p_2, and so on). Fractal landscapes are initially a continuously varying surface, much like a topographic relief map, which is useful for analyzing environmental gradients or resource distributions that are not discretely patchy (e.g., Palmer 1992). The fractal surface can be sectioned at a given "elevation" so that only a certain proportion is revealed (e.g., a two-dimensional contour map of a three-dimensional relief map). The resulting two-dimensional map is a binary landscape of habitat and nonhabitat, consistent with traditional lattice percolation models (With 1997a). Decreasing the elevation at which the section is made reveals more of the original fractal surface (i.e., more habitat is present on the landscape). The degree of spatial autocorrelation or contagion (H) among habitat cells can be adjusted to generate landscape patterns across a range of fragmentation (With 1999; With and King 1999a,b). Although NLMs are usually binary maps in the tradition of lattice percolation models, heterogeneous landscapes composed of multiple habitat types also can be generated (e.g., With and Crist 1995; Gustafson and Gardner 1996; With et al. 1997).

Neutral landscape models have served two main functions in ecology (With and King 1997): the generation and analysis of landscape patterns, and the analysis of the ecological consequences of landscape patterns. The prediction that landscape connectivity may become disrupted abruptly, as a nonlinear response to habitat loss, should concern conservation biologists because such thresholds in landscape connectivity are likely to have important ecological consequences. NLMs also have provided a general, spatially explicit framework for examining the effect of landscape pattern on a diverse array of ecological phenomena, and a variety of ecological processes have been modeled as a percolation process (With and King 1997).

7.2.4 Movement and Connectivity in Neutral Landscapes

Habitat connectivity in neutral landscapes is assessed by *neighborhood* or *movement rules,* which specify the distance across which sites are accessible to organisms by virtue of their dispersal or gap-crossing abilities. Movement rules are a direct extension of the neighborhood rules used in percolation theory. Thus, the critical threshold (p_c) at which the landscape becomes disconnected shifts to increasingly lower levels of habitat abundance (p) as the movement neighborhood becomes larger; that is, $p_c \rightarrow 0$ as the neighborhood size approaches m^2 (the size of the landscape). Habitat need not be adjacent to be connected if the organism is capable of crossing gaps of unsuitable habitat. Organisms with good dispersal abilities therefore perceive landscapes as connected across a greater range of habitat abundance than do those constrained to move only through habitat cells (Table 7.1). The use of different movement rules (or resource utilization scales; O'Neill et al. 1988) has permitted the identification of species' perceptions of landscape structure (Pearson et al. 1996). Defining connectivity from a species' perspective is the only relevant viewpoint if we are to develop meaningful and effective conservation strategies (Hansen and Urban 1992).

Determining the scale at which organisms are able to interact with the scale of the landscape pattern is thus the key to defining landscape connectivity (Doak et al. 1992; With and Crist 1995; Pearson et al. 1996). For species that lack gap-crossing abilities (neighborhood size = 4 cells; Rule 1 movement), habitat destruction at a fine scale (e.g., selective logging) disrupts connectivity sooner (at lower levels of habitat loss) than does coarse-scale habitat destruction (clear-cuts) in hierarchical random landscapes (Pearson et al. 1996). Interestingly, With (1999) demonstrated that gap-sensitive species (neighborhood size = 8 cells; Rule 2 movement) in fractal landscapes perceived highly fragmented landscapes (fine-scale habitat destruction, $H = 0.0$) as more connected than they did clumped landscape patterns (coarse-scale habitat destruction, $H = 1.0$) when 40% to 60% of the

TABLE 7.1. Percolation thresholds (p_c) for different neutral landscape models, as a function of the neighborhood size. The percolation threshold is defined here as the level of habitat abundance (p) at which the probability of having a percolation cluster, $P_{(p)}$, $\geq 50\%$.

Neutral landscape model	Neighborhood size	p_c	Source[a]
Random	4	0.59	1, 2
	8	0.41	1, 2
	12	0.29	1, 2
	24	0.17	1
	40	0.1	1
	60	0.07	1
Random with contagion[b]			
$Q_{11} = 0.48$	4	0.62	3
$Q_{11} = 0.7 - 0.8$		0.57	3
$Q_{11} = 0.97$		0.7	3
Fractal[c]			
$H = 1.0$	4	0.45	4
	8	0.33	5
$H = 0.0$	4	0.54	4
	8	0.38	5
	24	0.25	5
	80	0.15	5

[a] 1 = Plotnick and Gardner 1993; 2 = Pearson et al. 1996; 3 = Gardner and O'Neill 1991; 4 = With and King 1999a; 5 = With 1999.

[b] Contagion, Q_{11}, is the probability that neighboring sites are the same (i.e., habitat). Random landscapes with greater contagion ($Q_{11} > 0.8$) are less likely to percolate because habitat is aggregated in a few large patches, and thus a higher p is required before these coalesce to span the landscape.

[c] Fractal landscapes were generated by the midpoint displacement algorithm (Saupe 1988) in which H controls the degree of spatial autocorrelation (contagion) among sites.

habitat was destroyed (Figure 7.2a). A percolation cluster was more likely to form on fragmented fractal landscapes because the habitat was more dispersed, and thus individual habitat cells served as "stepping stones" to facilitate connectivity across the landscape (Figure 7.2b). Nevertheless, fragmented fractal landscapes had less suitable habitat (defined as connected habitat meeting the species' minimum area requirements; With 1999), and they therefore supported smaller populations that were more likely to go extinct (With and King 1999b). Landscape connectivity thus may not be sufficient for population persistence.

7.3 Recent Applications

Some potential applications of NLMs for biological conservation were recently presented in a review by With (1997a). In that paper, I identified several ways in which NLMs could contribute to the analysis of species' responses to habitat fragmentation, such as by providing a quantitative analysis of landscape connec-

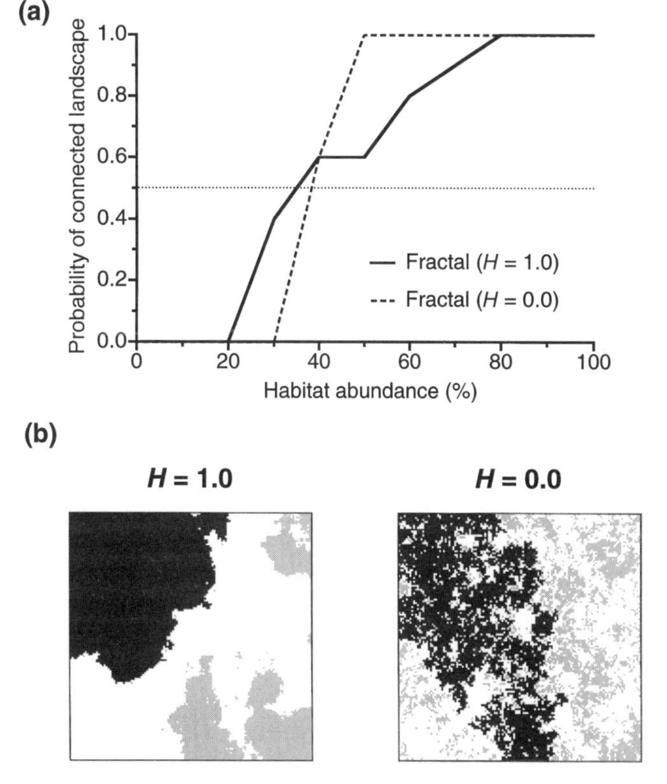

FIGURE 7.2. (a) Critical thresholds in connectivity for fractal landscapes (neighborhood size = 8 cells). (b) Fragmented fractal landscapes ($H = 0.0$) are more likely to be connected (percolation cluster present) than are clumped fractal landscapes ($H = 1.0$) at 50% habitat. The largest habitat cluster is highlighted in black in both landscapes.

tivity; permitting species' perceptions of landscape structure based on dispersal or gap-crossing abilities; contributing to the development of a general, spatially explicit framework for modeling population persistence; and facilitating the design of nature reserves by identifying critical landscape linkages necessary for maintaining connectivity. To what extent has percolation theory and its neutral landscape derivatives actually been applied in biological conservation?

7.3.1 Planning Application

Because percolation theory has only recently been presented to conservation biologists, it has not yet been adopted by management agencies concerned with maintaining landscape connectivity as a means of enhancing population persistence of target species or in the conservation of biodiversity. A search of the Internet and an informal survey of academic colleagues and staff scientists affiliated with federal resource and wildlife management agencies confirmed this initial impression. To date, the only application of percolation theory in a management context is the inclusion of a percolation-based analysis of habitat connectivity in the recovery plan for the Mexican Spotted Owl (Keitt et al. 1995). The Mexican Spotted Owl is federally listed as a threatened species and is genetically distinct from the other two spotted owl subspecies (Northern, *S. o. caurina;* California, *S. o. occidentalis*). The Mexican Spotted Owl has the most widespread distribution of the three subspecies, preferring forested slopes and canyons throughout its range in the southwestern United States and northern Mexico (Figure 7.3a).

This analysis represents an application of bond percolation, in which connectivity was assessed as the distance that could be traversed by owls between habitat sites. Percolation thus occurs at a critical mean dispersal distance instead of a critical habitat density as in lattice percolation. Connectivity (correlation length, C; Equation 7.2) of suitable owl nesting habitat, which included mixed-conifer and ponderosa pine (*Pinus ponderosa*) forest, exhibited a strong threshold between a dispersal range of 40–50 km (Figure 7.3b). Maximum dispersal distances do not capture the stochastic nature of dispersal, however, and thus a dispersal function was derived to assess the probability of dispersal among patches at a given distance. A sharp transition in connectivity was observed at an average dispersal distance of 15 km. Thus, forested habitat in the Southwest is connected if Mexican Spotted Owls can disperse 45 km over inhospitable habitat and have an average dispersal distance of at least 15 km. Information on the dispersal distance of Mexican Spotted Owls is limited, but there are reports of juvenile owls dispersing > 45 km and establishing new territories 10–20 km from their natal territory (reviewed in Keitt et al. 1995). Mexican Spotted Owls may therefore exist as a metapopulation in this region.

Apart from assessing habitat connectivity, this application of percolation theory also permitted the identification of critical habitat patches that are essential for the maintenance of landscape connectivity. Even with a maximum dispersal range of 45 km, the connectivity of forested habitat in this region is tenuous, at best, for Mexican Spotted Owls. At this critical dispersal distance, only one patch links the southwestern and northeastern parts of its range. Analysis revealed that

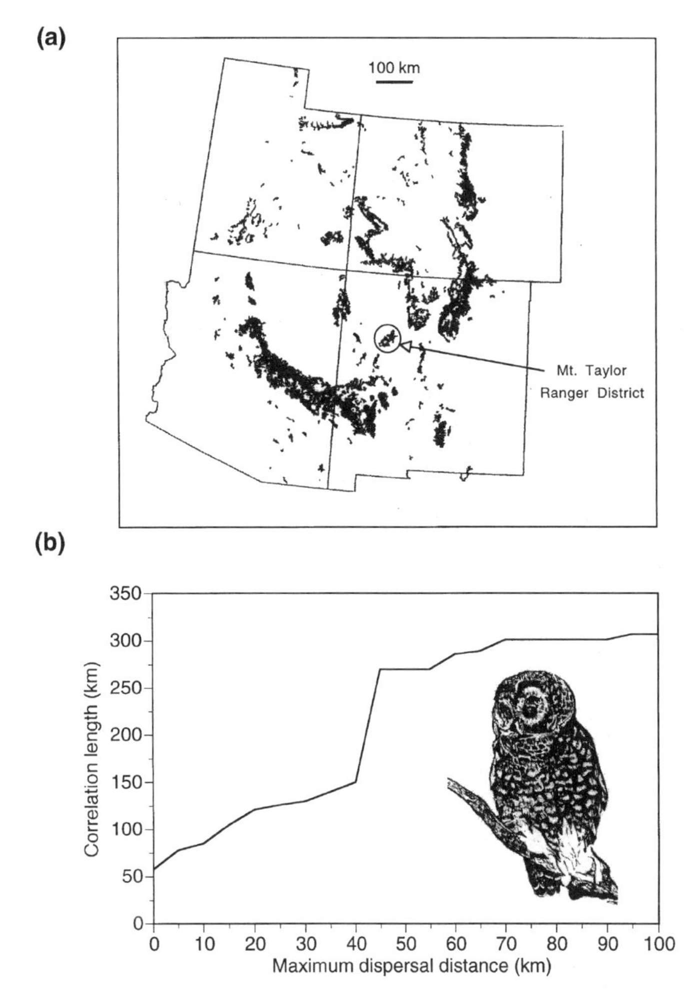

FIGURE 7.3. (a) Distribution of suitable breeding habitat for the Mexican Spotted Owl in the southwestern United States. (b) A critical threshold in the regional connectivity of suitable habitat occurs if the maximum dispersal distance of owls is less than 45 km. Modified from Keitt et al. (1995).

the U.S. Forest Service's Mount Taylor Ranger District in central New Mexico (Figure 7.3a) was a crucial stepping stone that connected owl populations in Arizona and New Mexico to those in Colorado and Utah. This was not obvious prior to this analysis of landscape connectivity, and in fact, the Mount Taylor District had previously been considered unimportant habitat for this species because it supported so few owls. Before the Mexican Spotted Owl was listed in 1993, commercial logging in the Mount Taylor District generated 12,000–19,200 m^3 yr^{-1} (5–8 million board feet per year) in saw-timber production (R. Suminski, personal communication). Commercial logging ceased following listing of the owl, and now specialty products such as small ponderosa pines harvested for vigas used in home construction, or the commercial sale of timber harvested incidental to road construction projects, generate only about 1200 $m^3 yr^{-1}$ (0.5 million board feet per year) in this district. Listing the owl was the main catalyst effecting management changes in the Mount Taylor Ranger District, and thus this percolation-based analysis was more of a planning application than a management application that was actually implemented. Nevertheless, the District's status as a critical habitat link, identified by this percolation-based analysis of habitat connectivity, was included in the species' recovery plan issued by the U.S. Fish and Wildlife Service (Keitt et al. 1995), and it therefore would have been critical information if more specialized guidelines had been required to protect this population.

7.3.2 Research Applications

Research applications have far outstripped management applications of percolation theory in biological conservation. Although percolation theory in the form of NLMs was first introduced to ecologists more than a decade ago, applications have only recently begun to "percolate" into conservation biology (e.g., Green 1994; With 1997a; Metzger and Décamps 1997). This lag may simply reflect the origins of NLMs, whose creators were all systems ecologists in the Environmental Sciences Division (ESD) at the Oak Ridge National Laboratory in Tennessee. The ESD has traditionally directed research in landscape ecology, environmental risk assessment, and ecotoxicology rather than in biological conservation (A. W. King, personal communication). In any case, it may take a decade for new ideas and approaches to become part of "mainstream" science, let alone be adopted in applied fields. Percolation-based applications may eventually gain a wider acceptance in fields such as conservation biology because of the obvious similarities between raster-based GIS applications and NLMs, making these theoretical applications more accessible to practitioners. This section highlights some recent research applications of percolation theory and their neutral landscape derivatives that have relevance for biological conservation: identifying thresholds in dispersal success, predicting population persistence in fragmented landscapes, managing keystone species, and assessing the effect of land-use change on sensitive species.

One of the more immediate consequences of a disruption in landscape connectivity should be a disruption in dispersal success, the ability of the organism to locate suitable habitat. Recent attempts to link measures of landscape connectivity, including percolation thresholds, with dispersal or colonization success have generally failed, however (Schumaker 1996; With and King 1999a). Thresholds in dispersal success occurred for simulated organisms dispersing on fractal landscapes at $p < 0.1$ (With and King 1999a). This dispersal threshold did not coincide with percolation thresholds, or any other patch-based metric used to quantify landscape pattern. Intuitively, the farther apart patches are, the less likely they are to be found or colonized by a dispersing organism. Thus, in terms of predicting fragmentation effects on dispersal success, a measure of the gap-size distribution of the landscape may be a more useful metric of landscape structure.

Lacunarity is a fractal index that can be used to describe the gap-size distribution of landscape patterns; it is basically a variance-to-mean ratio of gap sizes (Plotnick et al. 1993). Consider two landscapes with similar habitat coverages but that differ in the pattern of habitat destruction (Figure 7.2b). Gap sizes are more variable, and thus the lacunarity index (Λ) would be higher, in the clumped fractal landscape ($H = 1.0$) relative to the more fragmented landscape ($H = 0.0$), which has numerous small gaps that are more uniform in size. Interestingly, a threshold in the lacunarity index, Λ, was found at $p < 0.1$ in fractal landscapes across a range of scales, which is the same domain where dispersal thresholds occur. A similar threshold in the average distances between patches is also apparent for random landscapes below 10% habitat (see Figure 3 of Andrén 1994). Coincidentally, empirical work demonstrated thresholds in movement behavior for tenebrionid beetles in experimental random "microlandscapes" with <10% habitat (Wiens et al. 1997). This suggests that the gap structure of landscapes may be a more important determinant of dispersal success than is patch structure per se.

The quality of the matrix habitat through which the organism travels also should affect dispersal success and hence patch colonization rates. To assess the relative importance of matrix heterogeneity on patch colonization, Gustafson and Gardner (1996) simulated dispersal as a self-avoiding random walk (disperser was not permitted to return immediately to a previously visited cell) on random and hierarchical neutral landscapes, in which forest patches were embedded in a heterogeneous matrix composed of six, equally abundant habitat types that varied in their permeabilities to movement. Although distance between patches was the most important determinant of colonization success, matrix structure had the potential to induce asymmetrical flows between patches, where dispersal was more likely in one direction than in the other. Furthermore, regions of the heterogeneous matrix that facilitated movement were not discrete or obvious structures on the landscape. Thus, connectivity in heterogeneous landscapes must be assessed by quantifying movement responses to landscape structure and not *a priori* by identifying structural features of the landscape that appear to function as corridors.

If colonization success is affected by landscape structure, then this should have implications for population persistence. Habitat destruction may precipitate extinction thresholds (sensu Lande 1987), in which the equilibrium fraction of suitable habitat occupied by the population (p^*) crashes abruptly ($p^* \rightarrow 0$) at some critical level of available habitat (h_c). The maximum habitat occupancy by a species is determined by its demographic potential (k), which is a function of its dispersal ability and reproductive output (R'_o). For example, a species with a large demographic potential, $k = 0.9$, theoretically could occupy 90% of the available habitat when the entire landscape is suitable ($h = 1.0$). Thus, not all of the available habitat is expected to be occupied by the species even when the landscape is entirely suitable. A spatially implicit, generalized metapopulation model developed by Lande (1987) predicted the occurrence of extinction thresholds for species that differed in their demographic potential ($h_c = 1 - k$). Thus, a species with a demographic potential $k = 0.7$ is predicted to go extinct on the landscape when the availability of suitable habitat falls below 30% ($h_c = 1 - 0.7 = 0.3$), or alternatively, when 70% of the habitat has been destroyed. This generalized metapopulation model was parameterized for the Northern Spotted Owl of the Pacific Northwest to predict an extinction threshold of 21% for this species (Lande 1988).

A percolation version of this extinction threshold model was developed by Bascompte and Solé (1996), in which randomly distributed habitat patches (sites) could be colonized only if neighboring sites were occupied (i.e., neighborhood = 4 cells). This is equivalent to constraining dispersal to a local neighborhood, which effectively reduces the scale at which individuals are moving relative to the scale of landscape pattern and, thus, the probability that habitat patches will be colonized. Consequently, extinction thresholds occurred more precipitously, although generally at the same threshold of habitat destruction as in Lande's (1987) model (see Figure 6 of Bascompte and Solé 1996). The exception is for species with low demographic potentials ($k = 0.20$), for which extinction thresholds appeared to occur sooner (at lower levels of habitat destruction) than predicted by Lande (1987).

How does habitat fragmentation—the pattern of habitat loss—affect extinction thresholds? By coupling Lande's (1987) generalized metapopulation model with fractal neutral landscape models, With and King (1999b) demonstrated that extinction thresholds either did not occur, or they occurred later (at higher levels of habitat destruction) than predicted by Lande's model (Figure 7.4). The exception was, again, for species with limited demographic potentials, owing to a combination of poor dispersal abilities and limited reproductive output ($R'_o = 1.01$) in which extinction occurred sooner, and more precipitously, in all but the most clumped fractal landscapes ($H = 1.0$). Because limited demographic potential characterizes many of the species of concern to conservation biologists, this suggests that such species may be at greater risk of extinction from habitat loss and fragmentation than previously suspected. Even so, the absence of an extinction threshold should not be interpreted to mean that a species is not affected by habitat loss. A species that maintains near-maximum patch occupancy in the face of ongoing habitat destruction is still a species in decline. Occupying 70% of the landscape when it is entirely suitable is better than occupying 70% of it when only 20% is suitable; the species would occur in only 14% of the landscape in the latter scenario. Such reductions in population size can make species more vulnerable to stochastic demographic and environmental events, which would enhance the likelihood of extinction. Interestingly, dispersal had less of an effect on extinction thresholds than did reproductive output (compare effect of increasing m with increasing R'_o from 1.01 to 1.10, Figure 7.4). This implies that enhancing reproductive output, by the management of high-quality habitats, may ultimately have a greater effect on mitigating extinction risk than would enhancing connectivity of habitat to increase dispersal success.

Percolation theory also has been employed to evaluate the potential effects of habitat destruction and fragmentation on a keystone species, an army ant (*Eciton burchelli*) of the rainforests in Central and South America (Boswell et al. 1998).

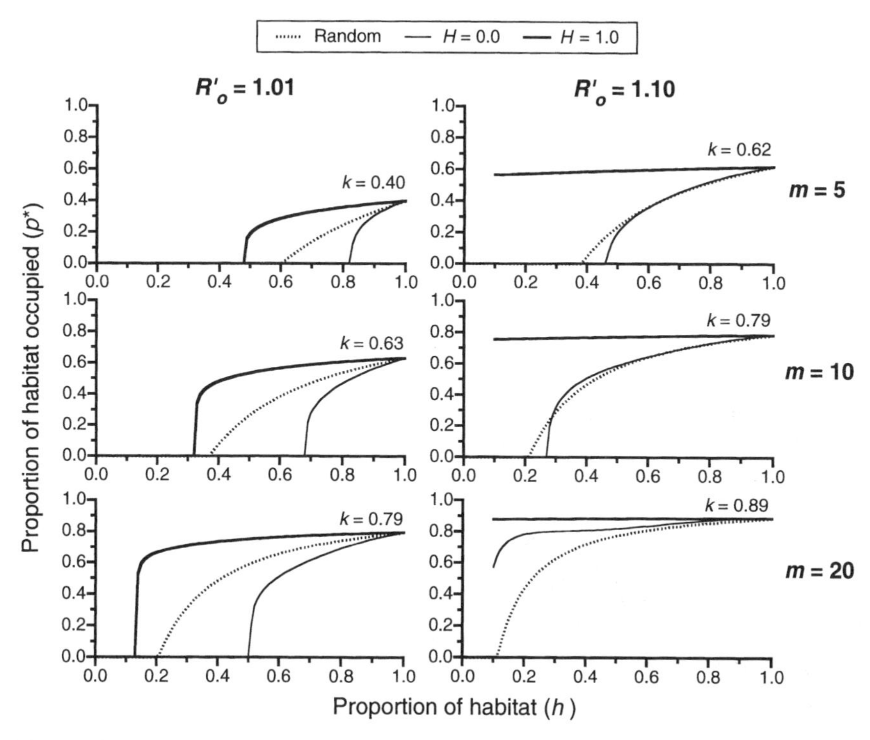

FIGURE 7.4. Extinction thresholds for species in fractal and random landscapes. Species are defined by their dispersal abilities (*m*, number of sites searched for a suitable unoccupied site) and reproductive output (R'_o). This gives rise to the demographic potential (*k*) of the species, which is the maximum fraction of sites that can be occupied when the landscape is entirely suitable. Curves for random landscapes were generated by Lande's (1987) model; results for fractal landscapes ($H = 0.0, 1.0$) are from With and King (1999b).

Army ants create a mosaic of disturbance as a consequence of their raiding activities, and the ensuing successional dynamics of these disturbed areas enhance local diversity. Furthermore, dozens of bird species are obligate followers of army ants and forage on the arthropod prey flushed from the leaf litter during their raids. The management of keystone species may therefore serve as an umbrella for the protection of other species.

Although extinction thresholds may not coincide with percolation thresholds in landscape connectivity (With and King 1999b), it is compelling that army ant colonies were predicted to go extinct when 45% of the rainforest was randomly cleared in small (180 × 180-m) blocks, which is sooner than the anticipated 60% when the landscape becomes disconnected (i.e., $p_c = 0.40$ for an 8-cell neighborhood on a random lattice). Destruction of larger blocks of habitat (720 × 720 m) actually increased the persistence of army ant colonies because contiguous habitat capable of supporting the large area requirements of this species remained.

Boswell et al. (1998) suggest that the removal of large blocks of habitat may be a more sustainable forestry practice, although they acknowledge the forest would take longer to regenerate. Habitat corridors are not viewed as a solution for restoring connectivity or enhancing the availability of habitat for army ants because colonies do not possess an "institutional memory." If individuals do not know how to exit a patch and are unable to locate corridors, a colony may become trapped in a pocket of rainforest and deplete local resources. Corridors would thus exacerbate fragmentation effects by producing habitat cul-de-sacs that enhance the extinction probability of individual colonies.

If the effects of land-use change can be predicted, then it might be possible to develop more sustainable land-management practices by directing land-use activities so as to maximize connectivity of suitable habitat for sensitive species. In an analysis of different land-use scenarios in the Brazilian state of Rondônia, Dale et al. (1994) quantified the availability of suitable habitat for a variety of rainforest species, based on their gap-crossing abilities and minimum area requirements. Species in this analysis were as diverse as jaguars (*Felis onca;* large area requirements and good gap-crossing abilities), three-toed sloths (*Bradypus variagatus;* small area requirement and poor gap-crossing abilities), and insects (Euglossine bees; large area requirements and poor gap-crossing abilities). Scenarios of land-use change were modeled using NLMs bracketing a range of agricultural practices in this region, which resulted in different rates and patterns of forest clearing. Available habitat for species under these various land-use scenarios was assessed by using their gap-crossing abilities to define whether habitat cells were connected. Clusters of connected habitat were then evaluated to determine whether they were of sufficient size to fulfill the minimum area requirements of the species. Gap-crossing ability tends to be positively correlated with area requirements for most, but not all, species. For example, Euglossine bees forage on orchids, which are widely dispersed, and thus, bees must search over many square kilometers. Nevertheless, bees are very sensitive to forest gaps, perhaps because they are an inhospitable environment to bees. Suitable habitat decreased at a much faster rate than did the amount of actual habitat loss for species like Euglossine bees whose gap-crossing abilities are smaller than are their minimum area requirements. Within 10 years, no suitable habitat would be available for such a species, even though nearly 50% of the landscape is still forested (Figure 7.5a).

Such estimates of habitat suitability are optimistic in any case. The minimum area required to sustain a viable population is much greater than that required for an individual's area requirements. Other ecological requirements may further modify assessment of habitat suitability for a given species. For example, an individual frog (*Chiasmocleis shudikarensis*) requires less than 0.001 ha, but about 500 ha of forest are required to support viable populations (Dale et al. 1994). Thus, only about half of the available forest cover would be suitable for frog populations after seven years of intense farming (Figure 7.5b). When other factors such as presence of pools for breeding and edge effects are included in the analysis, suitable habitat declined to only a third of the existing forested area. Again, habitat connectivity is not sufficient for population persistence. Nevertheless,

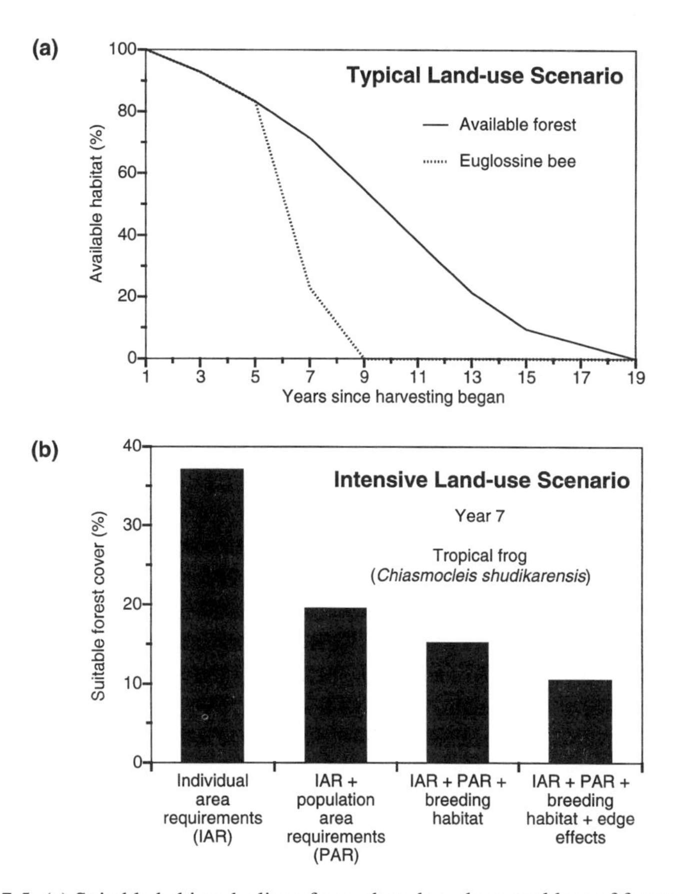

FIGURE 7.5. (a) Suitable habitat declines faster than does the actual loss of forest cover for rainforest species whose gap-crossing abilities are less than their area requirements (Euglossine bee). (b) Suitable habitat for a tropical frog declines precipitously when factors other than individual area requirements are assessed. These include habitat clusters of insufficient size to support viable populations, and a further reduction in available habitat due to edge effects and a lack of breeding pools. Modified from Dale et al. (1994).

such percolation-based analyses provide some preliminary guidelines for assessing how land-use change may affect sensitive species.

7.4 Principles for Applying Landscape Ecology

NLMs are a uniquely landscape ecological approach representing one of the few theoretical developments to emerge within this field. The potential utility of this approach for conservation has only recently been appreciated, however, and therefore principles derived from NLMs are necessarily based on ecological research rather than on demonstrated effects of successful management applications. Nevertheless, several principles of particular importance for conservation have emerged.

7.4.1 Principle 1: Disruption of Landscape Connectivity Can Occur as a Threshold Response to Habitat Loss

Habitat destruction can abruptly disconnect the landscape if it occurs near the threshold. It is the suddenness of thresholds that may make them difficult to detect empirically until after the threshold has been crossed. Conservation strategies are more likely to be successful and cost-effective if implemented before thresholds are crossed. Although landscape connectivity can be quantified in different ways, percolation theory can predict and quantify critical thresholds in connectivity. The challenge, however, is to determine where this threshold lies for a particular species in a given landscape, because the threshold is sensitive to the rule of patch connectivity used (i.e., a species' gap-crossing abilities) and the pattern of habitat destruction (fine vs. coarse scale, Figure 7.2). Thus, there is no single threshold of landscape connectivity, which may frustrate land managers. The oft-reported percolation threshold, $p_c = 0.59$, pertains only to the structural connectivity of random landscapes (Rule 1 movement; neighborhood size = 4 cells; Figure 7.1), which may not be an appropriate model for managed landscapes. Therefore, 59% should not be adopted as a general guideline for the management of landscape connectivity.

7.4.2 Principle 2: Landscape Connectivity Must Be Assessed From an Organismal Perspective

We must guard against viewing landscape structure in purely anthropocentric terms. Although it may be obvious when a landscape is structurally connected, it is usually not clear whether a fragmented landscape is actually disconnected. A fragmented landscape may be functionally connected for species with good dispersal or gap-crossing abilities. Successful conservation and land management will involve more than just identifying percolation clusters, however. Landscapes that have extensive areas of connected habitat may not percolate (e.g., fractal landscapes, $H = 1.0$; Figure 7.2), and landscapes that percolate may not have sufficient habitat to support viable populations (fractal landscapes, $H = 0.0$, Figure 7.2).

7.4.3 Principle 3: Habitat Connectivity Is Not Sufficient for Population Persistence

It is not enough that individuals can access suitable habitat; there also must be sufficient amounts of connected habitat to fulfill their minimum area requirements (territory or home range) and to support viable populations. Suitable habitat (connected habitat meeting the minimum area requirements of individuals) declines much faster than does the actual rate of habitat loss for gap-sensitive species with large area requirements (Dale et al. 1994). Fragmented landscapes may actually percolate at intermediate levels of habitat loss (40% to 60%), but these landscapes may be unable to support viable populations of such species (With 1999). Connectivity at one scale (overall landscape) may not guarantee connected habitat at other, finer scales (connected habitat meeting minimum area requirements). Connectivity thus needs to be assessed and maintained across multiple scales (Noss 1991).

7.4.4 Principle 4: Thresholds in Landscape Connectivity May Not Coincide With Other Ecological Thresholds

Although it is reasonable to expect that thresholds in landscape connectivity will have serious ecological consequences, this does not mean that thresholds in dispersal success or extinction will necessarily occur at this same point. Different scales of response to landscape structure are involved. Landscape connectivity is defined by individual gap-crossing abilities, but that does not guarantee that individuals will be successful in locating suitable habitat. The probability that the landscape percolates does not translate into dispersal thresholds, but dispersal thresholds do correlate with lacunarity thresholds, another element of landscape structure (With and King 1999a). In turn, extinction thresholds are a population-level response to habitat loss and fragmentation, and species may go extinct on a landscape long before the critical threshold in landscape connectivity is reached (Boswell et al. 1998; With and King 1999b; Figure 7.4). Management guidelines cannot be based solely on thresholds of landscape connectivity, therefore.

7.4.5 Principle 5: Corridors May Not Be Strictly Necessary to Enhance Connectivity

Regions of the landscape that facilitate movement (corridors) may in fact be diffuse and difficult to identify in heterogeneous landscapes (Gustafson and Gardner 1996). Furthermore, a functional definition of landscape connectivity does not require habitat to be structurally connected if species are capable of crossing areas of unsuitable habitat (gap-crossing abilities). Connectivity can be managed by identifying critical habitat "stepping stones" that are part of the percolation cluster (e.g., Keitt et al. 1997). From a management standpoint, these critical habitat nodes must be protected at the very least. This corresponds to identifying the *backbone* of the percolation cluster (Stauffer and Aharony 1991:91). The backbone consists of habitat cells that are ultimately responsible for landscape connectivity; as habitat is destroyed, it is the eventual loss of one of these cells that ultimately "breaks" the backbone of the percolation cluster and abruptly disrupts landscape connectivity. Nevertheless, it is not sufficient to maintain *only* these cells. Obviously, the more the percolation cluster dominates the landscape ($\Omega \rightarrow 1$), the more landscape connectivity is buffered from habitat loss. Furthermore, as discussed above, a minimally connected landscape (e.g., just the backbone present) may not have sufficient habitat to support viable populations.

7.4.6 Principle 6: Protection of High-Quality Habitat May Be as Important as Enhancing Connectivity

Protection or restoration of adequate amounts of habitat may ultimately be more important than mitigating fragmentation effects by enhancing connectivity (Fahrig 1997). Likewise, increasing reproductive output of species, such as through the management of sufficient amounts of high-quality habitat or predator-control programs, may have a greater effect on lowering extinction thresholds (the level of habitat abundance at which populations go extinct) than would enhancing dispersal success by generating more connected landscapes (With and King 1999b; Figure 7.4). Habitat connectivity *is* important, but management that focuses only on landscape connectivity will not be sufficient for the conservation of biodiversity.

7.5 Knowledge Gaps

Perhaps the greatest obstacle to the practical application of percolation theory in conservation biology is the perception by agency line officers and staff biologists that it is "too theoretical" (S. Rinkevitch, personal communication). Although the concepts of percolation theory may be understood and even appreciated by agency personnel, there nevertheless may be some reservations about applying results of percolation-based analyses or models that are not explicitly based on empirical data. Such skepticism by practitioners is not unique to applications of percolation theory, but it is characteristic of a more general apprehension regarding theory in conservation biology (Doak and Mills 1994; With 1997b). Nevertheless, additional theoretical work, discussed below, could enhance the relevance of percolation theory and its neutral landscape derivatives for conservation and land management. The following sections also indicate what empirical gaps need to be filled if percolation-based applications are to be useful to practitioners.

7.5.1 Theoretical Voids

Additional theoretical research needs to be done to expand percolation-based definitions of connectivity for the management of real landscapes. This includes adopting multilevel assessments of landscape connectivity, and defining connectivity in a dynamic landscape context. At present, we know little about how incorporating different scales of movement responses to landscape structure affects assessment of connectivity (Noss 1991; With 1999). This can best be explored in a theoretical context, particularly in the absence of available empirical information on how species respond to patch structure at multiple spatial scales (Section 7.5.2). Current percolation-based definitions of landscape connectivity emerge from the application of a particular movement rule (gap-crossing ability) or dispersal-distance function. There may not be a single characteristic scale at which a given species operates on the landscape, however. Species may exhibit different scales of movement corresponding to different behaviors (foraging vs. dispersal), different life-history stages, and in response to different scales of landscape patchiness (Kotliar and Wiens 1990). Overall landscape connectivity for a given species thus cannot be predicted from a single scale, but instead emerges as a result of movement responses to patch structure across a range of scales. Consequently, connectivity at one level (connected habitat meeting the minimum area requirements of individuals) may not guarantee connectivity at other levels (connected populations on a landscape).

Assessing connectivity within a dynamic framework is especially pertinent for land management, such as assessing how different timber harvest rotations can be developed to preserve percolation clusters on landscapes. Connectivity is generally assessed by comparing static NLMs across a range of disturbance, however (e.g., different levels of habitat abundance). Real landscapes, especially managed ones, are obviously not static and landscapes are being transformed faster than ever before in human history. The rate or frequency of habitat destruction may be as important as the spatial pattern of habitat loss. We need to determine how connectivity can be defined within a dynamic spatiotemporal framework.

7.5.2 Empirical Voids

The major empirical voids that must be overcome for the practical application of percolation theory are information on species' dispersal abilities (gap-crossing abilities) and empirical tests of the assumptions and predictions of percolation-based models. The former is required for adopting species-centered definitions of landscape connectivity, and the latter will facilitate practical applications of principles derived from percolation theory in conservation. Percolation theory, or rather, its neutral landscape derivatives, require information on species' gap-crossing or dispersal abilities. This is important for all assessments of landscape connectivity, and it is not unique to NLMs. Unfortunately, basic information on dispersal distances or movement responses to landscape structure is lacking for most vertebrate species (Lidicker and Koenig 1996), let alone for invertebrates and plants. Specifically, information is needed on dispersal-distance functions, gap-crossing abilities, and the different spatiotemporal scales at which dispersal occurs.

As with any theory, inherent assumptions and predictions generated by percolation-based applications need to be rigorously tested in the field before the principles emerging from these applications can be implemented in conservation. Unfortunately, the broad scale of managed landscapes generally precludes carefully controlled and well-replicated experiments. This explains why, to date, so much of the work has been theoretical, involving computer simulation modeling on neutral landscapes and percolation-based analyses of GIS data sets. Nevertheless, more empirical research validating model predictions is clearly warranted before we can put percolation theory into conservation practice. For example, do real organisms percolate? How do thresholds in landscape connectivity affect dispersal success of species in real landscapes? Theoretical models predict that lacunarity thresholds in landscape structure occur at <20% habitat, which coincides with thresholds in dispersal success. Does this hold for real organisms in fragmented landscapes? Extinction thresholds have been predicted for species in fragmented neutral landscapes. Is there empirical support for extinction thresholds? Assuming information on the demographic potential of the species (e.g., reproductive output and dispersal ability) can be obtained, do observed levels of patch occupancy or persistence match model predictions?

7.6 Research Approaches

7.6.1 Approaches for Theoretical Research

The main theoretical voids identified were the need to adopt a hierarchical, multilevel assessment of landscape connectivity and to define connectivity in a dynamic landscape context. A hierarchical assessment of landscape connectivity ideally would involve using different movement rules (gap-crossing abilities or neighborhood sizes) or dispersal-distance functions to evaluate habitat connectivity at various scales. In the absence of empirical information for a given species, reasonable approximations can be made using general movement rules or dispersal distances that bracket a range of scales to identify the domain of scale(s) at which thresholds in connectivity occur. Each scale imposes a "filter" on the availability of habitat at the next scale. For example, gap-crossing abilities define suitable habitat that fulfill individual area requirements at a fine scale. Habitat clusters capable of supporting viable populations (connected habitat capable of supporting a critical number of territories) can then be identified. Finally, average dispersal distances can be used to define connected populations (metapopulations) on the landscape. Maximum dispersal distances, representing infrequent long-distance dispersal events, may give some information on the regional connectivity of metapopulations at a broader scale.

To some extent, this type of analysis has been explored using hierarchical random NLMs. Complex landscape patterns can be generated by varying the availability of habitat at different levels, adjusting which level controls the availability of habitat at other levels (top-down vs. bottom-up), and readjusting the amount of habitat available at each level (Pearson et al. 1996; Pearson and Gardner 1998). Overall landscape connectivity is very sensitive to a disruption of habitat connectivity at a particular scale; that is, if habitat does not percolate at a fine or intermediate scale, then the overall landscape is unlikely to be connected (Lavorel et al. 1993; Pearson et al. 1996). Again, connectivity has been assessed by assuming a single scale of response (neighborhood size = 4 cells) across all levels of patch structure. Researchers should explore how different scales of movement interact with the different levels of patchiness to determine overall landscape connectivity. Mixed fractal landscapes (Russ 1994), in which the spatial contagion of habitat is adjusted at different scales, also could be generated and analyzed to provide a multiscale assessment of landscape connectivity.

To develop a dynamic definition of landscape connectivity, disturbances such as habitat destruction can be modeled as a spatiotemporal fractal distribution, in which both the spatial and temporal contagion of the disturbance is adjusted. Although habitat contagion has been adjusted in previous NLMs to generate different spatial patterns of disturbance (fragmentation), it also is possible to generate disturbances that are correlated in both time and space (e.g., frequent, fine-scale disturbances vs. rare, broad-scale disturbances). Collectively, this determines the *disturbance architecture* of the landscape, which is defined by the rate of disturbance, the size of individual disturbances, and the spatial and temporal correlation of disturbances (Moloney and Levin 1996). Landscape connectivity is expected to emerge as the complex interaction between disturbance architecture and species' dispersal abilities. This approach would permit an assessment of how thresholds in landscape connectivity shift in response to the frequency of habitat destruction, beyond just the amount of habitat destroyed. This would be particularly important for developing harvest rotation schedules that would maintain landscape connectivity.

7.6.2 Approaches for Empirical Research

As discussed previously, the main data needs are information on species' dispersal distances and empirical tests of the assumptions and predictions of percolation-based models. Information on species' responses to landscape structure is a research priority. Such information is absolutely essential for assessing habitat connectivity, if not by movement rules or gap-crossing abilities, then by dispersal-distance functions (as for bond percolation models; Keitt et al. 1997). Ideally, information on movement responses to landscape structure would be obtained across a range of temporal and spatial scales to permit a hierarchical assessment of habitat connectivity. Realistically, however, research on movement is laborious and time-intensive. Furthermore, most research on movement responses to landscape structure is done at relatively fine spatial scales over short time periods. At the other extreme, mark-recapture studies may document long-distance dispersal events, but do not provide information on how individuals interact with landscape patterns. Extrapolation of information on fine-scale movement responses to landscape structure across scales is a challenge, especially if organisms exhibit scale-dependent responses to patch structure.

Clever methodologies and recent advances in tracking technology are making studies on movement responses to landscape structure across a range of spatiotemporal scales more feasible, however. For example, the use of tape-recorded territorial songs or alarm calls to lure birds across habitat gaps of various sizes has been used to assess species' gap-crossing abilities (Desrochers and Hannon 1997; Rail et al. 1997; St. Clair et al. 1998), which can be related directly to the movement rules used in lattice-based neutral landscape applications. Other techniques that can be used to document species' responses to habitat edges and their gap-crossing abilities include dusting small mammals with flourescent powder, or affixing spools of string to either small mammals or amphibians (e.g., frogs; Heyer et al. 1994:153), which leave a trace of the animal's movement pathway. Long-term, broad-scale studies of animal movement across landscapes usually involve "tenacious tracking" of radio-collared individuals (e.g., Mladenoff et al. 1995; Beier 1996). Satellite tracking now permits the recording of nearly continuous georeferenced data on animal movements, which may facilitate the collection of fine-scale information over long periods of time, at least for larger animals. This provides the best opportunity for documenting scale-dependent shifts in movement responses to hierarchical patch structure.

Although well-replicated experiments at the traditionally defined landscape scale are not practical, "microlandscape" studies provide a means of manipulating habitat patterns at experimentally tractable scales (e.g., Wiens and Milne 1989). Although microlandscape studies have been criticized for not permitting a direct test of landscape connectivity per se (Beier and Noss 1998), they do serve as experimental model systems for quantifying movement responses to patch structure. Furthermore, landscape connectivity can be assessed in microlandscapes if appropriately scaled to the organism. Recent experiments founded on percolation theory involve the creation of neutral landscape patterns in the field, using random or fractal habitat distributions, in which the movement responses of terrestrial arthropods (tenebrionid beetles, crickets) were analyzed (Wiens et al. 1997; With et al. 1999). Hypotheses regarding the occurrence of thresholds in movement parameters were first generated by simulating movement as a percolation process on neutral landscapes across a range of habitat loss and fragmentation, and then compared with empirical movement pathways obtained from the microlandscape experiment (With et al. 1999). Experimentation is an underutilized protocol in conservation biology (With 1997b), and microlandscape studies have permitted initial testing of predictions generated by percolation-based applications, resulting in additional modification of theoretical models. Furthermore, these finer spatial scales may well be appropriate for investigating dispersal, population persistence, and community dynamics for organisms such as insects and plants. Developing scaling functions (e.g., Kunin 1998) that would permit the extrapolation of the results of these microlandscape studies to the broader scale of managed landscapes remains a research challenge, however.

7.7 Summary

Percolation theory offers a general quantitative approach for assessing landscape connectivity in a way that is meaningful for species of conservation concern. It provides a theoretical basis for predicting ecological thresholds, which have been called "a major unsolved problem facing conservationists" (Pulliam and Dunning 1994:193). Given the paucity of information on dispersal and life-history parameters for most species, percolation-based applications provide a first step toward identifying those species most susceptible to land-use change. Theoretical applications have outstripped management applications in conservation biology thus far. This is to be expected given that percolation theory and its neutral landscape derivatives have only recently been presented to conservation biologists; most practitioners are simply unaware of the potential applications of this approach. Practitioners may nevertheless be reluctant to implement principles derived from percolation theory unless they are empirically based. Future research should thus be directed at obtaining the necessary information on species' movement responses to landscape structure across a range of scales, and testing predictions generated by theoretical percolation models. This will allow the identification of critical information required for assessment.

Acknowledgments

I thank Robert Gardner and Anthony King for their comments on an earlier version of this chapter. My research on the applications of percolation theory and neutral landscape models in conservation biology has been supported by the National Science Foundation through grants DEB-9532079 and DEB-9610159. I thank Sarah Rinkevitch (Wildlife Biologist, U.S. Fish and Wildlife Service, Albuquerque, New Mexico) and Rita Suminski (Staff Biologist, Journey Level, U.S. Forest Service Mount Taylor Ranger District, New Mexico) for information on the application of percolation theory in the recovery plan of the federally listed Mexican Spotted Owl.

References

Andrén, H. 1994. Effects of habitat fragmentation on birds and mammals in landscapes with different proportions of suitable habitat: a review. *Oikos* 71:355–366.

Bascompte, J., and Solé, R.V. 1996. Habitat fragmentation and extinction thresholds in spatially explicit models. *J. Anim. Ecol.* 65:465–473.

Beier, P. 1996. Metapopulation models, tenacious tracking, and cougar conservation. In *Metapopulations and Wildlife Conservation,* ed. D.R. McCullough, pp. 293–323. Washington, DC: Island Press.

Beier, P., and Noss, R.F. 1998. Do habitat corridors provide connectivity? *Conserv. Biol.* 12:1241–1252.

Boswell, G.P., Britton, N.F., and Franks, N.R. 1998. Habitat fragmentation, percolation theory and the conservation of a keystone species. *Proc. R. Soc. London,* B 265:1921–1925.

Dale, V.H., Pearson, S.M., Offerman, H.L., and O'Neill, R.V. 1994. Relating patterns of land-use change to faunal biodiversity in the central Amazon. *Conserv. Biol.* 8:1027–1036.

Desrochers, A., and Hannon, S.J. 1997. Gap crossing decisions by forest songbirds during the post-fledging period. *Conserv. Biol.* 11:1204–1210.

Doak, D.F., Marino, P.C., and Kareiva, P.M. 1992. Spatial scale mediates the influence of habitat fragmentation on dispersal success: implications for conservation. *Theor. Pop. Biol.* 41:315–336.

Doak, D.F., and Mills, L.S. 1994. A useful role for theory in conservation. *Ecology* 75:615–626.

Fahrig, L. 1997. Relative effects of habitat loss and fragmentation on population extinction. *J. Wildl. Manage.* 61:603–610.

Forman, R.T.T., and Godron, M. 1986. *Landscape Ecology.* New York: John Wiley and Sons.

Gardner, R.H., Milne, B.T., Turner, M.G., and O'Neill, R.V. 1987. Neutral models for the analysis of broad-scale landscape patterns. *Landsc. Ecol.* 1:19–28.

Gardner, R.H., and O'Neill, R.V. 1991. Pattern, process, and predictability: the use of neutral models for landscape analysis. In *Quantitative Methods in Landscape Ecology,* eds. M.G. Turner and R.H. Gardner, pp. 289–307. New York: Springer-Verlag.

Green, D.G. 1994. Simulation studies of connectivity and complexity in landscapes and ecosystems. *Pac. Conserv. Biol.* 3:194–200.

Gustafson, E.J., and Gardner, R.H. 1996. The effect of landscape heterogeneity on the probability of patch colonization. *Ecology* 77:94–107.

Hansen, A.J., and Urban, D.L. 1992. Avian response to landscape pattern: the role of species' life histories. *Landsc. Ecol.* 7:163–180.

Heyer, R.W., Donnelly, M.A., McDiarmid, R.W., Hayek, L.C., and Foster, M.S., eds. 1994. *Measuring and Monitoring Biological Diversity: Standard Methods for Amphibians.* Washington, DC: Smithsonian Institution Press.

Keitt, T.H., Franklin, A., and Urban, D.L. 1995. Landscape analysis and metapopulation structure. In *Recovery Plan for the Mexican Spotted Owl, Volume II, Technical and Supporting Information.* Albuquerque, New Mexico: U.S. Department of the Interior, Fish and Wildlife Service.

Keitt, T.H., Urban, D.L., and Milne, B.T. 1997. Detecting critical scales in fragmented landscapes. *Conserv. Ecol.* [online] 1:4. Available from the Internet: www.consecol.org /vol1/iss1/art4.

Kotliar, N.B., and Wiens, J.A. 1990. Multiple scales of patchiness and patch structure: a hierarchical framework for the study of heterogeneity. *Oikos* 59:253–260.

Kunin, W.E. 1998. Extrapolating species abundance across spatial scales. *Science* 281:1513–1515.

Lande, R. 1987. Extinction thresholds in demographic models of territorial populations. *Am. Nat.* 130:624–635.

Lande, R. 1988. Demographic models of the Northern Spotted Owl (*Strix occidentalis caurina*). *Oecologia* 75:601–607.

Lavorel, S., Gardner, R.H., and O'Neill, R.V. 1993. Analysis of patterns in hierarchically structured landscapes. *Oikos* 67:521–528.

Lidicker, W.Z. Jr., and Koenig, W.D. 1996. Responses of territorial vertebrates to habitat edges and corridors. In *Metapopulations and Wildlife Conservation,* ed. D.R. McCullough, pp. 85–109. Washington, DC:Island Press.

Metzger, J.-P., and Décamps, H. 1997. The structural connectivity threshold: an hypothesis in conservation biology at the landscape scale. *Acta Œcologia* 18:1–12.

Mladenoff, D.J., Sickley, T.A., Haight, R.G., and Wydeven, A.P. 1995. A regional landscape analysis and prediction of favorable gray wolf habitat in the northern Great Lakes Region. *Conserv. Biol.* 9:279–294.

Moloney, K.A., and Levin, S.A. 1996. The effects of disturbance architecture on landscape-level population dynamics. *Ecology* 77:375–394.

Noss, R.F. 1991. Landscape connectivity: different functions at different scales. In *Landscape Linkages and Biodiversity,* ed. W. Hudson, pp. 23–39. Washington, DC:Island Press.

O'Neill, R.V., Gardner, R.H., and Turner, M.G. 1992. A hierarchical neutral model for landscape analysis. *Landsc. Ecol.* 7:55–61.

O'Neill, R.V., Milne, B.T., Turner, M.G., and Gardner, R.H. 1988. Resource utilization scales and landscape pattern. *Landsc. Ecol.* 2:63–69.

Palmer, M.W. 1992. The coexistence of species in fractal landscapes. *Am. Nat.* 139:375–397.

Pearson, S.M., and Gardner, R.H. 1998. Neutral models: useful tools for understanding landscape patterns. In *Wildlife and Landscape Ecology: Effects of Pattern and Scale,* ed. J.A. Bissonette, pp. 215–230. New York: Springer-Verlag.

Pearson, S.M., Turner, M.G., Gardner, R.H., and O'Neill, R.V. 1996. An organism-based perspective of habitat fragmentation. In *Biodiversity in Managed Landscapes,* eds. R.C. Szaro and D.W. Johnston, pp. 77–95. Oxford, United Kingdom: Oxford University Press.

Petit, L.J., Petit, D.R., and Martin, T.E. 1995. Landscape-level management of migratory birds: looking past the trees to see the forest. *Wildl. Soc. Bull.* 23:420–429.

Plotnick, R.E., and Gardner, R.H. 1993. Lattices and landscapes. *Lect. Math. Life Sci.* 23:129–157.

Plotnick, R.E., Gardner, R.H., and O'Neill, R.V. 1993. Lacunarity indices as measures of landscape texture. *Landsc. Ecol.* 8:201–211.

Pulliam, H.R., and Dunning, J.B., Jr. 1994. Demographic processes: population dynamics on heterogeneous landscapes. In *Principles of Conservation Biology,* eds. G.K. Meffe and C.R. Carroll, pp. 179–205. Sunderland, Massachusetts: Sinauer Associates.

Pulliam, H.R., and Dunning, J.B., Jr. 1997. Demographic processes: population dynamics on heterogeneous landscapes. In *Principles of Conservation Biology,* Second Edition, eds. G.K. Meffe and C.R. Carroll, pp. 203–232. Sunderland, Massachusetts: Sinauer Associates.

Pulliam, H.R., Dunning, J.B. Jr., and Liu, J. 1992. Population dynamics in complex landscapes: a case study. *Ecol. Appl.* 2:165–177.

Rail, J.-F., Darveau, M., Desrochers, A., and Huot, J. 1997. Territorial responses of boreal forest birds to habitat gaps. *Condor* 99:976–980.

Russ, J.C. 1994. *Fractal Surfaces.* New York: Plenum Press.

Salwasser, H. 1991. New perspectives for sustaining diversity in U.S. national forest ecosystems. *Conserv. Biol.* 5:567–569.

Saupe, D. 1988. Algorithms for random fractals. In *The Science of Fractal Images,* eds. H.-O. Petigen and D. Saupe, pp. 71–113. New York: Springer.

Schumaker, N. 1996. Using landscape indices to predict habitat connectivity. *Ecology* 77:1210–1225.

Stauffer, D., and Aharony, A. 1991. *Introduction to Percolation Theory,* Second Edition. London, United Kingdom: Taylor and Francis.

St. Clair, C.C., Bélisle, M., Desrochers, A., and Hannon, S. 1998. Winter responses of forest birds to habitat corridors and gaps. *Conserv. Ecol.* [online] 2:13. Available from the Internet: www.consecol.org/vol2/iss2/art13.

Taylor, P.D., Fahrig, L., Henein, K., and Merriam, G. 1993. Connectivity is a vital element of landscape structure. *Oikos* 68:571–573.

Wiens, J.A., and Milne, B.T. 1989. Scaling of 'landscapes' in landscape ecology, or, landscape ecology from a beetle's perspective. *Landsc. Ecol.* 3:87–96.

Wiens, J.A., Schooley, R.L., and Weeks, R.D., Jr., 1997. Patchy landscapes and animal movements: do beetles percolate? *Oikos* 78:257–264.

With, K.A. 1997a. The application of neutral landscape models in conservation biology. *Conserv. Biol.* 11:1069–1080.

With, K.A. 1997b. The theory of conservation biology. *Conserv. Biol.* 11:1436–1440.

With, K.A. 1999. Is landscape connectivity necessary and sufficient for wildlife management? In *Forest Fragmentation: Wildlife and Management Implications,* eds. J.A. Rochelle, L.L. Lehmann, and J. Wisniewski, pp. 97–115. Leiden, The Netherlands: Brill Academic Publishers.

With, K.A., Cadaret, S.J., and Davis, C. 1999. Movement responses to patch structure in experimental fractal landscapes. *Ecology* 80:1340–1353.

With, K.A., and Crist, T.O. 1995. Critical thresholds in species' responses to landscape structure. *Ecology* 76:2446–2459.

With, K.A., Gardner, R.H., and M.G. Turner. 1997. Landscape connectivity and population distributions in heterogeneous environments. *Oikos* 78:151–169.

With, K.A., and King, A.W. 1997. The use and misuse of neutral landscape models in ecology. *Oikos* 79:219–229.

With, K.A., and King, A.W. 1999a. Dispersal success on fractal landscapes: a consequence of lacunarity thresholds. *Landsc. Ecol.* 14:73–82.

With, K.A., and King, A.W. 1999b. Extinction thresholds for species in fractal landscapes. *Conserv. Biol.* 13:314–326.

Zallen, R. 1983. *The Physics of Amorphous Solids.* New York: John Wiley and Sons.

8

Landscape Connections and Genetic Diversity

Hugh B. Britten and Richard J. Baker

8.1 Introduction

Conceptual connections between ecology and population genetics have been an integral part of the field of conservation biology from its beginnings. The conservation of genes, the perpetuation of microevolutionary processes (e.g., natural selection, mutation, genetic drift, and gene flow) within populations, and the conservation of demographic and landscape-scale processes make up the three pillars of modern conservation biology (see Brussard 1991). This chapter reviews how population genetic processes interact with landscape processes in a conservation biology context and suggests practical ways that managers can address these interactions. Although extremely important in the preservation of biological diversity, these interactions are obscured by differences in scale that exist between genetic processes operating at the population level, and landscape processes operating at much broader spatial scales and significantly smaller temporal scales.

Section 8.2 provides historical background about the study of population genetics. We review the two major reasons that conservation geneticists study genetic variability in populations and some of the factors that affect levels of variability. We place these concepts in a landscape ecology context by discussing two emerging approaches in conservation genetics, phylogeography and the identification of conservation units within taxa. Section 8.3 discusses field and research applications of phylogeography and the genetic identification of conservation units. We stress that although research efforts in these areas have been considerable, few examples of field applications are available. Therefore, Section 8.4 on principles for applying landscape ecology to conservation genetics relies largely on research applications. We highlight the complexity of genetic processes at the landscape level and point out that landscape approaches may not always apply to questions about conservation genetics. Section 8.5 (Knowledge Gaps) stresses the need for greater understanding of the differences in scale in the operation of landscape and microevolutionary processes. Section 8.6 (Research Approaches) broadly outlines research approaches that are available in the rapidly developing fields of molecular biology, phylogenetics, and population genetics, and that have the potential to fill the theoretical and empirical gaps we have identified.

8.2 Concepts, Principles, and Emerging Ideas

Historically, two general arguments have been made for preserving genetic diversity in populations. First, evolution by natural selection cannot occur without heritable differences in fitness among individuals in populations. Populations can adapt to the selective pressure of a changing environment, if they have been able to gain and maintain the necessary genetic diversity. Second, the frequency and dispersion of genes within populations can record the history of populations over fairly long intervals of time. Patterns of isolation and changes in population sizes can often be reconstructed from present-day genetic data. These patterns can then inform management of populations in ways that are not possible with real-time ecological study.

These two arguments provide the rationale for managing populations to maximize *genetic diversity* (average heterozygosity and level of genetic polymorphism within populations). Although the preservation of genetic diversity may not be the direct goal of management, it is difficult to justify not taking the genetic implications of any given management plan into consideration.

8.2.1 Definitions and Principles

The genetic diversity of populations is the product of several mechanisms: gene flow, mutation, natural selection, inbreeding, and genetic drift. Perhaps the most critical connection between landscape ecology and population genetics is the important role that gene flow plays in maintaining genetic diversity. *Gene flow,* the movement of successfully breeding individuals between populations, is invaluable in maintaining genetic diversity and preserving a species' history. Gene flow also has strong geographic and ecological components. In general, individuals that are responsible for gene flow (i.e., those that move between local populations) must interact with the environment on a broader spatial scale than must those that remain in their original population. The rate of gene flow across the landscape over many generations determines *genetic population structure* (the pattern of genetic connections among subpopulations on a landscape or broader scale), and gene flow can be affected by management at the landscape scale. Genetic diversity can be reestablished in populations by gene flow from other populations. Mills and Allendorf (1996) demonstrated that a gene flow rate of about 10 reproductive migrants per generation can maintain genetic similarity.

We can define local populations, or *demes,* as groups of organisms that are essentially *panmictic* (i.e., all possible combinations of gametes have an equal probability of occurring within panmictic demes). Demes are genetically connected through gene flow. High rates of gene flow prevent genetic divergence between geographically separate demes to the point where they can be nearly panmictic. Subspecies, local races, and other geographic units within species are maintained by low levels (or the absence) of gene flow.

Groups of demes with gene flow between them compose a *metapopulation* (Hanski and Simberloff 1997). Several metapopulation subtypes, such as source-sink and mainland-island metapopulations, may be recognized (Hanski and Simberloff 1997). Virtually all of the examples we discuss below can be considered to have some form of metapopulation structure.

Management that increases rates of gene flow among local endemics may result in the *genetic swamping* (homogenization of gene pools) of unique populations. Deleterious effects of increased levels of gene flow between previously separated demes, known as *outbreeding depression,* have been noted in several species (Templeton 1986; Frankham 1995). Management actions that can lead to outbreeding depression include the removal of dispersal barriers, the creation of migration corridors, or both.

Hybridization, mating between individuals of two recognized taxa, is another manifestation of inappropriate levels of gene flow that can be caused by alteration of the landscape. Perhaps the best-known example of this is the red wolf (*Canis rufus*) from the southeastern United States. Agricultural conversion and other landscape alterations may have facilitated the hybridization of gray wolves (*Canis lupus*) and coyotes (*Canis latrans*), thereby forming a new species, the red wolf (Wayne and Jenks 1991). The red wolf is protected under the U.S. Endangered Species Act, but considerable controversy about its taxonomic status has surrounded this species (Dowling et al. 1992; Reich et al. 1999). A similar example is the hybridization of the coyote and gray wolf in the northeastern United States and southeastern Canada (Lehman et al. 1991). Deforestation for agriculture over the last two centuries has favored the coyote, a more ecological generalist, over the gray wolf in this area (Lehman et al. 1991). Gill (1997) reported hybridization of Blue-winged Warbler (*Vermivora pinus*) and Golden-winged Warbler (*V. chrysoptera*) populations in eastern Pennsylvania. Here, female Blue-winged Warblers have apparently dispersed onto farmland abandoned in the mid-1950s that is now becoming more forested. They breed with resident Golden-winged Warbler males, and eventually the hybrids and their progeny displace the pure Golden-winged Warbler population altogether (Gill 1997). In both cases, individuals dispersing into previously unsuitable habitat altered by human intervention breed with members of resident species and establish hybrid populations (Lehman et al. 1991; Gill 1997), and this causes the loss of genetically unique populations.

The ultimate source of genetic diversity is *mutation* (spontaneous changes in DNA), but this occurs too slowly to be useful in managing populations. *Natural selection* (differential survival and reproduction due to heritable traits) also affects genetic variability in demes. Because natural selection results in greater representation of genes from the most fit individuals in subsequent generations, it is extremely important that management actions preserve genetic diversity so that natural selection can continue.

Landscape change also can impact genetic diversity of demes through *inbreeding* (mating between close relatives). Inbreeding is likely a major cause of lowered reproductive capabilities and other problems observed in the cheetah (*Acinonyx jubatus*) and the federally protected Florida panther (*Felis concolor coryi*), both of which have experienced severe reductions in population size due to habitat loss and its associated problems (O'Brien et al. 1996). Similarly, Westemeier et al. (1998) document declines in a Greater Prairie Chicken (*Tympanuchus cupido pinnatus*) population that are linked to inbreeding resulting from extensive habitat fragmentation in Illinois.

Small populations that are isolated from gene flow also are subject to *genetic drift* (loss of genetic diversity due to the random sampling of genes that occurs each generation). Because the rate of loss of genetic variability due to drift is a function of generations, not years, this loss may not be apparent immediately after habitat perturbations in organisms with long generation times, such as greater one-horned rhinoceros (*Rhinoceros unicornis*; Dinnerstein and McCracken 1990). Although their habitat had been largely fragmented by agricultural practices during the 20th century, these animals have retained high levels of genetic variability because very few rhinoceros generations have actually passed since the fragmentation began in earnest (Dinnerstein and McCracken 1990). The loss of genetic variability due to genetic drift mimics the loss during inbreeding and can lead to problems similar to inbreeding depression.

The *founder effect* is a type of genetic drift that can be important in reintroduction efforts on the landscape. Founders (i.e., natural or human-mediated colonizers) represent a sample of the genetic variability present in the source population. Because founding populations are often small, they will have lower genetic diversity than will the source population. Their depauperate gene pools may make founding populations more vulnerable to local extirpation in the new habitat.

8.2.2 Emerging Ideas

Our understanding of the preceding microevolutionary processes enables us to reconstruct the history of taxa that may be of conservation concern, and to determine the genetic population structure of such taxa. This information can then be incorporated into the management of landscapes. Two new approaches that have recently emerged in conservation genetics (phylogeography and identification of conservation units) make use of advances in both theoretical population genetics and molecular biology.

Genetic population structure is species- and location-specific. The effects of landscape-scale habitat perturbations such as agricultural conversion, however, are not limited to a single species. The history of landscape change is recorded in the gene pools of all affected organisms that occupy a disturbed area. The basic approach in *phylogeography* is to use genetic data from a variety of species in an area or region to construct networks of genetic relatedness (or "trees") among the populations of each species. The trees for all of the species can be overlain on a map of the area under study to reveal geographic patterns common to all of the trees. Geographically concordant patterns of relationships in a number of species suggest that they shared a similar biogeographic history. The phylogeographic approach relies on the fact that species within a given area have experienced similar histories and often reveals landscape-scale ecological phenomena. Thus, Avise (1996) was able to detect a phylogeographic signature from the last glacial episode in the southeastern United States within terrestrial vertebrates, freshwater vertebrates, and marine organisms. Similarly, phylogeographic patterns within sand dune-obligate beetles from the Great Basin of North America can be traced to Pleistocene events (Epps et al. 1998).

There are several potentially useful ways that such an approach could be applied to management at the landscape scale. First, the identification of faunas that share a common biogeographic history can be important for the placement of preserves. If the goal is to capture as much biological diversity as possible in a given preserve, knowledge of local phylogeographic patterns would enable managers to delineate preserves that could include two or more branches within the regional phylogeny. Each branch would represent groups of species that experienced similar histories different from those in the other branch(es). Second, phylogeography can be used to identify groups of species that are phylogenetically distinct, but that have not received habitat protection. In this case, an area that supports a single branch within the regional phylogeny could be preserved for the phylogenetically unique species found there. Third, phylogeographic patterns can be used to keep genetically distinct populations separate. Here, the goal would be to prevent exchange between populations that have previously occupied different phylogeographic areas. Fourth, because species that have experienced the least human impact are the ones most likely to carry unaltered phylogeographic signals into the present, they are the best candidates for inclusion in phylogeographic studies (Avise 1996). This suggests that the concept of *umbrella species* (species targeted for management under the assumption that other sympatric species will be positively affected as well) may be modified in some cases to include phylogeographically, as well as ecologically, important species for which an area may be managed. Finally, the inherent hierarchical nature of regional phylogenies suggests that phylogeographic classification at the landscape scale may be possible. Thus, landscape subunits may be classified by biome, cover type, habitat type, and topological (branching pattern) features of regional phylogenies.

Numerous workers (e.g., Moritz 1994; Moritz et al. 1995) have pointed out that population genetics can be used to identify hierarchical units of conservation. The most inclusive level of the hierarchy delineates population segments that contain large portions of a species' evolutionary legacy and are termed *evolutionarily significant units* (ESUs; Ryder 1986). *Management units* (MUs) are sets of populations within ESUs that experience enough gene flow to keep them genetically linked, but the sets of populations are demographically independent. Thus, the total evolutionary potential of a species is found within all of its ESUs, whereas MUs define the genetic population structure of the species on the landscape.

Considerable discussion has occurred concerning the diagnosis of ESUs (e.g., Dizon et al. 1992; Vogler et al. 1993; Vogler and DeSalle 1993; Moritz 1994; Moritz et al. 1995; Waples 1995; Paetkau 1999). Many workers suggest that gene trees (phylogenies) based on neutral genetic markers are best for delineating ESUs. Others (Vogler and DeSalle 1993; Waples 1995) note that these phylogenies do not provide direct estimation of the distribution of adaptively important variability within species. For this reason, additional criteria for the diagnosis of ESUs may be considered, including "substantial reproductive isolation" (Waples 1995) and ecological factors (Vogler et al. 1993).

8.3 Recent Applications

Our search for examples of recent applications included an informal survey of several regional and international journals covering the time between the early 1980s to the present, and discussions with colleagues in management positions within U.S. and state natural resources agencies. These efforts, and our experiences, led us to conclude that population genetics theory has rarely been applied in landscape management, and ironically, it has had only a slight effect on actual management of populations in the wild (see Caughley 1994). There are three explanations for this observation. First, there is a general perception that demographic, environmental, and catastrophic variables, not genetic ones, determine the short-term persistence of populations (Lande 1988). Although this may be true for very small populations, there are numerous sound reasons why genetics should not be discounted in population management (Frankel and Soulé 1981; Hedrick and Miller 1992; Mills and Smouse 1994). Second, population genetics is a complicated field outside the realm of many managers' experiences and is often left to the "experts." Third, ascertaining the impact of a given management activity on the genetic structure of a population is often difficult.

The demographic results of management may be detected with a few years of monitoring, but the evolutionary consequences of management are impossible to observe. For example, viable Peregrine Falcon (*Falco peregrinus*) populations have been reintroduced into the midwestern United States, but the relative contribution of genetic variability to long-term persistence could not be determined with certainty (Moen and Tordoff 1993). Similarly, populations of the white-footed mouse (*Peromyscus leucopus*) on small islands in Norris Lake, Tennessee, once occupied continuous forest habitat. Flooding by the Tennessee Valley Authority left small populations isolated on islands that were former hilltops. There were no detectable differences in allozyme frequencies among these isolated populations some 50 years after the valleys were flooded (Britten 1985). In contrast, Gerlach and Musolf (2000) were able to detect genetic subdivision in a population of bank voles (*Clethrionomys glareolus*) caused by a 50-m-wide 25-year-old highway in Germany using microsatellite data. Numerous genetic studies of the microevolutionary effects of broad-scale climatic events in the Pleistocene clearly demonstrate the genetic effects of population isolation for a variety of taxa, including insects (e.g., Britten and Brussard 1992; Epps et al. 1998), freshwater fish (e.g., Vrijenhoek 1996), and terrestrial vertebrates and marine organisms (Avise 1992, 1996). Thus, microevolutionary processes may require somewhere between a few hundred years and 12,000 or more years to be detectable in most wild populations. Human-driven landscape-level disturbances such as urbanization, agricultural conversion, and pollution often occur in considerably shorter time frames.

Although we are unaware of any direct field applications of phylogeographic approaches, issues of genetic similarity are often considered in the context of single-species management. For example, plans for reintroducing Trumpeter Swans (*Cygnus buccanator*) into Wyoming must consider genetic differences between potential source populations in Alaska and Montana (C. Pelizza, personal communication). The Montana population is the only one that survived the species' decline in the lower 48 states, but it is genetically depauperate (Barrett and Vyse 1982). Although the extirpated Wyoming population may have been more closely related to the Montana population than it was to the Alaska one, managers must decide whether reconstructing a pattern of populations that more closely resembles the pre-extirpation distribution is more important than are the potential positive effects of a more varied gene pool as represented by Alaskan birds. Similarly, the decision to reintroduce gray wolves into Yellowstone National Park from Alberta, Canada, was made in part for genetic reasons (D. Smith, personal communication).

8.3.1 Field Applications—Units of Conservation

Under the U.S. Endangered Species Act (ESA), the National Marine Fisheries Service (NMFS) and the U.S. Fish and Wildlife Service (USFWS) have the authority to recognize species, subspecies, and in the case of vertebrates, Distinct Population Segments (DPSs) that do not necessarily have formal taxonomic recognition, as threatened or endangered. Both agencies use genetic data, in addition to geographical and ecological factors, to delineate DPSs. Waples (1991) proposed that both agencies use the ESU concept as a way of defining DPSs of endangered Pacific salmon (*Oncorhynchus* spp.) in the Pacific Northwest, USA, under the ESA. This became the official policy of the two agencies in 1996 (Federal Register 61 (26):4722–4725).

Salmonid conservation is complicated by migrations to the ocean undertaken by subadults, and return spawning runs to natal streams by mature adults several years later. The time of year that adults enter fresh water to begin spawning migrations can vary within a given salmonid population. For example, Waples (1995) noted that three *runs* (temporally isolated subpopulations in a stream-specific population), spring, summer, and fall, of Chinook salmon (*O. tshawytscha*) exist in the Snake River. A 1990 petition for protection of Snake River Chinook salmon under the ESA required that the NMFS determine whether each run constituted a reproductively isolated population (i.e., an ESU). Genetic and ecological data suggested two ESUs within the Snake River Chinook salmon population, a fall run and a spring–summer run. Both Snake River runs were found to be most similar to Columbia River runs that take place at these same times of year. Ecological differences, including different ocean distributions, suggested that fall-run Chinook salmon from the two rivers each constitute separate portions of the species' overall genetic heritage (i.e., each river's fall run is a separate ESU, Waples 1995). Similar data pointed to the distinctness of spring–summer runs of fish from the two rivers (Waples 1995). Thus, two Snake River Chinook salmon ESUs were listed as threatened in 1992. The spring–summer ESU underwent an emergency status change to endangered in 1994 (Waples 1995).

Broad-scale habitat disturbance brought about by the damming of rivers in the Pacific Northwest led to additional threatened listings of salmonid ESUs in the summer of 1999, including two ESUs of summer-run chum salmon (*O. keta*), the Ozette Lake sockeye salmon (*O. nerka*) ESU, and the Oregon Coast coho salmon (*O. kisutch*) ESU. Genetic data have been used extensively to identify these population segments for protection under the ESA. The goal is to maintain the genetic diversity of the protected species and allow microevolutionary processes to continue into the future (Allendorf and Waples 1996). For the reasons discussed above, no current means are available to assess the short-term success or failure of this approach.

The USFWS uses slightly different criteria than does NMFS in delineating DPSs (S. Lohr, personal communication); three general criteria are considered: "distinctness, significance, and conservation status." Distinctness refers to geographic disjunction or populations delimited by international boundaries, and it may include genetic separation. There are four elements used by USFWS to determine significance: persistence of the population in an unusual habitat, evidence that the loss of a particular population segment would result in a distributional gap within the taxon's range, presence of a discrete population within the natural range of the taxon with introduced populations persisting outside of the natural range, and genetic distinctness of the population segments (Federal Register 63 (111):31647–31674; 64(67):17110–17125). Conservation status refers to the present status of the taxon in relation to the standards for listing. Recent applications of these USFWS criteria include the listing of three bull trout (*Salvelinus confluentus*) DPSs from the Columbia, Jarbidge, and Klamath Rivers (Federal Register 63 (111):31647–31674; 64 (67):17110–17125). Landscape-scale disturbances that have contributed to the threatened status of bull trout DPSs include dams, forest management practices, livestock grazing, agricultural practices, road construction and maintenance, mining, and suburban sprawl (Federal Register 64 (67):17110–17125).

8.3.2 Research Applications—Phylogeography

Phylogeographic studies have burgeoned in recent years (Avise 1998). As a result, an entire issue of *Molecular Ecology* (1998, Volume 7, Number 4) was recently dedicated to this rapidly developing field. Studies by Avise (1992, 1996) were discussed above in presenting phylogeography as an emerging idea. Templeton and Georgiadis (1996) summarized phylogeographic results involving three bovids from eastern Africa and used evidence of specific barriers to gene flow, in addition to habitat patchiness, to explain the observed phylogeography. They stressed that genetic data can indicate demographic connectivity or disjunction of population segments on a landscape, whereas short-term observations of individual movements cannot (Templeton and Georgiadis 1996). Using genetic data from the marine copepod *Tigriopus californicus,* and examining phylogeographic studies of seven additional species whose distributions span Point Conception on the California coast, Burton (1998) assessed whether there was a purported major phylogeographic break in the study area. Burton (1998) concluded that no such break was detectable in the eight species' distributions.

The phylogeographic approach has been applied to a variety of organisms in single-species studies. Most are aimed at identifying unique or disjunct population segments that are potential targets for conservation efforts. For example, genetic data were used by Bermingham et al. (1996) to study Caribbean subspecies of the Bananaquit (*Coereba flaveola*), by Taberlet and Bouvet (1994) and Taberlet (1996) to study phylogeographic patterns in the European brown bear (*Ursus arctos*), and by Vogler et al. (1993) to study phylogeographic patterns and their possible conservation implications for Tiger Beetles (*Cicendela dorsalis*) on the Atlantic coast of the United States. Based on genetic data, Barrowclough et al. (1999) found that subspecies of the Spotted Owl (*Strix occidentalis*) in North America have been historically isolated from gene flow. Finally, Britten and Rust (1996) used genetic data to reveal phylogeographic patterns in the sand dune-obligate beetle *Eusattus muricatus* and their implications for dune management.

8.3.3 Research Applications—Units of Conservation

Single-species phylogeographic studies often lead to the identification of ESUs, MUs, or both. Thus, in their comprehensive review of conservation genetics in three species of marine turtles, Bowen and Avise (1996) used both phylogeography and identification of MUs in making conservation recommendations. Bowen and Avise (1996) found considerable genetic differentiation among rookeries, which implied strong demographic separation among female marine turtles on rookery beaches. The main conservation implication of this finding was that human-driven extirpation of rookeries through habitat alteration, turtle hunting, or egg gathering may result in the long-term loss of demographically independent rookery populations (Bowen and Avise 1996).

Britten et al. (1997) used genetic and morphological data to delineate MUs in the federally listed desert tortoise (*Gopherus agassizii*) in the Mojave Desert of southern Nevada and southwest Utah. The species recovery plan for the desert tortoise calls for the establishment of Desert Wildlife Management Areas (DWMAs) to ameliorate the effects of habitat fragmentation, urbanization, and disease on desert tortoise populations (USFWS Desert Tortoise Recovery Plan 1994). Significant genetic divergence, morphological differences, or both were used to delineate five desert tortoise MUs across the study area (Britten et al. 1997). MUs were overlain on a map of DWMAs to reveal gaps in the network of protected habitat for the tortoise. This exercise showed where additional DWMAs and corridor areas should be established to maintain the genetic population structure of desert tortoise (Britten et al. 1997).

We have provided a few examples of research applications for phylogeography and the identification of units of conservation. The broad geographic expanse of these studies (and others like them) makes it unlikely that the conservation recommendations they contain will ever be fully implemented. There is a clear negative relationship between the geographic scale of a study with conservation implications and the probability that those recommendations will be implemented. Research that identifies unique populations that may be worthy of conservation efforts tends to generate proactive conservation recommendations, often before the extirpation of the given population segment is clearly imminent. But, as with endangered salmonids in the Pacific Northwest, appropriate management on the landscape scale rarely takes place until species are on the brink of extinction and legal action requires amelioration of the situation.

8.4 Principles for Applying Landscape Ecology

One of the goals of this chapter is to suggest practical ways that managers can address the interactions between landscape-scale and population-genetic processes. We were not able to derive a set of principles for applying landscape ecology based primarily on the success or failure of previous field applications. This approach is not explicitly available to us when dealing with the interactions of microevolutionary processes and landscape ecology because of the paucity of actual applications in this context. We therefore suggest a set of principles for applying landscape ecology and population genetics in a conservation context based on available field applications and the body of theoretical and empirical work that addresses this issue.

8.4.1 Principle 1: Surveys Should Be Employed to Clarify Population Genetic Structure

The microevolutionary processes of mutation, natural selection, genetic drift, and gene flow occur simultaneously in populations across a managed landscape. Population genetics theory predicts how some of these processes may interact in general to reach equilibria (e.g., Hartl and Clark 1997), but due to the stochastic nature of several of these processes, we have predictive ability for only the shortest periods of time. Soulé and Mills (1998) present a simplified "extinction vortex" model that shows how some genetic and demographic processes can interact and lead to local extirpations. When time and other resources permit, genetic surveys aimed at estimating genetic population structure for species on managed landscapes are the only way to avoid potentially irreversible mistakes.

8.4.2 Principle 2: Landscape-Scale Disturbances Will Often Result in the Loss of Genetic Variability

Multispecies phylogeography is based on the premise that broad-scale changes (e.g., tectonic and global events) will similarly affect a large portion of the organisms living in the disturbed area (Avise 1996). In contrast, local-scale human disturbances to the landscape often affect species differentially. However, a general prediction is that at either scale, landscape disturbance will result in the loss of genetic variability within the affected populations. The main agent of this loss is genetic drift, which increases as habitat is further fragmented, isolating small populations on habitat islands and disrupting overall genetic population structure.

It is often unclear how short-term population viability is affected by the loss of genetic variability, but the maintenance of demographic links among disjunct populations can reinforce the natural genetic population structure on a landscape.

8.4.3 Principle 3: Management That Impacts Metapopulation Structure Will Also Affect Population Genetic Structure

In general, genes do not flow between populations without dispersing organisms to carry them. The intensity, frequency, and efficacy of gene flow among populations on a landscape define the genetic population structure of a metapopulation. However, genetic population structure is not necessarily obvious from observations of life histories, migration, and behavior. When management may impact metapopulation structure, managers should consider employing genetic studies to reveal historic and present-day genetic population structure not easily observed by other means. Knowledge of the disturbance history of an area from geologic, historic, or other independent sources can greatly facilitate the interpretation of these genetic data. Phylogenetic methods of analyses also can help one infer historic relationships among populations.

8.4.4 Principle 4: Management Impacts on Population Genetic Structure May Not Be Immediately Detectable

Changes to a landscape may result in readily observable demographic effects to populations within a few years, but microevolutionary changes may not be apparent for a large number of generations. However, models developed by Mills and Smouse (1994) for three representative mammalian life histories indicate that extinction probability may be significantly affected by inbreeding within 20 generations of severe population bottlenecks (where population size = 5, 20, and 80 in different simulations). This suggests that the "success" of management efforts should be assessed over the course of tens of generations post-manipulation for a set of species of interest.

8.4.5 Principle 5: Genetic Variability May Not Drive Population Viability at the Spatial and Temporal Scales Important in Landscape Management

The interplay among demography, genetic diversity, and population viability is complex. There is a well-supported evolutionary rationale for preserving genetic diversity: it provides genetic "capital" for future evolutionary change. However, Lande (1988) and others have argued that demographic concerns, particularly in small populations, should take management precedence in the short term to avoid compromising future management options. As suggested by Hedrick and Miller (1992), carefully conceived studies of genetic population structure and levels of genetic variability can make it possible to weigh the relative importance of demography and population genetics in species of conservation concern. Managers should be aware that both demographic and genetic parameters should be considered in evaluation of the long-term success of their management efforts.

8.4.6 Principle 6: Managers Should Conserve the Unique Genetic and Ecological Characteristics of Marginal Populations

It has long been recognized that populations on the margins of species' distributions may be genetically and ecologically unique (e.g., Brussard 1986). Insular populations are an easily recognized example, but population isolates in apparently continuous habitat also may be unique. This principle is implicitly applied within the population "distinctness" criterion of the USFWS definition of DPSs and within the "substantial reproductive isolation" criterion used by NMFS for delineating ESUs. In the United States, state endangered species statutes often function to protect these important populations where federal protection may be lacking.

8.4.7 Principle 7: Landscape Management Should Consider the Potential Range of Impacts on the Genetics of All Affected Species

As noted in Principles 1, 2, and 4, management of landscapes will differentially affect microevolutionary processes in species within the altered landscape. The effects may be species- or population-specific and sometimes unpredictable in terms of their temporal scale and impacts on population viability. Thus, the deliberate or inadvertent creation of corridors, buffers, and other landscape-scale features that may change population connectivity should be carefully considered. Some species (the largest and better studied) may benefit from such actions, if they are planned wisely. Others, such as more sedentary organisms, may be more adversely affected through inadvertent habitat fragmentation, outbreeding depression, or altered selection regimes.

8.5 Knowledge Gaps

A fairly large body of theory exists for microevolutionary processes and population genetics (e.g., Hartl and Clark 1997). But research aimed at the application of population genetics theory in conservation has not considered landscape-scale processes, perhaps because genetic variability is a property of populations (demes) that often do not have explicitly defined spatial limits. In this sense, populations to a geneticist are statistical abstractions. The effects of microevolutionary processes on gene pools may or may not have any connection to the landscape where the population actually exists. It is at the junction between theory and application that our gaps in knowledge are most obvious. Below, we discuss several of the most important theoretical and empirical gaps that inhibit the application of landscape ecology and associated genetic concepts in biological conservation. These gaps are important because without a better understanding of these relationships, landscape management will often produce unintended consequences that reduce microevolutionary options provided by genetic variability. Once the common scales are identified, a robust interface between these two important fields will have been achieved.

8.5.1 Theoretical Voids

The role of genetic variability in the short-term persistence of species is an area in need of further research. Do the genetic changes resulting from human-driven modification of the landscape affect species viability in the short term? It is widely assumed, as it has been in this chapter, that genetic variability is important to the long-term persistence of species. As a result, one of the major goals of conservation genetics is the maintenance of genetic variability in populations (Frankel and Soulé 1981; Shonewald-Cox et al. 1983; Hedrick and Miller 1992). The advent of molecular genetics techniques revolutionized the field of conservation genetics by enabling the estimation of variability at hundreds of neutral and noncoding genes. Because this variability is generally considered neutral to selection, its role in the short-term evolution of populations is unclear.

The effects of multiple, long-term landscape-scale human disturbances on microevolutionary processes are largely unknown, but generally include the loss of genetic variability. If we view evolution as a natural process worthy of preservation like other such processes, what are the long-term effects of continued human landscape-scale disturbances on the evolutionary potential of species? In contrast, what is the effect of human-mediated gene flow on formerly isolated populations? For example, it is possible that populations with the evolutionary potential to become species could be genetically swamped out of existence by human-mediated dispersers from other populations as the result of artificial dispersal corridors.

Very little is known about the genetic characteristics of populations at the landscape scale, although genetic population structure appears to be species- and location-specific. Thus, it is difficult to formulate general rules for genetic management of multiple species on a landscape. It is possible that characteristics such as generation length, body size, vagility, and geographic extent of demes are similar across a suite of species on a managed landscape. Identification of such similarities among species would aid in the application of landscape and microevolutionary principles in biological conservation.

We know little about the effects of source-sink dynamics on the genetic variability of both source and sink populations in changing environments. Individuals moving from inferior or destroyed habitat patches may affect the genetics of both the donor and the recipient populations (Porter 1999). Although recipient patches may be viewed as sinks during normal times, the loss of important source patches and the subsequent influx of refugees could convert a sink to a source, at least in the short term. The movement of viable individuals from a source population to a sink also may affect the source population's level of genetic variability. This could be important during periods in which specific genotypes within the source patch are predisposed to emigrate. These complex dynamics will require further theoretical development before their relevance to management becomes clear.

8.5.2 *Empirical Voids*

There is a conspicuous lack of genetic data for most populations of conservation concern. Landscape-scale management often takes place without consideration of genetic implications. In many cases, this is not because managers are unaware that genetic disturbances are possible, but because there are few genetic data for relevant species and little time or other resources to gather and analyze these data. A thorough genetic survey of several populations within a drainage basin, for example, could take two or more years and cost tens of thousands of dollars. Nevertheless, agencies should strive to gather genetic data and take genetic considerations into account with other factors when developing lists of priority species within their jurisdictions.

There also is usually a lack of genetic data from populations of different species within the same landscape. Populations of species least impacted by human disturbance may be the best at carrying phylogeographic signals into the present. Phylogeographic approaches have a number of potential advantages in landscape management, but few multispecies studies have been carried out. Furthermore, most phylogeographic studies that we are aware of have been carried out on species that have been impacted by broad-scale events at least 10,000 years ago. Apparently, there is a complete lack of phylogeographic data from multiple species subjected to human-driven landscape disturbances.

We have very few data on the potential existence or locations on the landscape of genetically significant "hotspots." Studies aimed at discerning phylogeographic patterns, and those designed to delineate ESUs and MUs, have demonstrated that species often contain genetically distinct units that may warrant special management attention. Large, landscape-scale disturbances, such as the agricultural conversion of the prairie in central North America or historic deforestation in northeastern North America, may have obscured or destroyed any genetic hotspots in these areas. Identification of hotspots in other areas could aid reintroductions and other management efforts.

8.6 Research Approaches

Below we outline research approaches needed to fill the gaps noted in Section 8.5. This section is not meant as a comprehensive recipe for research on these topics. Continued basic research into evolutionary questions, including phylogeography, phylogenetics, molecular biology, and quantitative genetics, are also of potential value to landscape ecologists.

8.6.1 Approaches for Theoretical Research

The theoretical gaps noted in Section 8.5.1 pose essentially historic questions. That is, the study of species that have survived landscape-scale disruptions in the short and long terms may reveal the genetic consequences of disturbance. This is clearly possible with landscape changes that occurred during the Pleistocene or a previous time, as demonstrated by many phylogeographic and historic biogeographic studies. Phylogeographic studies based on data from genetic markers of species within historically managed landscapes (e.g., agricultural areas) may reveal how these disturbances affect microevolutionary processes. Native plants and animals that are the least affected by human disturbance may retain accurate phylogeographical patterns better than would those brought to the brink of extinction by human activities. Phylogenetics can be used to indicate closely related, but undisturbed, taxa for comparative study.

The role of presumed neutral genetic variation in the short- and long-term persistence of species represents a theoretical gap that has a strong temporal component. Studies of "ancient DNA" isolated from museum specimens and other sources may provide a partial answer to this theoretical gap. For example, knowing how genetically similar wooly mammoths are from far-flung areas of the Palearctic may shed light on the landscape processes that brought about their extinction. These results may be applicable to the management of present-day boreal species. In addition, metapopulation modeling approaches are becoming increasingly sophisticated and may be applicable to questions concerning the genetic effects of human disturbances on microevolutionary processes (Hastings and Harrison 1994; Hedrick and Gilpin 1997).

The emphasis in this chapter has been on microevolutionary processes as revealed through the study of molecular markers like mitochondrial and nuclear DNA and proteins. The presumed neutrality of these marker molecules largely removes them from the study of adaptive evolutionary change. This is the realm of quantitative genetics and is largely beyond the scope of this chapter. However, as Lynch (1996) points out, quantitative genetics can help elucidate questions about the maintenance of adaptive potential that have relevance in conservation biology. This body of theory should be developed further with a goal of direct application to landscape ecology and conservation genetics.

The effects of source-sink dynamics on the genetic variability of both source and sink populations are also amenable to modeling approaches. For example, Porter (1999) recently modeled the "refugee effect," in which mobile organisms fleeing a destroyed habitat patch genetically augmented populations in the receiving patches. The refugee effect has several potentially important outcomes, including the obscuring of phylogenetic pattern among populations affected by the perturbations (Porter 1999).

Finally, basic research should focus on the identification of taxa that evolve at the landscape scale. This can be accomplished using some of the methods outlined above. Once identified, ecological, life-history, and other traits of these species should be compared among similar taxa from other disturbed landscapes to identify any common traits. The study of taxa that have survived mass extinction events (see Ridley 1996) may reveal traits that allow the survival of species in present-day landscapes. This research direction may eventually bring some predictive power to landscape-scale management.

8.6.2 Approaches for Empirical Research

All three of the empirical gaps noted in Section 8.5.2 (lack of genetic data for species of conservation concern, lack of genetic data for sympatric species within a landscape, and lack of data from potential genetic "hotspots") can be addressed with spatially explicit molecular genetic data. This approach requires accurate mapping of sample locations with global positioning systems, for example. Site-specific data of this type are essential before genetic considerations can be incorporated into any landscape-management plan.

The field of molecular genetics is evolving rapidly, with new techniques becoming available at a rate never before seen. Our increasing technical ability to detect genetic markers (e.g., nuclear and mtDNA sequencing, restriction fragment length polymorphisms, random amplified polymorphic DNA, amplified fragment length polymorphisms) makes it possible to study the role of genetic variability in the short-term persistence of species and to investigate the genetic population structure of species at increasingly finer geographic scales. Thus, it is becoming easier to investigate the numerous species that have become endangered due to human impacts in historic time. The successful application of most of these techniques requires a considerable investment in laboratory resources and training. We encourage landscape ecologists to develop some understanding in this area and to rely on collaborators with the requisite experience to add a genetic component to their research. Some basic knowledge of the utility of various techniques, and the molecules they involve, is essential to success in this endeavor, however. Mace et al. (1996) and Hedrick (1996) provide useful insight into these issues.

Acknowledgments

We thank P. Hedrick and R. Fredrickson for their insightful reviews of this work. We also thank P. Brussard, K. Helenurm, and L. Riley for their helpful discussions and ideas. The University of South Dakota Office of Research provided travel funds in support of this work. We thank K. Gutzwiller, the editor of this volume, for his patience during the development of this chapter.

References

Allendorf, F.W., and Waples, R.S. 1996. Conservation and genetics of salmonid fishes. In *Conservation Genetics: Case Histories from Nature,* eds. J.C. Avise and J.L. Hamrick, pp. 238–280. New York: Chapman and Hall.

Avise, J.C. 1992. Molecular population structure and the biogeographic history of a regional fauna: a case history with lessons for conservation biology. *Oikos* 63:62–76.

Avise, J.C. 1996. Toward a regional conservation genetics perspective: phylogeography of faunas in the southeastern United States. In *Conservation Genetics: Case Histories from Nature,* eds. J.C. Avise and J.L. Hamrick, pp. 431–470. New York: Chapman and Hall.

Avise, J.C. 1998. The history and purview of phylogeography: a personal reflection. *Molec. Ecol.* 7:371–379.

Barrett, V.A., and Vyse, E.R. 1982. Comparative genetics of three Trumpeter Swan populations. *Auk* 99:103–108.

Barrowclough, G.F., Gutierrez, R.J., and Groth, J.F. 1999. Phylogeography of Spotted Owl (*Strix occidentalis*) populations based on mitochondrial DNA sequences: gene flow, genetic structure, and a novel biogeographic pattern. *Evolution* 53:919–931.

Bermingham, E., Seutin, G., and Ricklefs, R.E. 1996. Regional approaches to conservation biology: RFLPs, DNA sequences, and Caribbean birds. In *Molecular Genetic Approaches in Conservation,* eds. T.B. Smith and R.K. Wayne, pp. 104–124. New York: Oxford University Press.

Bowen, B.W., and Avise, J.C. 1996. Conservation genetics of marine turtles. In *Conservation Genetics: Case Histories from Nature,* eds. J.C. Avise, and J.L. Hamrick, pp. 190–237. New York: Chapman and Hall.

Britten, H.B. 1985. *Genetics of Insular* Peromyscus leucopus *Populations at Norris Lake, Tennessee.* M.S. thesis. Knoxville, Tennessee: University of Tennessee.

Britten, H.B., and Brussard, P.F. 1992. Genetic divergence and the Pleistocene history of the alpine butterfly *Boloria improba* (Nymphalidae) and the endangered *Boloria acrocnema* (Nymphalidae) in western North America. *Can. J. Zool.* 70:539–548.

Britten, H.B., Riddle, B.R., Brussard, P.F., Marlow, R., and Lee, T.E. 1997. Genetic delineation of management units for the desert tortoise, *Gopherus agassizii,* in the northeastern Mojave Desert. *Copeia* 1997:523–530.

Britten, H.B., and Rust, R.W. 1996. Population structure of a sand dune-obligate beetle, *Eusattus muricatus,* and its implications for dune management. *Conserv. Biol.* 10:647–652.

Brussard, P.F. 1986. Geographic patterns and environmental gradients: the central-marginal model in *Drosophila* revisited. *Ann. Rev. Ecol. Syst.* 15:25–64.

Brussard, P.F. 1991. The role of ecology in biological conservation. *Ecol. Appl.* 1:6–12.

Burton, R.S. 1998. Intraspecific phylogeography across the Point Conception biogeographic boundary. *Evolution* 52:734–745.

Caughley, G. 1994. Directions in conservation biology. *J. Anim. Ecol.* 63:215–244.

Dinnerstein, E., and McCracken, G.F. 1990. Endangered greater one-horned rhinoceros carry high levels of genetic variation. *Conserv. Biol.* 4:417–422.

Dizon, A.E., Lockyer, C., Perrin, W.F., Demaster, D.P., and Sisson, J. 1992. Rethinking the stock concept: a phylogeographic approach. *Conserv. Biol.* 6:24–36.

Dowling, T.E., Minckley, W.L., Douglas, M.E., Marsh, P.C., and Demarais, B.D. 1992. Response to Wayne, Nowak, and Phillips and Henry: use of molecular characters in conservation biology. *Conserv. Biol.* 6:600–603.

Epps, T.M., Britten, H.B., and Rust, R.W. 1998. Historical biogeography of *Eusattus muricatus* (Coleoptera: Tenebrionidae) within the Great Basin, western North America. *J. Biogeog.* 25:957–968.

Frankel, O.H., and Soulé, M.E. 1981. *Conservation and Evolution.* Cambridge, United Kingdom: Cambridge University Press.

Frankham, R. 1995. Conservation genetics. *Ann. Rev. Genetics* 29:305–327.

Gerlach, G. and Musolf, K. 2000. Fragmentation of landscape as a cause for genetic subdivision in bank voles. *Conserv. Biol.* 14:1066–1074.

Gill, F.B. 1997. Local cytonuclear extinction of the Golden-winged Warbler. *Evolution* 51:519–525.

Hanski, I.A. and Simberloff, D. 1997. The metapopulation approach, its history, conceptual domain, and application to conservation. In *Metapopulation Biology: Ecology, Genetics, and Evolution,* eds. I.A. Hanski and M.E. Gilpin, pp. 5–26. San Diego: Academic Press.

Hartl, D.L., and Clark, A.G. 1997. *Principles of Population Genetics.* Third Edition. Sunderland, Massachusetts: Sinauer Associates.

Hastings, A., and Harrison, S. 1994. Metapopulation dynamics and genetics. *Ann. Rev. Ecol. Syst.* 25:167–188.

Hedrick, P.W. 1996. Conservation genetics and molecular techniques: a perspective. In *Molecular Genetic Approaches in Conservation,* eds. T.B. Smith and R.K. Wayne, pp. 459–477. New York: Oxford University Press.

Hedrick, P.W., and Gilpin, M.E. 1997. Genetic effective size of a metapopulation. In *Metapopulation Biology: Ecology, Genetics, and Evolution,* eds. I.A. Hanski and M.E. Gilpin, pp. 165–181. San Diego: Academic Press.

Hedrick, P.W., and Miller, P.S. 1992. Conservation genetics: techniques and fundamentals. *Ecol. Appl.* 2:30–46.

Lande, R. 1988. Genetics and demography in biological conservation. *Science* 241:1455–1460.

Lehman, N., Eisenhawer, A., Hansen, K., Mech, L.D., Peterson, R.O., Gogan, P.J.P., and Wayne, R.K. 1991. Introgression of coyote mitochondrial DNA into sympatric North American gray wolf populations. *Evolution* 45:104–119.

Lynch, M. 1996. A quantitative-genetic perspective on conservation issues. In *Conservation Genetics: Case Histories from Nature,* eds. J.C. Avise, and J.L. Hamrick, pp. 471–504. New York: Chapman and Hall.

Mace, G.M., Smith, T.B., Bruford, M.W., and Wayne, R.K. 1996. An overview of the issues. In *Molecular Genetic Approaches in Conservation,* eds. T.B. Smith and R.K. Wayne, pp. 3–24. New York: Oxford University Press.

Mills, L.S., and Allendorf, F.W. 1996. The one-migrant-per-generation rule in conservation and management. *Conserv. Biol.* 10:1509–1518.

Mills, L.S., and Smouse, P.E. 1994. Demographic consequences of inbreeding in remnant populations. *Am. Nat.* 144:412–431.

Moen, S.M., and Tordoff, H.B. 1993. *The Genetic and Demographic Status of Peregrine Falcons in the Upper Midwest.* St. Paul, Minnesota: Minnesota Department of Natural Resources Non-game Wildlife Program.

Moritz, C. 1994. Applications of mitochondrial DNA analysis in conservation: a critical review. *Molec. Ecol.* 3:401–411.

Moritz, C., Lavery, S., and Slade, R. 1995. Using allele frequency and phylogeny to define units for conservation and management. In *Evolution and the Aquatic Ecosystem: Defining Units in Population Conservation,* ed. J.L. Neilson, pp. 249–262. Bethesda, Maryland: American Fisheries Society.

O'Brien, S.J., Martenson, J.S., Miththapala, S., Janczewski, D., Pecon-Slattery, J., Johnson, W., Gilbert, D.A., Roelke, M., Packer, C., Bush, M., and Wildt, D.E. 1996. Conservation genetics of the Felidae. In *Conservation Genetics: Case Histories from Nature,* eds. J.C. Avise, and J.L. Hamrick, pp. 50–74. New York: Chapman and Hall.

Paetkau, D. 1999. Using genetics to identify intraspecific conservation units: a critique of current methods. *Conserv. Biol.* 13:1507–1509.

Porter, A.H. 1999. Refugees from lost habitat and reorganization of genetic population structure. *Conserv. Biol.* 13:850–859.

Reich, D.E., Wayne, R.K., and Goldstein, D.B. 1999. Genetic evidence for a recent origin by hybridization of red wolves. *Molec. Ecol.* 8:139–144.

Ridley, M. 1996. *Evolution.* Second Edition. Cambridge, Massachusetts: Blackwell Science.

Ryder, O.A. 1986. Species conservation and systematics: the dilemma of subspecies. *Trends Ecol. Evol.* 1:9–10.

Shonewald-Cox, C.M., Chambers, S.M., MacBryde, B., and Thomas, W.L., eds. 1983. *Genetics and Conservation: A Reference for Managing Wild Animal and Plant Populations.* Menlo Park, California: Benjamin/Cummings Publishing Co.

Soulé, M.E., and Mills, L.S. 1998. No need to isolate genetics. *Science* 282:1658–1659.

Taberlet, P. 1996. The use of mitochondrial DNA control region sequencing in conservation genetics. In *Molecular Genetic Approaches in Conservation,* eds. T.B. Smith and R.K. Wayne, pp. 125–142. New York: Oxford University Press.

Taberlet, P., and Bouvet, J. 1994. Mitochondrial DNA polymorphism, phylogeography, and conservation genetics of the brown bear (*Ursus arctos*) in Europe. *Proc. R. Soc. Lond. Biol.* 255:195–200.

Templeton, A.R. 1986. Coadaptation and outbreeding depression. In *Conservation Biology: The Science of Scarcity and Diversity,* ed. M.E. Soulé, pp. 105–116. Sunderland, Massachusetts: Sinauer Associates.

Templeton, A.R., and Georgiadis, N.J. 1996. A landscape approach to conservation genetics: conserving evolutionary processes in African Bovidae. In *Conservation Genetics: Case Histories from Nature,* eds. J.C. Avise and J.L. Hamrick, pp. 398–430. New York: Chapman and Hall.

Vogler, A.P., and DeSalle, R. 1993. Diagnosing units of conservation management. *Conserv. Biol.* 8:354–363.

Vogler, A.P., Knisley, C.B., Glueck, S.B., Hill, J.M., and DeSalle, R. 1993. Using molecular and ecological data to diagnose endangered populations of the Puritan Tiger Beetle *Cicindela puritana. Molec. Ecol.* 2:375–383.

Vrijenhoek, R.C. 1996. Conservation genetics of North American desert fishes. In *Conservation Genetics: Case Histories from Nature,* eds. J.C. Avise and J.L. Hamrick, pp. 367–397. New York: Chapman and Hall.

Waples, R.S. 1991. Pacific salmon, *Oncorhyncus* spp., and the definition of "species" under the Endangered Species Act. *Marine Fisheries Rev.* 53:11–22.

Waples, R.S. 1995. Evolutionarily significant units and the conservation of biological diversity under the Endangered Species Act. In *Evolution and the Aquatic Ecosystem: Defining Units in Population Conservation,* ed. J.L. Neilson, pp. 350–359. Bethesda, Maryland: American Fisheries Society.

Wayne, R.K. and Jenks, S.M. 1991. Mitochondrial DNA analysis implying extensive hybridization of the endangered red wolf *Canis rufus. Nature* 351:565–568.

Westemeier, R.L., Brawn, J.D., Simpson, S.A., Esker, T.L., Jansen, R.W., Walk, J.W., Kershner, E.L., Bouzat, J.L., and Paige, K.N. 1998. Tracking the long-term decline and recovery of an isolated population. *Science* 282:1695–1698.

9

Habitat Networks and Biological Conservation

RICHARD J. HOBBS

9.1 Introduction

This chapter examines the idea of habitat networks and how these relate to conservation. First, I explore the concepts behind the idea of habitat networks, particularly the need for placing reserves in a landscape context, the importance of connectivity, and the elements making up networks. Then I discuss examples of planning for, and implementation of, habitat networks, and I point out that although the network concept is well-established in a planning context, actual on-ground applications are still sparse. I then develop a set of principles for applying landscape ecology in the context of reserve networks and discuss gaps in our knowledge that hamper this application. I end with recommendations for an integrated research approach to redress these gaps.

9.2 Concepts, Principles, and Emerging Ideas

Previous chapters have indicated how landscape pattern, particularly in relation to habitat fragmentation, can influence persistence and movement of biota. It is generally accepted that habitat modification and fragmentation have numerous impacts on both the biotic and abiotic systems of a region (Saunders et al. 1991; Laurance and Bierregaard 1997; Schwartz 1997). These stem from the reduction in area of habitat, fragmentation of remaining habitat into smaller, more isolated patches, and impacts of changes in the surrounding matrix on these patches. These impacts become more severe and more widespread as human activity modifies and transforms increasing areas of Earth (Vitousek et al. 1997).

9.2.1 Reserves in Context

The challenge for biological conservation is to deal effectively with the threats to biota, both locally and regionally, to ensure the persistence of that biota. A range of conservation goals may be considered at both regional and local scales, ranging from the prevention of further loss of biota to reintroduction of particular biotic elements or restoration of entire communities (Chapter 20). However, conservation goals are often ill-defined and relate to "conservation of biodiversity," but they do not provide specific targets, time frames, or spatial scales. In the face of increased habitat modification and fragmentation, conservation efforts have often focused on the preservation of particular patches of habitat. These may be designated as nature reserves or preserves and may be managed by a statutory authority, a nongovernmental organization, or community group. The focus of this type of management is the composition and dynamics of biotic communities within a particular patch, and it may include such considerations as fire management and control of invasive plants and animals. In addition, in the case of habitat patches easily accessible to human populations, as in cities, a significant part of the management effort is directed at regulating access by, and dealing with problems caused by, human visitors.

This type of approach can be very successful in maintaining particular patches of habitat, and it can also enthuse large numbers of people who "adopt" and actively manage such patches. However, it is increasingly recognized that an approach dealing with individual reserves can only provide a partial solution to the problems facing conservation management. The reasons for this are numerous. First, a focus on individual fragments often ignores the fact that most of the factors impacting the fragment arise from the surrounding area or "matrix" (Hobbs 1993a; Hobbs 1994). Second, species operate at many different spatial scales, and although the requirements of some will be met entirely within individual patches, others need to move around the landscape to meet their habitat or resource requirements (Pearson et al. 1996; Hobbs 1998; Kolasa and Waltho 1998). Hence, management of individual patches in isolation may not guarantee the persistence of such species. Finally, there has been a move in conservation biology away from the ideas of island biogeography, relating species numbers to size and isolation of habitat "islands," toward the concept of *metapopulations*. This concept assumes that species' populations can exist as a series of interlinked subpopulations or metapopulations that persist in separate habitat patches but may be linked by interpatch movement (McCulloch 1996; Hanski and Gilpin 1997). More explicitly, the metapopulation approach assumes that "populations are spatially structured into assemblages of breeding populations and that migration among the local populations has some effect on local dynamics, including the possibility of population re-establishment following extinction" (Hanski and Simberloff 1997).

9.2.2 Connectivity

This array of considerations has given rise to the perceived need to consider more than just individual fragments or reserves, and from this has developed the concept of *habitat networks*. A habitat network can be defined as an interconnected set of habitat elements that together allow for movement of biota and enhance population survival probabilities. The basis of this concept is the assumption that patches of habitat are generally embedded in a matrix of "nonhabitat." The pervasive notion is that the matrix is hostile to the organisms within the relatively small fragments. This concept has been refined to more generally describe landscapes in terms of patch, corridor, and matrix (Forman 1995), with corridors representing narrow strips of habitat or, if not habitat, at least vegetation that allows biotic movement between patches (Figure 9.1). The universality of this model has been questioned recently, particularly in relation to the influence of the matrix, which may not always be completely hostile to all elements of the biota (McIntyre and Hobbs 1999). Although corridors have received widespread attention and are frequently part of conservation plans and activities, their utility is often debated, and they are in reality only one part of a broader picture (Hobbs 1992; Hobbs and Wilson 1998; Bennett 1999).

Probably a more constructive approach is to consider overall *landscape connectivity,* the extent to which different elements of the landscape are functionally connected from the viewpoint of particular biotic elements. Connectivity in a landscape depends on the relative isolation of habitat elements from one another and the extent to which the matrix represents a barrier to movement of organisms. There have been attempts to derive some generalities concerning connectivity from modeling simple geometric relationships arising from the distribution of habitat patches in landscapes with different proportions of habitat and nonhabitat (With and Crist 1995; Pearson et al. 1996; Wiens 1997; With 1997). These studies have indicated that there may be thresholds where small changes in the proportion of habitat present result in large changes in connectivity. However, the problem remains that different species will perceive the landscape differently, and landscape connectivity will depend on the mobility and habitat specificity of the species involved (Cale and Hobbs 1994; Pearson et al. 1996; Kolasa and Waltho 1998). Also, most modeling efforts consider straightforward habitat versus non-

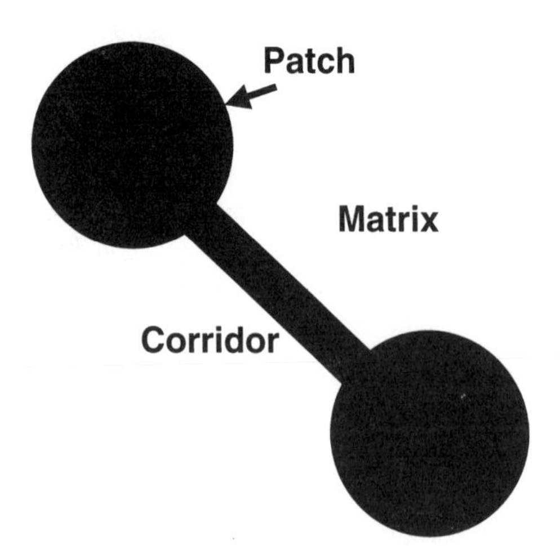

FIGURE 9.1. Simple patch-corridor-matrix model of landscapes, which dichotomizes landscape elements into habitat and nonhabitat (matrix), with corridors linking habitat patches. Derived from Forman (1995).

habitat dichotomies; they therefore do not deal with the possibility that the matrix may be more or less permeable or, conversely, resistant to movement. Hence, the role of all landscape elements in facilitating or inhibiting movement needs to be considered (Taylor et al. 1993; Wiens 1997; McIntyre and Hobbs 1999). Generalizing from that, it is apparent that a number of issues are of importance to conservation planning and management. If the conservation goal in a region is the maintenance of populations of target species, a number of factors are likely to increase the probability of population persistence, including the amount of habitat present, its distribution (or "clumpiness"), the distribution of patch sizes, and the connectivity of the matrix (Figure 9.2; Harrison and Fahrig 1995; Wiens 1996). In the case in which the matrix provides some habitat characteristics or is somewhat permeable to movement, a habitat versus nonhabitat model does not apply; hence, the final situation in Figure 9.2 indicates a "gray" matrix that helps increase the probability of population persistence.

9.2.3 Networks

Habitat networks thus have to be considered in the light of the above considerations. Realization that reserves cannot be considered in isolation from other habitat patches or from the surrounding matrix has led to ideas that conservation management should take place in an integrated way across the landscape (Hobbs et al. 1993a; Noss and Cooperrider 1994). This involves some consideration of the management of habitat patches for conservation and for other uses, and of how to manage the matrix in ways that are compatible with conservation. Such considerations are implicit in the concept of managing "*greater ecosystems,*" in which national parks are considered in the context of surrounding lands managed for production purposes, such as in the Yellowstone area in the United States (Povilitis 1993; Burroughs and Clark 1995). These considerations also are implicit in the

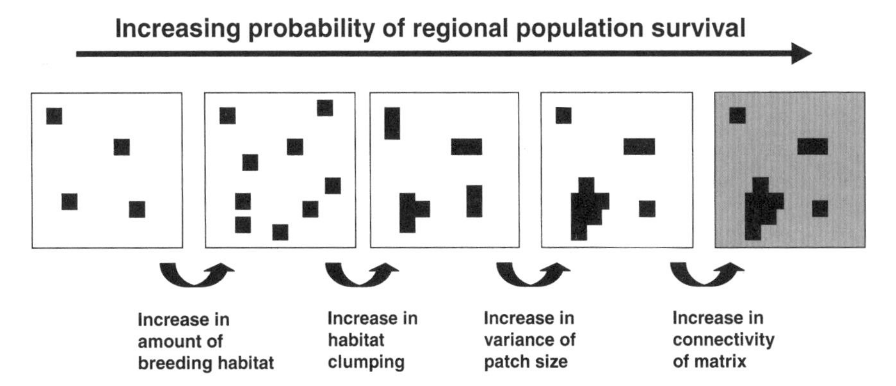

FIGURE 9.2. Hypothesized effects of changes in landscape characteristics on the probability of persistence of a regional (meta)population. Modified from Harrison and Fahrig (1995).

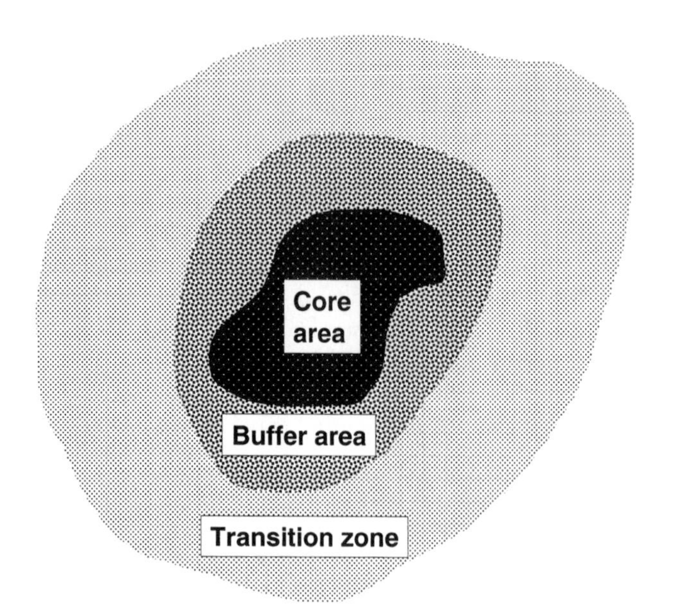

FIGURE 9.3. Idealized representation of the biosphere reserve concept, with a fully protected core conservation area surrounded by a limited-use buffer area and a transition zone containing land uses compatible with conservation objectives.

biosphere reserve concept, initiated under the Man and the Biosphere Program (UNESCO 1974; Dasmann 1988); this program sought to establish a global network of reserves with a zonation system consisting of a core reserve zone surrounded by a buffer zone and a transitional zone where other land uses were compatible with conservation within the reserve (Figure 9.3). Although both greater ecosystems and biosphere reserves sound good in theory, application of these ideas generally seems to fall short (Noss and Cooperrider 1994), with only a few examples around the world in which attempts have been made to operationalize the concepts (e.g., Watson et al. 1995; Harwell et al. 1996). Nevertheless, the concept of biosphere reserves was extended and generalized to a range of scales and management scenarios in the concept of the *multiple-use module* (MUM) (Harris 1984; Noss and Harris 1986). A MUM consists of a core reserve area surrounded by buffers containing increasing intensities of land use. The idea of zoning land uses into different degrees of protection also has been used in arid-zone ecosystems in Australia (Morton et al. 1995).

Building on these concepts, the idea developed that habitat networks should be established in which *nodes* of high-value habitat (i.e., key reserves or habitat patches) could be integrated into a system with buffer zones and connections. Such a network could potentially operate at a number of different spatial scales with, for instance, local networks of core reserves, buffers, and corridors linked into a regional network (Figure 9.4). In Europe, many plans for ecological net-

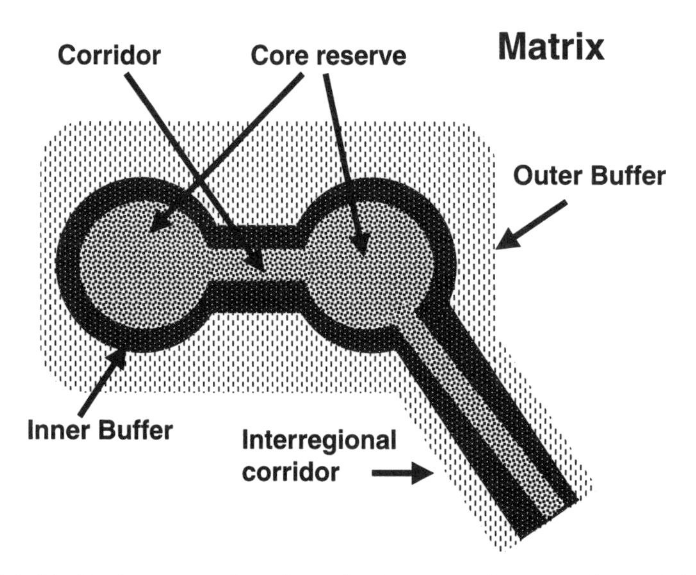

FIGURE 9.4. Idealized representation of a regional conservation network, consisting of a local network of core reserves and corridors surrounded by buffer zones, and linked to other local networks by regional corridors. Redrawn from Noss and Cooperrider (1994).

works are being developed (Arts et al. 1995a), and a variety of networks have been proposed for numerous locations in North America (Noss and Cooperrider 1994). *Greenways,* or vegetated linkages between public open spaces and parks, are one component of such networks. These are discussed more fully below.

9.3 Recent Applications

On-ground applications were sought from the published literature, contact with ecologists and planners in different areas, and conference presentations at a number of international meetings. From this search, it is apparent that although considerable attention has been given to placing reserves in a landscape context, to questions of connectivity, and to the development of habitat networks at a theoretical or research level, such considerations are only recently appearing in planning contexts and have rarely been applied in a meaningful way in on-ground activities. Here, I discuss three examples of where the concepts have reached the planning stage.

9.3.1 Ecological Network Planning in Europe

The extremely modified and fragmented nature of many European ecosystems has been recognized for some time, and serious attempts are being made at regional, national, and international levels to put plans in place that will formalize habitat networks that retain important elements of the European biota. At a local scale, such efforts have been underway for some time. For instance, the Biotopverbundsysteme (habitat network) has been actively promoted in Germany, resulting in the retention of habitat and corridors across the country (Jedicke 1994). Although this approach has been criticized as lacking rigor (Henle 1995), it does recognize the importance of the concepts listed above, particularly the importance of building on existing nodes and maintaining or enhancing connectivity. It also attempts to develop a network approach to conservation.

More generally, plans for habitat networks are being developed in many European countries (Arts et al. 1995a; Jongman 1995). These are being planned as "coherent spatial structures" consisting of cores, buffers, and corridors, and Arts et al. (1995b) concluded that "during the last decade, the nature conservation policies in many European countries have been based on landscape-ecological research, especially concerning the role of land use and landscape structure in the survival of species and in the protection of nature reserves. Plan proposals were made to establish ecological networks on local, regional and national scales." Articles in a special issue of *Landschap* (1995, Volume 12, Number 3) describe planning efforts in many different European countries, frequently with nested local, regional, and national plans, and they discuss the development of pan-European strategies. An example is the National Ecological Network in The Netherlands, which aims to protect, buffer, and link the core areas and nature development areas for the entire country and include linkages across national borders with Germany and Belgium (Opdam et al. 1994; Ahern 1995). Again, these planning activities recognize the concepts listed above, particularly the importance of core areas or nodes, connectivity, and networks.

Although these activities remain largely in the planning phase, they clearly indicate recognition of the need for local and regional conservation networks to be enshrined in policy. Actual on-ground implementation will undoubtedly pose significant challenges, especially in relation to configuring local and regional plans so that they are compatible (see Chapter 20 for further discussion on landscape and regional planning). The actual effectiveness of such activities cannot yet be assessed.

9.3.2 Greenways in Eastern North America

Considerable interest has developed recently in greenways (Ahern 1995; Fabos 1995; Linehan et al. 1995). These are developed as linkages between open spaces or parks and are primarily for human recreation or aesthetic enjoyment (Little 1990; Smith and Hellmund 1993), but they may nevertheless provide some benefits in terms of conservation (Hay 1991).

The idea of greenways builds on the concepts of connectivity and the development of networks, although it does not directly relate to habitat conservation in a landscape context. An example of the scope and vision for the development of greenways occurs in the State of Maryland in the United States. A statewide network of greenways, including a diverse array of linear habitats and linkages, such as stream valleys, ridgetops, and utility corridors, is envisaged as a major community conservation initiative (Therres et al. 1988; State of Maryland 1990). These plans include features relating to conservation such as the provision of a buffer zone along shorelines and streams to maintain the stream environment and protect riparian habitat, the protection of riparian forests, and the incorporation of wildlife corridors in development proposals. In this example, as with other planned greenways, the major challenge will be integrating conservation goals with the primary recreational goals.

As with the European networks, the greenway concept has been discussed widely in planning terms, but it has yet to be widely implemented. Nevertheless, the examples of implementation that are available (Fabos 1995 and other articles in the same volume) are cause for optimism that greenways provide the promise of community-supported networks that provide at least some conservation benefit, especially in densely populated areas.

9.3.3 Farm and Watershed Planning in Australia

In the agricultural regions of Australia, native ecosystems often remain only as relatively small patches, and there is clear evidence that considerable loss of biota has occurred and continues to occur (Hobbs et al. 1993b; Saunders and Ingram 1995). It is obvious that if we set a goal of retaining the existing biota within particular regions, there is a requirement not only for retention of most of the native vegetation that remains, regardless of its current status (i.e., designated reserve, private land, etc.), but also for reestablishment of extensive areas of native vegetation (Hobbs 1993b; Hobbs et al. 1993a). This involves recognition that the remaining fragments of native vegetation constitute a network and have to be managed in an integrated fashion. Strategic approaches to designing revegetation programs have been developed that build on the existing system of remnants, and that provide spatially explicit plans to landholders and community groups. This approach, discussed more fully in Chapter 20, applies the concepts of building networks using existing habitat as nodes and enhancing the connectivity of the landscape. Ryan (2000) indicates clearly the lack of evidence to date that carrying out such revegetation will actually do anything useful. Nevertheless, examples cited by him and Barrett and Davidson (2000) of the use of revegetated areas by a variety of bird species provide some hopeful signs that revegetation and regeneration do, in fact, result in conservation benefits in Australian agricultural areas.

9.4 Principles for Applying Landscape Ecology

The main point arising from the examples above is that the concept of habitat networks has been embraced at a planning level in many different instances, but relatively few of these plans have reached implementation. It remains to be seen whether this is just a matter of time frame, or whether there are fundamental constraints to the successful implementation of landscape and regional plans. Similarly, other concepts discussed above, such as greater ecosystem and biosphere reserve management, have gained considerable currency at both theoretical and applied levels, but when the actual examples of application are examined, they often fall short of expectations. For instance, Noss and Cooperrider (1994) discuss the example of the Greater Yellowstone Ecosystem, which purports to place Yellowstone National Park and other parks in a broader context of sympathetic management of surrounding lands. However, at least part of the boundary of the Park is clearly discernable from the air because of logging activities in the adjacent forest. On the other hand, de facto ecological networks are probably already in place in many parts of the world, and the main task is to recognize these networks and implement measures to ensure their protection or enhancement.

The following principles have been derived partially from the applications given in the previous section, and partially from a general theoretical and pragmatic consideration of what is required. The reason for this is that existing applications are limited in information about their on-ground effectiveness, and they may not be based on sound ecological or conservation principles.

9.4.1 Integrate the Management of Different Landscape Elements

The development of networks clearly depends on successful integration of the management of lands under various types of tenure and subject to many uses. At one extreme, there are calls for the development of dedicated conservation networks at the expense of other land uses (Noss 1992; Soulé 1995), whereas others see the need for developing conservation networks nested within a range of other land uses. The extent to which either of these is possible depends largely on social and political factors; the extent to which they are successful depends more on whether the developed networks actually achieve the goals espoused for them. To achieve network effectiveness, the application of landscape ecology principles is required.

9.4.2 Consider the Type of Landscape to Be Managed

If a series of conservation goals is developed for a region, how do we best design the habitat network to achieve these goals? This question has to be tackled from two angles. One must consider the type of landscape being managed, and the types of species being conserved. To date, a generalized response to these questions has been sought.

A set of general principles, derived from island biogeography theory, suggests that big patches are better than small patches, connected patches are better than unconnected patches, and so on. For fragments in agricultural landscapes, such principles can be translated into the need to retain existing patches (especially large ones) and existing connections, and to revegetate in such a way as to provide larger patches and more connections (Hobbs 1993b).

Important questions concern the sort of landscape-scale management and reconstruction that is appropriate for maintaining or developing habitat networks in different landscapes. If we can accept that priority actions involve, firstly, the protection of existing critical areas and the maintenance or redevelopment of landscape connectivity, we then need to set management priorities. The following questions need to be asked in any conservation planning process:

1. Which are the priority areas to retain?
2. Should we concentrate on retaining the existing fragments or on habitat reconstruction, and relatively how many resources (financial, manpower, etc.) should go into each?
3. How much reconstruction is required, and in what configuration?
4. When should we concentrate on protecting existing corridors or providing more corridors, versus protecting blocks of habitat or trying to provide additional habitat?

9.4.3 Develop a Generalized Landscape-Assessment Approach and Spatially Explicit Solutions

If we are to make a significant impact in terms of biological conservation, the above questions need to be addressed in a strategic way. Although generalizations on connectivity, metapopulation dynamics, and so forth are useful to a certain extent, most on-ground application will have to be related to the specifics of the landscape in question and the species involved. Thus, rather than looking for general principles, we need a generalized approach for assessing landscapes and species, and for determining specific requirements for habitat networks based on these assessments. I present two allied approaches to this that are currently under development.

McIntyre and Hobbs (1999, 2001) discuss the processes of habitat destruction and habitat modification, which can both be conceptualized as a continuum and are associated with the effects of human disturbance. Habitat destruction results in loss of all structural features of the vegetation and loss of the majority of species, as occurs during vegetation clearance. A gradient of habitat destruction is represented as a continuum, four levels of which are depicted in Figure 9.5a. McIntyre and Hobbs (1999, 2001) identified four broad types of landscapes along this gradient, with intact and relictual landscapes at the extremes, and two intermediate states, variegated and fragmented. In variegated landscapes, the habitat still forms the matrix, whereas in fragmented landscapes, the matrix is composed of "destroyed habitat." This allows for the possibility that in some landscape types (i.e., intact and variegated), the matrix retains some habitat value and contributes to a greater or lesser degree to habitat connectivity.

Each of the four levels described in Figure 9.5a is associated with a particular degree of habitat destruction, and this broad division can be regarded as a "first cut." Further investigation is required to test these categories, to clarify whether hypothesized thresholds are real or not, and to examine the need for additional subcategories (e.g., functionally different types of fragmented landscapes).

Habitat modification alters the condition of the remaining habitat and can occur in any of the situations illustrated in Figure 9.5a. Figure 9.5b illustrates this

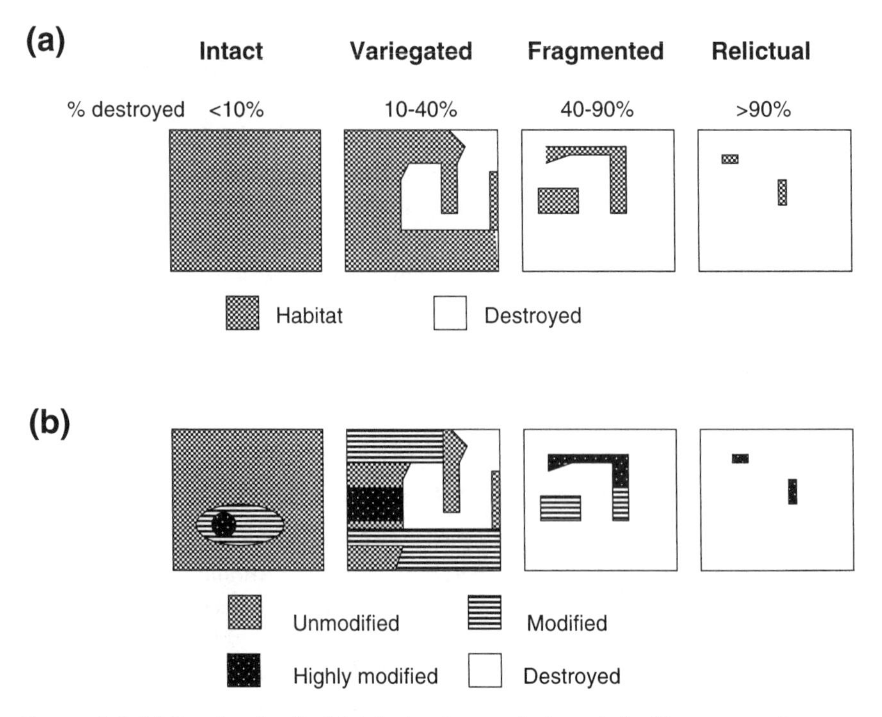

FIGURE 9.5. (a) Four levels of habitat destruction, each characterized by a range of proportions of habitat destroyed. (b) Pattern of habitat modification overlying landscape patterns of habitat destruction depicted in (a). Although any combination of destruction and modification levels is theoretically possible, those considered to be typical of different destruction levels are illustrated. Modified from McIntyre and Hobbs (1999, 2001).

by depicting three arbitrary levels of modification. Modification creates a source of variation in the landscape over and above the straightforward spatial patterning caused by vegetation destruction. If a modifying disturbance is intense, or protracted enough, it will eventually lead to habitat destruction.

The levels of modification depicted in Figure 9.5b reflect the tendency for habitats to become progressively more modified with increasing levels of destruction, owing to the progressively greater proportion of edge in remaining habitats. Theoretically, all combinations of habitat destruction and modification gradients could exist. In practice, it is highly unlikely to find a relictual, or even a fragmented, landscape containing unmodified remnants. This combination of level of destruction and degree of modification is an important consideration for management planning.

McIntyre and Hobbs (2001) suggested that the framework illustrated in Figure 9.5 can assist in deciding where on the landscape to allocate greater and lesser protection or management efforts. Three principles for applying landscape ecology to habitat management follow from this suggestion and are discussed in Sections 9.4.4–9.4.6.

9.4.4 *Maintain Existing Habitat*

Maintain existing condition of habitats by removing and controlling threatening processes. It is generally much easier to avoid the effects of degradation than it is to reverse them. The first priority is thus the maintenance of elements that are currently in good condition. These will be predominantly the vegetated matrix in intact and variegated landscapes and the remnants that remain in good condition in fragmented landscapes. Maintenance will involve ensuring continuation of population, community, and ecosystem processes that result in persistence of the species and communities present in the landscape. Note that maintaining fragments in good condition in a fragmented system may also require management activities in the matrix to control landscape processes, such as water flows.

9.4.5 *Improve Degraded Habitats*

Improve the condition of habitats by reducing or removing threatening processes. More active management may be needed to initiate a reversal of condition (e.g., removal of exotic species, reintroduction of native species) in highly modified habitats. The second priority is thus the improvement of elements that have been modified in some way. In variegated landscapes, buffer areas and corridors may be a priority, whereas in fragmented systems, improving the surrounding matrix to reduce threatening processes will be a priority, as indicated above. In relict landscapes, improving the condition of fragments will be essential for their continued persistence. Improvement may involve simply dealing with threatening processes such as stock grazing or feral predators, or it may involve active management to restore ecosystem processes, improve soil structure, encourage regeneration of plant species, or reintroduce flora or fauna species.

9.4.6 *Reconstruct Habitats as a Last Resort*

Reconstruct habitats where their total extent has been reduced below viable size using replanting and reintroduction techniques. Because this is so difficult and expensive, it is a last-resort action that is most relevant to fragmented and relictual landscapes. We have to recognize that restoration will not come close to restoring habitats to their unmodified state, and this reinforces the wisdom of maintaining existing habitat as a priority. Reconstruction is likely to be necessary only in fragmented and relictual areas. Primary goals of reconstruction will be to provide buffer areas around fragments, to increase connectivity with corridors, and to provide additional habitat (Hobbs 1993b).

In Table 9.1, the application of these principles of developing or maintaining habitat networks are explicitly linked to specific landscape components (matrix, fragments, patches, buffer areas, connecting areas) for which they would be most

TABLE 9.1. Management actions and location in the landscape where they may be most effective for the retention and development of an effective habitat network in landscapes with differing degrees of habitat destruction and modification. Actions and locations vary with the alteration state of particular landscapes. *Matrix*—the predominant part of the landscape, consisting of habitat (in intact and variegated landscapes) or destroyed habitat (in fragmented and relictual landscapes). *Fragments*—restricted areas of habitat surrounded by areas of destroyed habitat. *Patches*—least-modified habitat surrounded by areas of highly modified habitat. *Buffer areas*—occur around fragments or patches of least-modified habitat. *Connecting areas*—occur between fragments or patches of least-modified habitat. From McIntyre and Hobbs (2001).

	Landscape alteration level			
Action	Intact	Variegated	Fragmented	Relictual
Maintain	Matrix	Matrix, especially patches	Fragments in good condition	—
Improve	—	Buffer areas, connecting areas	Fragments	Fragments
Reconstruct	—	—	Buffer areas, connecting areas	Buffer areas

effective. Table 9.1 highlights the different priorities for management action in developing or maintaining habitat networks in different landscape types. These recommended actions reflect three basic principles, as follows:

1. Build on strengths of the remaining habitat by filling in gaps and increasing landscape connectivity.
2. Increase the availability of resources by rehabilitating degraded areas.
3. Expand habitat by revegetating to create larger blocks and restoring poorly represented habitats.

9.4.7 Develop Strategic Responses With Clear Goals

A general principle is that more efficient solutions to conservation problems can be developed if we take a strategic approach rather than a generalized one. This involves developing a clear set of conservation objectives rather than relying on vague statements of intent (Lambeck 1997). One set of objectives relates to the achievement of a comprehensive, adequate, and representative set of reserves or protected-area networks (see Chapter 20). Another complementary set of objectives relates to the adequacy of existing remnant vegetation (not only reserves). The process of setting conservation objectives in any given area can be simplified by identifying a set of key or "focal" species (Figure 9.6; Lambeck 1997). This approach is similar to a multispecies, indicator-species, or umbrella-species approach (Simberloff 1998), and it claims to link the best of single-species and landscape-scale approaches (Lambeck 1997; see also Chapter 20). For habitat networks to be effective, we need to be clear about what species we are providing the networks for, and how that relates more generally to other species.

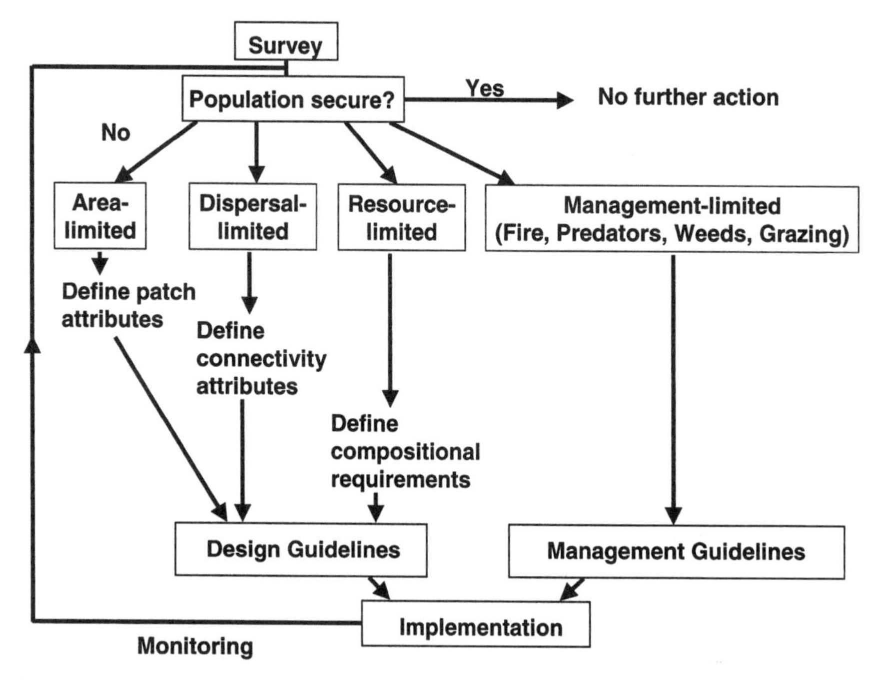

FIGURE 9.6. A process for selecting focal species in any given landscape, based on whether species are limited by area, movement, resources, or management, and resulting in landscape-design guidelines and management guidelines. Modified from Lambeck (1997).

To identify focal species, Lambeck (1997) recognized four distinct sets of species, each of which was likely to be limited or threatened by particular characteristics of the landscape. These were area- or habitat-limited species, whose numbers are limited by the availability of large enough patches of suitable habitat; movement-limited species, whose numbers are limited by the degree to which they can move between habitat patches; resource-limited species, whose numbers are limited by the availability of one or more critical resources; and management-limited species, whose numbers are limited by processes such as predation, disturbance, fire, and the like, which can be manipulated within particular sites.

Conservation objectives for an area can be discussed in terms of which species and communities are at risk, what the likely source of that risk is, and how prepared Society is to address the risk. The focal species approach could profitably be combined with the framework for categorizing landscapes suggested by McIntyre and Hobbs (1999, 2001) if the relative incidence of species in different categories could be linked to landscape configuration. A useful approach is thus to develop a set of principles or guidelines for habitat networks in a general way, for instance to decide the relative efforts needed in habitat protection or revegetation. More detailed guidelines then become necessary (in relation to goals for habitat networks for particular sets of species) to decide, for example, on the relative need for corridors versus provision of enlarged habitat patches. Lambeck (1997) has indicated how the identification of focal species and a rapid assessment of their habitat requirements can result in the production of quantitative guidance about how much habitat is needed, and in what configuration. Further development of this work involves being able to make spatially explicit recommendations about where reconstruction should occur (see Chapter 20).

9.5 Knowledge Gaps

9.5.1 Theoretical Voids

The concepts discussed in Section 9.2 arise from current theory on how species survive and move around in heterogeneous landscapes. As discussed previously, there has been a development in ideas away from island biogeographic theory toward metapopulation theory, and away from the relatively simple view of landscapes being composed of patches of habitat set in a matrix of nonhabitat to a more complex view of different landscape elements combining to provide patches of varying habitat and movement value. This theoretical development is ongoing, and there are numerous areas where further work is needed.

Metapopulation and population viability models have often been poorly validated, and their assumptions often remain untested. In addition, a variety of models is available, and decisions about their applicability in particular situations may be difficult. Metapopulation models also generally do not account for variations in habitat quality or condition.

Making such models spatially explicit, so that the impacts of changing landscape configurations can be tested, is a real challenge. Attempts to model the likely impact of changing landscape connectivity on the dynamics of particular species have only recently begun (Brooker et al. 1999). Assessing landscape connectivity in terms of both corridors and general permeability of different landscape elements also is an area that requires development (Wiens 1996, 1997). An effective theoretical framework that recognizes the potential importance of different landscape elements (including the matrix), and of variation in habitat quality or condition within patches, has yet to be fully developed. This is essential if we are to make headway in understanding and designing habitat networks.

9.5.2 Empirical Voids

Our ability to design habitat networks is limited by our knowledge of how species currently use the landscape and how they may respond to changes to the landscape. Moving from general principles derived from conservation biology to specific quantitative predictions of the outcome of particular actions is still problematic.

Maintenance and development of habitat networks requires that we can assess the current status of a particular landscape and decide what types of management actions are necessary. Our ability to categorize and describe landscapes remains restricted due to the failure to develop reliable and empirically tested landscape measures. Recent syntheses on measurement of landscape parameters have indicated significant drawbacks in many of the currently used landscape measures (Frohn 1997; Hargis et al. 1998; Trani and Giles 1999). Such measures not only need to be robust, but they also need to be relatively simple, easy to obtain, and easy to interpret. Describing landscape pattern in a meaningful way remains one of the biggest challenges in landscape ecology.

Once we have assessed the current landscape status, we then need to be able to detect biotic response to changes in the landscape resulting from management actions. These responses may include, for instance, population declines of a target species, indicating that the current habitat network is inadequate, or population persistence or increase resulting from the development of an improved network. In dealing with species responses to landscape change, we frequently have to deal with a sparse data set on species population parameters, especially where these species have small and declining populations. Given that these species are often going to be the focus of conservation activity, this presents real problems.

Assessing the conservation consequences of regional networks is difficult due to the limited development of such networks. Empirical data on species abundances and distributions are required before, during, and after the establishment of such networks, and such data are frequently either unavailable or difficult and expensive to collect. Often too, the consequences of retaining or establishing a network will not become apparent for a number of years and will need to be measured in terms of persistence or population increase of target species. It is not always the case that monitoring programs are either in place or designed to provide useful information. We need such data to assess the success of the network versus the impact of other factors.

9.6 Research Approaches

9.6.1 Approaches for Theoretical Research

From the above, it is clear that an integrated protocol involving modeling, monitoring, and empirical observations is required if we are to have any chance of assessing the likely performance of habitat networks or of improving their design and performance.

In terms of theoretical development, the primary requirement is for spatially explicit metapopulation modeling of population responses to landscape modification and alternative network designs. To achieve this, several components are required, as follows:

1. Refinement of metapopulation modeling approaches that factor in overall landscape structure, the impact of the landscape matrix, and the importance of within-patch habitat condition. This requires a move away from simple

metapopulation modeling based on patch/nonpatch habitat concepts to more generalized spatial population modeling.
2. Development of simple methods of landscape parameterization and empirical measures of biotic response to variation in these parameters. Careful assessment is required of whether measures actually reflect functionally important features of landscapes and can effectively detect differences between landscapes or changes in particular landscapes. Initial steps toward this include comparing parameter estimates against estimates of biotic response (e.g., population abundances or persistence probabilities).
3. Development of a simple theoretical framework that adequately describes the range of landscape configurations that are likely to occur, and that can then act as a basis for developing strategies for habitat network development and maintenance. McIntyre and Hobbs (1999, 2001) have provided a start for this, but more needs to be done.
4. Reassessment of the validity of different approaches and methods in landscape ecology, to ensure that useful information is garnered from a variety of approaches and is acceptable from theoretical and statistical viewpoints. Classic approaches to ecology, involving factorial experimentation and balanced designs, are virtually impossible to implement in the study of habitat networks. This involves developing a new research framework that can cope with the spatial extent and complexity of the features we wish to study.

9.6.2 Approaches for Empirical Research

To move from generalized principles to spatially explicit and locally relevant options, we need the following research components:

1. Assessment of species-specific habitat and movement requirements, using a combination of detailed study of individual species, other more rapid observational methods, and experimental and modeling (simulation and statistical) studies. The most useful data for this purpose are the most difficult to collect; detailed studies of individual species, conducted over a long enough time period (i.e., several generations of the species under consideration) to be reliable and to factor in natural environmental variation, are essential, but often costly and time-consuming. These need to be combined with more rapid assessment techniques, which focus on particular species or groups of species, and can be used as the basis for directing future detailed studies while providing input to current management decisions. Experimental studies yield invaluable data and a context for interpreting data derived from other sources (e.g., Laurance and Bierregaard 1997). Finally, these have to be combined effectively with the modeling efforts discussed above.
2. Quantitative assessment of the relationship between landscape patterns and biotic composition or population status and dynamics. The missing link in research on networks is consideration of the biotic response to changes in landscape configuration. This type of information will be exceptionally difficult to

obtain and, hence, will need to be derived from a variety of research approaches, as described in the preceeding paragraph.

Only through effective integration of a number of approaches can we hope to contribute significantly to our understanding of how habitat networks function and how to maintain and improve them in ways that will substantially improve the conservation of biodiversity. The challenge for landscape ecology is to be inclusive and synthetic enough for these integrated approaches to be both accepted scientifically and implemented on the ground.

Acknowledgments

I thank Rob Lambeck and Kevin Gutzwiller for helpful comments on the draft manuscript, and the numerous colleagues who have contributed to the ideas presented here, particularly Mike and Lesley Brooker, Peter Cale, Rob Lambeck, and Sue McIntyre.

References

Ahern, J. 1995. Greenways as a planning strategy. *Landsc. Urban Plan.* 33:131–155.

Arts, G.H.P., van Buuren, M., Jongman, R.H.G., Nowicki, P., Wascher, D., and Hoek, I.H.S., eds. 1995a. *Landschap–Special Issue on Ecological Networks* 12(3).

Arts, G.H.P., van Buuren, M., Jongman, R.H.G., Nowicki, P., Wascher, D., and Hoek, I.H.S. 1995b. Editorial. *Landschap* 12:5–9.

Barrett, G., and Davidson, I. 2000. Community monitoring of woodland habitats—the Birds on Farms Survey. In *Temperate Eucalypt Woodlands in Australia: Biology, Conservation, Management and Restoration,* eds. R.J. Hobbs and C.J. Yates, pp. 382–399. Chipping Norton, Australia: Surrey Beatty and Sons.

Bennett, A.F. 1999. *Linkages in the Landscape: The Role of Corridors and Connectivity in Wildlife Conservation.* Gland, Switzerland and Cambridge, United Kingdom: The World Conservation Union (IUCN) Forest Conservation Programme.

Brooker, L., Brooker, M., and Cale, P. 1999. Animal dispersal in fragmented habitat: measuring habitat connectivity, corridor use, and dispersal mortality. *Conserv. Ecol.* [online] 3:4. Available from the Internet: www.consecol.org/vol3/iss1/art4.

Burroughs, R.H., and Clark, T.W. 1995. Ecosystem management: a comparison of Greater Yellowstone and Georges Bank. *Environ. Manage.* 19:649–663.

Cale, P., and Hobbs, R.J. 1994. Landscape heterogeneity indices: problems of scale and applicability, with particular reference to animal habitat description. *Pac. Conserv. Biol.* 1:183–193.

Dasmann, R.F. 1988. Biosphere reserves, buffers and boundaries. *BioScience* 38:487–489.

Fabos, J.G. 1995. Introduction and overview: the greenway movement, uses and potentials of greenways. *Landsc. Urban. Plan.* 33:1–13.

Forman, R.T.T. 1995. *Land Mosiacs: The Ecology of Landscapes and Regions.* Cambridge, United Kingdom: Cambridge University Press.

Frohn, R.C. 1997. *Remote Sensing for Landscape Ecology: New Metrics for Monitoring, Modeling and Assessment of Ecosystems.* Boca Raton, Florida: Lewis.

Hanski, I.A., and Gilpin, M.E., eds. 1997. *Metapopulation Biology: Ecology, Genetics, and Evolution.* New York: Academic Press.

Hanski, I., and Simberloff, D. 1997. The metapopulation approach, its history, conceptual domain, and application to conservation. In *Metapopulation Biology: Ecology, Genetics, and Evolution,* eds. I.A. Hanski and M.E. Gilpin, pp. 5–26. New York: Academic Press.

Hargis, C.D., Bissonette, J.A., and David, J.L. 1998. The behavior of landscape metrics commonly used in the study of habitat fragmentation. *Landsc. Ecol.* 13:167–186.

Harris, L.D. 1984. *The Fragmented Forest: Island Biogeographic Theory and the Preservation of Biotic Diversity.* Chicago: University of Chicago Press.

Harrison, S., and Fahrig, L. 1995. Landscape pattern and population conservation. In *Mosaic Landscapes and Ecological Processes,* eds. L. Hansson, L. Fahrig, and G. Merriam, pp. 293–307. London: Chapman and Hall.

Harwell, M.A., Long, J.F., Bartuska, A.M., Gentile, J.H., Harwell, C.C., Myers, V., and Ogden, J.C. 1996. Ecosystem management to achieve ecological sustainability: the case of South Florida. *Environ. Manage.* 20:497–521.

Hay, K.G. 1991. Greenways and biodiversity. In *Landscape Linkages and Biodiversity,* ed. W.E. Hudson, pp. 162–175. Washington, DC: Island Press.

Henle, K. 1995. Biodiversity, people and a set of important connected questions. In *Nature Conservation 4: The Role of Networks,* eds. D.A. Saunders, J.L. Craig, and E.M. Mattiske, pp. 162–174. Chipping Norton, Australia: Surrey Beatty and Sons.

Hobbs, R.J. 1992. Corridors for conservation: solution or bandwagon? *Trends Ecol. Evol.* 7:389–392.

Hobbs, R.J. 1993a. Effects of landscape fragmentation on ecosystem processes in the Western Australian wheatbelt. *Biol. Conserv.* 64:193–201.

Hobbs, R.J. 1993b. Can revegetation assist in the conservation of biodiversity in agricultural areas? *Pac. Conserv. Biol.* 1:29–38.

Hobbs, R.J. 1994. Fragmentation in the wheatbelt of Western Australia: landscape scale problems and solutions. In *Fragmentation in Agricultural Landscapes,* ed. J. Dover, pp. 3–20. Garstang, United Kingdom: International Association for Landscape Ecology.

Hobbs, R.J. 1998. Managing ecological systems and processes. In *Ecological Scale: Theory and Applications,* eds. D. Peterson and V.T. Parker, pp. 459–484. New York: Columbia University Press.

Hobbs, R.J., Saunders, D.A., and Arnold, G.W. 1993a. Integrated landscape ecology: a Western Australian perspective. *Biol. Conserv.* 64:231–238.

Hobbs, R.J., Saunders, D.A., Lobry de Bruyn, L.A., and Main, A.R. 1993b. Changes in biota. In *Reintegrating Fragmented Landscapes: Towards Sustainable Production and Nature Conservation,* eds. R.J. Hobbs and D.A. Saunders, pp. 65–106. New York: Springer-Verlag.

Hobbs, R.J., and Wilson, A.-M. 1998. Corridors: theory, practice and the achievement of conservation objectives. In *Key Concepts in Landscape Ecology,* eds. J.W. Dover and R.G.H. Bunce, pp. 265–279. Preston, United Kingdom: International Association for Landscape Ecology.

Jedicke, E. 1994. *Biotopverbund: Grundlagen und Massnahmen einer neuen Natureschutzstrategie.* Stuttgart, Germany: Ulmer.

Jongman, R.H.G. 1995. Nature conservation planning in Europe: developing ecological networks. *Landsc. Urban Plan.* 32:169–183.

Kolasa, J., and Waltho, N. 1998. A hierarchical view of habitat and its relationship to species abundance. In *Ecological Scale: Theory and Applications,* eds. D. Peterson and V.T. Parker, pp. 55–76. New York: Columbia University Press.

Lambeck, R.J. 1997. Focal species: a multi-species umbrella for nature conservation. *Conserv. Biol.* 11:849–856.

Laurance, W.F., and Bierregaard, R.O., eds. 1997. *Tropical Forest Remnants: Ecology, Conservation and Management of Fragmented Communities.* Chicago: University of Chicago Press.

Linehan, J., Gross, M., and Finn, J. 1995. Greenway planning: developing a landscape ecological network approach. *Landsc. Urban Plan.* 33:179–193.

Little, C.E. 1990. *Greenways for America.* Baltimore, Maryland: The Johns Hopkins University Press.

McCulloch, D.R., ed. 1996. *Metapopulations and Wildlife Conservation.* Washington, DC: Island Press.

McIntyre, S., and Hobbs, R. 1999. A framework for conceptualizing human effects on landscapes and its relevance to management and research models. *Conserv. Biol.* 13:1282–1292.

McIntyre, S., and Hobbs, R.J. 2001. Human impacts on landscapes: matrix condition and management priorities. In *Nature Conservation 5: Nature Conservation in Production Environments,* eds. J. Craig, D.A. Saunders, and N. Mitchell, pp. 301–307. Chipping Norton, Australia: Surrey Beatty and Sons.

Morton, S.R., Stafford Smith, D.M., Friedel, M.H., Griffin, G.F., and Pickup, G. 1995. The stewardship of arid Australia: ecology and landscape management. *J. Environ. Manage.* 43:195–217.

Noss, R.F. 1992. The Wildlands Project land conservation strategy. *Wild Earth,* Special Issue: 10–25.

Noss, R.F., and Cooperrider, A.Y. 1994. *Saving Nature's Legacy: Protecting and Restoring Biodiversity.* Washington, DC: Island Press.

Noss, R.F., and Harris, L.D. 1986. Nodes, networks and MUMs: preserving diversity at all scales. *Environ. Manage.* 10:299–309.

Opdam, P., Foppen, R., Reijnen, R., and Schotman, A. 1994. The landscape ecological approach in bird conservation: integrating the metapopulation concept into spatial planning. *Ibis* 137:S139–S146.

Pearson, S.M., Turner, M.G., Gardner, R.H., and O'Neill, R.V. 1996. An organism-based perspective of habitat fragmentation. In *Biodiversity in Managed Landscapes: Theory and Practice,* eds. R.C. Szaro and D.W. Johnston, pp. 77–95. New York: Oxford University Press.

Povilitis, T. 1993. Applying the biosphere reserve concept to a greater ecosystem: the San Juan Mountain area of Colorado, New Mexico. *Nat. Areas J.* 13:18–28.

Ryan, P. 2000. The use of revegetated areas by vertebrate fauna in Australia: a review. In *Temperate Eucalypt Woodlands in Australia: Biology, Conservation, Management and Restoration,* eds. R.J. Hobbs and C.J. Yates, pp. 318–335. Chipping Norton, Australia: Surrey Beatty and Sons.

Saunders, D.A., Hobbs, R.J., and Margules, C.R. 1991. Biological consequences of ecosystem fragmentation: a review. *Conserv. Biol.* 5:18–32.

Saunders, D.A., and Ingram, J. 1995. *Birds of Southwestern Australia: An Atlas of Changes in the Distribution and Abundance of the Wheatbelt Fauna.* Chipping Norton, Australia: Surrey Beatty and Sons.

Schwartz, M.W., ed. 1997. *Conservation in Highly Fragmented Landscapes.* New York: Chapman and Hall.

Simberloff, D. 1998. Flagships, umbrellas, and keystones: is single species management passé in the landscape era? *Biol. Conserv.* 83:247–257.

Smith, D.S., and Hellmund, P.C., eds. 1993. *Ecology of Greenways: Design and Function of Linear Conservation Areas.* Minneapolis: University of Minnesota Press.

Soulé, M.E. 1995. An unflinching vision: networks of people defending networks of lands. In *Nature Conservation 4: The Role of Networks,* eds. D.A. Saunders, J.L. Craig, and E.M. Mattiske, pp. 1–8. Chipping Norton, Australia: Surrey Beatty and Sons.

State of Maryland 1990. *Greenways . . . A Bold Idea For Today, A Promise for Tomorrow.* State of Maryland: Department of Natural Resources.

Taylor, P.D., Fahrig, L., Henein, K., and Merriam, G. 1993. Connectivity as a vital element of landscape structure. *Oikos* 68:571–573.

Therres, G.D., McKegg, J.S., and Miller, R.L. 1988. Maryland's Chesapeake Bay Critical Area Program: implications for wildlife. *Trans. N. Am. Wildl. Nat. Resour. Conf.* 53:391–400.

Trani, M.K., and Giles, R.H. 1999. An analysis of deforestation: metrics used to describe pattern change. *For. Ecol. Manage.* 114:459–470.

UNESCO. 1974. *Task Force on Criteria and Guidelines for the Choice and Establishment of Biosphere Reserves. Man and the Biosphere Report Number 22.* Paris: United Nations Educational, Scientific, and Cultural Organization.

Vitousek, P.M., Mooney, H.A., Lubchenco, J., and Melillo, J. 1997. Human domination of Earth's ecosystems. *Science* 277:494–499.

Watson, J., Lullfitz, W., Sanders, A., and McQuoid, N. 1995. Networks and the Fitzgerald River National Park Biosphere Reserve, Western Australia. In *Nature Conservation 4: The Role of Networks,* eds. D.A. Saunders, J.L. Craig, and E.M. Mattiske, pp. 482–487. Chipping Norton, Australia: Surrey Beatty and Sons.

Wiens, J.A. 1996. Wildlife in patchy environments: metapopulations, mosaics, and management. In *Metapopulations and Wildlife Conservation,* ed. D.R. McCulloch, pp. 53–84. Washington, DC: Island Press.

Wiens, J.A. 1997. Metapopulation dynamics and landscape ecology. In *Metapopulation Biology: Ecology, Genetics, and Evolution,* eds. I.A. Hanski and M.E. Gilpin, pp. 43–62. New York: Academic Press.

With, K.A. 1997. The application of neutral landscape models in conservation biology. *Conserv. Biol.* 11:1069–1080.

With, K.A., and Crist, T.O. 1995. Critical thresholds in species' responses to landscape structure. *Ecology* 76:2446–2459.

10

Landscape Invasibility by Exotic Species

JOHN L. VANKAT AND D. GRAHAM ROY

10.1 Introduction

Species continually invade new areas, where—as exotics—they can impact ecosystem composition and function. The objective of this chapter is to examine concepts and applications of landscape ecology pertinent to the topic of landscape invasibility by exotic species. We begin by reviewing and developing concepts, principles, and emerging ideas. We then describe recent applications of landscape ecology determined by a survey of land managers. With this background, we develop principles for applying landscape ecology. We conclude by identifying theoretical and empirical knowledge gaps and describing possible research approaches to fill the gaps.

10.2 Concepts, Principles, and Emerging Ideas

10.2.1 Threat of Exotics

Exotic species across the world affect the composition and function of ecosystems they invade. The zebra mussel (*Dreissena polymorpha*) and the Amur honeysuckle shrub (*Lonicera maackii*) provide examples of the dramatic impacts of exotic species. The zebra mussel is native to the Caspian region of Central Asia, and in 1988 it was discovered for the first time in North America in Lake St. Clair, near Detroit, Michigan, and Windsor, Ontario (Hebert et al. 1989). In the following eight years, the mussel established in two provinces of Canada and 18 states in the eastern United States (Johnson and Padilla 1996), causing major problems in the Great Lakes and Mississippi River drainage systems. Zebra mussel reproduced in such great numbers that in addition to blocking water intake pipes, it replaced native mussels, thereby altering ecosystem composition. It also filter-fed on plankton so extensively that water became clearer, indicating a potential alteration in ecosystem function through changes in food-chain dynamics (Padilla et al. 1996).

Amur honeysuckle is native to northeastern Asia and was imported to the eastern United States in the late 1800s as an ornamental. It escaped from cultivation in the 1920s and became a prominent invasive in the 1950s (Luken and Thieret 1996). The current range of Amur honeysuckle includes 25 provinces and states in Canada and the United States (Trisel and Gorchov 1994). Dense stands of honeysuckle have developed in disturbed eastern forests where the species competes with native herbs and tree seedlings, altering ecosystem composition and, possibly, ecosystem function in terms of successional dynamics (Hutchinson and Vankat 1997; Collier et al. 2002).

10.2.2 *Stages of Invasion*

Successful invaders such as zebra mussel and Amur honeysuckle pass through four stages of invasion to enter the local species pool: introduction, dispersal, colonization, and establishment (Figure 10.1). Identification of these key stages of invasion is important to the development of principles connecting landscape ecology to landscape invasibility by exotic species.

In general, the process of invasion begins with a few individuals and concludes either when a species dies or becomes abundant (Vermeij 1996; Williamson and Fitter 1996). More specifically, the first stage of invasion is *introduction,* the addition of the exotic species into the regional species pool (Figure 10.1). The second stage is *dispersal,* the movement of individuals (usually few in number) from somewhere within the general region to the specific landscape. The third stage is *colonization,* the founding of a new population in the local landscape as the dispersed individuals grow, reproduce, or both. Colonizing populations are typically small in size and areal extent, as well as restricted to a particular habitat. The

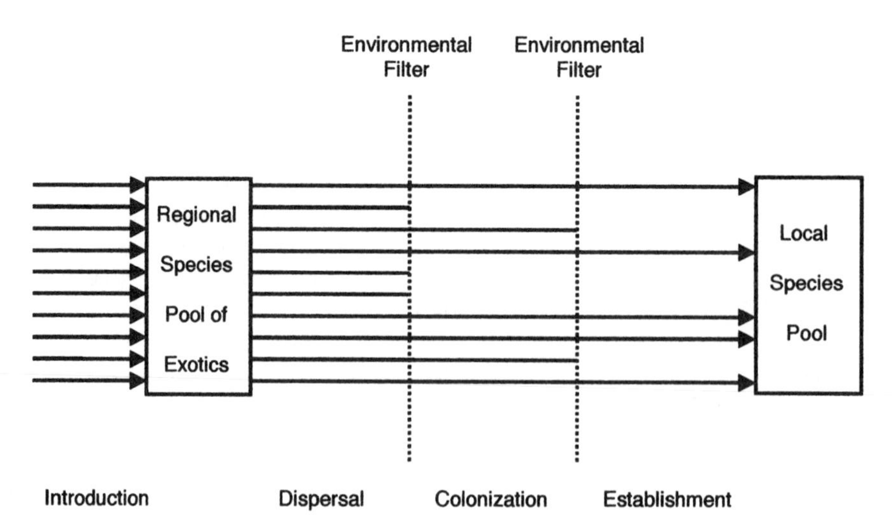

FIGURE 10.1. Model of the addition of exotic species to the local species pool. Some exotic species do not colonize or establish because of unfavorable environmental conditions illustrated as "environmental filters."

fourth stage is *establishment,* the spreading of reproducing populations into other areas of the landscape, often including other habitats. Of course, not all exotic species successfully pass through all four stages of invasion to become established in the local species pool, as unfavorable environmental conditions often combine with small population size to restrict exotics from passing beyond the second or third stages of invasion (cf., Mack 1996; Ewel et al. 1999).

It is important to recognize that this model of the invasion process involves different scales. As illustrated in Figure 10.1, the introduction stage of invasion involves arrival of species at a regional scale from the broader continental scale, and dispersal moves species from the regional scale to the local landscape scale where colonization and establishment occur. Our model also applies to broader scales. For example, introduction can involve arrival of species at a continental scale from the broader global scale, and dispersal would be movement from the continental scale to a regional scale where colonization and establishment occur. Similarly, our model can be applied at finer scales, such as individual patches.

10.2.3 The Role of Landscape Factors in Invasion

Few studies have addressed the invasion of exotic species at the landscape scale; nevertheless, these studies and anecdotal evidence indicate that landscape factors are important in invasion. For example, the zebra mussel rapidly spread through river and canal corridors in the Great Lakes and Mississippi River drainage systems, but it has only slowly invaded isolated inland lakes (Schneider et al. 1998). Amur honeysuckle, which colonizes forests (Hutchinson and Vankat 1997), spread more extensively through a landscape with abundant, well-connected forests than it did through an agricultural landscape with isolated forest patches (Hutchinson and Vankat 1998). There is a substantial need for additional studies to clarify the role of landscape factors in landscape invasibility (Ewel et al. 1999); however, some patterns of invasibility are evident from initial studies and observations. These can be summarized as three principles describing landscape structural elements most likely to affect the dispersal, colonization, and establishment stages of invasion of exotic species:

Corridors are routes for dispersal of exotic species.
Corridors and patches are sites of colonization of exotic species.
Corridors and patches invaded by exotic species facilitate the establishment of those species elsewhere in landscapes.

These principles presumably apply to any scale at which invasion occurs.

Corridors Are Routes for Dispersal of Exotic Species.

Not only is landscape ecology theory congruent with this principle (see proposal that the density of corridors influences the invasibility of the landscape, Forman 1995), but there also is direct and indirect empirical evidence. Direct evidence includes the findings that cars entering Kakadu National Park (Australia) along road corridors carried an average of 2.5 exotic plant seeds (Lonsdale and Lane 1994) and that horses entering Rocky Mountain National Park (Colorado) and using backcountry trail corridors deposited fecal material containing exotic species (Benninger 1989). Another example is the movement of zebra mussels along eastern North American river corridors, both as adults attached to boat hulls and as juveniles carried by river currents (Johnson and Padilla 1996), a dispersal mechanism used by exotic plants as well (Pyšek and Prach 1993; DeFerrari and Naiman 1994; Levine 2000). See Trombulak and Frissell (2000) for a review of the ecological effects of roads, including the spread of exotic species. Indirect evidence includes the presence of isolated populations of exotic species along corridors and in sites connected by corridors but otherwise isolated. For example, exotic species occurred along backcountry trail corridors in Rocky Mountain National Park (Benninger-Truax et al. 1992), in backcountry campsites connected only by trail corridors in the Selway-Bitterroot Wilderness (Idaho; Marcus et al. 1998), and along roads and streams in the Cascade Range of Oregon (Parendes and Jones 2000). Additional indirect evidence is that the bird-dispersed Amur honeysuckle spread more extensively where its forest habitat was connected, such as along stream corridors, than it did where forests were isolated (Hutchinson and Vankat 1998). The fact that the spread of Amur honeysuckle also was correlated with total area of forest habitat indicates that continuous habitat can mimic corridors as dispersal routes of exotic species.

Corridors and Patches Are Sites of Colonization of Exotic Species.

Corridors tend to be sites of colonization because (as routes of dispersal) they have propagules of exotic species, and because they often have disturbed habitat that favors exotics due to reduced competition from native species. Trail corridors in Rocky Mountain National Park and trail and road corridors in Glacier National Park (Montana) have been documented as sites of colonization of exotic species (Benninger-Truax et al. 1992; Tyser and Worley 1992). Also, watersheds in Washington have a higher percentage of exotic plants in riparian corridors than they do in upland areas (DeFerarri and Naiman 1994).

Patches can be sites of colonization of exotic species, not because of high propagule availability but because many landscape patches are formed by disturbance. In Washington, disturbed patches such as clear-cuts are more invasible than are undisturbed patches (DeFerrari and Naiman 1994). Also, the number of exotics in forests generally decreases with increasing time since disturbance (Hobbs and Atkins 1990; DeFerrari and Naiman 1994). Moreover, an experimental study involving gradients of disturbance and nutrient levels supported the hypothesis that disturbance favors exotic species—results attributed to exotics having faster growth rates than do native species (Burke and Grime 1996). Short generation times also can favor exotic species in sites that experience repeated disturbance, be they patches or corridors (Hobbs and Atkins 1990).

Human-dominated elements in the landscape are particularly susceptible to colonization by exotic species, as indicated by the presence of more exotic plants and animals in urban and agricultural landscapes than in more natural landscapes (Cowie and Werner 1993; Fensham and Cowie 1998). Humans inadvertently encourage exotics in at least three ways: by introducing species into the regional species pool, by dispersing individuals into specific landscapes, and by enhancing colonization and establishment with disturbance.

Corridors and Patches Invaded by Exotic Species Facilitate the Establishment of Those Species Elsewhere in Landscapes.

Because corridors and disturbance patches are highly susceptible to colonization, they can function as staging areas for the establishment of exotic species elsewhere in landscapes. Aerial photographs in Australia revealed that the exotic shrub *Acacia nilotica* initially invaded riparian areas disturbed by cattle and then increased in adjacent upland areas (Brown and Carter 1998). Also, this means of establishment may be inferred from a general pattern of exotic species being abundant near corridors and present in smaller numbers away from corridors, as observed for roads in Australia (Amor and Stevens 1976), Glacier National Park (Tyser and Worley 1992), and Massachusetts (Forman and Deblinger 2000), as well as for trails in Rocky Mountain National Park (Benninger-Truax et al. 1992). Presumably, disturbance patches would function similarly to corridors.

Edge environments adjacent to corridors and disturbance patches may be particularly important in the establishment of exotic species away from areas of initial colonization. Indeed, it is likely that edges function as corridors—at least they appear susceptible to colonization. For example, Argentine Ants (*Linepithema humile*) and exotic plants colonized edges of fragmented vegetation in San Diego, California (Suarez et al. 1998). Also, a general pattern of more exotics in patch edges than in interiors suggests that colonization begins at edges and is followed by establishment in interiors. For example, patches of deciduous forest in Indiana had a much higher frequency of exotic plants at edges than in interiors (Brothers and Spingarn 1992).

10.3 Recent Applications

Our search of the primary ecological literature and published government reports revealed little description of recent applications of landscape ecology principles to reduce invasibility by exotic species. Therefore, we decided to survey land managers about their current activities. In addition to seeking information on specific applications, we designed our survey to reflect general patterns. Although various limitations prevented a statistically valid survey, we suggest that our findings are representative of recent applications.

In designing our survey, we thought that the responses of land managers to our survey could be influenced by several factors. One of these was scale of the management unit. Therefore, we surveyed managers responsible for multiunit systems (regional scale) and managers responsible for large single units (landscape scale). We requested that multiunit managers restrict their responses to the scale of multiple units and not to consider single units within their system. A second possible factor was difference in land management objectives. Therefore, we surveyed managers responsible for nature preserves (e.g., national parks) and others responsible for multiple-use reserves (i.e., national forests and national grasslands). A third possible factor was biogeographic region, so we surveyed land managers from the eastern, central, and western United States (corresponding to regions of temperate deciduous forest, grassland, and coniferous forest biomes, respectively).

For each of the three biogeographic regions, we surveyed a multiunit manager from the U.S. National Park Service, the U.S. Forest Service, and The Nature Conservancy. We selected specific multiunit systems as representative of the region's primary biome. Also, for each biogeographic region, we surveyed three single-unit managers from the U.S. National Park Service and three from the U.S. Forest Service. The single units were selected for geographic breadth within each region, and to represent the region's primary biome. In total, we surveyed 27 managers. In each case, the specific individual interviewed was selected by telephoning the management unit and asking for the person most knowledgeable about the management of exotic species. Although we indicated interest in plants, animals, and microbial pathogens, most managers focused on plants.

We conducted the survey in Spring 1999. We used the telephone rather than the mail to ensure (1) the focus of the survey remained on invasibility by new exotics rather than on control of established exotics, (2) 100% response, thereby enabling us to compare management scales, objectives, and regions, and (3) thorough elaboration of questions and answers. We asked each land manager six primary questions (all but question 5 included a numeric rating scale):

1. Are exotic species a current problem in your management area?
2. Are exotic species a potential problem in your management area?
3. Have you used concepts of landscape ecology (such as the patch–corridor–matrix paradigm) to reduce the invasibility of your management area by exotic species?
4. Have these efforts been successful, and what scientific or ecological factors affected successes and failures?
5. What information would be most useful to you with regard to invasibility of your management area by exotic species?
6. What is your level of knowledge of landscape ecology, and how did you acquire it?

10.3.1 Are Exotic Species a Current and Potential Problem?

Before asking about applications of landscape ecology, we thought it was important to determine land managers' level of concern. When asked if exotics were a current problem in their management area, using a rating scale of 1 to 5 (very small to very large problem), responses averaged 3.7, i.e., close to "large problem" (Table 10.1).

TABLE 10.1. Responses (means and ranges) of United States land managers to our survey on landscape invasibility by exotic species. Evaluation scales ranged from 1 to 5, where 1 = very small or very infrequently, and 5 = very large or very frequently.

		Number of units		Type of unit		Region of U.S.		
	Total	Multi	Single	Preserve	Reserve	East	Central	West
Are exotic species a current problem in your management area?								
Mean	3.7	4.1	3.5	4.1	3.2	3.4	3.6	4.1
Range	1–5	2.5–5	1–5	2.5–5	1–5	1–5	2–5	3–5
n	27	9	18	15	12	9	9	9
Are exotic species a potential problem in your management area?								
Mean	4.2	4.6	4.0	4.8	3.5	4.1	3.9	4.7
Range	2–5	4–5	2–5	4–5	2–5	2–5	2–5	4–5
n	27	9	18	15	12	9	9	9
Have you used concepts of landscape ecology to reduce the invasibility of your management area by exotic species?								
Mean	2.3	1.8	2.6	2.5	2.1	1.8	2.4	2.7
Range	1–5	1–3	1–5	1–5	1–4	1–4	1–5	1–4
n	27	9	18	15	12	9	9	9
Have these efforts been successful?								
Mean	2.9	1.0	3.1	2.8	3.0	3.0	2.8	3.2
Range	1–4	1	2–4	1–4	2–4	3	1–4	2.5–4
n	10	1	9	6	4	1	6	3
What is your level of knowledge of landscape ecology?								
Mean	3.0	2.7	3.2	2.9	3.2	2.9	3.1	3.0
Range	1–5	1–5	2–5	1–4	2–5	1–5	2–4	1–5
n	27	9	18	15	12	9	9	9

When asked whether exotics were a potential problem, the mean increased to 4.2. Of the 19 managers who rated the current problem at less than the maximum, 14 (74%) expected problems with exotics to increase, four (21%) expected problems to remain constant, and one (5%) expected exotics would be less of a problem in the future. Of the eight managers who rated the current problem at the maximum, all expected the problem to remain at that level. Therefore, managers not only consider exotic species a large current problem, but nearly all also expect the problem to increase or remain very large in the future.

The responses of land managers appeared to be influenced by several factors. First, their concern about exotic species may have been heightened by an Executive Order by the President of the United States issued on 3 February 1999, just weeks before our survey. The order mandated prevention of invasion by new exotic species of federal lands such as national parks and national forests. Second, multiunit managers viewed exotics as a greater problem than did single-unit managers (4.1 vs. 3.5 currently and 4.6 vs. 4.0 potentially; Table 10.1). Third, managers of preserves rated exotics as more of a problem than did managers of reserves (4.1 vs. 3.2 currently and 4.8 vs. 3.5 potentially). Fourth, there was greater concern for exotics in the west than there was in the central and east regions (4.1 vs. 3.6 and 3.4 currently, and 4.7 vs. 3.9 and 4.1 potentially).

10.3.2 *Have You Applied Landscape Ecology Concepts?*

In sharp contrast to their high concern about exotic species, land managers indicated relatively little use of landscape ecology to reduce the invasibility of their management area. Rated on a scale of 1 to 5 (very infrequent to very frequent use), responses averaged 2.3, i.e., somewhat more than "infrequent" (Table 10.1). Several factors appeared to influence the ratings. For example, greater application of landscape ecology was reported by managers of single units than of multiple units (2.6 vs. 1.8), a difference likely reflective of the less complex landscape settings of single units. Second, somewhat greater application of landscape ecology was reported by managers of preserves than of reserves (2.5 vs. 2.1), paralleling greater concern about exotics among preserve managers. Third, greater use of landscape ecology was reported by managers in the west and central regions than by those in the east (2.7 and 2.4 vs. 1.8). One eastern manager stated that greater funding was available for management of exotics in the west and central regions where more land is in reserves and preserves and where exotic plants have long been a concern (especially on rangeland). Despite infrequent application of landscape ecology concepts, several managers, including nearly all responsible for multiple units, reported interest or plans to proceed with applying such concepts. Nearly always, these managers mentioned the need to change policy and planning.

Many managers recognized the need to develop policy and planning that no longer treated management units as isolated from the rest of the landscape. Managers spoke for cooperation with other agencies and private landowners in their landscapes. For example, at the time of the survey, the Southeast Region of the U.S. National Park Service was in initial stages of interagency cooperation to deal with exotics. Also, most single-unit managers reported examples of landscape-scale cooperation, including interagency identification of potential invasives, coordination with private landowners to control populations of exotics outside the management unit, and coordinated trapping of insect pests (lack of cooperation was a special problem reported by units that enclosed considerable private land owned by many individuals). Land managers also reported development of management plans applicable to all units within a region, a step toward homogeneity of objectives and application of management practices across a regional landscape. For example, all national forests in Colorado and Wyoming now require that hay for horses be free of weedy plants (because horses are thought to be a common seed dispersal vector along road and trail corridors).

Another policy and planning matter mentioned by several managers was switching from a policy of reactive management focusing on selected species to a policy of proactive management likely to reduce exotics in general. These managers expressed interest that landscape ecology could provide the foundation and framework for proactive management. For example, some envisioned applying landscape ecology concepts when planning new units or purchasing lands surrounding old units. Specifically, a land manager for The Nature Conservancy noted the possibilities of purchasing strategic buffer zones around preserves, minimizing preserve connectivity to increase isolation, and minimizing preserve edge length to reduce invasions beginning in edge habitat. Presumably, such ideas also can be applied when portions of units are designated for unique management, such as wilderness areas within national forests and national parks.

Many managers, especially those responsible for single units, reported going beyond policy and planning to include actual field application of concepts of landscape ecology in their land management. We describe these applications as they fit into our four-stage invasion process of introduction, dispersal, colonization, and establishment (Figure 10.1). The first of these, the introduction of the species into the regional species pool, has been examined by inventorying and mapping the regional distributions of exotic species to help managers identify exotics most likely to invade their unit.

Dispersal, the second stage of invasion, has been examined by mapping at finer scales. Detailed maps have assisted in correlating exotics with specific landscape features such as road corridors. In fact, some managers limit mapping to road and river corridors that are likely sites of dispersal and colonization. Several single-unit managers reported educating employees to increase their effectiveness in reporting the dispersal of exotics and in applying regulations designed to limit dispersal. For example, managers in Yellowstone National Park (Wyoming) trained about 150 employees, including staff who are stationed at gates where roads enter the Park, and law enforcement personnel who patrol roads within the Park. Managers in Boise National Forest (Idaho) provided weed identification training to construction crews and other employees whose jobs entail much road travel. Other activities to limit dispersal included pressure or stream cleaning of heavy equipment used for construction, logging, and so on, including privately owned equipment brought from outside the unit. Another regulation required road-fill materials to be certified as weed-free before being brought into the unit. In Mammoth Cave National Park (Kentucky), boats brought in by trailers are feared as possible vectors for dispersal of zebra mussels; however, no regulations are in effect because rivers flowing through the Park also are likely to disperse this exotic. Also, off-road vehicles are thought to be a likely dispersal vector for exotic plants to enter Daniel Boone National Forest (Kentucky), so use of these vehicles is now more restricted. Regulations regarding livestock as possible dispersal vectors are especially common. Many managers reported a requirement for weed-free feed. Others reported regulations that prohibit horses from roaming freely or that require the removal of manure. Boise National Forest has a regulation mandating a three-day quarantine of livestock brought from private land unless the land is known to be free of noxious weeds. Yellowstone National Park requires that horse trailers be examined and cleaned if necessary, and that trucks hauling hay must be certified as weed-free or be covered as they pass through the Park.

Colonization, the third stage of invasion, also has been the target of applications involving concepts of landscape ecology. Landscape-scale maps of exotic species have aided in determining the affinity of exotics for certain patch types. For example, managers at Glacier National Park have determined that prairie patches are more invasible than is the matrix of closed-canopy coniferous forest, and they therefore focus monitoring on these open patches. Often, sites of colonization are areas of disturbance. Therefore, several managers reported attempts to minimize the extent of disturbances and to monitor and control disturbed areas, as well as to accelerate succession on such sites. For example, in Thunder Basin National Grassland (Wyoming), areas of mine reclamation and other disturbances are replanted using native species in an attempt to reduce colonization by exotics. Similarly, Badlands National Park (South Dakota) uses a seed mix composed of native species for areas of road projects. Most of the land in Commanche National Grassland (Colorado) had been tilled prior to the establishment of the national grassland, so management focuses on increasing the cover of native herbs to reduce soil erosion that otherwise facilitates colonization by exotics. Disturbance and subsequent species colonization along road corridors have been such a concern in Daniel Boone National Forest that herbicides are applied along roadsides.

Other applications involving concepts of landscape ecology have focused on the establishment stage of species invasion by attempting to control new populations before they establish elsewhere in a management unit. Of course, ongoing monitoring of the landscape, or at least of likely dispersal routes and colonization sites, is essential for this approach to be successful. In Yellowstone National Park, all roads are walked at least once a year, a river corridor is surveyed by boat annually, and stream corridors are monitored when possible. With effective monitoring (which sometimes involves production of fine-scale maps), new outlier populations of exotics may be kept isolated and controlled at manageable levels, if not eradicated, before they establish elsewhere in the landscape. In Great Smokey Mountains National Park (Tennessee and North Carolina), a contract with a private company for road work required the company to return and remove exotic species should any appear after the work was finished. Approaches to prevent the establishment of newly colonized populations have included various methods used to control established populations, although these methods are likely to be applied more vigorously to new populations because they frequently are small in size and areal extent. In addition, some units manage activities within patches of known infestations in an attempt to reduce establishment of the exotic species elsewhere in the landscape. For example, livestock grazing is used to control newly established populations of exotic plants in Pawnee National Grassland (Colorado) and Mt. Hood National Forest, a particularly suitable approach because grazing is a normal land use in these reserves.

10.3.3 Have These Efforts Been Successful?

When asked to rate the frequency of success in applying concepts of landscape ecology, using a scale of 1 to 5 (very infrequently to very frequently successful), 17 of the 27 managers (63%) stated that they could not respond. The responses of the other 10 managers averaged 2.9, i.e., "intermediate" success (Table 10.1). Many managers stated that they did not respond or they chose a low rating because it was too early to determine whether their efforts had been successful.

Some also stated that they were unable to monitor applications thoroughly, properly enforce regulations, or both. Others commented that success at reducing invasibility was not as apparent as failure, but that the current situation would have been worse without attempts to reduce invasibility. Responses suggested a large disparity in success between managers of multiple versus single units. Only one of nine (11%) multiunit managers chose to rate their success, and that rating was a 1. This contrasts with responses from nine of 18 (50%) single-unit managers, whose ratings averaged 3.1. No major differences were found between managers of preserves versus multiple-use reserves (2.8 vs. 3.0), and differences among biogeographic regions were small except for the central versus the west (2.8 vs. 3.2, with 3.0 for the east).

Given that most respondents felt it was too soon to evaluate applications, the descriptions of successes and failures were limited. Some successes were attributed to good scientific information about the introduction stage of invasion, particularly information on which exotic species were in the regional species pool and the species' preferred habitat for colonization. This background information allowed for better control of the dispersal, colonization, and establishment stages of invasion. More specifically, ongoing monitoring coupled with early control of colonizing populations was most frequently mentioned as successful. For example, a manager from Yellowstone National Park reported success due to monitoring the introduction, dispersal, and colonization stages through partnerships with other agencies, Park staff assigned to the task, and education of other Park personnel, followed by early control of colonizing populations. Knowledge about specific ecological conditions pertinent to colonization was mentioned by several managers as important to success. For example, the dry climate in national grasslands of the central United States was reported to limit exotics to riparian areas, enabling monitoring and control of colonizing populations to be focused on only a relatively small portion of the landscape. Also, managers in Thunder Basin National Grassland determined the composition of natural communities on sites where disturbance was planned and then used this information in revegetation that successfully limited the number of colonizing exotic species. Other managers pointed out the importance of limiting the seed bank of exotics in sites scheduled for revegetation.

Only four managers provided insight on failures. Comments addressed policy and planning issues such as lack of a programmatic approach, insufficient funding, poor cooperation from personnel who did not recognize the impact of their activities on exotic species, as well as other broad issues such as insufficient expertise. In addition, one manager pointed out that exotic species that disperse and colonize along road and trail corridors are relatively easy to control in comparison to exotics that directly disperse into and colonize remote regions.

10.3.4 What Information Is Needed?

When asked what information would be most useful, multiunit managers focused on broad issues, whereas single-unit managers tended to focus on specific questions about specific species. Nevertheless, the responses overlapped. For example, in the area of policy and planning, multiunit and single-unit managers alike expressed the need for regional or national cooperation in communicating about exotic species, including sharing information on potential new invasives, current distributions of invasives, monitoring programs, and control methods. One manager also requested communication methods designed to gain cooperation of visitors and adjacent landowners. Other managers requested greater communication with scientists to better transfer basic research to applications, especially in applying concepts of landscape ecology to the management of exotic species. Others noted the need to have background information on exotic species to prioritize use of limited funds on species most likely to be invasive, have large impact, and be controlled. Similarly, a multiunit manager requested background information on each unit in the management area to prioritize efforts on those units most likely to be invaded and damaged.

Other needs were related to field applications. In terms of the first stage of the invasion process, introduction, a common request was for information on which exotic species posed potential threats. Managers wanted to know the current distributions of exotic species (including fine-scale distributions, not just general range maps) as well as potential ranges. Nearly all managers also noted the need for additional background information on the life histories of potential exotic species. With regard to the dispersal stage of invasion, managers requested background information on dispersal vectors of potential invasives, as well as patterns and rates of spread. In terms of the colonization stage, managers needed background information on habitat requirements and patterns of colonization of potential invasives. They also requested background information on land management practices and environmental conditions (especially as related to fire) that promote exotic species. Some managers mentioned the need for information on effective means of monitoring. Moreover, nearly all managers were interested in having more information on the control of exotic species, especially methods that worked well on small, colonizing populations. Of special interest were the use of biocontrol and the cost and time effectiveness of control methods. In terms of the establishment stage of invasion, managers requested background information on patterns of establishment, population dynamics, and impacts of potential invasives. Some managers requested information on means to slow the spread of exotics even when control was impossible. Some multiunit managers also mentioned the need for predictive models covering the entire invasion process from introduction to establishment.

No managers mentioned a need to have more information on the general environment within their management unit. This may not be surprising for single-unit managers because the environment of most national parks, national forests, and other single units is relatively well-known. However, it is surprising that multiunit managers—who deal with more complex, regional landscapes—did not express this need; perhaps this is related to the relatively low level of application of landscape ecology at this scale.

10.3.5 Knowledge of Landscape Ecology

An important question regarding applications is whether land managers have much knowledge of landscape ecology. When asked to evaluate their knowledge on a 1 to 5 scale (very little to very great), the average was 3.0, i.e., "intermediate" (Table 10.1). Managers of single units claimed greater knowledge than did managers of multiunit systems (3.2 vs. 2.7). Less difference was found between managers of reserves versus preserves (3.2 vs. 2.9), and regional differences were minor. In regard to sources of their knowledge of landscape ecology, eight managers (29%) indicated exposure to the field in their academic training. In addition, all but the two managers who claimed very little knowledge (i.e., 93%) acquired knowledge from various combinations of readings, workshops, training sessions, professional meetings, colleagues, and work experience.

10.4 Principles for Applying Landscape Ecology

Based primarily on our survey of land managers, we propose five action principles for reducing landscape invasibility by exotic species: (1) acquire—and share—background information; (2) monitor, monitor, monitor; (3) control dispersal vectors and routes; (4) reduce invasible habitat; and (5) eradicate colonizing populations. At this time, these should be considered to be emerging rather than well-established principles; the literature is sparse, and the number of successful field applications of landscape ecology reported by land managers is small. We present these action principles in Figure 10.2, superimposed on our general diagram of invasion (Figure 10.1). Our model is conceptually based on an assembly rules approach to modeling community composition (cf., Weiher and Keddy 1999).

10.4.1 Acquire—and Share—Background Information

It is important to know which taxa are invasive or potentially invasive and to have background information on their life histories—including patterns of reproduction, population demographics, vectors and pathways of dispersal, habitat preferences, and relationship with disturbance—as well as means of control. It also is important to have background information on the environment of the management unit, including its landscape structure (i.e., distribution of corridors and patches). Information on locations where disturbance promotes invasibility is essential. Although the environment is site-specific, species are not. Therefore, information on current and potential invasive species should be shared through regional (or broader) repositories established through partnerships among governmental agencies, private organizations, and possibly academic institutions.

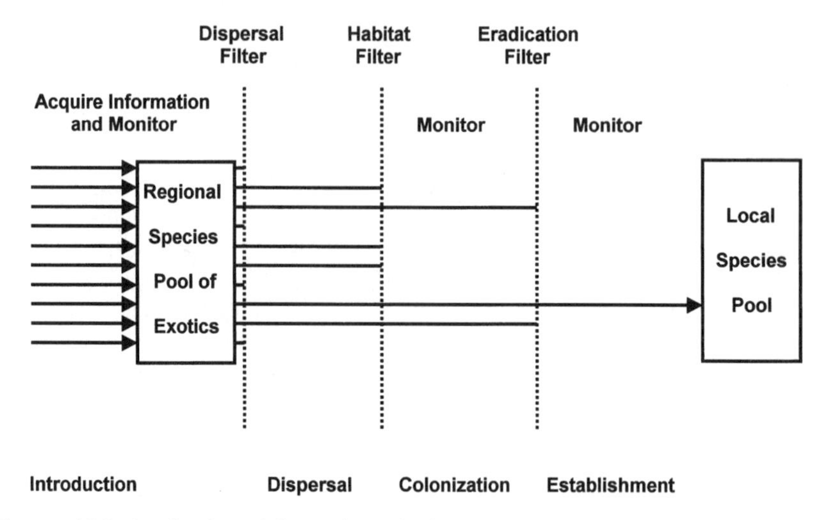

FIGURE 10.2. Application of five action principles to the model of the addition of exotic species to the local species pool (Figure 10.1). The principles are: acquire—and share—background information; monitor, monitor, monitor; control dispersal vectors and routes; reduce invasible habitat; and eradicate colonizing populations. The "habitat" and "eradication" filters supplement (and are superimposed over) the environmental filters shown in Figure 10.1.

10.4.2 Monitor, Monitor, Monitor

Monitoring is suggested at three different stages of the invasion process (Figure 10.2). First, the regional species pool of exotics must be monitored to know which species are potentially invasive. Second, habitat must be monitored to identify colonizations. Third, the effects of eradication efforts must be monitored to avoid establishments. Thorough monitoring may be difficult and expensive, but even limited monitoring can be very useful, especially if focused on landscape features likely to have invasives, such as corridors, edges, and disturbance patches. Monitoring staff can be augmented at minimal cost by educating other employees (e.g., road maintenance workers) and the public.

10.4.3 Control Dispersal Vectors and Routes

Effective reduction of landscape invasibility involves control of dispersal vectors and dispersal routes. Such efforts are illustrated in Figure 10.2 as a "dispersal filter" that prevents some exotics from dispersing. Although some vectors such as wind and native animals may be impossible to control, human dispersal along human-produced dispersal routes such as road corridors is more manageable. Methods of control that focus on dispersal vectors include cleaning vehicles and other equipment, as well as using weed-free road fill and weed-free livestock feed. Methods that focus on dispersal routes include reducing corridors such as roads, horse and off-road-vehicle trails, and power- and pipe-line corridors, especially in areas that are otherwise isolated.

10.4.4 Reduce Invasible Habitat

The most complex of our action principles also has the greatest potential for applying concepts of landscape ecology. Efforts to reduce invasible habitat are illustrated in Figure 10.2 as a "habitat filter" (which supplements an environmental filter), preventing some exotics from colonizing. Many invasive species are initially most successful in disturbed areas. The obvious recommendation is to minimize human disturbance. Managers of nature preserves, where logging, livestock grazing, and other human disturbances normally are not allowed, can follow this recommendation relatively easily because most human disturbances can be relegated to road corridors and patches with visitor facilities. Therefore, managers of preserves may be able to develop a fine-gauge habitat filter that effectively eliminates many exotic species. In contrast, managers of reserves, where logging, grazing, and so on, are allowed, must deal with widespread, ongoing disturbances that can favor exotic species. Therefore, managers of reserves may be able to develop only a relatively coarse-gauge habitat filter that eliminates relatively few species.

The effectiveness of a habitat filter, at least those associated with planned human disturbances (e.g., logging), can be maximized by several steps. First, sites can be selected for isolation or habitat conditions that will result in relatively low invasibility by exotic species after disturbance. Second, disturbance can be designed or managed to minimize landscape invasibility, such as by keeping disturbance intensity low, reducing disturbance duration, and restricting the length and width of edge habitat favored by exotic species. Third, disturbance sites can be kept as isolated as possible by, for example, eliminating road corridors connecting them to the rest of the landscape. Fourth, strategies such as accelerating succession via revegetation with native species can be developed that minimize the time period disturbed sites remain highly invasible.

Although such steps may be effective where human disturbances are planned as part of land management, unplanned human disturbances can be a larger problem. Moreover, natural disturbances that promote invasion by exotic species pose a special problem. A strategy of reducing natural disturbance can be counterproductive in that it may lead to increased disturbance later (e.g., suppression of naturally occurring low-intensity fires has increased probabilities of high-intensity fires in many areas of western North America). Furthermore, reducing natural disturbance is inappropriate for nature preserves because it alters natural processes. Instead of reducing natural disturbance, a more productive and appropriate strategy would be to focus on the other four action principles we propose.

10.4.5 Eradicate Colonizing Populations

Implementation of dispersal and habitat filters is unlikely to prevent all invasive species from initiating the colonization stage of invasion. Fortunately, colonizing populations are generally small in size and areal extent, in contrast with established species. Therefore, eradication of invasives at the colonization stage of invasion is likely to be less difficult, less expensive, less damaging to the environment, and more effective than after the establishment stage. Such efforts are represented in Figure 10.2 as an "eradication filter" that supplements an environmental filter and prevents some species from establishing.

10.5 Knowledge Gaps

The low level of application of landscape ecology concepts to reduce invasibility by exotic species (see Section 10.3.2) results from lack of basic knowledge rather than lack of transfer of basic knowledge. In this section, we identify the most important knowledge gaps. These gaps restrict development of predictive capabilities and limit understanding of underlying biological mechanisms. As a consequence, these gaps severely constrain the use of landscape ecology concepts in controlling invasive species.

10.5.1 Theoretical Voids

A primary reason for little application of landscape ecology is the lack of unifying theory relating invasions to landscape conditions. No broad-based theory incorporates landscape principles to explain how the structure of the landscape, with its network of corridors and patches, influences the distribution and success of exotic species. This void limits understanding of how landscapes affect invasions, precludes an organizing framework for empirical research on invasive species and landscapes, and prevents land managers from developing predictive tools to reduce landscape invasibility.

10.5.2 Empirical Voids

Applications of landscape ecology to control landscape invasibility are limited not only by the lack of unifying theory, but also by the paucity of empirical studies that have examined invasions of exotic species at the landscape scale. With regard to the introduction stage of invasion, empirical research on life-history characteristics of individual invasive species is too limited. We need to know more thoroughly the suites of characteristics that predispose species to be invasive, and whether invasive characteristics are similar or different among taxonomic groups. A greater void is that past empirical research on life histories of exotic species largely has failed to consider how traits such as reproductive ability, dispersal methods, and habitat preferences interact with landscape conditions (Lodge 1993).

Even less is known about how exotic species, including prominent invasives, move through landscapes (i.e., disperse, colonize, and establish). Unfortunately, most studies of invasions of exotics have focused on generalized descriptions of range expansions at the scale of large geographic regions rather than of landscapes (cf., Mack 1986; Frean et al. 1997; Richardson 1998). Moreover, many such studies have been based on historical records rather than on field observations of the actual invasion. Such research has value at continental or subcontinental scales, but land managers need to understand movement of exotic species at regional and landscape scales. Few empirical studies indicate how landscape features such as corridors and patches serve as dispersal routes, provide areas of colonization, and promote establishment (Brown and Carter 1998), or how managers can apply this information to reduce landscape invasibility. For example, little is known about establishment rates away from colonized corridors and patches and what factors are most effective in reducing those rates. Although many land managers attempt to reduce invasions by exotic species, little evidence verifies the effectiveness of these efforts. In addition, empirical research is needed on whether edges function as corridors, on how to reduce and isolate invasible habitat, and on the relationship between natural disturbance and invasibility. Methods to eradicate established populations need to be investigated for their applicability to colonizing populations.

10.6 Research Approaches

10.6.1 Approaches for Theoretical Research

Simulation modeling is needed for development of unifying theory that explains and predicts invasions. Model development needs to integrate species' life-history characteristics, including population demographics and dispersal modes, with landscape characteristics such as connectivity, corridor and edge density, and distribution and extent of suitable habitat. Models (and the theory they help develop) should be applicable from local to continental scales, as well as across a range of organisms from animals to plants to microbes. Our linkage in this chapter—of individual stages of invasion, specific landscape factors, and action principles for applying concepts of landscape ecology—represents a step in the development and application of such models. Further development may usefully incorporate established ecological models such as island biogeography theory. Of course, well-designed empirical studies will be critical for verification of newly developed models.

10.6.2 Approaches for Empirical Research

We recommend a two-step process for studies of life-history characteristics related to the introduction stage of invasion. First, the need for greater assessment of characteristics that predispose species to be invasive can be addressed by surveying the life-history characteristics of known invasive species (especially in comparison to exotic species that are not invasive; Ewel et al. 1999). Such surveys should examine large groups of organisms of broadly similar form, function, or both (e.g., short-lived herbaceous plants) with the objective of developing a profile of life-history characteristics associated with invasiveness, which would aid in predicting which species are likely to be invasive (cf., Rejmánek and Richardson 1996; Reichard and Hamilton 1997). Groups that prove largely similar should be combined, and target species should be selected. This should be followed by a second type of study, one that examines how key life-history characteristics of these target species interact with landscape features to determine species' success in landscapes. Such research should adapt protocols used in basic life-history studies to fit a landscape perspective. For example, studies of habitat requirements must determine not only what habitats are most suited to the species, but also examine how the abundance and configuration of these habitats in the landscape influence species' success (cf., Hutchinson and Vankat 1998). The selection of which landscape features to examine can be aided by additional observational studies correlating the abundance and distribution of exotic species with various landscape features (cf., Cowie and Werner 1993; DeFerrari and Naiman 1994; Gilfedder and Kirkpatrick 1998).

For understanding species' movement through landscapes—the dispersal, colonization, and establishment stages of invasion—we recommend observational studies of the movement of exotic species as invasion actually occurs, presumably at the front of species' range expansions. Individual populations of target species should be systematically monitored at the scale of single patches, with data compiled to produce maps at the landscape scale. This would enable determination of dispersal rates, distances, and patterns (i.e., spatial and temporal patterns of colonization and establishment). Data should be entered into a geographic information system (GIS) database and related to spatial data on landscape features such as corridor and edge density and human disturbance. Models developed from GIS analysis could describe and predict the movement of species across landscapes and link with data acquired at other scales, such as species' range expansion at continental and subcontinental scales. Models should be verified with experimental studies, such as research in which landscape invasibility is manipulated by varying such factors as patch isolation, edge habitat, and extent and duration of disturbance. The efforts of land managers to reduce invasion by exotic species provide numerous research opportunities. The efficiency and effectiveness of these reduction efforts can be examined by comparing the presence and abundance of exotic species in managed and unmanaged landscapes.

In conclusion, theoretical and empirical research will provide land managers with scientifically justified management prescriptions that are proactive in preventing exotic species from establishing (cf., Hobbs and Humphries 1995). Regional- or broader-scale repositories of results from empirical studies are needed to enable widespread, efficient use of information. With these repositories, findings from empirical research would be readily available to land managers, facilitating, for example, prioritization of areas and species (cf., Hiebert and Stubbendieck 1993) for application of action principles such as ours (Section 10.4). In addition, increased availability of empirical research will contribute to the development of unifying theory that explains and predicts invasions. Such theory will stimulate additional empirical research and lead to greater use of landscape ecology concepts in efforts to reduce landscape invasibility by exotic species.

Acknowledgments

We thank Drs. Richard T. T. Forman, Ronald D. Hiebert, and Laura F. Huenneke for suggestions on an early draft of this chapter. We also thank the 27 land managers who participated in our survey of recent applications of landscape ecology concepts.

References

Amor, R.L., and Stevens, P.L. 1976. Spread of weeds from a roadside into sclerophyll forests at Dartmouth, Australia. *Weed Research* 16:111–118.

Benninger, M.C. 1989. *Trails as Conduits of Movement for Plant Species in Coniferous Forests of Rocky Mountain National Park, Colorado.* M.S. thesis. Oxford, Ohio: Miami University.

Benninger-Truax, M., Vankat, J.L., and Schaefer, R.L. 1992. Trail corridors as habitat and conduits for movement of plant species in Rocky Mountain National Park, Colorado, USA. *Landsc. Ecol.* 6:269–278.

Brothers, T.S., and Spingarn, A. 1992. Forest fragmentation and alien plant invasion of central Indiana old-growth forests. *Conserv. Biol.* 6:91–100.

Brown, J.R., and Carter, J. 1998. Spatial and temporal patterns of exotic shrub invasion in an Australian tropical grassland. *Landsc. Ecol.* 13:93–102.

Burke, M.J.W., and Grime, J.P. 1996. An experimental study of plant community invasibility. *Ecology* 77:776–790.

Collier, M.H., Vankat, J.L., and Hughes, M.R. 2002. Diminished plant richness and abundance below *Lonicera maackii*, an invasive shrub. *Am. Midl. Nat.* 147: 60–71.

Cowie, I.D., and Werner, P.A. 1993. Alien plant species invasive in Kakadu National Park, tropical northern Australia. *Biol. Conserv.* 63:127–135.

DeFerrari, C.M., and Naiman, R.J. 1994. A multi-scale assessment of the occurrence of exotic plants on the Olympic Peninsula, Washington. *J. Veg. Sci.* 5:247–258.

Ewel, J.J., O'Dowd, D.J., Bergelson, J., Daehler, C.C., D'Antonio, C.M., Gomez, L.D., Gordon, D.R., Hobbs, R.J., Holt, A., Hopper, K.R., Hughes, C.E., LaHart, M., Leakey, R.R.B., Lee, W.G., Loope, L.L., Lorence, D.H., Louda, S.M., Lugo, A.E., McEvoy, P.B., Richardson, D.M., and Vitousek, P.M. 1999. Deliberate introductions of species: research needs. *BioScience* 49:619–630.

Fensham, R.J., and Cowie, I.D. 1998. Alien plant invasions on the Tiwi Islands. Extent, implications and priorities for control. *Biol. Conserv.* 83:55–68.

Forman, R.T.T. 1995. *Land Mosaics: The Ecology of Landscapes and Regions.* Cambridge, United Kingdom: Cambridge University Press.

Forman, R.T.T., and Deblinger, R.D. 2000. The ecological road-effect zone of a Massachusetts (U.S.A.) suburban highway. *Conserv. Biol.* 14:36–46.

Frean, M., Balkwill, K., Gold, C., and Burt, S. 1997. The expanding distributions and invasiveness of *Oenothera* in southern Africa. *South African J. Bot.* 63:449–458.

Gilfedder, L., and Kirkpatrick, J.B. 1998. Factors influencing the integrity of remnant bushland in subhumid Tasmania. *Biol. Conserv.* 84:89–96.

Hebert, P.D.N., Muncaster, B.W., and Mackie, G.L. 1989. Ecological and genetic studies on *Dreissena polymorpha* (Pallas): a new mollusc in the Great Lakes. *Can. J. Fish. Aquat. Sci.* 46:1587–1591.

Hiebert, R.D., and Stubbendieck, J. 1993. *Handbook for Ranking Exotic Plants for Management and Control.* National Park Research Report NPS/NRMWRO/NRR-93/08. Denver, Colorado: U.S. Department of the Interior, National Park Service, Natural Resources Publication Office.

Hobbs, R.J., and Atkins, L. 1990. Fire-related dynamics of a *Banksia* woodland in southwestern Western Australia. *Australian J. Bot.* 38:97–110.

Hobbs, R.J., and Humphries, S.E. 1995. An integrated approach to the ecology and management of plant invasions. *Conserv. Biol.* 9:761–770.

Hutchinson, T.F., and Vankat, J.L. 1997. Invasibility and effects of Amur honeysuckle in southwestern Ohio forests. *Conserv. Biol.* 11:1117–1124.

Hutchinson, T.F., and Vankat, J.L. 1998. Landscape structure and spread of the exotic shrub *Lonicera maackii* (Amur honeysuckle) in southwestern Ohio forests. *Am. Midl. Nat.* 139:383–390.

Johnson, L.E., and Padilla, D.K. 1996. Geographic spread of exotic species: ecological lessons and opportunities from the invasion of the zebra mussel *Dreissena polymorpha. Biol. Conserv.* 78:23–33.

Levine, J.M. 2000. Species diversity and biological invasions: relating local process to community pattern. *Science* 288:852–854.

Lodge, D.M. 1993. Biological invasions: lessons for ecology. *Trends Ecol. Evol.* 8:133–137.

Lonsdale, W.M., and Lane, A.M. 1994. Tourist vehicles as vectors of weed seeds in Kakadu National Park, northern Australia. *Biol. Conserv.* 69:277–283.

Luken, J.O., and Thieret, J.W. 1996. Amur honeysuckle, its fall from grace. *BioScience* 46:18–24.

Mack, R.N. 1986. Alien plant invasion into the Intermountain West: a case history. In *Ecology of Biological Invasions of North America and Hawaii,* eds. H.A. Mooney and J.A. Drake, pp. 191–213. New York: Springer-Verlag.

Mack, R.N. 1996. Biotic barriers to plant naturalization. In *Proceedings of the IX International Symposium on Biological Control of Weeds,* eds. V.C. Moran and J.H. Hoffman, pp. 39–46. Stellenbosch, South Africa: University of Cape Town.

Marcus, W.A., Milner, G., and Maxwell, B. 1998. Spotted knapweed distribution in stock camps and trails of the Selway-Bitterroot Wilderness. *Great Basin Nat.* 58:156–166.

Padilla, D.K., Adolph, S.C., Cottingham, K.L., and Schneider, D.W. 1996. Predicting the consequences of dreissenid mussels on a pelagic food web. *Ecol. Model.* 85:129–144.

Parendes, L.A., and Jones, J.A. 2000. Role of light availability and dispersal in exotic plant invasion along roads and streams in the H.J. Andrews Experimental Forest, Oregon. *Conserv. Biol.* 14:64–75.

Pyšek, P., and Prach, K. 1993. Plant invasions and the role of riparian habitats: a comparison of four species alien to central Europe. *J. Biogeog.* 20:413–420.

Reichard, S.H., and Hamilton, C.W. 1997. Predicting invasions of woody plants introduced into North America. *Conserv. Biol.* 11:193–203.

Rejmánek, M., and Richardson, D.M. 1996. What attributes make some plant species more invasive? *Ecology* 77:1655–1661.

Richardson, D.M. 1998. Forestry trees as invasive aliens. *Conserv. Biol.* 12:18–26.

Schneider, D.W., Ellis, C.D., and Cummings, K.S. 1998. A transportation model assessment of the risk to native mussel communities from zebra mussel spread. *Conserv. Biol.* 12:788–800.

Suarez, A.V., Bolger, D.T., and Case, T.J. 1998. Effects of fragmentation and invasion on native ant communities in coastal southern California. *Ecology* 79:2041–2056.

Trisel, D.E., and D.L. Gorchov. 1994. Regional distribution, ecological impact, and leaf phenology of the invasive shrub, *Lonicera maackii*. *Bull. Ecol. Soc. Am.* 75:231–232.

Trombulak, S.C., and Frissell, C.A. 2000. Review of ecological effects of roads on terrestrial and aquatic communities. *Conserv. Biol.* 14:18–30.

Tyser, R.W., and Worley, C.A. 1992. Alien flora in grasslands adjacent to road and trail corridors in Glacier National Park, Montana (U.S.A.). *Conserv. Biol.* 6:253–262.

Vermeij, G.J. 1996. An agenda for invasion biology. *Biol. Conserv.* 78:3–9.

Weiher, E., and Keddy, P.A., eds. 1999. *Ecological Assembly Rules: Perspectives, Advances, Retreats*. Cambridge, United Kingdom: Cambridge University Press.

Williamson, M., and Fitter, A. 1996. The varying success of invaders. *Ecology* 77:1661–1666.

Section III

Landscape Change

11

Conservation in Human-Altered Landscapes: Introduction to Section III

KEVIN J. GUTZWILLER

Mankind's influence on natural landscapes is readily apparent in most sectors of the globe, and the frequency and spatial intensity of these effects are increasing. Many of the associated changes are detrimental to indigenous species. And many of the impacts occur at rates that overwhelm the scientific expertise and the socioeconomic and political means available to control them. Applying the science of landscape ecology is one strategy by which we may improve scientific approaches for coping with this crisis. But, if we are to be successful in using landscape ecology to help save large parts of living systems, a broader and deeper knowledge base is necessary. Specifically, we must develop a better understanding of the nature and extent of human-induced landscape changes, how landscape alterations affect organism distribution and persistence, how landscape ecology can be used to avoid or minimize detrimental effects, and what can be done to increase such applications of landscape ecology. These are the subjects of Section III.

11.1 Landscape Transformation

Environmental modifications associated with agriculture, urbanization, and timber extraction often degrade or destroy natural landscapes. Such transformations occur through five main spatial processes that have distinct ecological effects (Forman 1995). *Perforation* involves creation of a hole or gap (e.g., clear-cut, housing development) in the original land cover. Subdivision of land cover into sections by equal-width alterations (e.g., roads, powerline rights-of-way) is known as *dissection.* The well-studied process of *fragmentation* entails the splitting apart of large areas of land cover (e.g., contiguous forest) into smaller ones; the resulting parts are separated unevenly and often more widely than are those generated by dissection. *Shrinkage* refers to size reduction of a given land-cover element (corridor or patch). The complete loss of an element, *attrition,* is most common for small patches.

Although these processes overlap in time, perforation and dissection peak during initial stages of land transformation, fragmentation and then shrinkage dominate during intermediate phases, and attrition is prominent at the end when original land cover is almost nonexistent (Forman 1995). One or more of these five processes are involved in the human conversion of terrestrial habitats (addressed in Chapter 12), landscape transformation by roads (considered in Chapter 13), and timber extraction (covered in Chapter 14). Because habitat for native species is frequently lost as a result of such human activities, these three chapters address some of the most important factors affecting biological conservation today.

11.2 Animal Behavior

Habitat fragmentation can affect population size, dynamics, and persistence (Meffe and Carroll 1997). These effects result in part from behavioral responses to altered habitat conditions. Fragmentation can create barriers that limit dispersal or foraging movements, and it can create edge conditions that exacerbate predation or nest parasitism (e.g., Saunders et al. 1987; Robinson et al. 1995). Many species abandon and avoid areas that are subject to these influences. Knowledge about behavioral responses to landscape conditions (e.g., Dale et al. 1994; Lima and Zollner 1996), and development of approaches for applying that information, will improve the success of conservation efforts. Chapter 15 addresses conservation implications of animal behavior in fragmented landscapes.

11.3 Impacts on Aquatic Species

Because terrestrial and aquatic systems are inextricably linked, human alteration of landscapes frequently has impacts on aquatic organisms (e.g., Verhoeven et al. 1993). Often, the effects stem from transported materials (e.g., sediments, pesticides, excess nutrients) and reduction of vegetative cover (e.g., loss of shade over streams, loss of filtering vegetation near wetlands). Human-degraded landscapes can detrimentally affect the physical and chemical components of aquatic ecosystems (discussed in Chapter 16) and, consequently, aquatic biodiversity and biointegrity (examined in Chapter 17). An understanding of these effects and how landscape ecology can be applied to reduce or prevent them is indispensable for the holistic approach needed to conserve both terrestrial and aquatic species.

11.4 Time Lags in Responses to Landscape Change

In many conservation situations, it is valuable to be able to predict the effects of landscape change on organisms. But, because the full consequences of landscape alteration may not become evident for long periods (Noss and Csuti 1997), and interactions between natural patch dynamics and human-induced changes can obscure actual human effects, useful predictions can be difficult or impossible to obtain. Knowledge about the causes, lengths, and consequences of lags in species' responses to landscape change is valuable for realistic prediction of population viability, persistence time, extinction risk, habitat occupancy, and other conservation-related factors. Chapter 18 discusses time lags in metapopulation responses to landscape change. The need to understand this subject and apply associated principles in both terrestrial and aquatic systems is becoming increasingly imperative as unchecked landscape transformation continues to break apart contiguous populations and degrade the viability of naturally occurring metapopulations.

References

Dale, V.H., Pearson, S.M., Offerman, H.L., and O'Neill, R.V. 1994. Relating patterns of land-use change to faunal biodiversity in the central Amazon. *Conserv. Biol.* 8:1027–1036.

Forman, R.T.T. 1995. *Land Mosaics: The Ecology of Landscapes and Regions.* Cambridge, United Kingdom: Cambridge University Press.

Lima, S.L., and Zollner, P.A. 1996. Towards a behavioral ecology of ecological landscapes. *Trends Ecol. Evol.* 11:131–134.

Meffe, G.K., and Carroll, C.R., eds. 1997. *Principles of Conservation Biology.* Second Edition. Sunderland, Massachusetts: Sinauer Associates.

Noss, R.F., and Csuti, B. 1997. Habitat fragmentation. In *Principles of Conservation Biology,* Second Edition, eds. G.K. Meffe and C.R. Carroll, pp. 269–304. Sunderland, Massachusetts: Sinauer Associates.

Robinson, S.K., Thompson, F.R., III, Donovan, T.M., Whitehead, D.R., and Faaborg, J. 1995. Regional forest fragmentation and the nesting success of migratory birds. *Science* 267:1987–1990.

Saunders, D.A., Arnold, G.W., Burbidge, A.A., and Hopkins, A.J.M., eds. 1987. *Nature Conservation: The Role of Remnants of Native Vegetation.* Chipping Norton, Australia: Surrey Beatty and Sons.

Verhoeven, J.T.A., Kemmers, R.H., and Koerselman, W. 1993. Nutrient enrichment of freshwater wetlands. In *Landscape Ecology of a Stressed Environment,* eds. C.C. Vos and P. Opdam, pp. 33–59. London: Chapman and Hall.

12

Human Conversion of Terrestrial Habitats

PETER AUGUST, LOUIS IVERSON, AND JARUNEE NUGRANAD

12.1 Introduction

In this chapter, we describe how human activities change the abundance and quality of terrestrial habitats and discuss the ecological implications of these changes for biota. We begin by identifying fundamental principles associated with human conversion of terrestrial habitats and how fauna and flora respond to habitat conversion. We present a number of examples of how landscape ecologists and conservation biologists use these basic principles of land-cover change to develop management strategies to minimize ecological impacts from habitat loss. Next, we discuss principles for applying landscape ecology. We identify major voids in ecological theory and existing data that need to be filled for land managers to be better prepared to apply the principles of landscape ecology to biological conservation. Finally, we suggest research approaches that may be used to fill knowledge gaps. Although social and economic considerations are fundamental to land-cover change dynamics (Riebsame et al. 1994), detailed discussion of these factors is beyond the scope of this book; therefore, we focus our remarks on the ecological aspects of human conversion of terrestrial habitats.

12.2 Concepts, Principles, and Emerging Ideas

Human disturbance is the most significant contemporary agent of change in terrestrial ecosystems (Forman 1995). The rates with which natural habitats are lost, disturbed landscapes are created, species go extinct, and ecosystem processes are altered are higher now than they have been since the last cataclysmic event that impacted the planet 50 million years ago (Fastovsky and Weishampel 1996). Transformation of habitat can occur as a result of either natural or human activities (Table 12.1). The outcomes of some natural phenomena (e.g., fire, hurricane, flooding; Turner et al. 1997) are greatly modified by human influences. For example, dams will modify the rate and extent of flooding; forest removal and fragmentation will modify the effects of windstorms and overall climatic conditions.

TABLE 12.1. Possible types of land transformation by human influences (adapted from Forman 1995).

Cause of transformation	Type of human influence
Forest cutting	Direct
Urbanization	Direct
Corridor construction (road, rail, irrigation, utility)	Direct
Agriculture	Direct
Wetland drainage	Direct
Reforestation/restoration	Mostly direct, some indirect
Desertification	Mostly indirect, some direct
Chronic air pollution	Direct
Burning	Mostly direct, some indirect
Flooding	Mostly direct, some indirect
Bombing	Direct
Removal, mining	Direct
Herbicide	Direct
Release of non-native species	Direct and indirect

Human impacts occur at all scales, from site-specific to global, and they can occur rapidly or slowly (Figure 12.1).

When, where, and how rapidly natural landscapes are changed by human activities depends on a number of factors (Figure 12.2). Landscapes with significant biophysical constraints to development such as steep slopes, extreme climate, low soil fertility, or seasonal inundation are less prone to extensive human settlement than are landscapes without such impediments (Iverson 1988; Fuentes 1990; LaGro 1994). Reliable, safe access is a prerequisite to human development of landscapes. Some conduits for transportation are naturally occurring (e.g., rivers and sea) and have long provided human access to remote landscapes; other transportation conduits are products of human engineering (e.g., roads and rail) and permit access to undeveloped landscapes (Forman and Alexander 1998). The creation of a reliable access system is frequently the precursor to sudden and dramatic landscape conversions in both historical and modern times (LaGro and DeGloria 1992; Greene 1997; Pedlowski et al. 1997).

Human conversion of natural landscapes is precipitated by social, economic, and political factors, but it is bounded by environmental constraints (Ojima et al. 1994). Human population growth is a fundamental driving force in land conversion; as population numbers increase, the need to produce or extract more food, fuel, and fiber, and to develop infrastructure to support homes and commerce increases (Meyer and Turner 1992; Figure 12.3). Economic development is frequently the cause of broad-scale and rapid conversion of landscapes. This is clearly seen in the rapid loss of tropical forests because of logging, or the creation of large residential communities in suburban expansion (Browder et al. 1995; Scheer and Mintcho 1998).

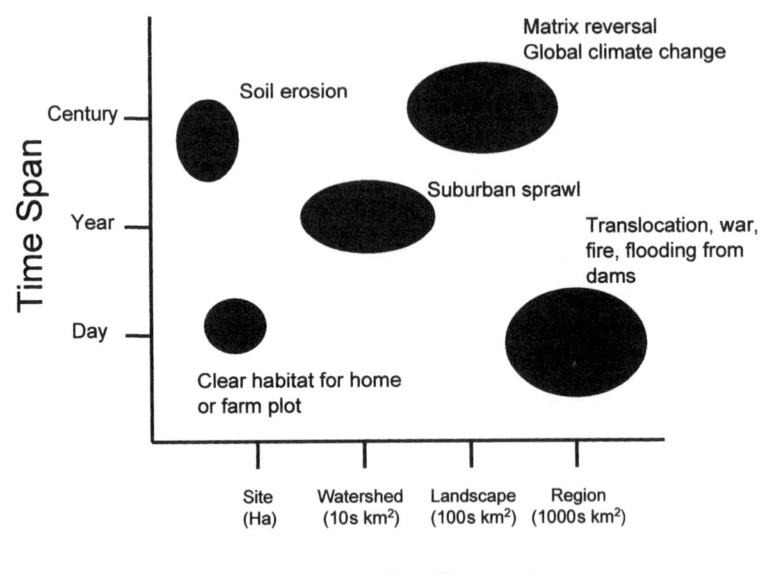

FIGURE 12.1. Human disturbance of terrestrial ecosystems occurs across a spectrum of time and spatial scales. Adapted from Forman (1995).

Anthropogenic disturbance of natural landscapes results from harvesting of environmental resources (e.g., forest, mineral, animal), farming, creation of residential opportunities, and commerce. The rate and extent of conversion can be rapid and extensive (Turner et al. 1998). For example, much of the native forest and prairie ecosystems of the northeastern and midwestern United States were rapidly transformed to pasture and cropland in the 18th and 19th centuries (Iverson 1988; Foster 1995). Massachusetts had an average deforestation rate of 1.37% per year during the period 1845–1875, whereas the Illinois deforestation rate was about 1% per year from 1850 to 1924 (Figure 12.4). During the period 1850–1900 in the United States, the human population tripled to 76 million, while the area of cropland increased over four times, from 31 million to 129 million ha (MacCleery 1992).

Suburban sprawl (also known as urban sprawl, edge cities, or metropolitan fringe) is the creeping of residential neighborhoods and light commerce into rural or natural landscapes surrounding urban centers (Browder et al. 1995; Scheer and Mintcho 1998). Sprawl can occur slowly (over decades) and consist of a gradual incursion of homes and residential developments into natural habitats, or it can occur extremely fast (over years) with the rapid building of large residential compounds that consist of hundreds of homes. In America, sprawl and commercial development are significant causes for the loss of rural and agricultural landscapes (Ilbery and Evans 1989; LaGro 1994) and are consistently cited as top-ranking threats to biodiversity (Flather et al. 1998; Wilcove et al. 1998).

In many cases, landscapes that have been modified by human activities are in a continual process of change. Some human-dominated landscapes remain in a dis-

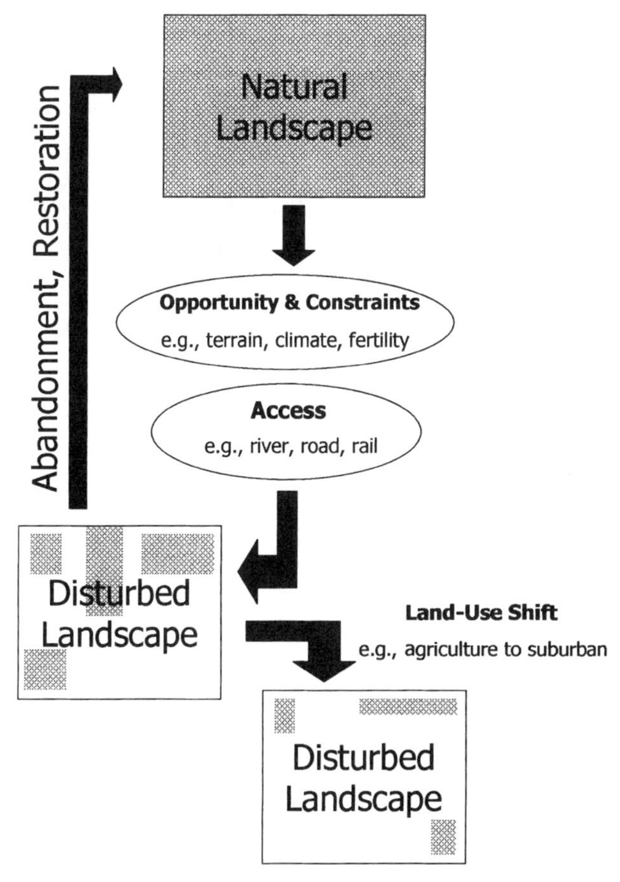

FIGURE 12.2. General sequence of events in the conversion of natural landscapes to disturbed landscapes and their continued disturbance or return to a natural or seminatural condition. Crossed-hatched areas in boxes symbolize the extent and pattern of undisturbed terrestrial habitats.

turbed state as evidenced by the large metropolitan centers of the world, other landscapes shift to another form of human disturbance, and some landscapes revert back to a natural condition through succession (Figure 12.2; Turner and Ruscher 1988; Odum and Turner 1990). Once abandoned, agricultural landscapes tend to revert back to a forested or prairie condition depending on the surrounding matrix and the long-term impacts of agricultural practices on soils, nutrients, and hydrological patterns (O'Keefe and Foster 1998).

From the above, we can deduce two basic principles that come into play when human activities convert natural habitats to a disturbed condition: *Principle 1—A safe, reliable, low-cost (time, energy) system (road, river, rail) to access landscapes is a prerequisite to human conversion of terrestrial habitats. Principle 2—Barring*

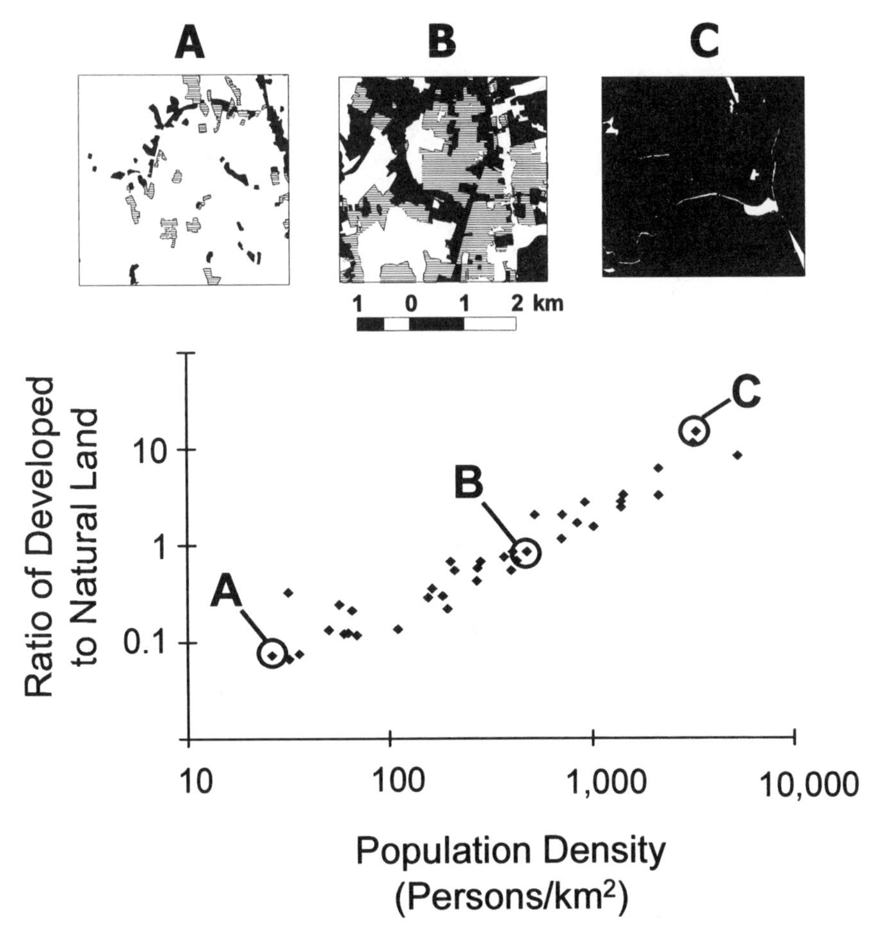

FIGURE 12.3. Relationship among human population density, natural and disturbed habitats, and the landscape matrix among Rhode Island (RI), USA towns spanning a continuum of human population density (1990 census data). The ratio of disturbed habitat (residential, commercial, and industrial [shaded black]) to natural habitat (forests, water, wetlands, and brushlands [shaded white]) is derived from 1988 land-cover data. Agricultural lands are hatched. Insets show a representative landscape for (A) a low-density rural community (West Greenwich, RI), (B) a medium-density community (Portsmouth, RI), and (C) a high-density community (Providence, RI). Note the matrix shift from natural (white shade) to developed (black shade) habitats.

profound changes to the soil and hydrological properties of a landscape that may occur during human modification, terrestrial habitats are resilient and can revert back to a natural condition if left undisturbed.

The transformation of terrestrial landscapes to landscapes dominated by human uses results in measurable changes to the composition and pattern of habitats, and the fauna and flora that occur in them (Table 12.2). Undisturbed landscapes typi-

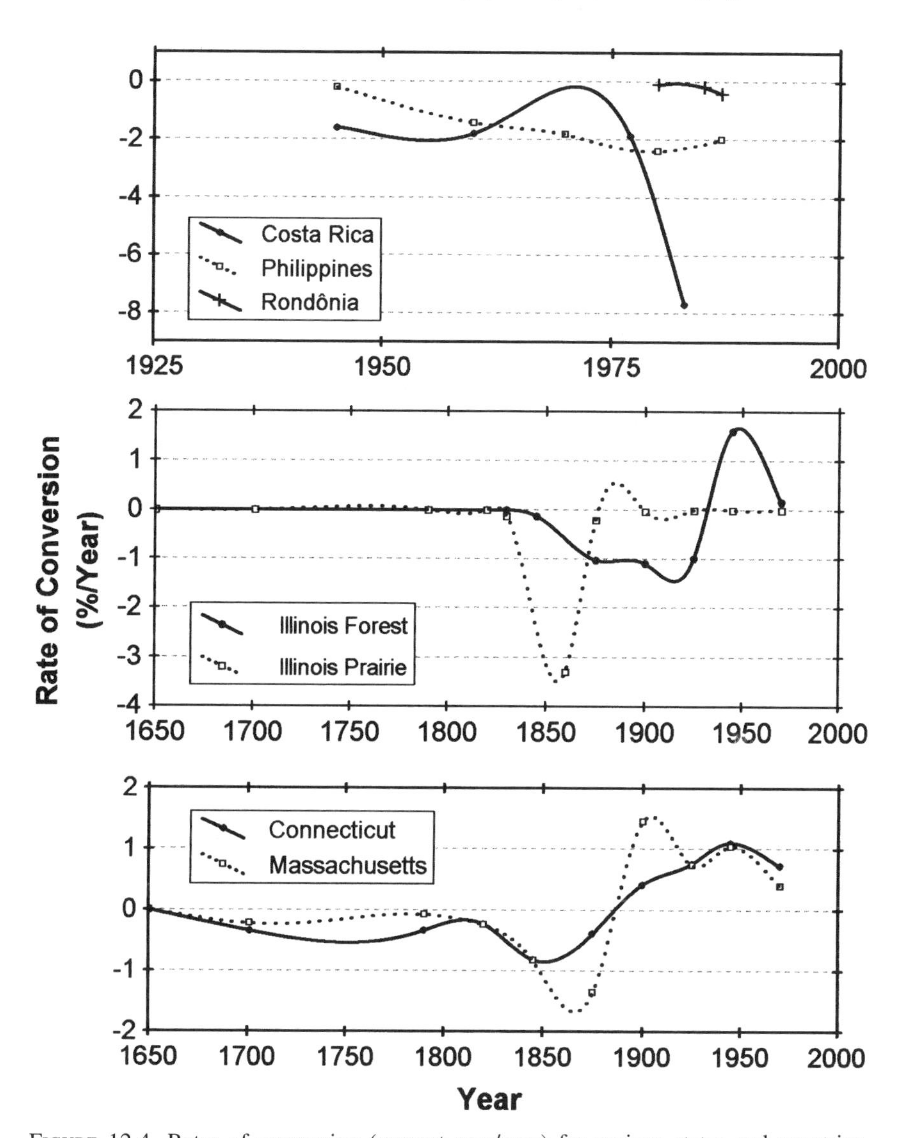

FIGURE 12.4. Rates of conversion (percent area/year) for various states and countries. Conversion rates are based on the amount of remaining forest (and prairie for Illinois) at the beginning of the time interval. Negative values indicate conversion from natural to disturbed habitats. Positive rates of conversion indicate a reversion from disturbed to natural condition (some smoothing of the trends has been done). Data are from Foster (1995, Connecticut and Massachusetts), Iverson (1991, Illinois forest and prairie), Sader and Joyce (1988, Costa Rica), Kummer and Turner (1994, Philippines), and Malingreau and Tucker (1988, Rondônia, Brazil).

TABLE 12.2. Typical changes in landscape and biological characteristics during the conversion of natural lands to human-dominated landscapes.

Landscape and biological characteristics	Landscape condition			Explanation	References†
	Natural	Transition	Disturbed		
Native/core habitat	Extensive	Medium	Uncommon	Native vegetation is lost or replaced with non-native species	1, 10, 11 12, 13, 15
Patch size	Large	Medium	Small	Large contiguous tracts of habitat are broken up	1, 2, 10, 13, 16
Patch shape	Complex	Simple	Simple	Human-modified edges tend to be rectilinear	7, 14, 17
Patch density	Low	Low to medium	Low to high	As fragmentation occurs, the landscape becomes more subdivided	10, 13, 16, 18
Edge density	Low	Moderate	High	Edge habitats become more common as conversion increases	2, 10, 13, 18
Connectedness	High	High to medium	Low	Patches become isolated	4, 10, 13
Landscape heterogeneity	Medium	High	Low to medium	Moderate disturbance can create a diverse landscape	2, 10, 16,
Roads	Few	Some to many	Many	Roads provide access for development or resource extraction	11, 12, 15, 18
Human population density	Low	Low to moderate	Low to high	Highly disturbed landscapes can have low human populations (e.g., farmlands, commercial forests) or high density (cities)	11, 17
Richness of native species	High	Moderate	Low	Native species are frequently replaced with non-native taxa as development proceeds	3, 4, 5
Richness of introduced/weedy species	Low	Moderate to high	Moderate to high	Generalist or non-native species frequently dominate heavily disturbed landscapes	2, 3, 18
Neotropical migratory birds	High	Moderate	Low	Migrants are often highly sensitive to disturbance	2, 3, 5
Predators	Normal	Depressed or elevated	Depressed or elevated	Domestic pets can increase predator density; natural predators usually decline	4, 6, 9, 12, 18
Wide-ranging species	Normal	Moderate	Depressed	As patches become smaller and isolated, less contiguous habitat is available for wide-ranging taxa	9
Core forest species	High	Moderate	Low	Species requiring large tracts of undisturbed habitat are impacted as landscapes become fragmented	1, 2, 5, 18
Total species diversity	Medium	High	Low to medium	Intermediate levels of disturbance often create complex landscapes that support high species diversity	1, 2, 4

Native species density	Medium to high	High	Low to medium	Native species are often outcompeted by invasives in disturbed habitats	4, 5, 10
Probability of extinction	Low	Moderate	High	Native taxa have a high probability of extinction as habitat patches become small and isolated	4
Edge-related mortality	Low	Moderate to high	High	Edge habitat increases mortality and decreases productivity for some species, such as ground-nesting birds	6, 8

† [1] Franklin and Forman (1987), [2] Miller et al. (1997), [3] Germaine et al. (1998), [4] Bolger et al. (1997a), [5] Hanowski et al. (1997), [6] Bayne and Hobson (1997), [7] O'Neill et al. (1988), [8] Esseen and Renhorn (1998), [9] Woodroffe and Ginsberg (1998), [10] Pearlstine et al. (1997), [11] Pedlowski et al. (1997), [12] LaGro (1994), [13] Dunn et al. (1990), [14] Turner and Ruscher (1988), [15] LaGro and DeGloria (1992), [16] Ambrose and Bratton (1990), [17] Iverson (1988), [18] Nilon et al. (1995).

cally have large amounts of native habitat, and landscape diversity is maintained by geomorphological properties of the area and naturally occurring disturbance such as fire, flood, landslide, and so on (Turner et al. 1997; Nichols et al. 1998). In contrast, human-disturbed patches are often simpler in shape than are naturally occurring patches; landscape edges defined by human uses tend to be rectilinear, whereas edges defined by natural processes are curvilinear. In heavily disturbed settings, fragments of original habitat become disconnected from one another and become isolated islands. Edge habitats can increase considerably in disturbed landscapes. In cases of extreme human disturbance, there can be a total shift in matrix habitat (Figure 12.3).

The results of human conversion of landscapes—loss of native habitat, small isolated remnant patches of native habitat, and increased edge—have profound effects on the fauna and flora of the region (Table 12.2). As native habitat decreases in extent, populations of resident plants and animals decrease in size and become more vulnerable to local extinction (Lande 1988, 1995; Andrén 1994). The sequence of impacts to local biota is largely scale-dependent. Large predators and wide-ranging taxa are first affected by habitat loss and fragmentation. Specialist taxa, such as species requiring large homogeneous expanses of native habitat, are frequently lost, whereas generalist taxa that can exist in natural or disturbed settings may dominate the biota. Species of plants and animals that are not native to the landscape are brought into the ecological setting as human disturbance continues. For example, domestic cats (*Felis cattus*) that accompany residential communities represent a new predator to transitional ecosystems and can have a profound impact on bird and small mammal populations. Horticultural and invasive plants frequently outcompete native flora and result in significant changes to habitats (Stohlgren et al. 1999). Loss and fragmentation of native habitat increases the amount of edge habitat. Some species, such as deer (*Odocoileus* sp.), Wild Turkeys (*Meleagris gallopavo*), and Brown-headed Cowbirds (*Molothrus ater*), benefit from edge, but others are negatively impacted (Paton 1994; Hartley and Hunter 1998). The changes in climate and ecology that occur in habitat edges can extend far into patches of native habitat. Edge contrast (sharp or gradual) and the types of adjacent land uses are important parameters in assessing the ecological impact of edge (Forman 1995).

A growing body of evidence shows that the ecological context (i.e., surrounding matrix habitat) of patches is as important (or sometimes more so) to resident biota as are the ecological conditions within the patch. Bird communities in plantation forests (Hanowski et al. 1997), oak (*Quercus* spp.) woodland (Sisk et al. 1997), and remnant deciduous forests patches in Chile (Estades and Temple 1999) all showed greater sensitivity to habitats surrounding occupied patches than they did to the composition and structure of vegetation within the patches.

In addition to the two previously mentioned principles that relate to landscape conversion, three principles summarize the ecological implications of such disturbances: *Principle 3—Human disturbance reduces the amount and alters the spatial properties (patch size, amount of edge, connectedness) of native habitat. Principle 4—Loss of native habitat results in the loss of resident species of plants and animals; non-native species frequently invade and can sometimes dominate disturbed landscapes. Principle 5—Habitat context is a very important characteristic of patches, and the biota within a patch are often affected by the habitat(s) surrounding it.*

12.3 Recent Applications

The principles of landscape ecology that bear on human conversion of terrestrial habitats can play a significant role in reducing the negative impact of human disturbance. However, our review of primary literature and government documents, as well as contacts with colleagues, indicated that the principles identified in the preceding section have rarely been applied on the ground to ameliorate these impacts. In contrast, the principles have been very important in directing the focus of recent ecological research. It is this type of use of the principles that we discuss in this section. These principles have been used in landscape planning; recent examples of such applications are discussed in Section IV of this book but not in this chapter. Here, we focus on research associated with three basic classes of anthropogenic land-cover change—forest and farmland conversion, suburban sprawl, and human resettlement. These three classes of conversion account for much of the historical and current human disturbance to natural ecosystems on the planet (Houghton 1994).

12.3.1 Forest and Farmland Conversion

Cutting forests for wood products or clearing native vegetation to allow for the planting of crops or creation of pasture are, by areal extent, the most significant human activities resulting in the loss of natural habitat world-wide (Myers 1980). The contemporary landscapes of many developed countries, for example much of western Europe and the United States, are defined by historical conversions for farming and forestry. Broad-scale conversions are now occurring over much of the tropical world in a fashion similar to what occurred in the United States 150 to 300 years ago (Figure 12.4) and in Europe before then. If suburban sprawl is viewed as a gradual nibbling away of natural or rural landscapes, conversion for forestry or farming must be considered a large and sudden bite. The time course for conversion and the details of the ecological impacts vary by region and land use. Swanson et al. (1990) found that extensive cutting of native forests in the Pacific Northwest of the United States resulted in the loss of older-age-class forests, created a mosaic of forest patches of different seral stages, increased sedimentation and alteration of coarse woody debris, and led to significant changes in local hydrology and stream ecosystem health. Similarly, conversion of natural landscapes for farming and pasture resulted in major losses of native habitat, as evidenced by the near total loss of tall-grass prairie in America (Iverson 1988), and it created a mosaic in which remnant patches of forest or native vegetation were isolated or loosely connected by narrow hedgerows (Merriam 1991).

For a variety of social and economic reasons, farming and forestry are declining in many regions of the world, and landscapes that were under cultivation or pasture are reverting back to a more natural ecological condition. Research by David Foster and colleagues in New England has chronicled the land-use history of this region. Their research demonstrates the long-term changes that occur in ecosystem patterns and processes, and the resiliency of natural habitats when human impacts are curtailed (Foster 1992, 1995; Foster et al. 1992, 1998). The precolonial (16th and early 17th centuries) landscape of New England was dominated by forest and wetland. Aboriginal human occupants cultivated lowland habitats in close proximity to rivers and streams, and they used fire to maintain fields and clear understory vegetation to facilitate travel and enhance the quality of habitat for game animals (MacCleery 1992). Occupation of the New England landscape by European colonists occurred in the 18th and 19th centuries. In approximately 200 years, much of the native forest of the region was cut and the land cleared for crops or pasture (Figure 12.4). In the 19th century, the Industrial Revolution precipitated a major change in land use and human settlement; farms were abandoned, cities grew, and industry developed along the major waterways (a source of energy and a transportation conduit). The abandoned farms reverted to forest in the 20th century (Figure 12.4). The pattern and species composition of contemporary forests has, however, been significantly altered when compared with precolonial forests. For example, many of the oak-hickory (*Quercus* spp., *Carya* spp.) forests in the eastern United States were established during a time of high fire and forest clearing (Abrams 1994). Today, red maple (*Acer rubrum*) is emerging as an increasing dominant over much of this region (Abrams 1998).

The landscape patterns observed by Foster and colleagues in their research mimic patterns found in other parts of the world. For example, in the southeastern United States and western Europe, lands that were intensively farmed in the 19th and early 20th centuries have been replaced by forest (Turner and Ruscher 1988; Preiss et al. 1997). In other areas, the farmland has remained relatively constant or is being converted to residential areas, a shift from one form of intensive human impact to another (Browder et al. 1995; Greene 1997). These research studies of land-cover change over long time scales clearly demonstrate the ecological resiliency of terrestrial habitats.

12.3.2 Suburban Sprawl

The principles described in Section 12.2 have provided direction for a number of excellent research projects on the ecological impacts of suburban sprawl. Sprawl destroys native habitat and creates isolated patches of remnant native habitat (Figure 12.3). Germaine et al. (1998) found that non-native horticultural plants that accompany residential development can expand into natural or seminatural habitats on the fringe of sprawl. Luken and Thieret (1996) found that ecological communities can shift from being dominated by native taxa to communities infiltrated by aggressively superior introduced species or species tolerant of human disturbance. The composition of ecological communities shifts as wide-ranging taxa and large predators are replaced by large numbers of smaller predators, both native and introduced (e.g., domestic cats; Soulé et al. 1988). Studies in the United Kingdom have shown that house cats have a significant impact on nearby bird and small-mammal communities, especially at the fringe of villages and neighborhoods that abut natural or seminatural landscapes (Churcher and Lawton 1987; May 1988). Some species of large predators may continue to exist in the face of human intrusion. For example, interactions between mountain lions (*Felis concolor*) and humans are becoming more common as sprawl encroaches on largely undisturbed habitats in the western United States (Torres et al. 1996; Torres 2000).

The studies by Bolger, Soulé, and colleagues (Soulé et al. 1988; Bolger et al. 1997a,b) have done an excellent job of measuring the ecological impacts of habitat fragmentation due to suburban sprawl in coastal southern California. The native flora of the region is coastal sage scrub and chaparral (Davis et al. 1995), and this landscape has exceptional levels of habitat diversity, species diversity, and endemism. The authors found that the size of remnant patches of shrub habitat and how long they had been isolated from other patches of habitat were critical variables in explaining variation in the number and diversity of chaparral-specialist birds and rodents; the smaller the patch and the longer the isolation, the fewer the resident species. Distance to the closest edge of residential habitat was a significant predictor of the distributions of many bird species. Some taxa were tolerant of human disturbance and occurred most often in edge habitats, whereas others were intolerant of disturbance and only occurred in core habitat.

12.3.3 Human Resettlement

As a means to relieve overpopulation pressures in urban centers, a number of national governments have initiated ambitious human resettlement programs that strive to relocate large numbers of citizens to underdeveloped or underutilized landscapes. Two recent examples are the Indonesian Transmigration from the densely populated island of Java to the sparsely populated islands of Sumatra and Kalimantan (MacKinnon 1996), and the Brazilian settlement of Rondônia in the Amazon Basin. Ecologists have studied the land-cover changes in Rondônia for a decade, and this research has been very important in clarifying the patterns and impacts of human invasion of undisturbed habitats.

The Brazilian government has had a policy of settling the Amazon Basin region of the country for over a century (Dale et al. 1993; Dale et al. 1994a,b; Pedlowski et al. 1997). The area is large (3.3 million km^2) and is rich in natural resources. In 1970, a road was built to provide reliable overland transportation from Rondônia to neighboring Brazilian states. During the next two decades, the human population density increased by an order of magnitude (from 100,000 to over 1,000,000 people). Most of the colonists cleared small patches of forest and planted crops and pasture. New patches were cleared every two years, and previously farmed patches were sold to cattle ranchers. Cattle ranchers removed any remnant forest habitat on the previously farmed patches and grazed cattle on the cleared lands. The changes in the Rondônian landscape have been systematically monitored with AVHRR and TM satellite data (Malingreau and Tucker 1988; Skole and Tucker 1993). From 1970 to 1993, over 52,500 km^2 of forest were destroyed. Much of the forest loss occurred adjacent to roads, the total length of which increased at a phenomenal rate (1,400 km in 1979, 25,000 km in 1988; Dale et al. 1994a,b). Similar conversion rates have historically been seen elsewhere in the world (Figure 12.4), usually as a result of a rapid influx of people and new access to the land.

The rapid settlement of Rondônia and concomitant habitat changes created an exceptional opportunity to study the ramifications of many of the principles described in Section 12.2. Dale et al. (1994b) modeled the probability of extinction of nine groups of animals from the Brazilian Amazon under three scenarios of land-use conversion over a 40-year period. Animals were characterized by their gap-crossing ability and area requirements. If a species could readily cross gaps, its fate was proportional to the amount of remaining forest and was not sensitive to fragmentation per se. If, however, a species had large area requirements and would only cross small gaps, the rate of habitat loss for this species was disproportionately greater than was the rate of forest clearing. Tropical frogs fared the worst of all species, with their habitat shrinking to as little as 39% of the remaining forest land after seven years of land conversion.

12.4 Principles for Applying Landscape Ecology

The sciences of landscape ecology and conservation biology provide many important insights into the management and monitoring of human conversion of terrestrial habitats. These ecological insights are the basis for the application principles we discuss in this section. In Figure 12.5, we illustrate the various points in the conversion sequence where land managers can intercede to eliminate or minimize adverse ecological impacts to biodiversity. The scientific principles described in this chapter guide planning and regulatory activities to prevent (or reduce) land uses that adversely impact ecosystems, help determine the most important measurements necessary to monitor the rate and extent of human disturbance, assess the response of populations and communities of native biota to change, and assist in the development and implementation of restoration activities to return impacted ecosystems to their natural form and function.

A critically important venue for minimizing the ecological impact of human activities is the institution where local land-use controls are established (LaGro 1994; Rookwood 1995). Depending on the form of governance and the ownership patterns of the land, venues may be town or county planning boards, or federal agencies such as the U.S. Forest Service or the U.S. Bureau of Land Management. The topic of ecologically sustainable environmental planning is taken up in detail in Section IV of this book, so our discussion is brief and focuses on

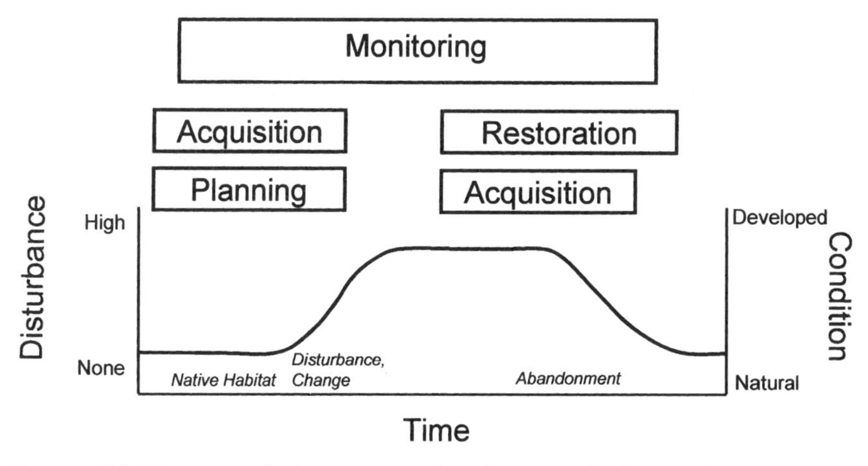

FIGURE 12.5. Time course for human conversion of terrestrial habitats and intervention opportunities (in boxes) for land managers.

how ecological principles described in Section 12.2 can be incorporated into the planning process.

Land-use managers and planners at regional and local levels are critically important players in the conservation process, and the principles of landscape ecology provide guidance to many of the decisions they must make. Carefully constructed zoning regulations can prevent destructive land uses in ecologically sensitive areas. For example, we have shown that a reliable method of access frequently precedes rapid settlement of undisturbed habitats and that creating or enhancing safe and reliable access to rural or wilderness landscapes is a precursor to future land-use change (see Section 12.2); therefore, transportation planners should assess the ecological impacts their projects will likely induce. This fundamental principle has been applied in wilderness regions of the United States where human access is restricted to nonmotorized methods of transportation (e.g., hiking, canoeing; The Wilderness Act of 1964 [16 U.S. C 11 21]).

Native species composition, population sizes, and community structure are severely impacted when native habitat is reduced to small, isolated patches in close association with highly disturbed land uses (Section 12.2). Therefore, two simple design principles for planners and land managers are as follows: maintain connections (corridors) among tracts of native habitat (Bueno et al. 1995), and ensure that remnant patches of natural habitat are as large as possible. Corridors that are naturally defined on the landscape, such as stream networks, should be maintained in an undisturbed condition to permit natural flux of species, energy, and materials (Harris 1984; Ahern 1995; Rosenberg et al. 1997). The amount and form of edge impacts the quality of habitat within a patch. In most cases, land management goals should be to minimize edge-to-core ratios. Edge contrast can be an important landscape characteristic and, when possible and appropriate, gradual edges rather than abrupt transitions from natural to impacted land uses should be designed. Gradual contrast can be achieved by surrounding natural regions of the landscape with buffer zones having only slight or moderate disturbance (Koslowski and Vass-Bowen 1997). Patch size is nearly always the most important variable defining the level of impact to species and communities; small patches are heavily impacted, and large patches are less-impacted. Therefore, environmental planners should minimize the extent of high-disturbance land uses adjacent to or near existing natural habitats.

Ecological monitoring is necessary to measure the rate and extent of human conversion of terrestrial habitats. Through monitoring, we quantify habitat loss, changes in landscape structure (patchiness, isolation of habitats), and the loss and gain of species and communities. Furthermore, continual ecological monitoring is the only way to determine if planning, regulatory, or conservation programs and policies to control human disturbance actually work (Goldsmith 1991). With adequate monitoring, we can shift our thinking and management strategies in a timely manner as new knowledge is acquired, a process called adaptive management (Stanford 1996).

Perhaps the most effective tool that planners and land managers have to stop human conversion of terrestrial habitats is to obtain ownership or development rights to the land. This strategy can occur at all scales, from continental initiatives (Soulé and Terborgh 1999) to municipal-scale programs (Peck 1998). Landscape ecology and the basic principles describing the changes that occur in species and populations that have experienced extensive habitat loss (see Section 12.2) have much to offer the process of creating and managing bioreserves. Some simple design or management objectives are as follows: refuges should be as large as possible and be connected with nearby refuges by protected corridors of habitat, round refuges offer the greatest core area of protected land relative to edge habitats, land adjacent to refuges should be managed as a buffer zone with only limited opportunity for major anthropogenic impacts, and disturbance regimes within refuges may have to be managed to maintain landscape diversity. Indeed, these simple principles are becoming entrenched within environmental planning and bioreserve design protocols (Noss et al. 1997; Peck 1998; Baydack et al. 1999).

The long-term resiliency of terrestrial habitats is an important principle to bear in mind while developing conservation plans or designing refuge systems. Habitats that are presently degraded, but have the potential to revert to a successional stage that has value for native fauna and flora (MacCleery 1992), may be important targets for conservation (Nichols et al. 1998). The cost to acquire degraded lands may be less than that of pristine habitats, and their geographic position in the landscape can fill important gaps in conservation networks.

12.5 Knowledge Gaps

Several areas of theoretical and empirical research must be advanced if we are to reconcile the need to accommodate human needs for space and natural resources with the need to protect biodiversity. Here we identify knowledge gaps that we feel are the most crucial to fill because of their constraining influence on our ability to convert principles of landscape ecology into hands-on planning, management, and restoration programs.

12.5.1 Theoretical Voids

Integration of Ecological Theory and Land-Use Planning

There is a lack of communication between scientists and land-use planners. Each group has a strong tradition of research and application within their respective disciplines, but there is little integration of science and practical land-use management. This lack of interdisciplinary interaction is perhaps the most significant gap in our ability to apply the principles of landscape ecology to on-the-ground natural resource management. We need to develop a means to integrate ecological theory with the land-use planning process. Likewise, we must integrate planning theory as a significant driving force into the ecological assessment of landscapes and regions. Planners and economists have developed tools to assess the social and financial impacts of development scenarios (Wegener 1994; Landis 1995), but ecologists have just begun to provide planners the same ability to assess ecological impacts (Linehan et al. 1995). Until a better connection is made between ecology and planning, land managers will have limited knowledge of the real impact of their decisions on fauna and flora within their jurisdiction.

Detecting Anthropogenic Change

If we are to manage human impacts at landscape scales, it is essential that we know rates and patterns of habitat conversion. Measuring and monitoring human impact of terrestrial ecosystems is a large and complex task; the variables measured, the technologies used, and the indices computed differ markedly depending on the size of the target landscape. Geographic information systems (GIS) and remote sensing technologies are fundamental tools for measuring the extent and pattern of land-use change (Iverson et al. 1989; Johnson 1990; August et al. 1996; Jensen 1996). Critical data for monitoring projects may be aerial photography or orthophotography for small areas, or satellite-based imagery for large regions (Dunn et al. 1990). Technological advancement in Earth imaging science is proceeding faster than is the research community's ability to identify the best suite of sensors, classification and change detection protocols, and summary indices to define accurately human impact on the Earth.

Predictive Mapping of Biota and Biodiversity

If we are to understand the impacts of land-use conversion and other human disturbances on biota, we must have the ability to measure or predict the distribution and abundance of fauna and flora. With the onset of better GIS capability, better landscape data on the environment, and geographically referenced records of biota, this area of research is rapidly changing (e.g., Franklin 1998; Wiser et al. 1998). Franklin (1995) has reviewed the field of biodiversity mapping and the many methods now available for such work. The Gap Analysis Program (Scott et al. 1993) uses a GIS approach to determine suitable habitat for a suite of animals. Our own work has enabled the prediction of species richness (Iverson and Prasad 1998a; Nichols et al. 1998), and particular tree species importance at local (Iverson et al. 1997) and regional (Iverson and Prasad 1998b) scales of analysis. Predictive mapping of various animal species is also undergoing rapid growth (e.g., Dettmers and Bart 1999; Mladenoff et al. 1999). However, it is common to find that the predictive models explain a relatively small portion of the variance. Additional research on what constitutes the critical input data to improve these models is required.

Land-Conversion Models

There is a need to develop better models to determine the most appropriate locations on the landscape for habitat conversions to occur and where they are most likely to occur (Wegener 1994). Modeling the spread of urban areas and other land-use changes more than 30 years into the future is rich in uncertainty, and more research is needed in this area (Landis 1995). Consideration of landscape ecological principles in land-use planning, via model development and outputs, is needed to ensure that planners identify spatial configurations of landscape development that permit sustainable ecological goods and services.

Conversion-Impact Models

We do not have the ability to model the impact of habitat conversion of the distribution and abundance of biota. Initial attempts to model impacted populations and communities are site-specific and not yet transferable to other areas (Friesen et al. 1995; Blair 1996; White et al. 1997; Rottenborn 1999). Research is needed to model the integrated effects of projected land-use conversion on the fauna and flora of the impacted area (Naiman et al. 1993; Lawton 1999).

12.5.2 Empirical Voids

Role of Habitat Corridors

Despite decades of research, we still do not know when and where corridors should be used to connect patches of habitat. Furthermore, we do not have a set of design guidelines to determine the optimal corridor width or habitat composition. It will be difficult or impossible for land managers and ecological planners to be effective if ecologists remain uncertain about the long- and short-term benefits of connected habitats (Simberloff et al. 1992; Rosenberg et al. 1997).

Fragmentation Versus Habitat Abundance

We do not yet know the relative importance of habitat loss versus fragmentation with respect to impacts on biota. In many cases, the two factors are correlated; as habitat is lost, remaining patches become increasingly fragmented. However, there is increasing evidence (e.g., McGarigal and McComb 1995; Trzcinski et al. 1999) that habitat abundance, rather than patchiness, is the dominant landscape feature controlling biotic integrity. If true, this tendency will have far-reaching implications in the way we integrate human activities into natural habitats at the landscape scale.

Role of Landscape Matrix

We do not understand the interactions that occur between habitat patches and their surrounding matrix. Much of our thinking about the relationships between patches and matrix has it roots in the theory of island biogeography advanced by MacArthur and Wilson (1967). The importance of island (patch) size, shape, and distance from source habitats is fundamental to landscape ecology. However, island biogeography is based on the notion that the matrix within which islands lie (e.g., oceanic islands in the sea, mountaintops in the desert) is totally unsuitable for resident biota. In terrestrial landscapes, this may not be the case because the matrix can have varying degrees of suitability for species occupying patches. Recent research suggests that the ecological context of patches of native habitat is critically important to the biota of the patch (Hanowski et al. 1997; Sisk et al. 1997; Estades and Temple 1999).

12.6 Research Approaches

12.6.1 Approaches for Theoretical Research

Integration of Ecological Theory and Land-Use Planning

Research is needed to develop a process to incorporate ecological thinking into land-use planning, and vice-versa. Central to this is knowing how human conversion of terrestrial habitats impacts species and communities. Long-term research in transitional landscapes, such as that conducted in urban long-term ecological research projects (Grove and Burch 1997), can provide knowledge needed for the integration of ecological theory and land-use planning. Publication of integrative treatises that promote unification of ecological science and land management are essential. Important also are the efforts of professional societies (e.g., International Association for Landscape Ecology) that provide forums for communication and interaction among planners and scientists.

Detecting Anthropogenic Change

Only recently have satellite-based data that have high spatial resolution (<5-m pixel size) and multispectral information content (e.g., Ikonos imagery) become publicly available. These data, in theory, provide an excellent basis to monitor detailed changes in land use over small areas (Li 1998). Satellite-based data sources (e.g., Spot, Landsat, AVHRR) that produce imagery at coarser scales (10–1,000-m pixel size) are valuable sources of data for measuring change over extensive regions (Skole and Tucker 1993; Laporte et al. 1998). GIS and remote sensing technology enable us to measure changes in the extent of natural habitats or levels of disturbance, patch characteristics, edge metrics, connectedness, landscape diversity, and other landscape metrics (O'Neill et al. 1997). What constitutes the best suite of landscape metrics to measure is, however, still an open question (O'Neill et al. 1988; Gustafson 1998).

Besides using wall-to-wall imagery to monitor human impacts, there remains immense value to plot-level assessments to understand ecological trends occurring spatially and temporally at ground level. Three national examples exemplify these data. The Forest Inventory and Analysis data of the U.S. Forest Service has data for over 100,000 forested plots across the eastern United States (Scott 1998). The National Resources Inventory of the U.S. Natural Resources Conservation Service is a longitudinal survey of soil, water, and related environmental resources designed to assess conditions and trends every five years on nonfederal United States lands (Nusser and Goebel 1997). The North American Breeding Bird Survey of the U.S. Fish and Wildlife Service provides field survey data on bird distributions for much of North America (Flather and Sauer 1996).

Predictive Mapping of Biota and Biodiversity

Improvement of models to predict the extent and abundance of species and communities can be achieved by several actions. The models will only ever be as good as the data from which they are built. First and foremost, local and regional biota must be inventoried and monitored (National Research Council 1993). Extensive and intensive surveys are needed, and the information should be entered into readily available databases. The National Biological Information Infrastructure (Anonymous 1998) provides one source for biological data to be distributed in clearinghouse fashion. Global positioning devices should be standard for any survey operations so that exact locations can be recorded in coordinates that can be easily entered into GIS databases (August et al. 1994). Ecologically meaningful environmental data should be acquired at a fine resolution with wall-to-wall coverage. Satellite imagery provides much of this information, but additional data are critical. Digital elevation models, at a scale of at least 1:24,000, are needed for modeling the variation in the abundance of flora and fauna across specific landscapes. Soils information is critical in identifying particular texture, depth, pH, drainage, or nutrient regimes especially related to specific organisms.

Land-Conversion Models

Land-use-change models have, for the most part, been based on or calibrated by historical trends. As such, there is a great dependency on adequate spatial data that describes the land use of an area at several previous intervals. To be most useful, these data should be in digital format over wide spatial extents. Unfortunately, these types of data sets are scarce; they need to be built from historical photographs and maps, and this is a slow and laborious process. The incorporation of less-used variables into land-use-change models should be explored. Variables related to human attitudes, quality of life, potential future transportation systems, and job locations in the age of telecommunications may become relatively more important than traditional land-capability, road network, and human population density variables in land-use-change models.

Conversion-Impact Models

There is no shortage of empirical research showing the ecological effects of habitat conversion at landscape scales (Friesen et al. 1995; Blair 1996; Rottenborn 1999). However, ecologists are only now trying to consolidate this knowledge into predictive models that strive to assess future impacts to biodiversity based on various land-use-change scenarios (Sisk et al. 1994; Freemark 1995; Lawton 1999). An excellent example of this kind of endeavor is the work of White et al. (1997), who developed a model to measure the gain or loss of species under six land-use scenarios ranging from complete build-out (maximum development of all developable lands) to immediate protection of all undeveloped lands in Monroe County, Pennsylvania. Their model attempts to predict the fate of individual species of mammals, birds, reptiles, and amphibians based on species-specific habitat requirements and the fate of those habitats under the different development scenarios. Their analysis is a benchmark beginning for the development of predictive planning tools to assess human impacts to local and regional patterns of biodiversity and should be extended to different landscapes and different land-use change scenarios. Metapopulation theory also offers considerable potential insight into the ecological impacts of land-use conversion. For example, the models developed by Wahlberg et al. (1996) have been used successfully to identify critically important patches of habitat for Fritillary Butterfly (*Melitaea diamina*) populations in moist meadow patches on the Finnish landscape.

12.6.2 Approaches for Empirical Research

Role of Habitat Corridors

Controlled experiments, such as those of Haddad (1999), need to be conducted across various landscapes and with various organisms to determine the overall importance of corridors. "Created" landscapes consisting of isolated and connected patches of habitat have been a productive method to measure the importance of corridors (Rosenberg et al. 1997). Organisms should be classified as to their requirements for corridor type and width, and a database that summarizes the work should be built. Only then can the modeling community begin to predict potential outcomes of conversions, which leave corridors as the primary protection against loss of biodiversity, at a species or guild level. Dale et al. (1994b) used an approach like this to evaluate the effects of tropical deforestation on biodiversity. Experimental determination of what species (or classes of species) benefit from corridors and what design principles should be followed when creating corridors are essential baseline data.

Fragmentation Versus Habitat Abundance

Much of the research on the ecological effects of habitat loss and fragmentation has been based on observational studies. Few well-controlled experimental studies exist to clarify cause-and-effect relationships between habitat change and concomitant changes in biota. The urban–rural landscape gradient is an interesting setting for these studies because it offers a continuum of patch-mosaic configurations along a relatively short distance (e.g., Bolger et al. 1997b; Germaine et al. 1998). The results of this research need to be categorized by organism and by guild, in a database, so that the modeling community can better use the work. The long-term studies of Lovejoy and colleagues (Bierregaard et al. 1992) in the Amazonian forest islands project constitute an excellent model for such experiments.

The scale of habitat patches relative to movement and space needs of the resident species is an important parameter to consider in analysis of habitat abundance versus patchiness (Fahrig 1997). If native species will tolerate levels of fragmentation that were previously considered excessive, planners and land managers will have greater flexibility in establishing development plans and policies that do not negatively impact the ecological integrity of the landscape. Carefully controlled studies comparing biodiversity among different geometric (patchy versus contiguous) configurations of landscapes should provide insight into the factors most important in maintaining viable populations and ecological communities.

Role of Landscape Matrix

Realization of the importance of the matrix in determining species' fates after disturbances is increasing, even for those species that spend very little time in the matrix. Landscapes with patches occurring in a variety of matrix habitats, for example wetland patches surrounded by disturbed habitats (residential areas) and natural matrix (upland forest), will be useful settings for this research. The fundamental question to answer is how much of the total variation in population and community characteristics is explained by the landscape ecological context of patches versus the habitat content of patches. This can be achieved observationally or experimentally. In an observational approach, community composition can be measured in a large number of landscape patches in a diversity of matrix settings. These data can be analyzed using multivariate statistical procedures to estimate the total variation explained by patch content and matrix type (McGarigal and McComb 1995). In an experimental approach, the fate of communities and populations can be monitored in landscape patches that have been created with varying levels of internal habitat complexity in a diversity of matrix settings.

Acknowledgments

Rolf Pendall, Marshall Feldman, and Marty Fujita were very helpful in connecting us with the literature on community and regional planning. Art Gold and Peter

Paton provided thoughtful reviews of the manuscript. Kevin Gutzwiller exercised incredible counsel and patience in helping us create a chapter that meshed with the rest of the book. This is contribution number 3723 of the Rhode Island Agricultural Experiment Station.

References

Abrams, M.D. 1994. Fire and the development of oak forests. *BioScience* 42:346–353.

Abrams, M.D. 1998. The red maple paradox. *BioScience* 48:355–364.

Ahern, J. 1995. Greenways as a planning strategy. *Landsc. Urban Plan.* 33:131–155.

Ambrose, J.P., and Bratton, S.P. 1990. Trends in landscape heterogeneity along the borders of Great Smoky Mountains National Park. *Conserv. Biol.* 4:135–143.

Andrén, H. 1994. Effects of habitat fragmentation on birds and mammals in landscapes with different proportions of suitable habitat: a review. *Oikos* 71:355–366.

Anonymous. 1998. What is the NBII? In *Access: Newsletter of the National Biological Information Infrastructure,* ed. R. Sepic, p. 2. Reston, Virginia: National Biological Information Infrastructure National Program Office.

August, P.V., Baker, C., LaBash, C., and Smith, C. 1996. Geographic information systems for the storage and analysis of biodiversity data. In *Measuring and Monitoring Biological Diversity: Standard Methods for Mammals,* eds. D. Wilson, F.R. Cole, J.D. Nichols, R. Rudran, and M.S. Foster, pp. 235–246. Washington, DC: Smithsonian Institution Press.

August, P.V., Michaud, J., LaBash, C., and Smith, C. 1994. GPS for environmental applications: accuracy and precision of positional data. *Photogramm. Eng. Rem. Sens.* 60:41–45.

Baydack, R.K., Campa, H., III, and Haufler, J.B. 1999. *Practical Approaches to the Conservation of Biological Diversity.* Washington, DC: Island Press.

Bayne, E.M., and Hobson, K.A. 1997. Comparing the effects of landscape fragmentation by forestry and agriculture on predation of artificial nests. *Conserv. Biol.* 11:1418–1429.

Bierregaard, R.O., Lovejoy, T.E., Kapos, V., Dos Santos, A.A., and Hutchings, R.W. 1992. The biological dynamics of tropical rainforest fragments. *BioScience* 42:859–866.

Blair, R.B. 1996. Land use and avian species diversity along an urban gradient. *Ecol. Appl.* 6:506–519.

Bolger, D.T., Alberts, A.C., Sauvajot, R.M., Potenza, P., McCalvin, C., Tran, D., Mazzoni, S., and Soulé, M.E., 1997a. Response of rodents to habitat fragmentation in coastal southern California. *Ecol. Appl.* 7:552–563.

Bolger, D.T., Scott, T.A., and Rotenberry, J.T. 1997b. Breeding bird abundance in an urbanizing landscape in coastal southern California. *Conserv. Biol.* 11:406–421.

Browder, J.O., Bohland, J.R., and Scarpaci, J.L. 1995. Patterns of development on the metropolitan fringe: urban fringe expansion in Bangkok, Jakarta, and Santiago. *J. Am. Plan. Assoc.* 61:310–327.

Bueno, J.A., Tsihrintzis, V.A., and Alvarez, L. 1995. South Florida greenways: a conceptual framework for the ecological reconnectivity of the region. *Landsc. Urban Plan.* 33:247–266.

Churcher, P.B., and Lawton, J.H. 1987. Predation by domestic cats in an English village. *J. Zool.* 212:439–455.

Dale, V.H., O'Neill, R.V., Pedlowski, M.A., and Southworth, F. 1993. Causes and effects of land-use change in Central Rondônia, Brazil. *Photogramm. Eng. Rem. Sens.* 59:997–1005.

Dale, V.H., O'Neill, R.V., Southworth, F., and Pedlowski, M.A. 1994a. Modeling effects of land management in the Brazilian settlement of Rondônia. *Conserv. Biol.* 8:196–206.

Dale, V.H., Pearson, S.M., Offerman, H.L., and O'Neill, R.V. 1994b. Relating patterns of land-use change to faunal biodiversity in the central Amazon. *Conserv. Biol.* 8:1027–1036.

Davis, F.W., Stine, P.A., Stoms, D.M., Borchert, M.I., and Hollander, A.D. 1995. GAP analysis of the actual vegetation of California 1: the southwestern region. *Madrono* 42:40–78.

Dettmers, R., and Bart, J. 1999. A GIS modeling method applied to predicting forest songbird habitat. *Ecol. Appl.* 9:152–163.

Dunn, C.P., Sharpe, D.M., Guntenspergen, G.R., Stearns, F., and Yang, Z. 1990. Methods for analyzing temporal changes in landscape pattern. In *Quantitative Methods in Landscape Ecology: The Analysis and Interpretation of Landscape Heterogeneity,* eds. M.G. Turner and R.H. Gardner, pp. 173–198. New York: Springer-Verlag.

Esseen, P, and Renhorn, K. 1998. Edge effects on an epiphytic lichen in fragmented forests. *Conserv. Biol.* 12:1307–1317.

Estades, C.F., and Temple, S.A. 1999. Deciduous forest bird communities in a fragmented landscape dominated by exotic pine plantations. *Ecol. Appl.* 9:573–585.

Fahrig, L. 1997. Relative effects of habitat loss and fragmentation on population extinction. *J. Wildl. Manage.* 61:603–610.

Fastovsky, D., and Weishampel, D.B. 1996. *The Evolution and Extinction of the Dinosaurs.* New York: Cambridge University Press.

Flather, C.H., Knowles, M.S., and Kendall, I.A. 1998. Threatened and endangered species geography. *BioScience* 48:365–376.

Flather, C.H., and Sauer, J.R. 1996. Using landscape ecology to test hypotheses about large-scale abundance patterns in migratory birds. *Ecology* 77:28–35.

Forman, R.T.T. 1995. *Land Mosaics: The Ecology of Landscapes and Regions.* Cambridge, United Kingdom: Cambridge University Press.

Forman, R.T.T., and Alexander, L.E. 1998. Roads and their major ecological effects. *Ann. Rev. Ecol. Syst.* 29:207–231.

Foster, D.R. 1992. Land-use history (1730–1990) and vegetation dynamics in central New England, USA. *J. Ecol.* 80:753–772.

Foster, D.R. 1995. Land use history and four hundred years of vegetation change in New England. In *Global Land Use Change: A Perspective From the Colombian Encounter,* eds. B.L. Turner, A. Gomez Sal, F. Gonzalez Bernaldez, and F. di Castri, pp. 253–319. Madrid: Consejo Superior de Investigaciones Cientifica.

Foster, D.R., Motzkin, G., and Slater, B. 1998. Land-use history as long-term broad-scale disturbance: regional forest dynamics in central New England. *Ecosystems* 1:96–119.

Foster, D., Zebryk, T., Schoonmaker, P., and Lezberg, A. 1992. Post-settlement history of human land-use and vegetation dynamics of a hemlock woodlot in central New England. *J. Ecol.* 80:773–786.

Franklin, J. 1995. Predictive vegetation mapping: geographic modeling of biospatial patterns in relation to environmental gradients. *Prog. Phys. Geog.* 19:494–519.

Franklin, J. 1998. Predicting the distribution of shrub species in southern California from climate and terrain-derived variables. *J. Veg. Sci.* 9:733–748.

Franklin, J.F., and Forman, R.T.T. 1987. Creating landscape pattern by forest cutting: ecological consequences and principles. *Landsc. Ecol.* 1:5–18.

Freemark, K.E. 1995. Assessing effects of agriculture on terrestrial wildlife: developing a hierarchical approach for the US EPA. *Landsc. Urban Plan.* 31:99–115.

Friesen, L.E., Eagles, P.F.J., and Mackay, R.J. 1995. Effects of residential development on forest-dwelling Neotropical migrant songbirds. *Conserv. Biol.* 6:1408–1414.

Fuentes, E.R. 1990. Vegetation change in Mediterranean-type habitats of Chile: patterns and processes. In *Changing Landscapes: An Ecological Perspective,* eds. I.S. Zonneveld and R.T.T. Forman, pp. 165–190. New York: Springer-Verlag.

Germaine, S.S., Rosenstock, S.S., Schweinsburg, R.E., and Richardson, W.S. 1998. Relationships among breeding birds, habitat, and residential development in greater Tucson, Arizona. *Ecol. Appl.* 8:680–691.

Goldsmith, F.B. 1991. *Monitoring for Conservation and Ecology.* London: Chapman and Hall.

Greene, R. 1997. The farmland conversion process in a polynucleated metropolis. *Landsc. Urban Plan.* 36:291–300.

Grove, J.M., and Burch, W.R., Jr. 1997. A social ecology approach and applications of urban ecosystem and landscape analyses: a case study of Baltimore, Maryland. *Urban Ecosyst.* 1:259–275.

Gustafson, E.J. 1998. Quantifying landscape spatial pattern: what is the state of the art? *Ecosystems* 1:143–156.

Haddad, N.M. 1999. Corridor and distance effects on interpatch movements: a landscape experiment with butterflies. *Ecol. Appl.* 9:612–622.

Hanowski, J.M., Niemi, G.J., and Christian, D.C. 1997. Influence of within-plantation heterogeneity and surrounding landscape composition on avian communities in hybrid poplar plantations. *Conserv. Biol.* 11:936–944.

Harris, L. 1984. *The Fragmented Forest: Island Biogeography and the Preservation of Biological Diversity.* Chicago: University of Chicago Press.

Hartley, M.J., and Hunter, M.L., Jr. 1998. A meta-analysis of forest cover, edge effects, and artificial nest predation rates. *Conserv. Biol.* 12:465–469.

Houghton, R.A. 1994. The worldwide extent of land-use change. *BioScience* 44:305–313.

Ilbery, B.W., and Evans, N.J. 1989. Estimating land loss on the urban fringe: a comparison of the agricultural census and aerial photograph/map evidence. *Geography* 74:214–221.

Iverson, L.R. 1988. Land-use changes in Illinois, USA: the influence of landscape attributes on current and historic land use. *Landsc. Ecol.* 1:45–61.

Iverson, L.R. 1991. Forest resources of Illinois: what do we have and what are they doing for us? *Illinois Nat. Hist. Surv. Bull.* 34:361–374.

Iverson, L.R., Dale, M.E., Scott, C.T., and Prasad, A. 1997. A GIS-derived integrated moisture index to predict forest composition and productivity in Ohio forests. *Landsc. Ecol.* 12:331–348.

Iverson, L.R., Graham, R.L., and Cook, E.A. 1989. Applications of satellite remote sensing to forested ecosystems. *Landsc. Ecol.* 3:131–143.

Iverson, L.R., and Prasad, A.M. 1998a. Predicting abundance of 80 tree species following climate change in the eastern United States. *Ecol. Monogr.* 68:465–485.

Iverson, L.R., and Prasad, A.M. 1998b. Estimating regional plant biodiversity with GIS modeling. *Divers. Distrib.* 4:49–61.

Jensen, J. 1996. *Introduction to Digital Image Processing: A Remote Sensing Perspective.* Englewood Cliffs, New Jersey: Prentice Hall.

Johnson, L.B. 1990. Analyzing spatial and temporal phenomena using geographical information systems: a review of ecological applications. *Landsc. Ecol.* 4:31–43.

Koslowski, J., and Vass-Bowen, N. 1997. Buffering external threats to heritage conservation areas: a planner's perspective. *Landsc. Urban Plan.* 37:245–267.

Kummer, D.M., and Turner, B.L., III. 1994. The human causes of deforestation in southeast Asia. *BioScience* 44:323–328.

LaGro, J.A., Jr. 1994. Population growth beyond the urban fringe: implications for rural land use policy. *Landsc. Urban Plan.* 28:143–158.

LaGro, J.A., Jr., and DeGloria, S.D. 1992. Land use dynamics in an urbanizing nonmetropolitan county in New York State (USA). *Landsc. Ecol.* 7:275–289.

Lande, R. 1988. Genetics and demography in biological conservation. *Science* 241:911–927.

Lande, R. 1995. Mutation and conservation. *Conserv. Biol.* 9:782–791.

Landis, J. 1995. Imagining land-use futures: applying the California urban futures model. *J. Am. Plan. Assoc.* 61:438–457.

Laporte, N.T., Goetz, S.J., Justice, C.O., and Heinicke, M. 1998. A new land-cover map of central Africa derived from multi-resolution, multi-temporal AVHRR data. *Int. J. Rem. Sens.* 19:3537–3550.

Lawton, J.H. 1999. Are there general laws in ecology? *Oikos* 84:177–192.

Li, R. 1998. Potential of high-resolution satellite imagery for national mapping products. *Photogramm. Eng. Rem. Sens.* 59:1165–1170.

Linehan, J., Gross, M., and Finn, J. 1995. Greenway planning: developing a landscape ecological network approach. *Landsc. Urban Plan.* 33:179–193.

Luken, J.O., and Thieret, J.W. 1996. Amur honeysuckle, its fall from grace. *BioScience* 46:18–24.

MacArthur, R.H., and Wilson, E.O. 1967. *The Theory of Island Biogeography.* Princeton: Princeton University Press.

MacCleery, D.W. 1992. *American Forests, A History of Resiliency and Recovery.* FS-540. Washington, DC: USDA Forest Service.

MacKinnon, K. 1996. *The Ecology of Kalimantan.* Hong Kong: Periplus Editions.

Malingreau, J.P., and Tucker, J.C. 1988. Large-scale deforestation in the southeastern Amazon Basin of Brazil. *Ambio* 17:49–55.

May, R.M. 1988. Control of feline delinquency. *Nature* 332:392–393.

McGarigal, K., and McComb, W.C. 1995. Relationships between landscape structure and breeding birds in the Oregon Coast Range. *Ecol. Monogr.* 65:235–260.

Merriam, G. 1991. Corridors and connectivity: animal populations in heterogeneous environments. In *Nature Conservation 2: The Role of Corridors,* eds. D.A. Saunders and R.J. Hobbs, pp. 133–142. Chipping Norton, Australia: Surrey Beatty and Sons.

Meyer, W.B., and Turner, B.L. 1992. Human population growth and global land-use/cover change. *Ann. Rev. Ecol. Syst.* 23:39–61.

Miller, J.N., Brooks, R.P., and Croonquist, M.J. 1997. Effects of landscape pattern on biotic communities. *Landsc. Ecol.* 12:137–153.

Mladenoff, D.J., Sickley, T.A., and Wydeven, A.P. 1999. Predicting gray wolf landscape recolonization: logistic regression models vs. new field data. *Ecol. Appl.* 9:37–44.

Myers, N. 1980. *Conversion of Moist Tropical Forests.* Washington, DC: National Academy of Sciences.

Naiman, R.J., Décamps, H., and Pollock, M. 1993. The role of riparian corridors in maintaining regional biodiversity. *Ecol. Appl.* 3:209–212.

National Research Council. 1993. *A Biological Survey for the Nation.* Washington, DC: National Academy Press.

Nichols, W., Killingbeck, K.T., and August, P.V. 1998. The influence of geomorphological heterogeneity on biodiversity: II. a landscape perspective. *Conserv. Biol.* 12:371–379.

Nilon, C.H., Long, C.N., and Zipperer, W.C. 1995. Effects of wildland development on forest bird communities. *Landsc. Urban Plan.* 32:81–92.

Noss, R.F., O'Connell, M.A., and Murphy, D. 1997. *The Science of Conservation Planning: Habitat Conservation Under the Endangered Species Act.* Washington, DC: Island Press.

Nusser, S.M., and Goebel, J.J. 1997. The National Resources Inventory: a long-term multi-resource monitoring program. *Environ. Monit. Assess.* 4:181–204.

Odum, E.P., and Turner, M. 1990. The Georgian landscape: a changing resource. In *Changing Landscapes: An Ecological Perspective,* eds. I.S. Zonneveld and R.T.T. Forman, pp. 137–164. New York: Springer-Verlag.

Ojima, D.S., Galvin, K.A., and Turner, B.L., III. 1994. The global impact of land-use change. *BioScience* 44:300–304.

O'Keefe, J.F., and Foster, D. 1998. An ecological history of Massachusetts forests. In *Stepping Back to Look Forward: A History of Massachusetts Forests,* ed. C.H. Foster, pp. 19–66. Cambridge, Massachusetts: Harvard University Press.

O'Neill, R.V., Hunsaker, C.T., Jones, K.B., Riitters, K.H., Wickham, J.D., Schwartz, P.M., Goodman, I.A., Jackson, B.L., and Baillargeon, W.S. 1997. Monitoring environmental quality at the landscape scale. *BioScience* 47:513–519.

O'Neill, R.V., Krummel, J.R., Gardner, R.H., Sugihara, G., Jackson, B., DeAngelis, D.L., Milne, B.T., Turner, M.G., Zygmunt, B., Christensen, S.W., Dale, V.H., and Graham, R.L. 1988. Indices of landscape pattern. *Landsc. Ecol.* 1:153–162.

Paton, P.W.C. 1994. The effect of edge on avian nest success: how strong is the evidence? *Conserv. Biol.* 8:17–26.

Pearlstine, L.G., Brandt, L.A., Mazzotti, F.J., and Kitchens, W.M. 1997. Fragmentation of pine flatwood and marsh communities converted for ranching and citrus. *Landsc. Urban Plan.* 38:159–169.

Peck, S. 1998. *Planning for Biodiversity: Issues and Examples.* Washington, DC: Island Press.

Pedlowski, M.A., Dale, V.H., Matricardi, E.A.T., and Pereira da Silva Filho, E. 1997. Patterns and impacts of deforestation in Rondônia, Brazil. *Landsc. Urban Plan.* 38:149–157.

Preiss, E., Martin, J.-L., and Debussche, M. 1997. Rural depopulation and recent landscape changes in a Mediterranean region: consequences to the breeding avifauna. *Landsc. Ecol.* 12:51–61.

Riebsame, W.E., Parton, W.J., Galvin, K.A., Burke, I.C., Bohren, L., Young, R., and Knop, E. 1994. Integrated modeling of land use and cover change. *BioScience* 44:350–356.

Rookwood, P. 1995. Landscape planning for biodiversity. *Landsc. Urban Plan.* 31:379–385.

Rosenberg, D.K., Noon, B.R., and Meslow, E.C. 1997. Biological corridors: form, function, and efficacy. *BioScience* 47:677–687.

Rottenborn, S.C. 1999. Predicting the impacts of urbanization on riparian bird communities. *Biol. Conserv.* 88:289–299.

Sader, S.A., and Joyce, A.T. 1988. Deforestation rates and trends in Costa Rica, 1940 to 1983. *Biotropica* 20:11–19.

Scheer, B.C., and Mintcho, P. 1998. Edge city morphology: a comparison of commercial centers. *J. Am. Plan. Assoc.* 64:298–310.

Scott, C.T. 1998. Sampling methods for estimating change in forest resources. *Ecol. Appl.* 8:228–233.

Scott, J.M., Davis, F.W., Csuti, B., Noss, R., Butterfield, B., Groves, C., Anderson, H., Caicco, S., D'Erchia, F.D., Edwards, T.C., Ulliman, J., and Wright, R.G. 1993. Gap analysis: a geographic approach to protection of biological diversity. *Wildl. Monogr.* 123:1–41.

Simberloff, D., Farr, J.A., Cox, J., and Mehlman, D.W. 1992. Movement corridors: conservation bargains or poor investments? *Conserv. Biol.* 6:493–504.

Sisk, T.D., Haddad, N.M., and Ehrlich, P. 1997. Bird assemblages in patchy woodlands: modeling the effects of edge and matrix habitats. *Ecol. Appl.* 7:1170–1180.

Sisk, T.D., Launer, A.E., Switky, K.R., and Ehrlich, P.R. 1994. Identifying extinction threats. *BioScience* 44:592–604.

Skole, D., and Tucker, C. 1993. Tropical deforestation and habitat fragmentation in the Amazon: satellite data from 1978–1988. *Science* 260:1905–1910.

Soulé, M.E., Bolger, D.T., Alberts, A.C., Wright, J., Sorice, M., and Hill, S. 1988. Reconstructed dynamics of rapid extinctions of chaparral-requiring birds in urban habitat islands. *Conserv. Biol.* 2:75–92.

Soulé, M.E., and Terborgh, J., eds. 1999. *Continental Conservation: Scientific Foundations of Regional Reserve Networks.* Washington, DC: Island Press.

Stanford, J.A. 1996. A protocol for ecosystem management. *Ecol. Appl.* 6:741–744.

Stohlgren, T.J., Binkley, D., Chong, G.W., Kalkham, M.A., Schell, L.D., Bull, K.A., Otsuki, Y., Newman, G., Bashkin, M., and Son, Y. 1999. Exotic plant species invade hot spots of native plant diversity. *Ecol. Monogr.* 69:25–46.

Swanson, F.J., Franklin, J.F., and Sedell, J.R. 1990. Landscape patterns, disturbance, and management in the Pacific Northwest, USA. In *Changing Landscapes: An Ecological Perspective,* eds. I.S. Zonneveld and R.T.T. Forman, pp. 190–213. New York: Springer-Verlag.

Torres, S.G. 2000. Counting cougars in California. *Outdoor Calif.* 61:7–9.

Torres, S.G., Mansfield, T.M., Foley, J.E., Lupo, T., and Brinkhaus, A. 1996. Mountain lion and human activity in California: testing speculations. *Wildl. Soc. Bull.* 24:451–460.

Trzcinski, M.K., Fahrig, L., and Merriam, G. 1999. Independent effects of forest cover and fragmentation on the distribution of forest breeding birds. *Ecol. Appl.* 9:586–593.

Turner, M.G., Carpenter, S.R., Gustafson, E.J., Naiman, R.J., and Pearson, S.M. 1998. Land use. In *Status and Trends of the Nation's Biological Resources,* eds. M.J. Mac, P.A. Opler, C.E. Puckett Haecker, and P.D. Doran, pp. 37–61. Reston, Virginia: U.S. Department of the Interior, U.S. Geological Survey.

Turner, M.G., Dale, V.H., and Everham, E.H., III. 1997. Fires, hurricanes, and volcanoes: comparing large disturbances. *BioScience* 47:758–768.

Turner, M.G., and Ruscher, C.L. 1988. Changes in landscape patterns in Georgia, USA. *Landsc. Ecol.* 1:241–251.

Wahlberg, N., Moilanen, A., and Hanski, I. 1996. Predicting the occurrence of endangered species in fragmented landscapes. *Science* 273:1536–1538.

Wegener, M. 1994. Operational urban models: state of the art. *J. Am. Plan. Assoc.* 60:17–29.

White, D., Minotti, P.G., Barczak, M.J., Sifneos, J.C., Freemark, K.E., Santelmann, M.V., Steinitz, C.F., Kiester, A.R., and Preston, E.M. 1997. Assessing risks to biodiversity from future landscape change. *Conserv. Biol.* 11:349–360.

Wilcove, D.S., Rothstein, D., Dubow, J., Phillips, A., and Losos, E. 1998. Quantifying threats to imperiled species in the United States. *BioScience* 48:607–615.

Wiser, S.K., Peet, R.K., and White, R.S. 1998. Prediction of rare-plant occurrence: a southern Appalachian example. *Ecol. Appl.* 8:909–920.

Woodroffe, R., and Ginsberg, J.R. 1998. Edge effects and the extinction of populations inside protected areas. *Science* 280:2126–2128.

13

Impacts of Landscape Transformation by Roads

LAURIE W. CARR, LENORE FAHRIG, AND SHEALAGH E. POPE

13.1 Introduction

The transformation of landscapes by roads has spread rapidly with little consideration of the ecological consequences. Between 1986 and 1994, the number of vehicles increased 18%, to over 6 billion world-wide (United Nations 1997). In this chapter, we outline the ecological impacts of roads on biological conservation, from a landscape ecology perspective. In Section 13.2 we outline the relevant landscape ecological concepts, in Section 13.3 we review measures to compensate for negative impacts of roads on wildlife, and in Section 13.4 we outline principles for successful mitigation of landscape-level road effects. Finally, in Sections 13.5 and 13.6, we discuss existing knowledge gaps and suggest research approaches for filling these gaps.

13.2 Concepts, Principles, and Emerging Ideas

Landscape composition, configuration, and connectivity are key descriptors of landscape structure that determine the influence of the landscape on population persistence (Dunning et al. 1992; Taylor et al. 1993). *Composition* is the number and extent of different landscape elements (e.g., habitat types) in a landscape. *Configuration* describes the spatial arrangement of landscape elements, and *connectivity* is the degree to which the landscape facilitates animal movement (Taylor et al. 1993).

13.2.1 Road Effects on Landscape Composition

Roads affect landscape composition through loss of habitat, habitat introduction, and changes in habitat quality. Road construction results in conversion of existing habitats to pavement and road verges. Roads through wooded areas also increase the amount of forest edge, resulting in additional loss of habitat for forest interior species (Ranney et al. 1981). Road construction also can result in the indirect destruction of surrounding habitat through siltation of streams and drying of wetlands, due to interrupted water flow. The loss of habitat to roads is far from inconsequential. For example, Adams and Geis (1983) estimated that the 6.3 million km of roads and their associated rights-of-way occupied 8.1 million ha in the United States.

Roads also change landscape composition by introducing novel "habitats." Pavement affords resources to few species (thermoregulation for ectotherms and roadkill as a food source being notable exceptions) and, therefore, barely deserves the moniker "habitat." However, vegetated verges and medians provide resources such as forage, shelter, and nest sites to many species (Bennett 1991). Roadside habitat is often characterized by grassland species, generalist species, disturbance-tolerant species, and exotic species (Adams and Geis 1983; Forman and Alexander 1998).

In addition to habitat loss and introduction, roads alter habitat quality through a direct increase in wildlife mortality (Forman and Alexander 1998). For example, during a two-year period, 32,000 amphibians, reptiles, birds, and mammals were found as roadkill along the 3.6-km Long Point Causeway adjacent to Big Creek Wetland, Lake Erie, Canada (Ashley and Robinson 1996). Ehmann and Cogger (1985) estimated that 5.48 million reptiles and frogs are killed in Australia each year by road traffic. The amount of traffic mortality is influenced by traffic volume and movement behavior of the organism (Fahrig et al. 1995; Carr and Fahrig 2001).

Roadside habitats become *ecological traps* when the mortality rate of individuals using this habitat is higher than that of individuals using an alternative habitat. For example, mortality of Florida Scrub-Jays (*Aphelocoma coerulescens*) was almost 100% greater in populations bordering a two-lane country road than it was for populations living away from the road (Mumme 1994, cited in Lidicker and Koenig 1996). Mortality was greatest for immigrants, implying that the roadside habitat was luring birds away from other habitat.

The extent to which road mortality negatively affects local populations depends on the intrinsic growth rate of the organism and the degree to which the killed individuals were surplus to the local population (e.g., juveniles without local territories). Road mortality can be a serious cause of decline for small populations. For example, Harris and Scheck (1991) reported that vehicle collisions were the largest source of mortality for all of Florida's large rare and endangered vertebrates, including panther (*Felis concolor coryi*), black bear (*Ursus americanus floridanus*), key deer (*Odocoileus virginianus clavium*), American crocodile (*Crocodylus acutus*), and Bald Eagle (*Haliaeetus leucocephalus*).

Roads also can create suboptimal habitat through *edge effects* (Murcia 1995). Road edge effects result when adjacent habitat is exposed to road effects such as: runoff of chemicals, particulate matter, and water; noise; and changes in microclimate (Garland and Bradley 1984; Forman 1995; Reijnen 1995). These can have direct biological consequences. For example, an increase in traffic noise is correlated with a decrease in abundance of farmland birds in The Netherlands (Reijnen 1995). Roads also indirectly alter habitat quality by providing human access (e.g., hunting, development) to previously remote areas.

Suboptimal roadside habitat may play a role as *sink habitat* for nonbreeders in *source-sink population dynamics.* Source-sink dynamics occur when a local source population exceeds the number of breeding sites available, and nonbreeders disperse to a poorer quality sink habitat. In the sink habitat, mean annual reproduction is insufficient to balance mean annual mortality. Immigration from a source population is required to maintain the sink population over the long term. However, sink populations can benefit the overall regional population by temporarily "storing" individuals that can move back into a source habitat, should numbers decline there (Pulliam 1988). If roads convert source habitat to sink habitat, persistence of the regional population can be jeopardized. For example, subordinate and some adult female grizzly bears (*Ursus arctos*) are forced to use roadside habitat in Bow River Valley, Alberta, when adult males and resident females take the highest-quality habitat. Most grizzly bears are unwilling to use habitats with a high human presence (Gibeau and Herrero 1998).

The *road-effect zone* is the area (extending outward from the road) over which the road has significant ecological impacts (Forman and Deblinger 2000). It combines the road effects on habitat loss and habitat quality. The width of the zone depends on the intensity of the road effects and the sensitivity of organisms to them.

13.2.2 Road Effects on Landscape Configuration and Connectivity

Roads affect landscape configuration and connectivity by introducing barriers and corridors into the landscape. These affect landscape processes, including metapopulation dynamics, landscape supplementation, and landscape complementation.

Many species avoid roads in response to the noise, light, altered microclimate, and reduced cover (Bennett 1991). Movement across roads by wood mice, carabid beetles, and lycosid spiders is reduced compared with movement over similar distances in adjacent habitat (Mader 1984; Mader et al. 1990). Oxley et al. (1974) found that road width was the main determinant of reduced crossing by small forest animals. Garland and Bradley (1984) found that desert rodents avoided roads, despite the small difference between the road and the desert environment. Individuals may also be prevented from crossing a road by a physical barrier such as a median barrier, roadside fence, or bank (Andrews 1990).

In addition to the *barrier effect* caused by road avoidance, high road mortality can act as a barrier to movement in the landscape (e.g., van Gelder 1973; Ehmann and Cogger 1985). Road mortality therefore has two effects: (1) it reduces local habitat quality by increasing mortality (Section 13.2.1), and (2) it decreases the rate of movement through the landscape (Figure 13.1).

As barriers, roads can cause *habitat fragmentation,* or the "breaking apart" of habitat, which reduces local population sizes. Smaller populations are at a greater risk of local extinction, due to stochastic demographic, genetic, and environmental events (Wilcox and Murphy 1985). Isolated populations also have a lower chance of survival without the demographic and genetic input of immigrants, and a lower chance of recolonization after local extinctions (Lande 1988).

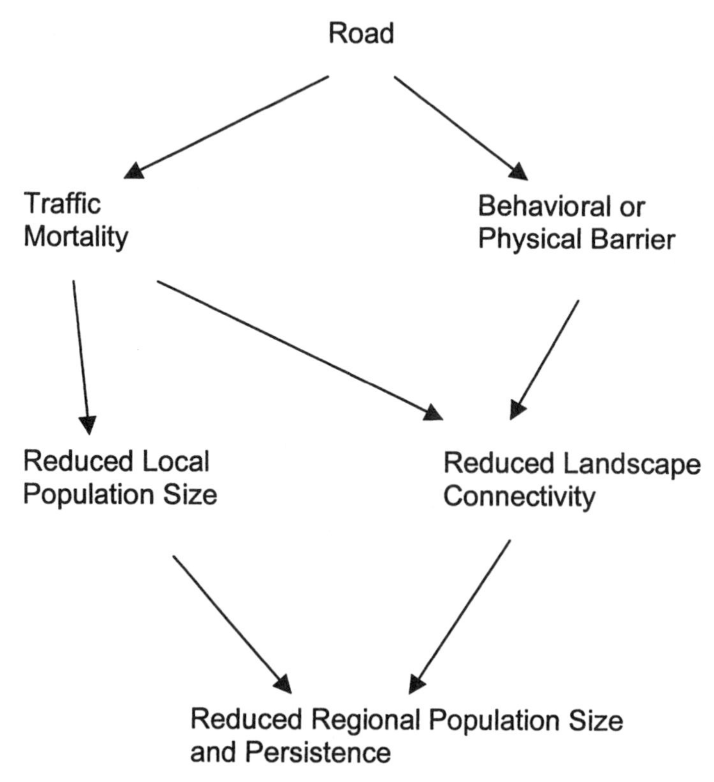

FIGURE 13.1. Roads affect regional persistence of wildlife populations through (1) directly increasing mortality and (2) directly and indirectly (through mortality) decreasing landscape connectivity.

A *corridor* is a linear landscape feature that facilitates movement between resource patches, and it may increase the interchange of individuals between local populations. This may increase the persistence of local and regional populations (see metapopulation dynamics and rescue effect below). On the negative side, corridors can facilitate the spread of exotics, disease, and disturbances (Wiens 1996). There is little evidence that animals are more likely to move along roads or roadsides than through other habitats. Unpaved, narrow roads may be used by large mammals such as lions (*Panthera leo*), cheetahs (*Acinonyx jubatus*), dingos (*Canis dingo*), coyotes (*Canis latrans*), foxes (*Vulpes sp.*), wolves (*Canis lupus*), and tasmanian devils (*Sarcophilus harrisi*) (Forman 1995). Pocket gophers (Geomyidae), meadow voles (*Microtus pennsylvanicus*), and heathland ground beetles appear to have expanded their ranges by using roadside verges as dispersal routes (Vermeulen 1994; Forman 1995). Probably the best evidence of roadsides as corridors is the dispersal of disturbance-adapted plant species. Sludge from a carwash in Canberra, Australia, contained viable seeds from 259 plant species, indicating movements of up to 100 km along roads based on the nearest possible seed sources (Wace 1977).

A *metapopulation* is composed of a group of local populations linked by movement. By definition, local population dynamics are asynchronous, such that not all local populations undergo extinction at the same time—a process known as *spreading of risk* (Den Boer 1968). Movement rate between local populations permits recolonization of local extinctions, but this rate is not so high that the local population dynamics become synchronized (Hanski 1989). Interpatch movement is the defining characteristic of metapopulations, and therefore metapopulations depend on landscape connectivity for persistence. Roads can disrupt connectivity, thereby reducing metapopulation persistence. Movement between populations also reduces local extinction probability by rescuing populations from low numbers (the *rescue effect*; Brown and Kodric-Brown 1977). By reducing connectivity, roads reduce the potential of the rescue effect to support regional populations.

In addition to affecting between-population movements, landscape structure also affects the ability of an individual animal to move among different landscape elements to meet its life-history demands. The scale at which such movements occur is defined by the organism's *ecological neighborhood,* the region within which it is active or has influence (Addicott et al 1987). High connectivity of habitat patches containing nonsubstitutable resources (*landscape complementation*) and substitutable resources (*landscape supplementation*) enables an organism to collect resources efficiently, resulting in lower mortality, higher reproduction, and ultimately larger populations (Dunning et al. 1992; Pope et al. 2000). When roads act as barriers to movement, they reduce landscape complementation and supplementation by reducing accessibility of resources.

13.3 Recent Applications

We surveyed the primary literature and the secondary literature (mainly conference proceedings), and we informally surveyed individuals at the 1999 International Association for Landscape Ecology World Congress meeting in Colorado, looking for examples of applications of landscape ecology that address road impacts. For the most part, the concepts outlined in Section 13.2 have not been explicitly applied to road effects on biological conservation. Nevertheless, various measures to compensate for the negative impacts of roads ("mitigation measures") have been applied and can be considered in a landscape ecological context. Such measures are intended to mitigate the effects of roads on either (1) landscape composition, by reducing road effects on habitat loss and quality, or (2) landscape connectivity, by increasing successful movement of organisms across roads. Due to the recent nature of most mitigation projects, existing performance evaluations measure only whether individual organisms reacted positively. The success of these measures at the population level has yet to be assessed.

The mitigation measures we highlight were developed based on the assumption that roads are needed. However, the most effective methods to reduce road impacts are to remove roads, or to reduce the physical size and traffic volume of roads (e.g., through better public transportation, increased gas taxes, and reduced commuting distances).

When roads must be built, the proper placement of a road is the best mitigation measure. This requires knowledge of the landscape and the distributions and movement patterns of the wild populations of concern. A good example is the decision to extend the Florida Turnpike (USA) so that it bypasses rather than bisects two state forests. Placing the road through high-quality wildlife habitat would have resulted in habitat loss and fragmentation, edge effects, and mortality of the low-flying, threatened Red-cockaded Woodpecker (*Picoides borealis*) (Gilbert 1998). Plans also are being formulated to prevent secondary impacts from development pressures caused by the Turnpike (Gilbert 1998).

Measures for mitigation against habitat loss involve limiting rights-of-way and clearing activities to as small an area as possible (Gilbert 1998). For example, eliminating debris, erosion, and sedimentation in waterways crossed by roads has been successful in maintaining intact mussel habitat in North Carolina, USA (Savidge 1998). Creating alternative, accessible habitat has shown some success with amphibian spawning grounds in western Europe (Langton 1989). The Florida Department of Transportation has created and manages six regional mitigation parks to "offset" development impacts to threatened species (Gilbert 1998). To truly mitigate habitat loss, the new habitat must be placed within the ecological neighborhood of existing populations of the species of interest. Designating already existing habitat as "protected" or creating isolated habitats does not replace lost habitat.

Various measures have been implemented to mitigate effects of roads on the quality of roadside habitat. Educating roadside workers on low-disturbance work practices and the value of roadside habitat has had success in Australia (Straker 1998). Roadside vegetation screens and earthen berms can be used to reduce visual and noise disturbance in habitat adjacent to roadways (Tewes and Blanton 1998). Disturbance of ungulates by highways was lower when shielding was present (Ward 1973, cited in Singer and Doherty 1985). Eliminating or reducing light pollution by directing all lighting toward the road surface may be necessary before cougars (*Felis concolor*) will use roadside habitat (Beier 1995). Reduced habitat quality due to disruption of water flow by roads can be mitigated using properly placed culverts. The open-bottomed culvert and bridge of relatively large size allow fish passage and normal flow of water, and they reduce debris accumulation (Thomas 1998).

The main effect of roads on habitat quality is through increases in wildlife mortality. Faunal traffic mortality can be reduced by modifying either wildlife behavior or human behavior. Some animals are attracted to roads because roads offer food in the form of roadkill, basking animals, spilled grain, and road salt (Forman and Alexander 1998). Removing these sources of food reduces the likelihood that attracted individuals will be on the road and vulnerable to traffic (Straker 1998). An effective deterrent to wildlife traffic mortality is to erect a physical barrier such as a fence (Jackson and Griffin 1998). The fence must be sufficiently tall, long, solid, and deep in the ground to stop animals from going over, around, through, or under it. An improperly installed fence that permits animals to enter the road may trap the animals on the road, thus increasing mortality (J. Duquesnel, personal communication).

Creating a complete barrier limits the risk of traffic mortality but also reduces landscape connectivity, which could have negative consequences for the regional population. Measures permitting animals to cross roads safely simultaneously mitigate direct effects of mortality on local populations and indirect effects on landscape connectivity (see Figure 13.1). One such measure is the placement of light reflectors, which are designed so that light from car headlights hits a reflector and is reflected perpendicular to the road and, in theory, into the eyes of approaching wildlife. The light is assumed to frighten animals and cause them to retreat from the road edge when a car is on the road. Reflectors have shown mixed success for deer in North America and small mammals in Australia (Armstrong 1992; Straker 1998). Animals appear to become habituated to the light and eventually disregard it (A. P. Clevenger, personal communication). In addition, where there is an embankment along the road, the light may shine over or under the head of the approaching animal. Reduction in traffic mortality after installation of reflectors in Australia was probably due to increased driver awareness rather than deterrence of animals (Straker 1998).

Human behavior can be modified by raising awareness of wildlife traffic mortality and limiting human use of roads. The most common measure is the installation of wildlife crossing signs to raise driver awareness. However, this is of limited value, especially when the animals are rarely seen (Pafko and Kovak 1996). Reducing speed limits is more effective. Yellowstone National Park (USA) instituted a speed limit of 73 km/h (45 mph) when they determined that vehicle speed was the primary factor contributing to vehicle-wildlife collisions. They found that road design, such as uneven surfaces, more curves than standard highway specifications, and narrow pavement width, was more effective at reducing vehicular speed than were posted signs (Gunther et al. 1998). An effective measure that works on a site-by-site basis is road closures during times of high wildlife activity. On Phillip Island, Australia, nighttime curfews and restricted access to a subdivision during peak penguin activity resulted in a 10-fold reduction in Fairy Penguin (*Eudyptula minor*) roadkill (Straker 1998).

Other methods for increasing permeability of roads to wildlife involve installation of wildlife crossing structures, i.e., underpasses or overpasses. Underpasses are spaces created under the road and include tunnels, culverts, and areas under bridges. Modifications to existing culverts and bridges so they can be used by animals include adding a raised platform in culverts to provide dry passage, and extending bridges on land so that animals can travel along the bank of the waterway. Overpasses are small-scale land bridges that extend over the road. Overpasses can be vegetated to allow for seminatural crossing conditions.

Research on wildlife use of crossing structures has focused largely on ungulates (deer, *Odocoileus virginianus;* elk, *Cervus canadensis;* mountain goats, *Oreamnos americanus*), panthers, and amphibians (e.g., Singer and Doherty 1985; Langton 1989; Foster and Humphrey 1995; Clevenger 1998). Ungulates and panthers have successfully used underpasses, but results are mixed for amphibian tunnels. In the Foster and Humphrey (1995) and Clevenger (1998) studies, several other mid-sized and large animals such as bobcats (*Felis rufus*), coyotes, and raccoons (*Procyon lotor*) were found to use underpasses frequently. Occasional users included alligators (*Alligator mississippiensis*), black bears, wolves, and cougars. Smaller underpasses (<1.2 m high) are used by small mammals such as mice and rabbits, amphibians and reptiles, and mid-sized carnivores such as badgers, foxes, and wildcats (*Felis sylvestris*) (Yanes et al. 1995; Bekker and Canters 1997). Rodriguez et al. (1996) found that the most important factor determining vertebrate use of culverts under a high-speed railway was the proximity of the passage to the species' habitat.

Clevenger and Waltho (1999) investigated the effects of culvert design and placement on crossing frequencies of eight small- and medium-sized mammals. They found that species differed in their responses. For example, coyotes and martens (*Martes americana*) had a tendency to use older culverts, whereas weasels (*Mustela* sp.), snowshoe hares (*Lepus americanus*), and red squirrels (*Tamiasciurus hudsonicus*) used newer culverts. Culvert use by red squirrels and voles (Arvicolinae) was negatively correlated with culvert openness, whereas use was positively correlated with openness for coyotes. Amount of forest cover nearby, distance to cover, or both affected culvert use by some species. The higher the traffic volume, the greater the use of culverts by martens, weasels, hares, red squirrels, and voles, whereas coyotes used culverts less in high traffic density situations.

Vegetated overpasses may be more successful than underpasses because they are quieter; are less constricting; maintain more natural conditions through the presence of soil, vegetation, and small ponds; and have ambient conditions of rainfall, temperature, and light (Jackson and Griffin 1998). Overpasses appear to accommodate a greater range of organisms, because they can serve as both passageways and as habitat for small mammals, reptiles, amphibians, and invertebrates. Width of overpasses is a crucial element for effectiveness; narrow (<50 m) overpasses are associated with fewer wildlife movements (Keller and Pfister 1997).

Crossing structures do not need to be spaced as closely together for large mobile species as they do for smaller animals. Also, Rodriguez et al. (1996) and Clevenger and Waltho (1999, 2000) found that some species preferentially use large culverts, whereas others prefer small ones. Therefore, a variety of sizes of structures is likely to represent a cost-effective strategy that will provide for both large and small species. Roadside fencing that funnels animals to the structures and prevents them from crossing the road at other places is considered important for the success of most crossing structures (Jackson and Griffin 1998). Despite the clear benefits of crossing structures for facilitating faunal movement, there have been no empirical tests of their effect on population abundance or persistence. To date, there are no mitigation measures to address the negative impacts of "road corridors," such as the spread of exotics.

13.4 Principles for Applying Landscape Ecology

As discussed above, the potential landscape-scale effects of roads on natural populations are significant and widespread. Roads affect many species at once, but to varying degrees and in different ways depending on how the species' movement and demographic characteristics interact with landscape structure, which includes the road network. Principles for successful mitigation require that the task be simplified without compromising the effectiveness of the mitigation. Based on these criteria, we suggest three principles for successful mitigation.

First, the species most vulnerable to road effects should be identified. Several life-history and population traits contribute to the vulnerability of faunal populations to road effects. Large carnivores are inherently vulnerable because they exist in low densities, have low population growth rates and large home ranges, and require large amounts of interconnected habitat (Ruediger 1998). Forman and Alexander (1998) suggest that sustainable populations of large carnivores are possible only in landscapes containing road densities below about 0.6 km/km^2. The average road density in the United States is 1.2 km/km^2 (Forman and Deblinger 1998), and average road density in many European countries is over 2 km/km^2 (European Commission 1999). Generally, species that occur in low densities and have low reproductive rates and long generation times (many threatened and endangered species) are poor at recovering from additional mortality. Highly vagile species, with large home ranges or life cycles that require shifts in habitat, are more likely to encounter roads and be subjected to their effects. Species that depend on more than one habitat (landscape complementation) are particularly susceptible to habitat loss and reduced landscape connectivity. For example, Carr and Fahrig (2001) found that leopard frog (*Rana pipiens*) populations were more vulnerable to traffic mortality than were green frog (*Rana clamitans*) populations, possibly because of the leopard frog's higher vagility and greater reliance on landscape complementation.

A species' vulnerability also depends on the state of its habitat. Species in landscapes with low amounts of habitat are likely already close to an extinction threshold and, therefore, are vulnerable to additional habitat loss and mortality from roads (Fahrig 2001). Fragmentation effects of roads also will be more severe in landscapes containing low amounts of habitat (Fahrig 1998).

The second principle is that mitigation measures should target the most vulnerable species. For example, removing food trees along roadsides except at an underpass as a funneling measure may work for the koala (*Phascolarctos cinerus*) (Straker 1998) but is detrimental to other species that depend on those trees. Installing culverts may allow the flow of aquatic species but not terrestrial species. Species that exhibit behavioral avoidance due to noise will not use overpasses unless traffic noise is reduced in the area. If most traffic mortality occurs during seasonal migrations, then closing roads during these migration periods can reduce mortality.

The third principle is that mitigation measures should be evaluated based on their effectiveness to reduce population-level impacts for a variety of species and landscapes simultaneously. Here we attempt a ranking of measures in this context, based on current knowledge. First, the most general and influential measure is to reduce the number of roads (existing and new) and vehicles. This would result in the greatest reduction of road impacts for every species. No amount or combination of other mitigation measures can attain the effectiveness of eliminating roads. If a road must be built, the most effective measure is to build the road completely below ground as a tunnel or (less effective) above ground as a viaduct. However, due to the cost of such construction, these measures are likely only feasible over short distances to avoid critical habitats (see below). A more practical solution is to place the road where it is least likely to have an effect on wildlife. When the objective of a new road is primarily to accommodate increased traffic volume to existing destinations, new roads should be placed in already developed areas near other roads, or existing roads should be expanded to increase traffic volume. This would result in a smaller additional effect than would building a new road through undeveloped natural areas. Expansion of an existing road also may result in a road-effect zone that is smaller than is the total road-effect zone of two separate roads (but see Section 13.5).

These measures may not be acceptable options when the objective of a new road is to create a more direct link between locations. In this case, construction of the new road will likely result in unavoidable loss and fragmentation of habitat. Where this occurs, mitigation measures should increase the ability of animals to move successfully across (under or over) the roadway. The most successful of such measures are those that maintain critical stretches of habitat intact. This is most effectively done by building sections of the road to avoid the habitat (i.e., build a road bridge or tunnel), as opposed to placing the road through the habitat and then reconstructing the broken link (i.e., wildlife overpass or tunnel). If the road is already in place, wildlife overpasses, wildlife tunnels, and fences should be used.

13.5 Knowledge Gaps

Little or no research has been conducted explicitly to study the effects of roads on landscape processes, and mitigation measures have been applied largely without consideration of landscape-scale effects. Therefore, the scope for research in this area is almost limitless. However, some knowledge gaps are particularly problematic. The most critical gap is in the area of design principles for landscape planners. To date, landscape-scale ecological considerations in design of road networks have been nonexistent at worst and post hoc, ad hoc, or both, at best. Well-designed studies are needed to form the basis for such design principles. Theoretical work is needed to derive design principles, whereas empirical work is needed to provide realistic input assumptions for modeling studies as well as to test resulting theoretical predictions.

13.5.1 Theoretical Voids

Currently, the only theories linking roads and landscape ecology concepts are the verbal arguments in Section 13.2. Quantitative theory is needed to address the following three questions.

First, under what conditions is the impact of one large road less than that of several smaller roads? The answer to this depends on the road-effect zone, which varies among species with different behaviors and vulnerabilities to road impacts. Expanding an existing road rather than building a new road may be more costly, given higher land prices in already developed areas, and it may result in less-efficient human travel. Justification of this action therefore requires evidence that there is an overall benefit to regional wildlife populations.

Second, are there general indices of landscape structure that can predict impacts of different road pattern scenarios for different species guilds? Here, a guild is defined as a set of species with similar life-history and movement attributes. The effect of roads on population survival depends on species' traits, the road pattern, and the landscape structure. Predicting such effects for many species in a particular landscape is therefore likely to require a complex multispecies analysis. It would be of great value to know whether there are robust measures of landscape structure (including the road pattern) that can be used to draw reasonable conclusions without requiring highly complex multispecies analyses.

Third, can one derive an optimal mitigation strategy, such that landscape connectivity can be restored for the largest number of species at the lowest possible cost? The goal would be an algorithm for predicting the optimal distribution of types, sizes, and spacing of mitigation measures for a given set of species and a given road pattern. The objective is to move away from ad hoc solutions for individual species, but also (as above) to avoid the need for building a complex multispecies landscape-scale simulation model for each road development.

13.5.2 Empirical Voids

Many of the relationships between road effects and landscape ecology concepts (Section 13.2), as well as much of the collective wisdom regarding mitigation measures, are based on assumptions and inferences rather than on empirical evidence. Empirical research is critically needed in at least five areas.

First, studies documenting the population-level effects of roads are needed. Counts of traffic mortality victims can reach the thousands (Ashley and Robinson 1996). However, only a few studies have demonstrated significant population-level impacts of roads (e.g., Reh and Seitz 1990; Fahrig et al. 1995; Reijnen 1995). Transportation agencies may require this kind of evidence to justify expensive mitigation efforts (D. McAvoy, personal communication).

Second, empirical studies are needed to determine the size of the road-effect zone. Forman and Deblinger (1998) provide a rough estimation that 15% to 20% of the land in the United States is directly affected ecologically by roads. Indirect effects of roads on individual species or guilds can extend out to at least 1 km (Findlay and Houlahan 1996; Carr and Fahrig 2001). Such information is needed for a wide range of taxa to build an understanding of the cumulative scale of roads on the landscape.

Third, the responses of wildlife individuals and populations to traffic may include threshold phenomena. For example, there may be a threshold in traffic volume above which the probability of safe crossing drops to near zero. Similarly, there are likely threshold effects of traffic mortality on the regional population, in which a small increase in traffic volume has a large effect on persistence of the regional population (Fahrig 2001). If such thresholds occur but their locations (in terms of traffic volume) are unknown, then gradually increasing traffic volume could lead to precipitous crashes of wildlife populations when the threshold is (unknowingly) crossed. Knowledge of such thresholds is therefore critical.

Fourth, studies are needed to quantify the separate effects of traffic mortality on local population persistence and landscape connectivity. Traffic mortality eliminates individuals that would have contributed to both local population dynamics and regional population survival through movement. If loss of individuals from the local population causes the greatest impact, then blocking the population from crossing roads is a feasible mitigation option. For example, Henein and Merriam (1990) showed that decreasing structural connectivity of a landscape can increase population persistence if dispersal mortality is high. If traffic mortality is a significant factor at both the local and regional levels, then reducing mortality and enhancing landscape connectivity are both needed for mitigation.

Fifth, few studies quantitatively test and compare effectiveness of different mitigation structures and identify the factors important for their effectiveness. Clevenger and Waltho (1999, 2000) conducted such analyses for wildlife underpasses. However, many widely held assumptions, such as that wildlife overpasses are generally more effective than are wildlife underpasses, remain untested.

13.6 Research Approaches

13.6.1 Approaches for Theoretical Research

The theoretical research questions identified in Section 13.5.1 could be addressed by conducting simulation experiments using a generalized landscape-scale spatial model of animal population dynamics and movement. Such a model would not be tailored to a particular species or landscape, but would be general and flexible enough to represent different species guilds and landscape structures by altering parameter values. The approach would be to develop general principles by conducting many simulation runs in an experimental design, in which parameter values are changed systematically between runs. Principles would then be derived from the relationships between the parameter values and population response variables such as population abundance and persistence (Fahrig 1991).

This approach could be used for all three theoretical research questions. To compare effects of one large versus several small roads, simulations would be conducted for a range of landscape structures and species guilds to determine how these factors affect the tradeoff between size and number of roads. Indices of landscape structure for predicting road impacts could be developed by conducting simulations in which landscape structure (including road pattern) is systematically altered between simulation runs, and the effects on regional abundance and persistence of species guilds are determined. The results would then be analyzed to determine whether there are indices of landscape structure that can predict the simulated road impacts (e.g., Schumaker 1996). The question of optimal mitigation strategies would require the added step of incorporating economic costs of each mitigation measure. Different simulation runs would be conducted with different mitigation strategies. Model output would be the population response and the cost of the mitigation strategy. The optimal mitigation strategy for a given road pattern would be the one that produced acceptable abundance and persistence levels for wildlife populations (using predefined criteria) at the lowest cost.

13.6.2 Approaches for Empirical Research

Success of these theoretical studies would depend on availability of realistic parameter values, generated from empirical studies. Critical empirical information for model input includes (1) road-effect zones for different species, (2) threshold values of species' responses to roads, and (3) information on responses of various species to different mitigation structures.

Two general sampling designs are appropriate for empirical studies of road effects on wildlife populations. The first is road-centered, in that the road is the focal point of study and the wildlife population is sampled in the surrounding landscape (e.g., Reijnen 1995; Fahrig et al. 1995). The second design is habitat-centered, in that a habitat type of interest is the focal point of the study. The wildlife population is assessed within the habitat patch, and the predictor variable is the length of road or amount of traffic in the landscape surrounding the patch (e.g., Findlay and Houlahan 1996; Carr and Fahrig 2001).

The road-effect zone could be determined using either approach. Using the road-centered approach, several transects could be run perpendicular to the road(s). Along each transect, response variables, such as abundance of individual species or species richness, would be measured. By plotting the mean response against distance from the road, one should detect a maximum distance at which there is no change in the response. This distance defines the road-effect zone.

The habitat-centered approach also could be used to estimate the road-effect zone. Road (or preferably traffic) density would be calculated in concentric circular landscapes of increasing diameter around patches of the habitat type of interest (e.g., wetlands: Findlay and Houlahan 1996; ponds: Carr and Fahrig 2001). The response variable (e.g., population density, species richness) would be measured in each focal habitat area, and the predictor variable would be road density in the circular landscapes around each focal habitat. If traffic volume information is available for all roads in the landscapes, traffic density rather than road density would be used because traffic density integrates the length of road in the landscape with the traffic volume on those roads (e.g., Carr and Fahrig 2001). A separate analysis would be conducted for each landscape diameter. The strength of the relationship (e.g., R^2 value) between road density and the response variable(s) should increase with increasing diameter of the landscape, up to a critical diameter, which determines the road-effect zone.

Either the road-centered or the habitat-centered approach could be used to identify thresholds in the relationship between traffic volume and population densities in adjacent habitat. For the road-centered approach, the sample would need to include roads with a wide range of traffic volumes. For the habitat-centered approach, one would need habitats situated in various landscapes whose traffic densities varied widely. The relationship between population density and either traffic volume (road-centered approach) or traffic density (habitat-centered approach) would then be examined for evidence of a sharp drop or threshold in population density at a given traffic volume or traffic density.

Local population and regional population (Figure 13.1) effects of road mortality could be separated using a road-centered, multilandscape experiment. In each landscape, three treatments would be applied, each to a different local population. In the first treatment, a barrier (e.g., a fence) would be installed on the other side of the road from the population. In the second treatment, a barrier would be installed on the same side of the road as the population. And for the third treatment, no barrier would be installed. Therefore, the first population would experience both loss of individuals (local impact) and the barrier effect (regional impact), the second would experience only the barrier effect, and the third would experience only loss of individuals. Abundances of all populations would be monitored. If the results showed no difference in mean abundance between populations experiencing both mortality and the barrier effect and populations only experiencing the barrier effect, then one would conclude that local mortality is of little import and impact of roads on landscape connectivity is the main concern. If there is no difference in mean abundance between populations experiencing both mortality and the barrier effect and populations only experiencing mortality, then one would conclude that local mortality has a relatively greater impact than does the barrier effect and connectivity is not the main concern. Note that this experimental design would work regardless of the direction of animal movement.

Empirical studies to compare different mitigation measures could be conducted using the road-centered approach. The study design would depend on the degree of control the researcher has over installation of the mitigation structures. In the ideal situation, a multilandscape design would be used in which, in each landscape, each type of mitigation structure would be built on different segments of the same road. Wildlife populations near the road would be monitored before and after construction both near the structures and in control locations where no mitigation structure was built. Statistical analyses (e.g., analysis of variance) would be used to test for differences among mean population responses due to the different structures. In reality, however, this level of control is unlikely to be available to the researcher. To compare different types of mitigation measures using existing structures, populations near the structures (and control populations near the roads in locations where no structure exists) could be sampled for a very large number of structures. The large sample size would be necessary so that uncontrolled variables, such as length of time since a structure has been in place, could be included in the statistical analyses.

13.7 Conclusions

Addressing the impact of roads on the landscape is a daunting task. Roads affect landscape structure and species survival in many ways, through traffic mortality, habitat fragmentation, habitat loss, and changes in the quality of the surrounding habitat. Although the most visible impact, traffic mortality, brought the issue of roads to public attention, closer examination of the issue has brought some important questions to light. What is the overall impact of roads, especially when the landscape is already under stress? To what extent are roads responsible for species' declines? Answering these questions is an important task, as preliminary studies indicate that roads could be a far greater force in population declines than previously thought.

Although mitigation measures are being developed and perfected to deal with some of the impacts of roads, they rely on the assumption that roads are a necessary evil. However, most mitigation measures have a limited potential in their ability to reduce impacts for a large number of species over large areas. The best method from a biological conservation viewpoint is to alter human values and behavior such that our dependency on roads is reduced.

Acknowledgments

We thank Tony Clevenger, Richard Forman, and Kringen Henein for comments on an earlier draft of this chapter. This work was supported by a Natural Sciences and Engineering Council of Canada (NSERC) scholarship to L. Carr and a NSERC grant to L. Fahrig.

References

Adams, L.W., and Geis, A.D. 1983. Effects of roads on small mammals. *J. Appl. Ecol.* 20:403–415.

Addicott, J.F., Aho, J.M., Antolin, M.F., Padilla, D.K., Richardson, J.S., and Soluk, D.A. 1987. Ecological neighborhoods: scaling environmental patterns. *Oikos* 49:340–346.

Andrews, A. 1990. Fragmentation of habitat by roads and utility corridors: a review. *Australian Zool.* 26:130–141.

Armstrong, J.J. 1992. *An Evaluation of the Effectiveness of Swareflex (TM) Deer Reflectors.* Downsview, Ontario, Canada: Ontario Ministry of Transportation.

Ashley E.P., and Robinson, J.T. 1996. Road mortality of amphibians, reptiles and other wildlife on the Long Point Causeway, Lake Erie, Ontario. *Can. Field-Nat.* 110:403–412.

Beier, P. 1995. Dispersal of juvenile cougars in fragmented habitat. *J. Wildl. Manage.* 59:228–237.

Bekker, H.G.J., and Canters, K.J. 1997. The continuing story of badgers and their tunnels. In *Habitat Fragmentation and Infrastructure,* ed. K.J. Canters, pp. 344–353. Delft, The Netherlands: Ministry of Transport, Public Works and Water Management.

Bennett, A.F. 1991. Roads, roadsides and wildlife conservation: a review. In *Nature Conservation 2: The Role of Corridors,* eds. D.A. Saunders and R.J. Hobbs, pp. 99–118. Chipping Norton, Australia: Surrey Beatty and Sons.

Brown, J.H., and Kodric-Brown, A. 1977. Turnover rates in insular biogeography: the effect of immigration on extinction. *Ecology* 58:445–449.

Carr, L.W., and Fahrig, L. 2001. Impact of road traffic on two amphibian species of differing vagility. *Conserv. Biol.* 15:1071–1078.

Clevenger, A.P. 1998. Permeability of the Trans-Canada Highway to wildlife in Banff National Park: importance of crossing structures and factors influencing their effectiveness. In *Proceedings of the International Conference on Wildlife Ecology and Transportation, FL-ER-69-98,* eds. G.L. Evink, P. Garret, D. Zeigler, and J. Berry, pp. 109–119. Tallahassee, Florida: Florida Department of Transportation.

Clevenger, A.P., and Waltho, N. 1999. Dry drainage culvert use and design considerations for small- and medium-sized mammal movement across a major transportation corridor. In *Proceedings of the Third International Conference on Wildlife Ecology and Transportation, FL-ER-73-99*, eds. G.L. Evink, P. Garret and D. Zeigler, pp. 263–277. Tallahassee, Florida: Florida Department of Transportation.

Clevenger, A.P., and Waltho, N. 2000. Factors influencing the effectiveness of wildlife underpasses in Banff National Park, Alberta, Canada. *Conserv. Biol.* 14:47–56.

Den Boer, P.J. 1968. Spreading of risk and stabilization of animal numbers. *Acta Biotheor.* 18:165–194.

Dunning, J.B., Danielson, B.J., and Pulliam, H.R. 1992. Ecological processes that affect populations in complex landscapes. *Oikos* 65:169–175.

Ehmann, H., and Cogger, H. 1985. Australia's endangered herpetofauna: a review of criteria and policies. In *Biology of Australasian Frogs and Reptiles,* eds. G. Grigg, R. Shine, and H. Ehmann, pp. 435–447. Sydney: Surrey Beatty and Sons and Royal Zoological Society of New South Wales.

European Commission. 1999. *EUROSTAT: Regions: Statistical Yearbook 1999.* Luxembourg: European Commission.

Fahrig, L. 1991. Simulation methods for developing general landscape-level hypotheses of single species dynamics. In *Quantitative Methods in Landscape Ecology,* eds. M.G. Turner and R.H. Gardner, pp. 417–442. New York: Springer-Verlag.

Fahrig, L. 1998. When does fragmentation of breeding habitat affect population survival? *Ecol. Model.* 105:273–292.

Fahrig, L. 2001. How much habitat is enough? *Biol. Conserv.* 100:65–74.

Fahrig, L., Pedlar, J.H., Pope, S.E., Talyor, P.D., and Wegner, J.F. 1995. Effect of road traffic on amphibian density. *Biol. Conserv.* 74:177–182.

Findlay, C.S., and Houlahan, J. 1996. Anthropogenic correlates of species richness in southeastern Ontario wetlands. *Conserv. Biol.* 11:1000–1009.

Forman, R.T.T. 1995. *Land Mosaics: The Ecology of Landscapes and Regions.* Cambridge, United Kingdom: Cambridge University Press.

Forman, R.T.T., and Alexander, L.E. 1998. Roads and their major ecological effects. *Ann. Rev. Ecol. Syst.* 29:207–231.

Forman, R.T.T., and Deblinger, R.D. 1998. The ecological road-effect zone for transportation planning, and a Massachusetts highway example. In *Proceedings of the International Conference on Wildlife Ecology and Transportation, FL-ER-69-98,* eds. G.L. Evink, P. Garret, D. Zeigler, and J. Berry, pp. 78–96. Tallahassee, Florida: Florida Department of Transportation.

Forman, R.T.T., and Deblinger, R.D. 2000. The ecological road-effect zone of a Massachusetts (U.S.A.) suburban highway. *Conserv. Biol.* 14:36–46.

Foster, M.L., and Humphrey, S.R. 1995. Use of highway underpasses by Florida panthers and other wildlife. *Wildl. Soc. Bull.* 23:92–94.

Garland, T., Jr., and Bradley, W.G. 1984. Effects of highway on Mojave desert rodent populations. *Am. Midl. Nat.* 111:47–56.

Gibeau, M.L., and Herrero, S. 1998. Roads, rails, and grizzly bears in the Bow River Valley, Alberta. In *Proceedings of the International Conference on Wildlife Ecology and Transportation, FL-ER-69-98,* eds. G.L. Evink, P. Garret, D. Zeigler, and J. Berry, pp. 104–108. Tallahassee, Florida: Florida Department of Transportation.

Gilbert, T. 1998. Technical assistance and agency coordination on wildlife and habitat conservation issues associated with highway projects in Florida. In *Proceedings of the International Conference on Wildlife Ecology and Transportation, FL-ER-69-98,* eds. G.L. Evink, P. Garret, D. Zeigler, and J. Berry, pp. 209–213. Tallahassee, Florida: Florida Department of Transportation.

Gunther, K.A., Biel, M.K.J., and Robison, H.L. 1998. Factors influencing the frequency of road-killed wildlife in Yellowstone National Park. In *Proceedings of the International Conference on Wildlife Ecology and Transportation, FL-ER-69-98,* eds. G.L. Evink, P. Garret, D. Zeigler, and J. Berry, pp. 32–42. Tallahassee, Florida: Florida Department of Transportation.

Hanski, I. 1989. Metapopulation dynamics: does it help to have more of the same? *Trends Ecol. Evol.* 4:113–114.

Harris, L.D., and Scheck, J. 1991. From implications to applications: the dispersal corridor principle applied to the conservation of biological diversity. In *Nature Conservation 2: The Role of Corridors,* eds. D.A. Saunders and R.J. Hobbs, pp. 189–220. Chipping Norton, Australia: Surrey Beatty and Sons.

Henein, K., and Merriam, G. 1990. The elements of connectivity where corridor quality is variable. *Landsc. Ecol.* 4:157–170.

Jackson, S.D., and Griffin, C.R. 1998. Toward a practical strategy for mitigating highway impacts on wildlife. In *Proceedings of the International Conference on Wildlife Ecology and Transportation, FL-ER-69-98,* eds. G.L. Evink, P. Garret, D. Zeigler, and J. Berry, pp. 17–22. Tallahassee, Florida: Florida Department of Transportation.

Keller, V., and Pfister, H.P. 1997. Wildlife passages as a means of mitigating effects of habitat fragmentation by roads and railway lines. In *Habitat Fragmentation and Infrastructure,* ed. K.J. Canters, pp. 70–80. Delft, The Netherlands: Ministry of Transport, Public Works and Water Management.

Lande, R. 1988. Genetics and demography in biological conservation. *Science* 241:1455–1460.

Langton, T.E.S., ed. 1989. *Amphibians and Roads.* Shefford, Bedfordshire, United Kingdom: ACO Polymer Products Ltd.

Lidicker, W.Z., and Koenig, W.D. 1996. Responses of terrestrial vertebrates to habitat edges and corridors. In *Metapopulations and Wildlife Conservation,* ed. D.R. McCullough, pp. 85–109. Washington, DC: Island Press.

Mader, H.J. 1984. Animal habitat isolation by roads and agricultural fields. *Biol. Conserv.* 29:81–96.

Mader, H.J., Schell, C., and Kornacker, P. 1990. Linear barriers to arthropod movements in the landscape. *Biol. Conserv.* 54:209–222.

Murcia, C. 1995. Edge effects in fragmented forests: implications for conservation. *Trends Ecol. Evol.* 10:58–62.

Oxley, D.J., Fenton, M.B., and Carmody, G.R. 1974. The effects of roads on populations of small mammals. *J. Appl. Ecol.* 11:51–59.

Pafko, F., and Kovach, B. 1996. Experience with deer reflectors. In *Trends in Transportation Related Wildlife Mortality, FL-ER-58-96,* eds. G.L. Evink, D. Zeigler, and J. Berry. Tallahassee, Florida: Florida Department of Transportation.

Pope, S.E., Fahrig, L., and Merriam, H.G. 2000. Landscape complementation and metapopulation effects on leopard frog populations. *Ecology* 81:2498–2508.

Pulliam, H.R. 1988. Sources, sinks, and population regulation. *Am. Nat.* 132:652–661.

Ranney, J.W., Bruner, M.C., and Levenson, J.B. 1981. The importance of edge in the structure and dynamics of forest islands. In *Forest Island Dynamics in Man-Dominated Landscapes,* eds. R.L. Burgess and D.M. Sharpe, pp. 67–92. New York: Springer-Verlag.

Reh, W., and Seitz, A. 1990. The influence of land use on the genetic structure of populations of the common frog *Rana temporaria. Biol. Conserv.* 54:239–249.

Reijnen, R. 1995. *Disturbance by Car Traffic as a Threat to Breeding Birds in The Netherlands.* Ph.D. dissertation. Wageningen, The Netherlands: DLO Institute for Forestry and Nature Research.

Rodriguez, A., Crema, G., and Delibes, M. 1996. Use of non-wildlife passages across a high speed railway by terrestrial vertebrates. *J. Appl. Ecol.* 33:1527–1540.

Ruediger, B. 1998. Rare carnivores and highways—moving into the 21st century. In *Proceedings of the International Conference on Wildlife Ecology and Transportation, FL-ER-69-98,* eds. G.L. Evink, P. Garret, D. Zeigler, and J. Berry, pp. 10–16. Tallahassee, Florida: Florida Department of Transportation.

Savidge, T. 1998. Management of protected freshwater mussels with regard to North Carolina Department of Transportation highway projects. In *Proceedings of the International Conference on Wildlife Ecology and Transportation, FL-ER-69-98,* eds. G.L. Evink, P. Garret, D. Zeigler, and J. Berry, pp. 143–150. Tallahassee, Florida: Florida Department of Transportation.

Schumaker, N.H. 1996. Using landscape indices to predict habitat connectivity. *Ecology* 77:1210–1225.

Singer, F.J., and Doherty, J.L. 1985. Managing mountain goats at a highway crossing. *Wildl. Soc. Bull.* 13:469–477.

Straker, A. 1998. Management of roads as biolinks and habitat zones in Australia. In *Proceedings of the International Conference on Wildlife Ecology and Transportation, FL-ER-69-98,* eds. G.L. Evink, P. Garret, D. Zeigler, and J. Berry, pp. 181–188. Tallahassee, Florida: Florida Department of Transportation.

Taylor, P.D., Fahrig, L., Henein, K., and Merriam, G. 1993. Connectivity is a vital element of landscape structure. *Oikos* 68:571–572.

Tewes, M.E., and Blanton, D.R. 1998. Potential impacts of international bridges on ocelots and jaguarundis along the Rio Grande Wildlife Corridor. In *Proceedings of the International Conference on Wildlife Ecology and Transportation, FL-ER-69-98,* eds. G.L. Evink, P. Garret, D. Zeigler, and J. Berry, pp. 135–139. Tallahassee, Florida: Florida Department of Transportation.

Thomas, A.E. 1998. The effects of highways on western cold water fisheries. In *Proceedings of the International Conference on Wildlife Ecology and Transportation, FL-ER-69-98,* eds. G.L. Evink, P. Garret, D. Zeigler, and J. Berry, pp. 249–252. Tallahassee, Florida: Florida Department of Transportation.

United Nations. 1997. *Statistical Yearbook 42nd Issue.* New York: United Nations.

van Gelder, J.J. 1973. A quantitative approach to the mortality resulting from traffic in a population of *Bufo bufo* L. *Oecologia* 13:93–95.

Vermeulen, H.J.W. 1994. Corridor function of a road verge for dispersal of stenotopic heathland ground beetles. *Biol. Conserv.* 69:339–349.

Wace, N.M. 1977. Assessment of dispersal of plant species—the car-borne flora in Canberra. *Proc. Ecol. Soc. Australia* 10:167–186.

Wiens, J.A. 1996. Wildlife in patchy environments: metapopulations, mosaics, and management. In *Metapopulations and Wildlife Conservation,* ed. D.R. McCullough, pp. 53–85. Washington, DC: Island Press.

Wilcox, B.A., and Murphy, D.D. 1985. Conservation strategy: the effects of fragmentation on extinction. *Am. Nat.* 125:879–887.

Yanes, M., Velasco, J.M., and Suárez, F. 1995. Permeability of roads and railways to vertebrates: the importance of culverts. *Biol. Conserv.* 71:217–222.

14

Landscape Pattern, Timber Extraction, and Biological Conservation

ERIC J. GUSTAFSON AND NANCY DIAZ

14.1 Introduction

Timber extraction has become controversial because it is perceived to be harmful to forested and riparian ecosystems. This perception is based on aesthetic concerns and the belief that current timber management practices are neither sustainable nor compatible with a functional level of biodiversity. Reduction in biodiversity is believed to result from reduction in the area of certain habitat conditions and by changes in the spatial configuration of ecosystems that disrupt ecosystem interdependencies. In this chapter, we will (1) review the concepts, principles, and emerging ideas about how the spatial configuration of timber extraction activity affects biodiversity, (2) describe recent applications of landscape ecology to forest management, (3) define principles for applying landscape ecology to timber management to conserve biodiversity, (4) describe knowledge gaps that hamper our ability to manage more effectively, and (5) propose research approaches for filling those gaps.

14.2 Concepts, Principles, and Emerging Ideas

14.2.1 Reciprocal Effects of Pattern and Process

Landscape ecology emphasizes relationships between the spatial pattern of landscape elements and ecological processes (Turner 1989). *Landscapes* are usually described as mosaics of various land uses arranged in some spatial pattern (Forman 1995). These land uses are usually distinguished by differences in vegetation. For example, human-dominated landscapes have some areas devoid of vegetation, whereas other areas are cropped or have vegetation that is intensely managed (e.g., parks, golf courses). In more natural landscapes, the *mosaic* (interspersed land-cover types) is described by differences in the vegetative communities found there (e.g., hardwood forest, early successional forest, grassland). The spatial arrangement of the landscape mosaic is believed to have important consequences for ecological processes such as nutrient cycling, animal and plant dispersal, and predator-prey interactions (Turner 1989).

Landscape ecology recognizes the dynamic nature of landscapes, sometimes described as a *shifting mosaic* (Bormann and Likens 1981; Wiens 1994). Change in the spatial pattern of landscapes occurs by *disturbance* (destruction of vegetation) and the gradual change in the structure and composition of ecological communities by succession. However, succession may be prevented in intensely managed or developed areas. Disturbance can be natural or anthropogenic, and it usually results in some change to the vegetation in the disturbed area. The ecological structure seen today was produced by past ecological processes, including disturbance (Forman and Godron 1986). Obviously, the structure, function, and disturbance of the present are forming future ecological systems. Changes in the pattern of vegetative communities through time have consequences on ecological phenomena such as dispersal, energy and nutrient flows, and faunal community composition. Community dynamics are essential to the functioning of healthy ecosystems because all species do not have the same habitat requirements. Although many species require late-successional habitat, many other species depend on recently disturbed habitat. Managing community dynamics to provide a continuous supply of this variety of habitat conditions is a complex challenge.

All disturbance, including that by timber extraction, produces a shifting mosaic of successional habitats that has spatial and temporal dynamics to which ecological communities respond. A specific habitat condition may be destroyed by harvest, but suitable conditions for other species temporarily replace it. Succession often reproduces the original habitat condition, although the possibility of alternative stable communities as endpoints of succession (Gilpin and Case 1976) adds to the complexity of forest management. Change in ecological spatial pattern also is related to the rate of recovery from disturbance. When recovery is quick, disturbance effects are more transient. The rate of recovery from timber harvest may vary widely depending on a number of factors, most notably climate (precipitation and temperature), soil conditions, intensity of harvest disturbance, and subsequent silvicultural treatments such as thinning and planting (Oliver 1981). For this reason, the persistence of landscape change varies in different parts of the world. To understand the effect of timber extraction on biodiversity, we must understand how ecological communities are adapted to a spatially and temporally dynamic mosaic of habitat conditions across a landscape.

Vegetative management, usually by silvicultural extraction of timber, is the primary determinant of landscape pattern in managed forests (Franklin and Forman 1987; Li et al. 1993). Disturbance has two primary effects on vegetative communities. It may change (1) the relative abundance of community types and successional stages, and (2) the spatial configuration (juxtaposition and scale of patchiness) of those communities across a landscape. *Forest management* is the management of disturbance and succession to achieve specific vegetation and ecological conditions that in turn support the products and benefits sought by the manager. Disturbance by timber harvest is in some ways similar to disturbance by natural forces such as fire, hurricanes, or more local wind events (Attiwill 1994). The effects of timber harvest, as in natural disturbance, are usually ephemeral—the forest is eventually replaced through succession. Both types of disturbance can be characterized by the frequency, severity, and size of disturbance events, and these characteristics occur across a continuum of temporal and spatial scales (Attiwill 1994). However, timber harvest activity rarely corresponds exactly to natural disturbances. For example, clearcutting may be as destructive to vegetation as are severe natural fire events, but the size of harvest units may be smaller and harvest may occur more frequently than do severe fires. Some partial cutting systems produce residual forests similar in structure to those produced by wind events, but these systems may be implemented in addition to naturally occurring wind disturbance. Even low-intensity harvest methods such as single-tree selection leave less biomass on a site than do windthrow disturbance events. These differences add complexity that confounds our attempts to understand the effects of timber management in the context of natural disturbance.

Patches are contiguous areas of similar conditions, and they may be defined using one of many ecological criteria such as dominant vegetation type, successional stage, or canopy closure. The spatial configuration (*patchiness*) of a landscape mosaic is the result of the interaction of past disturbance and the heterogeneity of the abiotic environment (Turner 1989). Disturbance (including by timber harvest) usually produces new patches by changing one (or many) of these various criteria. A change in the disturbance regime has the potential to alter significantly the scale of patchiness of the landscape mosaic (Wu and Loucks 1995).

The practice of dispersing harvests has been implicated as a major contributor to the reduction in forest interior habitat and the increase in linear edge (Franklin and Forman 1987), which negatively impacts species that are sensitive to the size of forested blocks (Blake and Karr 1987). One potential solution for managers of large ($>10^4$ ha) forest holdings is a *dynamic-zoning strategy* (Gustafson 1996). Here, timber production is concentrated in a relatively small portion (zone) of the timber production land base for a limited amount of time, and this zone is moved sequentially to other portions of the timber production land base at 30–50-year intervals (variant of progressive cutting; Li et al. 1993).

Ecological resilience is the ability of an ecological system to maintain its functions and ecological relationships in the face of change or disturbance (Holling 1973). A resilient system is *stable* when it tends to return to its original dynamics after being disturbed, rather than reorganizing into an alternative state. Ecologists believe resilience is related to diversity because species perform diverse ecological functions such as regulating biogeochemical cycles (Vitousek 1990) and modifying the physical environment (Jones et al. 1994). It is not clear why greater diversity usually makes ecosystems more resilient, but most hypotheses are related to redundancy of functional species groups. Recently, Peterson et al. (1998) proposed that ecological resilience is a consequence of diverse, but overlapping functional groups within a scale and (apparently) redundant species that operate at different spatial scales. If this is correct, the scale and pattern of ecosystem patchiness should be an important component of ecological resilience.

14.2.2 Principles of Landscape Ecology Relevant to Forest Management

Landscape ecology is the study of the reciprocal effects between ecological pattern and ecological function. Although the discipline is relatively young, a theoretical basis has emerged. The following ideas are relevant to timber extraction and biological diversity issues.

Landscape Structure and Function Principle

Landscapes that differ structurally (i.e., vary in amount and spatial arrangement of patch types) will differ in ecological function (ecological processes such as nutrient cycling, population dynamics, resistance to disturbance) (Forman and Godron 1986). For example, we know that birds nesting in forests embedded in agricultural lands experience higher rates of nest predation and brood parasitism by cowbirds than do birds nesting in heavily forested landscapes (Robinson 1992). It is less clear how birds are affected by timber harvest activities, but the spatial pattern and relative proportion of seral stages in a managed forest have important impacts on Neotropical birds (Thompson et al. 1992) and some keystone species such as deer (Augustine and Jordan 1998). It also is likely that the spread of disturbance in forests is related to the spatial pattern of the forest mosaic (Turner et al. 1989).

Ecological Edge Principle

There are unique ecological interactions that occur near *forest edges* (forest that is in proximity to an opening or nonforest type). Vegetation is affected by changes in light and microclimate that may extend tens of meters into the forest (Chen et al. 1992). Some species prefer edge habitat, and their numbers respond positively to the creation of edge (Litvaitis 1993). Conversely, some species (e.g., certain Neotropical migratory landbirds) appear to be negatively impacted by edge habitat, either by avoidance (King et al. 1997) or through reproductive losses in edge habitats (Paton 1994).

Hierarchy Theory

Dynamics of ecological systems are driven by events occurring at finer and faster scales, and they are constrained by events occurring at broader and slower scales. For example, changes in forest stands are the cumulative result of tree growth and mortality, nutrient cycling, and the like, which exhibit dynamics at the scale of trees measured on a time scale of years and decades. The effects of these phenomena are constrained by climate and regional patterns of land use and disturbance, phenomena that are dynamic at much broader scales and measured on a time scale of centuries (Urban et al. 1987). An important consequence of the hierarchical structure of ecosystems is that nonequilibrium dynamics at a particular scale may produce stable dynamics at another (Rahel 1990). For example, cherry

(*Prunus* spp.) readily invades forest openings but is quickly replaced by longer-lived shade-tolerant species. At the stand level, cherry is ephemeral, but the abundance of cherry remains nearly constant over the landscape (Urban et al. 1987).

Nonequilibrium Theory of Community Composition

Community composition is determined primarily by disturbance (Huston 1994), and there may be multiple possible successional endpoints (Gilpin and Case 1976). Nonequilibrium theories of community structure suggest that the diversity of species and the coexistence of functionally similar species in communities are generated by disturbance and the resulting opportunity for recruitment of new species into the community (Huston 1994). The long-term role of disturbance is sometimes minimized when considering the preservation of "natural ecosystems" (Attiwill 1994). For example, in the Central Hardwood Region (USA), there appears to be a trend toward the conversion of native oak-hickory (*Quercus-Carya*) communities to beech-maple (*Fagus-Acer*) communities, which is thought to be the result of fire suppression (Lorimer 1985).

14.2.3 Emerging Ideas

Forest management is usually implemented with the goal of achieving multiple uses and benefits from forests. With increasing emphasis on the benefits provided by biological diversity, the spatial pattern of managed landscapes has become an important management consideration (Diaz and Apostol 1992). Management plans may contain desired future conditions for landscape patterns. These conditions fall on a continuum based on mimicking natural patterns, ranging from being narrowly focused on producing natural patterns to having a broader range of pattern objectives that may include designed reserves or other land allocations. Typical objectives may be to provide connectivity, protect unstable landforms, locate harvest activities away from sensitive areas, and provide for human settlements. The common assumption is that ecological systems are adapted to past disturbance regimes and can be sustained under dynamics similar to past dynamics (Attiwill 1994).

Approaches to timber-extraction planning that primarily focus on mimicking natural disturbance patterns are illustrated by the following three examples. An analysis of historic landscape composition and pattern was used to develop objectives for landscape patterns in the Blue Mountains of Oregon (Shlisky 1994). The analysis described patterns found in landscapes prior to fire suppression and identified management tools that could be used to restore those patterns. Bergeron and Harvey (1997) proposed a silvicultural system for boreal forests based on natural ecosystem dynamics. They argue that the current cyclical rotation of softwood types causes regeneration difficulties, more insect outbreaks, and loss of site productivity. They advocate cutting systems that result in a succession from hardwoods to mixedwoods to conifers that resembles the sequence observed after natural fire. They also provide direction for a target proportion of various forest types in landscapes, but they provide little direction about the size and spatial distribution of managed stands. Mladenoff et al. (1993) compared adjacent managed and undisturbed landscapes in sub-boreal forests to describe patterns associated with the natural disturbance regime. They advocate management to mimic the complexities and characteristics of natural landscapes.

Other approaches expand the focus beyond natural disturbance patterns to include additional spatial configuration objectives. For example, the island-archipelago approach was proposed to conserve old-growth-dependent communities in the face of timber extraction (Harris 1984). This approach requires a system of habitat islands forming a network connected by organism dispersal through corridors and through the intervening matrix. The network must be designed so that timber harvest neither disrupts the ability of dispersers to reach habitat islands nor degrades conditions within islands so that they lose their habitat quality. Temporal planning must ensure that these criteria are met through time, as forest stands are harvested and regenerate. See Mladenoff et al. (1994) for an application of these ideas to integrate old-growth protection and timber extraction in a northern Great Lakes (USA) landscape.

Landscape design has two defining features that proactively help to guide landscape change: (1) it is spatially explicit (i.e., it depicts the spatial arrangement of pattern elements), and (2) it models landscape change through time in one way or another. In contrast, *landscape planning* typically involves establishing objectives (sometimes spatially explicit) in a narrative fashion, and it may even allocate particular sectors of landscapes to a suite of uses or goals, but it stops short of actually mapping out in detail the resulting landscape patterns in either two- or three-dimensional space.

14.3 Recent Applications

A review of recent landscape ecology literature (e.g., *Landscape Ecology, Ecological Applications*) was conducted to find recent applications of ecological landscape planning. Examples also were sought among planning documents for public lands in the Pacific Northwest and Eastern Canada. Case studies demonstrating an intentional ecosystem approach to managing landscape change and timber extraction are growing in number as conceptual and technological barriers begin to fall. However, actual on-the-ground results are almost nonexistent for two reasons: (1) such techniques have only recently (i.e., the past decade) been developed and applied, and (2) the implementation of landscape designs in forests requires time for trees to grow. Although it is possible to describe various approaches for designing dynamic landscapes that have a conservation biology orientation, it is *not* possible to demonstrate quickly the long-term outcome.

A widely embraced planning model is Landscape Analysis and Design (LAD) described in detail by Diaz and Apostol (1992). This process can be used to facilitate implementation of any of the approaches described in Section 14.2, or a combination of approaches. The process allows for a systematic analysis of existing landscape structure and function and a determination of conditions that would prevail under a natural disturbance regime. It includes factors such as multiscale connections to the larger landscape and land shape (topography), which are ignored by some approaches. LAD also offers spatial design strategies for designing landscape patterns on three-dimensional landscapes to meet specific landscape objectives.

The case studies presented below represent some of the best examples of diverse approaches to creating landscape change using an ecological framework. These examples also reflect the element of landscape design, in contrast to landscape planning.

14.3.1 Landscape Design Applications

In this section, we describe four examples of landscape design that illustrate various approaches.

The Augusta Creek Planning Area effort (Cissel et al. 1998) represents a synthesis of a variety of approaches, but it strongly emphasizes producing landscape patterns that might have arisen from natural disturbance processes (in this case, fire). Augusta Creek is a 7,600-ha planning area dominated by conifer forests on the western flank of the Cascade Mountains in Oregon (USA). The area has been impacted by road building, a system of dispersed clear-cuts, tree plantations, and fire suppression. All of the land is in federal ownership and is managed under a conservation strategy of reserves and intervening managed areas for protection of late-successional forest and aquatic values.

The objectives of the Augusta Creek project were to design a landscape that would maintain populations of native species, ecosystem processes and structures, and ensure long-term productive capacity of the forest, allowing for a substantial amount of timber harvest within nonreserve areas. The design attempted to mimic landscape patterns within the range of that expected under a natural fire regime. An intensive study of fire, based on dendrochronology over the past 500 years, provided the foundation for this work. Based on the frequency, intensity, and size of fires, nine separate fire-regime categories were described. The nine categories were mapped, primarily on the basis of topographic and climatic factors, yielding a depiction of sectors of the landscape within which a particular balance of patch sizes, ages, and stand tree retention levels would have likely occurred. The planners then superimposed a set of aquatic reserves, distributing no-harvest zones (both blocks and stream buffers) across the variety of topographic positions and habitat types throughout the planning area. The intent of these reserves was to protect overall hydrologic stability and aquatic values within the watershed, and to provide for habitat connectivity across the landscape. The resulting conceptual design is shown in Figure 14.1. Within this framework, a harvest scheduling process was applied to individual landscape blocks (20–140 ha), resulting in a specific progression of treatments over 200 years (Figure 14.2) implementing the conceptual design of Figure 14.1.

A key strength of this approach is that it clearly links the broader (i.e., landscape or watershed) and finer (i.e., stand) scales. Timber harvest and other man-

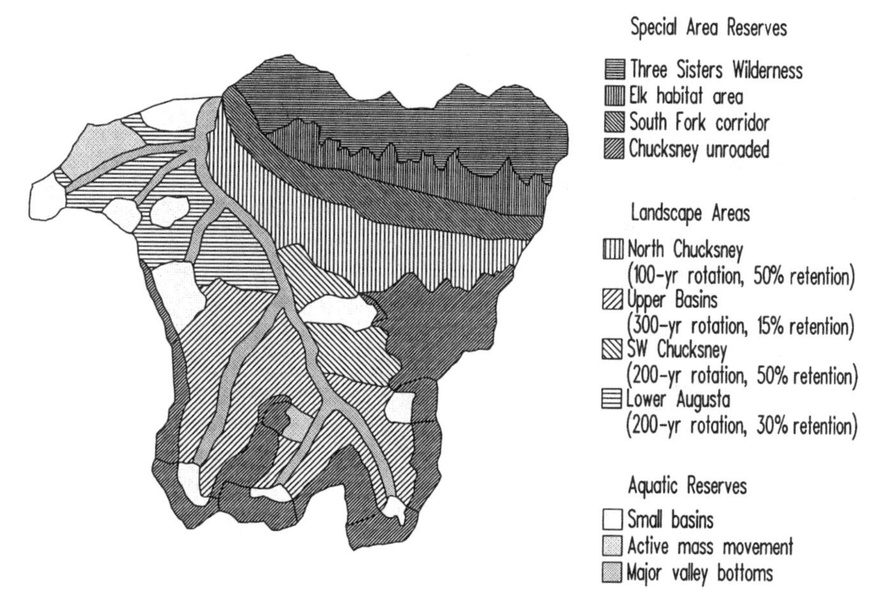

FIGURE 14.1. Boundaries of the management categories for the Augusta Creek Landscape Plan. Each category specifies the reserve status or vegetation management strategy to be applied within the boundaries of each category.

agement activities have a clear road map leading to desired conditions at multiple scales through time. The Augusta Creek example differs from some of the case studies described below in the rather narrow range of issues it addressed, and in its heavy reliance on fire history (in contrast to a broader interpretation of landform position or disturbance, and vegetation potential) to delineate treatment blocks. Also, it is less explicitly driven by conservation biology objectives and less focused on sustainability than are some other efforts.

The Sutherland Brook area is a 5,100-ha tract of Crown Land in central Nova Scotia (Canada) (Murray and Singleton 1995). It is an area of rolling to low relief, with a diverse mix of conifer, mixedwood, and hardwood forests, containing numerous lakes, bogs, and small streams. The area has high value for timber production, fisheries, game and nongame wildlife, and recreation. It also has a high proportion of undisturbed forest relative to the rest of the region. The Forest Ecosystem Design Project for Sutherland Brook was carried out through a partnership between industry, natural resource management agencies, local governmental and nongovernmental organizations, and individual forest users. Goals of sustainability, utility, efficiency, and beauty were primary in the design process.

As in the Augusta Creek project, the Sutherland Brook planners developed a spatially explicit conceptual design for the landscape in which timber harvest and other activities were to be implemented, but the means they used were substantially different. The Sutherland Brook design process combined techniques described by

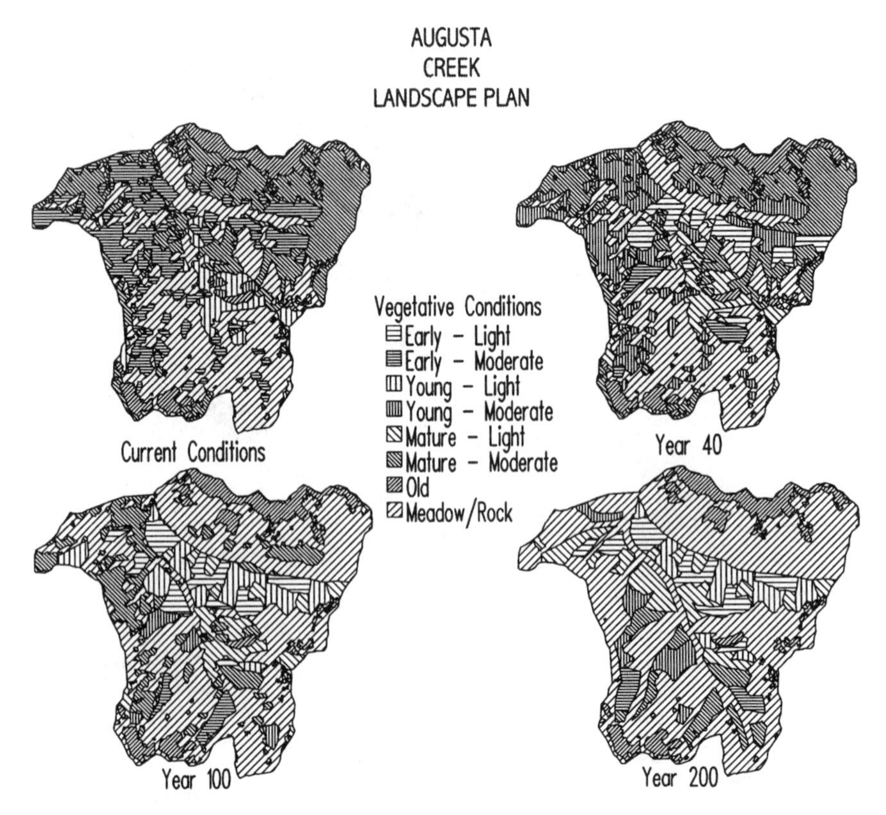

FIGURE 14.2. Future landscape conditions under the Augusta Creek Landscape Plan. Each category (i.e., early-light, etc.) refers to the seral stage and canopy retention level projected for each management unit.

Bell (1994) and Diaz and Apostol (1992), in which a systematic assessment of the composition, structure, and function of the landscape was completed. Existing patch types, landforms, and overall pattern were qualitatively evaluated for their effect on landscape flows and processes. At the same time, the effects of disturbance processes (fire, wind, and biological) and forest succession were analyzed to determine the range of conditions that would likely exist under a natural disturbance regime in various sectors of the landscape.

The development of the conceptual design for Sutherland Brook revolved around a skeleton of critical habitats, blocked reserve areas, and connectivity corridors designed to accomplish multiple objectives of protecting (1) rare or key forest types, (2) aquatic or riparian values, (3) deer wintering grounds, and (4) bird migration and nesting habitats. Around this framework of critical habitats, blocks within which timber harvest would occur were delineated with the objectives of emulating natural disturbance patterns and enhancing visual diversity (aesthetics). A unique feature of Bell's (1994) approach to this step is heavy reliance on the shape and grain of landforms to guide cutblock delineation (Diaz and Bell 1997).

Because landform determines both ecological processes and human visual perception, this approach can synthesize a multitude of seemingly unrelated goals.

The major strengths of the approach used at Sutherland Brook were the wide range of social, biological, and economic issues it addressed, the inclusion of a wide spectrum of interested groups, and the comprehensive approach to analysis of the landscape. The focus on interactions among landform, process, and pattern resulted in a highly integrated ecological design. The Sutherland Brook project was not based on as much biological and disturbance data as would have been desirable (Murray and Singleton 1995), and thus it did not have the scientific rigor found in the historic fire analysis of Augusta Creek.

The Little Applegate Watershed encompasses roughly 28,800 ha of mountain and valley topography in southwestern Oregon. It is a more ecologically and culturally diverse landscape than are Augusta Creek and Sutherland Brook, including agricultural fields, bunchgrass prairies, oak woodlands, chaparral, pine-oak (*Pinus-Quercus*) savanna, conifer forests, and subalpine parklands. About two-thirds of the area is in federal ownership, and these lands are managed for a variety of uses, including timber, water, wildlife, recreation, and scenery. The remaining private holdings include ranches and farms, industrial forests, and residential areas. The area has had an extensive history of human use that has left an imprint on the structure and function of the ecosystem. Extirpation of the beaver, large-scale hydraulic mining, irrigation, logging, grazing, farming, and fire suppression have significantly altered both landscape pattern and ecological processes in the Little Applegate Watershed.

The landscape design team undertook a landscape analysis of structure, pattern, processes, and flows similar to that for Sutherland Brook, based on the Diaz and Apostol (1992) LAD process (M. Sinclair and D. Apostol, unpublished manuscript). An additional step was to zone the landscape into ecological landscape units based on broad categories of potential natural vegetation, landform, and natural disturbance regime (Figure 14.3). Within these ecological landscape units, a list of most probable landscape structural elements was compiled, based on knowledge of disturbances and the way they play out in different sectors of the landscape (Table 14.1). Conceptually, this step was similar to the derivation of landscape blocks and target structure types in Augusta Creek, but here it was done at a coarser scale.

Because of the extreme amount of natural diversity, existing land-use patterns, and history of human manipulation, the landscape design team could not consider natural disturbance as the sole template for designing future landscape conditions, as was done with Augusta Creek. Instead, objectives such as providing a connected late-successional forest network, building a more fire-resilient landscape (to protect watershed and economic values), watershed restoration, and maintaining economic opportunities (farming, ranching, timber, etc.) were used to sort among the ecologically possible pattern types within each ecological landscape unit. The final conceptual design (Figure 14.4) portrays a mosaic of target landscape pattern types that achieves a blending of these objectives, and it plans for orderly inclusion of non-natural structural elements (e.g., pastures, managed forests) that exist in such a modified landscape.

Notable omissions from the Little Applegate project are a detailed schedule of management activities for stands and a depiction of interim landscape conditions

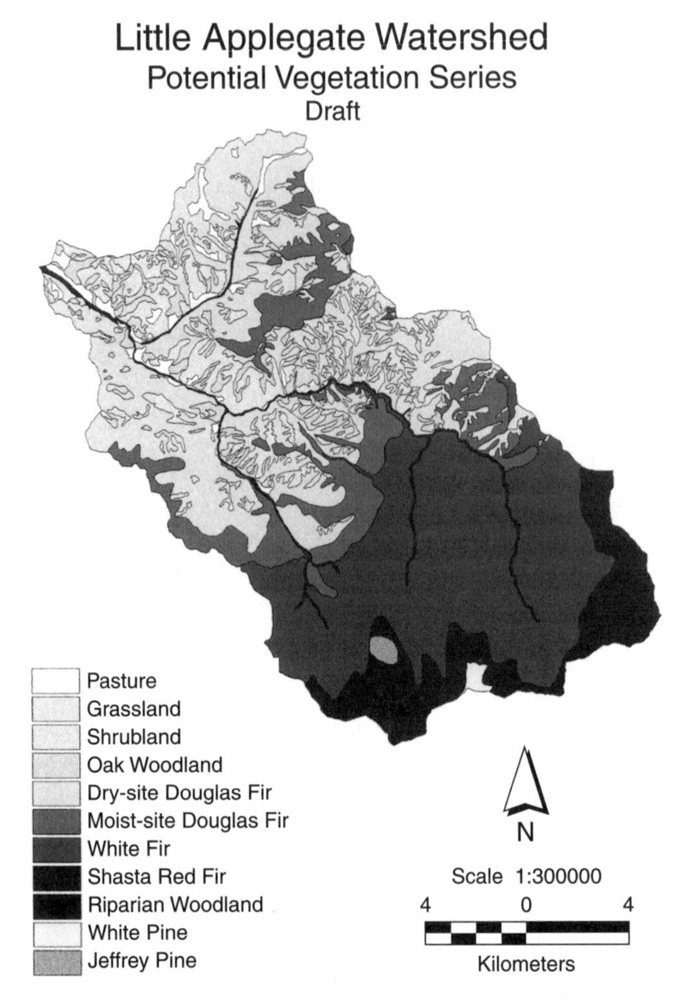

FIGURE 14.3. Potential natural vegetation, landform, and natural disturbance regimes of the Little Applegate Watershed. (See Color Plate I.)

during design implementation. However, because the project is ongoing, such a step may occur in the future. Such testing of how the conceptual design will change through time is essential to identify potential flaws in the final product, and to project economic benefits and costs. Major strengths of the Little Applegate project include the vast number of social, economic, and ecological factors it integrates, the large amount of community participation in the design process that encouraged people to work for the success of the project, and the clear connection between landscape capability and the conceptual design.

TABLE 14.1. Landscape structures by potential natural vegetation series, Little Applegate Watershed (see Figure 14.3).[a]

Landscape Structures	Shasta Fir/Mt. Hemlock Series	White Fir Series	Moist Douglas-Fir Series	Dry Douglas-Fir Series	Oak Wood-land	Shrub-land	Grass-land	Riparian Wood-land	White Pine	Jeffrey Pine
Fields and Pastures				+	+	+	+			
Perennial Grasslands					+	+	+			
Brush Mosaic				+	+	++	+			
Oak Woodland			+	+	++					
Pine/Oak Savanna		+	+	++	++	+				
Patchy Forest		++	+	+					+	+
Two-Layer Forest		+	+	+	+					
Perforated Forest	++	+	+	++					++	++
Multilayer Forest	+	++	++	+	+					
Dry Late Seral Forest	++			+						
Late Seral Forest		++	++						+	+
Subalpine Forest/Meadow Mosaic	++				+	++	+			
Ultramafic Vegetation	+	+	+	+						
Riparian Woodland				+	+	+	+	++		
Stem Exclusion Forest	+	++	++	+						
Madrone Patches			+	+	+					
Plantation Forests		+	+	+						
Residential Areas			+	+	+	+	+			

+ = may occur; ++ = very likely to occur.

[a]Scientific names of plant species: Shasta fir (*Abies magnifica* var. *shastensis*), mountain hemlock (*Tsuga mertensiana*), white fir (*Abies concolor*), Douglas-Fir (*Pseudotsuga menziesii*), white pine (*Pinus monticola*), Jeffrey pine (*Pinus jeffreyi*), madrone (*Arbutus menziesii*).

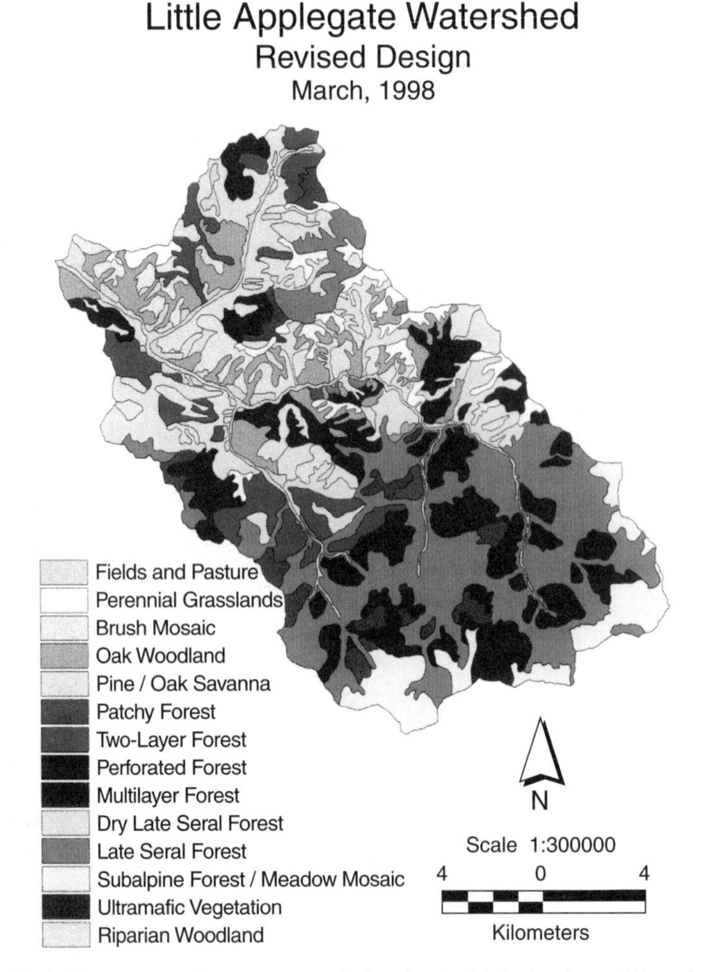

FIGURE 14.4. The conceptual management design for the Little Applegate Watershed. (See Color Plate II.)

14.3.2 Planning Applications

The key habitat-corridor model identifies areas where threatened and other species of interest occur, and it manages those areas to enhance those species (Fries et al. 1998). The key habitats are supported by corridor connections and are surrounded by buffer and restoration zones to mitigate negative effects from the matrix. This model has been applied in a number of places, including Sweden (Fries et al. 1998) and the Pacific Northwest (USA) (Forest Ecosystem Management Assessment Team 1993).

The supportive-features model was developed to aid landscape design in the cultural landscape of southern Sweden (Fries et al. 1998). This approach uses the existing landscape as the starting point for planning for multiple uses by identifying existing natural and cultural features that are valued by Society or landowners. Silvicultural activity is guided by its effect on the maintenance or enhancement of these landscape assets.

For the entire Brazilian State of Para (1,248,042 km^2), a spatial design for the placement of timber-extraction zones and nature reserves was developed (Verissimo et al. 1998). The spatial distribution of forest, land ownership, log-processing facilities, roads, and areas with high biodiversity was analyzed to develop a timber development plan. The criteria used included economic accessibility, current legal protection status, vegetation cover, and conservation priorities. This plan illustrates the value of spatial planning to mitigate impacts on biodiversity from timber extraction, and it considers the subsequent settlement of lands made accessible by logging-road development.

14.3.3 Research Application

Simulation studies have shown that it is possible to increase timber production *and* reduce forest fragmentation across the landscape by using a dynamic-zoning strategy that clusters harvests (Gustafson 1996). In fact, a dynamic zoning strategy produced amounts of forest interior and edge comparable to that of the dispersed alternative with half the rate of harvest (Figure 14.5).

14.4 Principles for Applying Landscape Ecology

From these examples and our experience, we propose the following principles for applying landscape ecology to the extraction of timber to mitigate impacts on biological diversity.

14.4.1 Principle 1: Effective Forest Management for Biological Diversity Requires a Broad Spatial View

It is imperative to consider complete ecological systems when seeking to maintain biodiversity—that is, the mix of habitat conditions, the spatial pattern of those conditions, and the nature of the disturbance regime(s) that maintain those conditions. Managers must consider what is outside of the boundaries of their management unit. Specific landscape considerations include (1) the ecological functions to be maintained (e.g., nutrient, water and animal flows, nutrient and energy cycles), (2) the forces of change and disturbance (e.g., land-use change, fire), and (3) the context within the ecological hierarchy. Because no management unit will completely encompass all of the elements relevant to some ecological processes, effective landscape management necessitates a broad view (Diaz and Apostol 1992). Timber harvest can be a valuable forest management tool for producing disturbance events that maintain biodiversity and for yielding wood products, but forest managers must consider ecological processes beyond those operating at the stand level (Hansen et al. 1991).

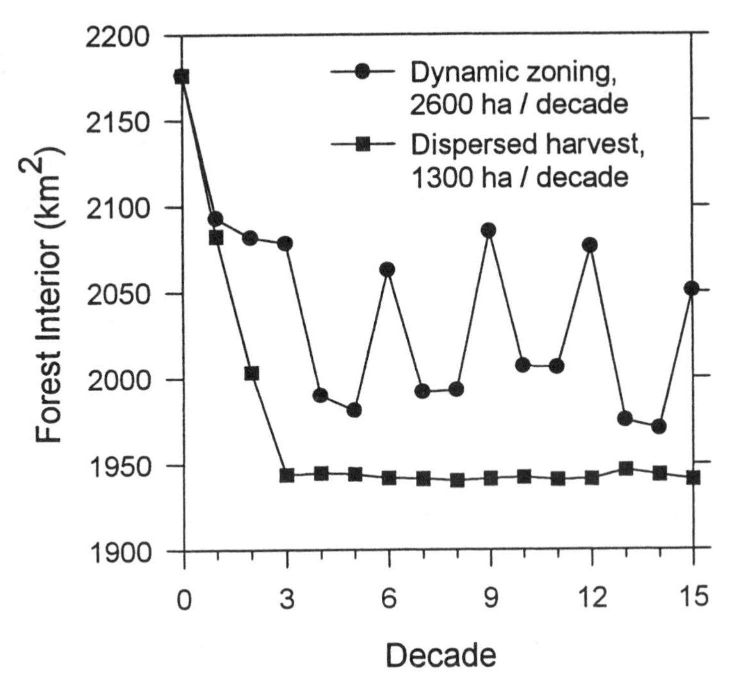

FIGURE 14.5. Amount of forest interior produced over simulated time by a dynamic-zoning alternative at high-intensity harvest (2600 ha/decade) and a dispersed alternative at low-intensity harvest (1300 ha/decade) on the Hoosier National Forest (Indiana, USA). The dynamic zoning alternative rotated timber harvest activity among five cutting zones every 30 years, whereas the dispersed alternative had no spatial clustering. After Gustafson (1996). Reprinted from Forest Ecology and Management Volume 87, Gustafson, E. J., Expanding the scale of forest management: allocating timber harvests in time and space, pp. 27–39, Copyright 1996, with permission from Elsevier Science.

14.4.2 Principle 2: Managers Must Recognize That Landscapes Are by Nature Dynamic

Because forested ecosystems are by nature dynamic, it is useful to think about both desired future dynamics and desired future conditions. It has been argued that if ecological dynamics are within natural bounds, then the ecological conditions at any given point in time will also be within an acceptable range (Cissel et al. 1998). It is sometimes difficult for humans to perceive the normalcy of ecological dynamics that occur at scales different from those of normal human experience. For example, in an ecosystem that normally burns every 50 years, the 50 years without fire come to be perceived as the norm, and the year with the fire is perceived as a catastrophe. Ecologically, the fire year is as normal as the other 49 nonfire years. An understanding that infrequent events are normal is important for both managers and those interested in the outcome of forest management. Because biota are adapted to change, some of it catastrophic, it is necessary that forest managers and users embrace (or at least accept) change (Rykiel 1998). Change is inevitable, but management can help avoid undesirable ecological consequences in the future.

14.4.3 Principle 3: Managers Should Guide Timber Extraction With Spatial and Temporal Plans

Landscape management must be intentional, meaning that the spatial distribution of management activities should be guided by a spatial plan to achieve objectives for landscape conditions. Certain management objectives can be enhanced through manipulation of the spatial and temporal configuration of timber extraction. For example, the creation of large contiguous blocks of habitat can be achieved by clustering harvests or increasing the size of harvest units, thereby limiting the perforating effects of harvest to a portion of the landscape (Li et al. 1993; Gustafson 1996). A diversity of seral stages in a locality can be achieved by periodically harvesting in the area. Some early stages of certain forest types require significant time to develop, requiring advance planning to ensure their presence on the landscape in the future. To achieve sustainable forest ecosystems, it is necessary to consider the spatial and temporal configuration of disturbance management and timber extraction, and the multiple scales at which ecological processes occur. All of the examples described above use timber extraction to create a desired forest mosaic through time.

14.5 Knowledge Gaps

Critical gaps in our knowledge impede the ability of land managers to apply confidently the principles of landscape ecology for biological conservation. Based on our experience, we identify here the most significant of those gaps.

14.5.1 Theoretical Voids

Because ecological systems are knit together with functional relationships that stretch across scales, there can be no successful management of ecological systems without improved understanding of how ecological phenomena are related across scales. Although the problem of scale has been a focus of study for some time, we still do not understand how the interactions of these processes produce the dynamic patterns and behaviors that are observed at every level of biological organization (Levin 1992; Rykiel 1998). This void makes it very difficult to relate processes that occur on certain scales of space and time to ecological patterns and dynamic behaviors observed on other scales of space and time.

Theoretical landscape pattern models tend to be formulated in two dimensions. But in the real world, land shape exerts a significant effect on landscape processes and the resulting patterns of vegetation. The degree of vertical relief, slope length, scale of dissection, and orientation to prevailing winds and solar input are examples of topographic attributes that determine how disturbances and other processes behave at broader scales. Landscape analyses that ignore these factors run the risk of generating completely spurious results. For example, an assessment of seasonal elk (*Cervus elaphus*) migration routes based solely on vegetation patterns and habitat connectivity resulted in a completely erroneous picture of animal movements (Diaz and Apostol 1992). When topography was considered, the predicted pathways matched the actual observations. Not surprisingly, elk apparently take the path of least resistance vertically as well as horizontally. Because topography is not considered in much of landscape ecological theory, models based on this theory have diminished predictive power in areas of high relief.

Current thinking about the effects of habitat spatial pattern is almost wholly derived from the theory of island biogeography (MacArthur and Wilson 1967). However, habitat patches usually exist in a complex landscape mosaic, and dynamics within a patch are affected by external factors related to the structure of the mosaic (Wiens 1994). The development of a mosaic theory is needed to improve the scientific basis for conservation and management (Wiens 1994).

It is possible that ecosystem stability represents a critical phenomenon, whereby the ability to recover from disturbance is rapidly diminished when ecological redundancy is reduced below some threshold (Kaufman et al. 1998). Are ecosystems that develop under anthropogenic disturbance regimes (such as timber management) sufficiently diverse and stable to withstand extreme disturbance events? Some have suggested that increased temporal variability (e.g., Boulinier et al. 1998) may signal that the system is approaching a threshold. However, there is an insufficient theoretical basis for managers to discover what the threshold is, and it is difficult to judge objectively the sustainability of managed ecosystems.

14.5.2 Empirical Voids

Corridors have been widely embraced by managers as a means to mitigate the landscape fragmentation effects of forest management (Harris 1984). However, the effectiveness of corridors to sustain wildlife populations remains the subject of debate. Corridors are thought to enhance immigration and genetic diversity (Harris 1984), but some argue that increased transmission of disease more than offsets these benefits (Hess 1994). Empirical evidence to settle this debate would substantively improve forest-management planning.

Advances have been made in our understanding of how species respond to landscape pattern in managed-forest landscapes (e.g., McGarigal and McComb 1995). However, a single-species approach to biological conservation has proved unsatisfactory because the number of species and lack of information is overwhelming (Franklin 1994). Models to predict how entire communities will respond to vegetation management are urgently needed. Such models must be capable of integrating many species' responses to disturbance to predict community composition and stability.

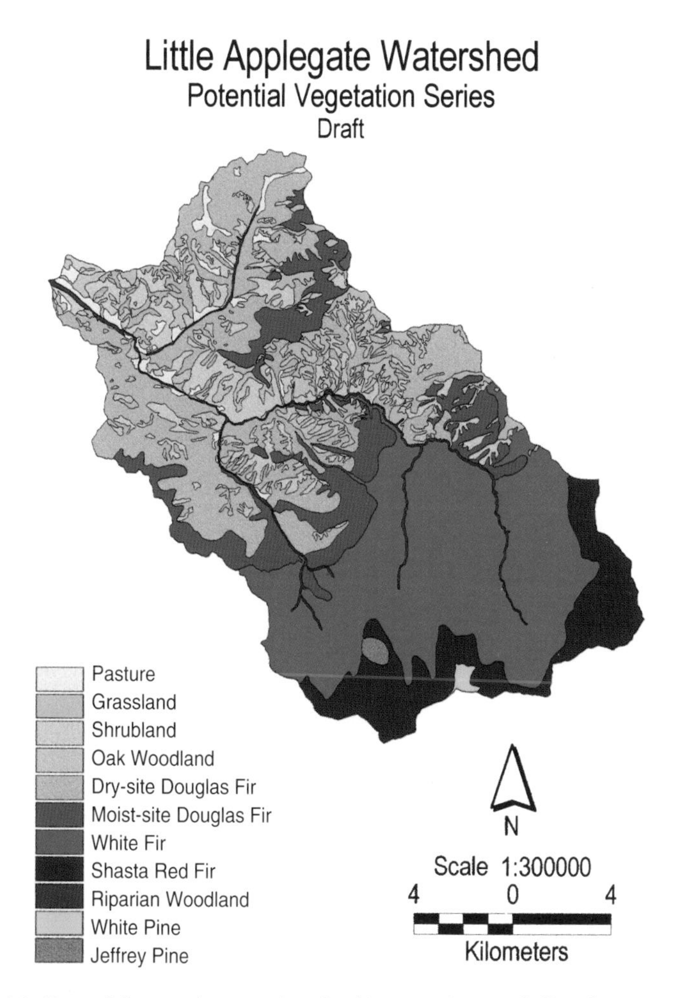

FIGURE 14.3. Potential natural vegetation, landform, and natural disturbance regimes of the Little Applegate Watershed.

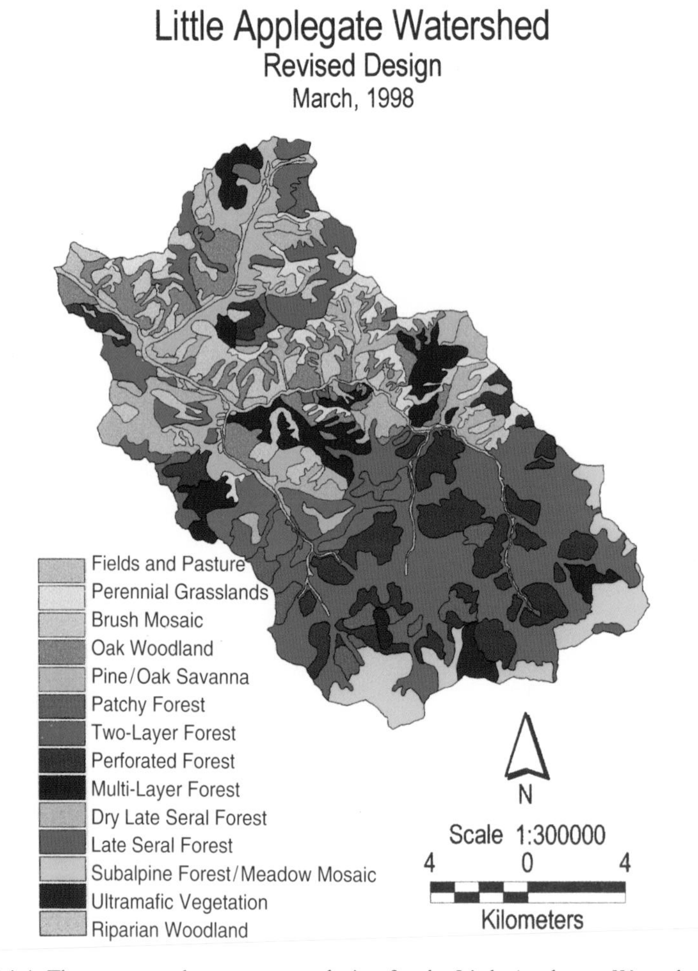

FIGURE 14.4. The conceptual management design for the Little Applegate Watershed.

14.6 Research Approaches

14.6.1 Approaches for Theoretical Research

Traditional studies conducted at single scales of space, time, or organizational complexity must be complemented by cross-scale studies (Levin 1992). Such studies should include scale as an independent variable (Turner et al. 1989). Ecological studies should explicitly consider scale effects in their design and interpretation. Simulation studies that explicitly model scale effects can be used to generate testable hypotheses about how management activity affects ecosystems at various scales.

Application of existing theory from other disciplines such as geology (Miall 1989) for understanding three-dimensional systems may jump-start development of a theoretical framework for adding the vertical dimension to ecological models. Theoretical models could include topographic factors as independent variables.

Habitat and ecological process models could be improved by representing the landscape more explicitly as a mosaic, rather than as a simple binary system (habitat vs. nonhabitat). For example, habitat can be spatially represented as a continuous surface rather than as a categorical map (i.e., not patch based) (Gustafson 1998). Alternatively, fuzzy habitat mosaics can be described where habitat value is determined using fuzzy logic (Roberts 1996). These approaches may be useful to describe better the ecotones in landscapes. Perhaps habitats can be represented as hierarchically scaled objects having explicit properties that vary across hierarchies of scale. For example, habitat for an Ovenbird (*Seiurus aurocapillus*) could be described by availability of nest-concealment objects at the microsite scale, vegetation composition and structure at the stand level, size of the forest patch at the landscape scale, and position relative to the edge of the Ovenbird's range at the regional scale.

Hierarchy theory predicts that systems approaching a critical threshold will take longer to recover (O'Neill et al. 1989). This prediction can be tested in microcosms (experimental model systems; Wiens et al. 1993) with experiments that allow control of spatial variables (e.g., Banks 1998). Simulation experiments can be used to illuminate the expected behavior of a system operating as theory describes it. These insights can then be used to guide empirical research and adaptive management. A simulation approach requires sophisticated community and vegetation-management (timber-harvest) models (e.g., LANDIS; Gustafson et al. 2000). Simulations of changes in forest composition in response to timber harvest, or how animal communities respond to changes in forest habitat pattern, could be explored to understand better the relationships between redundancy and resilience.

14.6.2 Approaches for Empirical Research

One approach to assess the effectiveness of corridors for sustaining wildlife populations is to study animal behavior at habitat boundaries to determine if corridors are used actively or passively, and to determine what is perceived as a corridor from the individual's perspective (Haddad 1999). Another approach is to collect detailed information on individual movement paths to determine the role of corridors for interpatch movements. Genetic diversity and incidence of disease can be compared between patches connected by corridors and isolated patches.

Landscape-scale experimental manipulation is required to develop models about the effects of landscape pattern and timber extraction on community composition and stability. Unfortunately, controlled broad-scale experiments are difficult to implement, replicate, and interpret, and many questions require long-term studies. Confounding factors include ecological land type and disturbance history (Foster et al. 1997), ownership patterns, road and infrastructure patterns, and landscape context (Wiens 1994). A shortcoming of many descriptive landscape studies is that their results are difficult to generalize to other landscapes. Expensive landscape manipulation studies must be designed as a rigorous test of mechanistic hypotheses about the relationships between spatial pattern and community responses. Promising approaches include simulating a disturbance process (e.g., hurricane; Foster et al. 1997) and manipulating landscape pattern (e.g., Haddad and Baum 1999). Assembling multidisciplinary teams to study the effects of the manipulations on community response can maximize knowledge and improve the cost-effectiveness of implementing the treatments.

A partial solution to the cost and logistical barriers to conducting experimental landscape research is adaptive management. Adaptive management seeks to implement management activity in a way that maximizes scientific knowledge. The typical strategy is to implement a management action with a clear statement of potential outcomes, monitor the outcomes, modify similar management actions in the future based on what was learned, and repeat the process. Adaptive management is most effective when a long-term partnership is established between scientists and managers (Rogers 1998). This could be a realistic component of landscape research on how timber management affects community responses and biological diversity.

Acknowledgments

We thank Patrick Zollner and David Shriner for critical reviews that helped us improve the manuscript.

References

Attiwill, P.M. 1994. The disturbance of forest ecosystems: the ecological basis for conservative management. *For. Ecol. Manage.* 63:247–300.

Augustine, D.J., and Jordan, P.A. 1998. Predictors of white-tailed deer grazing intensity in fragmented deciduous forests. *J. Wildl. Manage.* 62:1076–1085.

Banks, J.E. 1998. The scale of landscape fragmentation affects herbivore response to vegetation heterogeneity. *Oecologia* 117:239–246.

Bell, S. 1994. *Forest Landscape Design: A Handbook for the Maritime Provinces of Canada.* Fredricton, New Brunswick, Canada: Maritime Forest Ranger School.

Bergeron, Y., and Harvey, B. 1997. Basing silviculture on natural ecosystem dynamics: an approach applied to the southern boreal mixedwood forest of Quebec. *For. Ecol. Manage.* 92:235–242.

Blake, J.G., and Karr, J.R. 1987. Breeding birds of isolated woodlots: area and habitat relationships. *Ecology* 68:1724–1734.

Bormann, F.H., and Likens, G.E. 1981. *Pattern and Process in a Forested Ecosystem.* New York: Springer-Verlag.

Boulinier, T., Nichols, J.D., Hines, J.E., Sauer, J.R., Flather, C.H., and Pollock, K.H. 1998. Higher temporal variability of forest breeding bird communities in fragmented landscapes. *Proc. Nat. Acad. Sci. USA* 95:7497–7501.

Chen, J., Franklin, J.F., and Spies, T.A. 1992. Vegetation responses to edge in old-growth Douglas-fir forests. *Ecol. Appl.* 2:387–396.

Cissel, J.H., Swanson, F.J., Grant, G.E., Olson, D.H., Gregory, S.V., Garman, S.L., Ashkenas, L.R., Hunter, M.G., Kertis, J.A., Mayo, J.H., McSwain, M.D., Swetland, S.G., Swindle, K.A., and Wallin, D.O. 1998. *A Landscape Plan Based on Historical Fire Regimes for a Managed Forest Ecosystem: The Augusta Creek Study.* PNW-GTR-422. Portland, Oregon: USDA Forest Service, Pacific Northwest Research Station.

Diaz, N., and Apostol, D. 1992. *Forest Landscape Analysis and Design: A Process for Developing and Implementing Land Management Objectives for Landscape Patterns.* R6 ECO-TO-043-92. Portland, Oregon: USDA Forest Service, Pacific Northwest Region.

Diaz, N.M., and Bell, S. 1997. Landscape analysis and design. In *Creating a Forestry for the 21st Century: The Science of Ecosystem Management,* eds. K.A. Kohm and J.F. Franklin, pp. 255–269. Washington, DC: Island Press.

Forest Ecosystem Management Assessment Team. 1993. *Forest Ecosystem Management: An Ecological, Economic, and Social Assessment.* 1993–793–071. Washington, DC: U.S. Government Printing Office.

Forman, R.T.T. 1995. *Land Mosaics: The Ecology of Landscapes and Regions.* Cambridge, United Kingdom: Cambridge University Press.

Forman, R.T.T., and Godron, M. 1986. *Landscape Ecology.* New York: John Wiley and Sons.

Foster, D.R., Aber, J.D., Melillo, J.M., Bowden, R.D., and Bazzaz, F.A. 1997. Forest response to disturbance and anthropogenic stress. *BioScience* 47:437–445.

Franklin, J.F. 1994. Preserving biodiversity: species in landscapes—response. *Ecol. Appl.* 4:208–209.

Franklin, J.F., and Forman, R.T.T. 1987. Creating landscape patterns by forest cutting: ecological consequences and principles. *Landsc. Ecol.* 1:5–18.

Fries, C., Carlsson, M., Dahlin, B., Lamas, T., and Sallnas, O. 1998. A review of conceptual landscape planning models for multiobjective forestry in Sweden. *Can. J. For. Res.* 28:159–167.

Gilpin, M.E., and Case, T.J. 1976. Multiple domains of attraction in competition communities. *Nature* 261:40–42.

Gustafson, E.J. 1996. Expanding the scale of forest management: allocating timber harvests in time and space. *For. Ecol. Manage.* 87:27–39.

Gustafson, E.J. 1998. Quantifying landscape spatial pattern: what is the state of the art? *Ecosystems* 1:143–156.

Gustafson, E.J., Shifley, S.R., Mladenoff, D.J., He, H.S., and Nimerfro, K.K. 2000. Spatial simulation of forest succession and harvesting using LANDIS. *Can. J. For. Res.* 30:32–43.

Haddad, N.M. 1999. Corridor use predicted from behaviors at habitat boundaries. *Am. Nat.* 153:215–227.

Haddad, N.M., and Baum, K.A. 1999. An experimental test of corridor effects on butterfly densities. *Ecol. Appl.* 9:623–633.

Hansen, A.J., Spies, T.A., Swanson, F.J., and Ohmann, J.L. 1991. Conserving biodiversity in managed forests. *BioScience* 41:382–392.

Harris, L.D. 1984.*The Fragmented Forest: Island Biogeography Theory and the Preservation of Biotic Diversity.* Chicago: University of Chicago Press.

Hess, G.R. 1994. Conservation corridors and contagious disease: a cautionary note. *Conserv. Biol.* 6:256–262.

Holling, C.S. 1973. Resilience and stability of ecological systems. *Ann. Rev. Ecol. Syst.* 4:1–23.

Huston, M.A. 1994. *Biological Diversity: The Coexistence of Species on Changing Landscapes.* Cambridge, United Kingdom: Cambridge University Press.

Jones, C.G., Lawton, J.H., and Shachak, M. 1994. Organisms as ecosystem engineers. *Oikos* 69:373–386.

Kaufman, J.H., Brodbeck, D., and Melroy, O.R. 1998. Critical biodiversity. *Conserv. Biol.* 12:521–532.

King, D.I., Griffin, C.R., and DeGraaf, R.M. 1997. Effect of clearcut borders on distribution and abundance of forest birds in northern New Hampshire. *Wilson Bull.* 109:239–245.

Levin, S.A. 1992. The problem of pattern scale in ecology. *Ecology* 73:1943–1967.

Li, H., Franklin, J.F., Swanson, F.J., and Spies, T.A. 1993. Developing alternative forest cutting patterns: a simulation approach. *Landsc. Ecol.* 8:63–75.

Litvaitis, J.A. 1993. Response of early successional vertebrates to historic changes in land use. *Conserv. Biol.* 7:866–873.

Lorimer, C.G. 1985. The role of fire in the perpetuation of oak forests. In *Proceedings of Challenges in Oak Management and Utilization,* ed. J.E. Johnson, pp. 8–25. Madison: University of Wisconsin-Extension.

MacArthur, R.H., and Wilson, E.O. 1967. *The Theory of Island Biogeography.* Princeton: Princeton University Press.

McGarigal, K. and McComb, W.C. 1995. Relationships between landscape structure and breeding birds in the Oregon Coast Range. *Ecol. Monogr.* 65:235–265.

Miall, A.D. 1989. Can there be life after facies models? The development of a framework for the quantitative description of complex, three-dimensional facies architectures. In *Quantitative Dynamic Stratigraphy,* ed. T.A. Cross, pp. 601–615. New York: Prentice-Hall.

Mladenoff, D.J., White, M.A., Crow, T.R., and Pastor, J. 1994. Applying principles of landscape design and management to integrate old-growth forest enhancement and commodity use. *Conserv. Biol.* 8:752–762.

Mladenoff, D.J., White, M.A., Pastor, J., and Crow, T.R. 1993. Comparing spatial pattern in unaltered old-growth and disturbed forest landscapes. *Ecol. Appl.* 3:294–306.

Murray, T., and Singleton, J. 1995. *Sutherland Brook Forest Ecosystem Design Pilot Project.* Fredricton, New Brunswick, Canada: Canadian Forest Service.

Oliver, C.D. 1981. Forest development in North America following major disturbances. *For. Ecol. Manage.* 3:153–168.

O'Neill, R.V., Johnson, A.R., and King, A.W. 1989. A hierarchical framework for the analysis of scale. *Landsc. Ecol.* 3:193–205.

Paton, P.W.C. 1994. The effect of edge on avian nest success: how strong is the evidence? *Conserv. Biol.* 8:17–26.

Peterson, G., Allen, C.R., and Holling, C.S. 1998. Ecological resilience, biodiversity, and scale. *Ecosystems* 1:6–18.

Rahel, F.J. 1990. The hierarchical nature of community persistence: a problem of scale. *Am. Nat.* 136:328–344.

Roberts, D.W. 1996. Landscape vegetation modelling with vital attributes and fuzzy systems theory. *Ecol. Model.* 90:175–184.

Robinson, S.K. 1992. Population dynamics of breeding Neotropical migrants in a fragmented Illinois landscape. In *Ecology and Conservation of Neotropical Migrant Landbirds,* eds. J.M. Hagen and D.W. Johnston, pp. 408–418. Washington, DC: Smithsonian Institution Press.

Rogers, K. 1998. Managing science/management partnerships: a challenge of adaptive management. *Conserv. Ecol.* [online] 2(2):R1. Available from the Internet: www.consecol.org/vol2/iss2/resp1.

Rykiel, E.J., Jr. 1998. Relationships of scale to policy and decision making. In *Ecological Scale: Theory and Applications,* eds. D.L. Peterson and V. Thomas, pp. 485–497. New York: Columbia University Press.

Shlisky, A.J. 1994. Multi-scale ecosystem analysis and design in the Pacific Northwest region: the Umatilla National Forest restoration project. In *Eastside Forest Ecosystem Health Assessment - Vol. 2. Ecosystem Management: Principles and Applications,* eds. M.E. Jensen and P.S. Bourgeron, pp. 267–275. GTR-PNW-318. Portland, Oregon: USDA Forest Service, Pacific Northwest Forest Experiment Station.

Thompson, F.R., Dijak, W.D., Kulowiec, T.G., and Hamilton, D.A. 1992. Breeding bird populations in Missouri Ozark forests with and without clearcutting. *J. Wildl. Manage.* 56:23–30.

Turner, M.G. 1989. Landscape ecology: the effect of pattern on process. *Ann. Rev. Ecol. Syst.* 20:171–197.

Turner, M.G., Dale. V.H., and Gardner, R.H. 1989. Predicting across scales: theory development and testing. *Landsc. Ecol.* 3:245–252.

Turner, M.G., Gardner, R.H., Dale, V.H., and O'Neill, R.V. 1989. Predicting the spread of disturbance across heterogeneous landscapes. *Oikos* 55:121–129.

Urban, D.L., O'Neill, R.V., and Shugart, H.H. 1987. Landscape ecology. *BioScience* 37:119–127.

Verissimo, A., Junior, C.S., Stone, S., and Uhl, C. 1998. Zoning of timber extraction in the Brazilian Amazon. *Conserv. Biol.* 12:128–136.

Vitousek, P.M. 1990. Biological invasions and ecosystem processes: towards an integration of population biology and ecosystem studies. *Oikos* 57:7–13.

Wiens, J.A. 1994. Habitat fragmentation: island v landscape perspectives on bird conservation. *Ibis* 137:S97–S104.

Wiens, J.A., Stenseth, N.C., Van Horne, B., and Ims, R.A. 1993. Ecological mechanisms and landscape ecology. *Oikos* 66:369–380.

Wu, J., and Loucks, O.L. 1995. From balance of nature to hierarchical patch dynamics: a paradigm shift in ecology. *Q. Rev. Biol.* 70:439–466.

15

Animal Behavior in Fragmented Landscapes

RICHARD H. YAHNER AND CAROLYN G. MAHAN

15.1 Introduction

Concepts from island-biogeography theory (MacArthur and Wilson 1967) have been instrumental in testing and understanding current patterns of faunal distribution in fragmented landscapes. The species-area curve, for instance, has enabled us to examine the relationship between patch size and species richness in various landscapes (e.g., Robbins et al. 1989). The effects of fragmented landscapes on community structure (e.g., Friesen et al. 1995; Yahner 1997), reproductive success (e.g., Robinson et al. 1995), and population demographics (e.g., Wauters et al. 1994; Litvaitis and Villafuerte 1996; Wolff et al. 1997; Mahan and Yahner 1998) have been examined. These previous studies are vital to natural resource managers and conservation biologists concerned with biological conservation and ecosystem management. However, an important and largely untapped area of research is the effects of fragmented landscapes on animal behavior (Yahner and Mahan 1997a).

The behavior of an animal is constrained by species-specific (e.g., mobility) and extrinsic (e.g., resource availability) factors, and we contend that both types of factors can be influenced by landscape patterns. We begin this chapter by discussing concepts, principles, and emerging ideas that relate to animal behavior in fragmented landscapes. We then describe field, research, and policy applications of this information; provide principles for applying behavior and landscape ecology concepts in biological conservation; and identify theoretical and empirical knowledge gaps that constrain such applications. We conclude by describing fundamental research approaches for determining how fragmented landscapes affect animal behavior.

15.2 Concepts, Principles, and Emerging Ideas

The distribution of individual animals in a landscape tends to be nonrandom and has a characteristic pattern that may in part be dictated by responses in behavior to environmental or social factors (Crook et al. 1976). This characteristic pattern of distribution is referred to as the *social structure* for a species and includes features of a species, such as group size, duration of pair bonds, and the spacing of individuals of a species in the landscape. This spatial arrangement of members of a population within a habitat is referred to as the *dispersion pattern* of a population and may be random, clumped, or uniform (Brower and Zar 1984).

A *pair bond* is a short- or long-term relationship with a single member of the opposite sex for the purpose of reproducing or raising of young (Gill 1990). Many birds and mammals have at least some degree of sociality and form a social unit sometime during their life history, with the basic unit being the *parent-young relationship* (Eisenberg 1966). A good example of a parent-young relationship is the familiar matriarch group formed by an adult female white-tailed deer (*Odocoileus virginianus*), fawns of the year, and occasionally yearlings from the previous year (Hirth 1977). *Parental behavior* is an example of a behavior that mediates social structure within a species. Parental behavior is exhibited by adult animals when rearing young and, in passerine birds, includes all the activities that an adult must conduct to meet the brooding and feeding demands of young birds (Gill 1990).

Social structure, dispersion patterns, and all behavioral interactions among individuals of a species are mediated by various *communication signals,* such as auditory, olfactory, or visual cues, which generate the particular social structure characteristic of that species (Crook et al. 1976). This social structure is maintained by spacing mechanisms, e.g., territoriality in eastern chipmunks (*Tamias striatus*) or red squirrels (*Tamiasciurus hudsonicus*), or dominance hierarchies in gray wolf (*Canis lupus*) packs or male white-tailed deer during the rutting season (Smith 1968; Mech 1970; Yahner 1978; Hesselton and Hesselton 1982). Social structures ensure vital life-history events, including mating, rearing of young, resource exploitation, and predator avoidance (Crook et al. 1976). The totality of interactions that arise among individuals of a species (e.g., intraspecific behavior, social units, spacing systems) can be referred to as the *social organization* (or social system) of the species (Eisenberg 1966; Crook et al. 1976).

Habitat fragmentation occurs when contiguous land cover (e.g., forest) is converted into one or more smaller tracts by processes such as agricultural development, timber harvest, or urbanization (Harris and Silva-Lopez 1992; Yahner 2000) (Figure 15.1). During fragmentation, the original habitat not only becomes smaller in total area, but the remaining tracts become isolated by a dissimilar land use. Fragmentation may also be caused by the creation of *habitat gaps,* which are openings (nonhabitat land use) in an otherwise continuous land-cover type (e.g., forest or prairie).

Conservation biologists have often advocated the use of *corridors* to connect otherwise isolated habitat patches and to minimize the effects of habitat fragmentation (Meffe and Carroll 1997). These strips of habitat may occur naturally (e.g., riparian corridors) or may be constructed (e.g., tunnels that direct animals under highways). *Connectivity* can be defined as the degree to which animals can make movements between isolated habitat patches (Hunter 1996). The degree of connectivity will often depend on a particular species of animal. For example, woodlots separated by a 25-m agricultural field will have a lower degree of connectivity

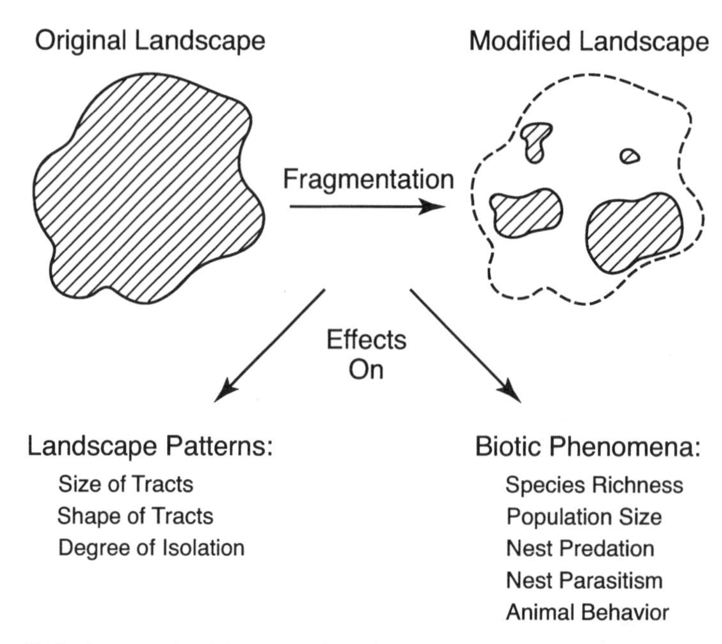

FIGURE 15.1. As a result of fragmentation, the original land cover is reduced to one or more smaller tracts, and these tracts become more isolated. Landscape ecologists are primarily interested in the effects of fragmentation on landscape patterns, whereas conservation biologists focus more on the effects of fragmentation on biotic phenomena (modified from Yahner 2000).

to a woodland salamander than to a black bear (*Ursus americanus*). Corridors that are large and broad may be used by animals that prefer continuous tracts of habitat (Noss 1991). Corridors also may help facilitate animal *dispersal,* which consists of movements made by young animals away from their natal site (Hunter 1996). *Landscape resistance* is the degree to which animal movement across landscapes is impeded by barriers such as highways, developed areas, and gaps in habitat. (Harms and Opdam 1990; Knaapen et al. 1992).

15.3 Recent Applications

To document how concepts related to animal behavior and landscape fragmentation have been applied, we reviewed the primary literature as well as reports and manuscripts compiled by natural resource agencies. In addition, our ongoing research in cooperation with the U.S. National Park Service, U.S. Fish and Wildlife Service, U.S. Forest Service, and numerous state agencies has enabled us to discuss with natural resource managers how an understanding of animal behavior has been applied in fragmented habitats. We found that some of the associated concepts have been applied in field, research, and policy contexts.

15.3.1 Field Applications

Artificial corridors may be needed in fragmented landscapes to connect isolated populations of some rare species. Mansergh and Scotts (1989) constructed a tunnel that successfully connected two breeding areas of the endangered mountain pygmy-possum (*Burramys parvus*) in Australia. The breeding population of this species had been isolated by roads and development associated with a ski resort; individuals seldom moved across these developed areas, thereby dramatically reducing the size of the effective breeding population. Highway underpasses constructed in southern Florida enabled alligators (*Alligator mississipiensis*), raccoons (*Procyon lotor*), black bears, bobcats (*Lynx rufus*), Florida panthers (*Puma concolor coryi*), and white-tailed deer to move across fragmented landscapes and to avoid vehicular collisions (Foster and Humphrey 1995).

15.3.2 Research Applications

Corridors facilitate movements by animals across inhospitable landscapes. For example, wooded corridors enable species adapted to woodlands to move between woodlots in an agricultural landscape (Swihart and Yahner 1982; Henein and Merriam 1990; Haas 1995). Based on empirical models, butterfly species tend to use wider corridors (≥50 m) more often than they use narrow corridors (Hadded 1999). In forested landscapes, birds used 100-m-wide wooded buffer strips as dispersal and movement corridors adjacent to clear cut stands (Machtans et al. 1996). Larger organisms (e.g., white-tailed deer), however, may require dispersal corridors that are at least 2 km wide (Harrison 1992).

Animals may avoid using certain types of corridors that are found in fragmented landscapes. Logging roads in a fragmented landscape, for instance, can negatively affect dispersal and mortality of many species (e.g., Ambrose and Bratton 1990; Gibbs 1998). Grizzly bears (*Ursus arctos*) avoided areas within 100 m of resource-extraction roads in British Columbia, presumably because of disturbance by humans (McLellan and Shackleton 1988), and their behavioral response reduced the amount of available habitat by 8.7%. Trombulak and Frissell (2000) provide an excellent overview of the effects of roads on animal behavior. Landscape resistance has been examined theoretically and experimentally in various animal groups (Harms and Opdam 1990).

In some cases, roads may facilitate dispersal and foraging behavior of mammalian carnivores. Roads built for oil exploitation in Alaska have enabled red fox (*Vulpes vulpes*) to colonize previously unoccupied areas (Rudzinski et al. 1982). The presence of logging roads conceivably could increase predation rates on avian nests located near roads compared with those farther away from roads. However, some evidence suggests that logging roads may not enhance foraging efficiency of nest predators in forested landscapes (Yahner and Mahan 1997b).

Human land uses may act as barriers to movements of organisms between habitats. For instance, forest clearcutting reduces (or eliminates) overstory trees, which in turn alters the vegetative structure, floristic composition, and microclimate of the original forest (Saunders et al. 1991; Yahner 2000). As a consequence, clear cut stands impede movements of southern flying squirrels (*Glaucomys volans*) that rely on overstory trees for gliding while foraging or escaping predators (Bendel and Gates 1987). Woodland amphibians, such as wood frogs (*Rana sylvatica*) and spotted salamanders (*Ambystoma maculatum*), tend to avoid clear cut stands or other discontinuities created by forest fragmentation (deMaynadier and Hunter 1999).

An examination of the relationship between landscape fragmentation or pattern and animal behavior is not new. However, the number of published studies focusing on this issue is limited (Yahner and Mahan 1997a). One classic study of wildlife response to landscape pattern involves a comparison of social behavior of white-tailed deer at the E.S. George Reserve in Michigan (USA) and the Rob and Bessie Welder Wildlife Refuge in Texas (USA) (Hirth 1977). The George Reserve is characterized by expansive wooded areas interspersed with small (<25 ha) open fields (total open areas = 22%). The Welder Refuge, on the other hand, has fewer wooded areas amid large (some >400 ha) open areas (total open area = 57%). Deer at the more open Welder Refuge, which was fragmented by agricultural land use, were consistently found in larger social units than were deer at the less-fragmented George Reserve (Figure 15.2). Larger social units at the

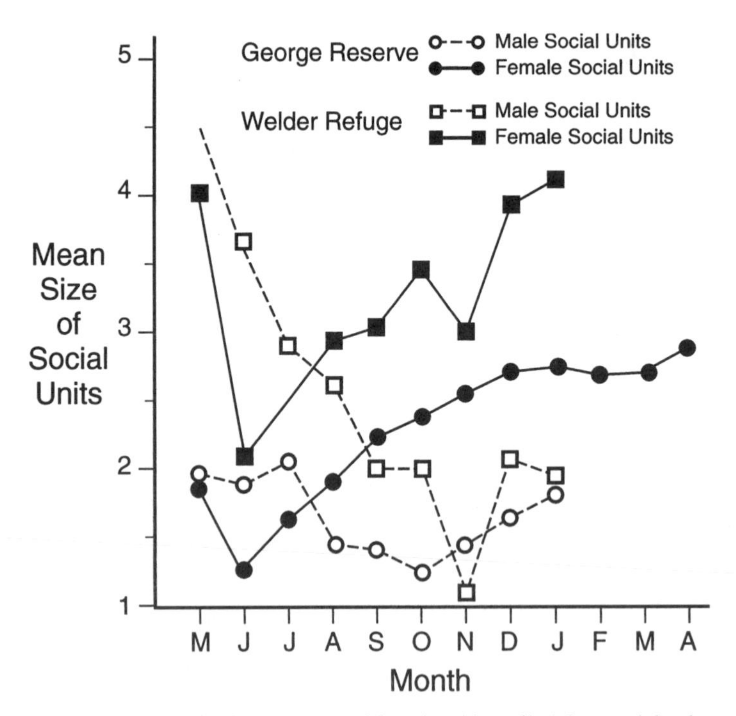

FIGURE 15.2. Mean monthly size of male and female white-tailed deer social units at the E. S. George Reserve, Michigan (USA), and the Rob and Bessie Welder Wildlife Refuge, Texas (USA) (modified from Hirth 1977).

Welder Refuge were adaptive, presumably by reducing predation risks—the larger the deer herd, the more "eyes" available to detect danger. Furthermore, rates of aggression among deer within a social unit were much higher at the George Reserve than they were at the Welder Refuge, thereby facilitating the formation of larger social units at the Welder Refuge.

Another example of how behavior and landscape concepts have been applied in a research setting involves a study of song rates of forest songbirds in relation to size and isolation of forested tracts (McShea and Rappole 1997). Song rates of two mature-forest species, Wood Thrushes (*Hylocichla mustelina*) and Ovenbirds (*Seiurus aurocapillus*), were higher in tracts of contiguous forest (>50 ha) than they were in isolated forested tracts (4–24 ha). On the other hand, song rates of an early-successional forest species, Northern Cardinals (*Cardinalis cardinalis*), were lower in contiguous forest than they were in isolated forested tracts. These differences in song rates between tract types did not affect the ability of investigators to detect individual birds, but they did have an effect on estimates of population densities.

Landscape fragmentation per se may not necessarily be the proximate stimulus affecting animal behavior. Instead, fragmentation may directly influence the availability of critical resources, such as food or mates, and, hence, indirectly affect animal behavior. (e.g., see McLean 1997) (Figure 15.3). As an example, butterflies are abundant in human-modified landscapes during summer, provided that wildflowers are available as nectar sources (Yahner 1996, 1998). In managed

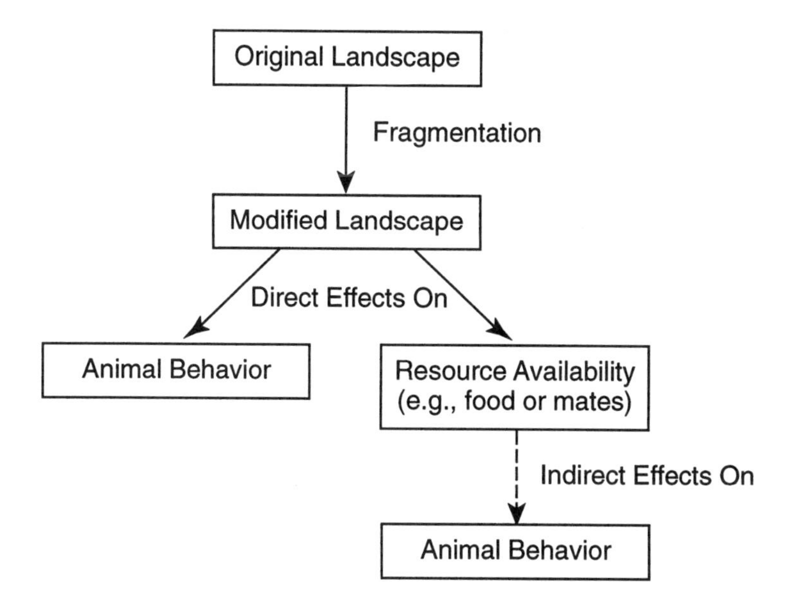

FIGURE 15.3. Landscape fragmentation may directly affect animal behavior. Alternatively, fragmentation may directly influence the availability of critical resources, such as food and mates, thereby indirectly affecting animal behavior.

forested landscapes, butterflies occur along unmowed logging roads containing wildflowers, but they are relatively scarce in clear cut or mature forest stands lacking wildflowers. In agricultural landscapes, wildflowers are common along herbaceous strips left unmowed at the interface of woodlands and crop fields. If wildflowers are mowed with other vegetation along logging roads or herbaceous strips, butterflies immediately forage elsewhere.

Another example of the effects of food resources on behavior in fragmented landscapes involves a prairie wildflower, the royal catchfly (*Silene regia*), and a major pollinator, the Ruby-throated Hummingbird (*Archilochus columbris*) (Menges 1991). In some Midwestern (USA) states, the range of the royal catchfly has been reduced to isolated, small populations because of landscape fragmentation. As a consequence, hummingbirds visit relatively small populations (e.g., <100 individuals) of royal catchfly less often than they visit larger populations (e.g., >150 individuals). In turn, smaller populations of royal catchfly often are less viable and persist for shorter times compared with larger populations because of significantly lower germination rates and higher rates of inbreeding in smaller versus larger populations.

Another important resource that may be affected directly by landscape fragmentation is mate availability, which in turn may indirectly affect animal behavior. Two recent studies have shown that the probability of pair bonding in Ovenbirds is considerably lower in smaller than in larger forested tracts (e.g., Gibbs and Faaborg 1990; Villard et al. 1993). For instance, 76% and 25% of male Ovenbirds were unpaired in small (9–150 ha) versus large (>500 ha) tracts, respectively. Reasons given for this differential pairing success in Ovenbirds range from male-biased sex ratios to greater susceptibility of ground-nesting female Ovenbirds to predators in smaller tracts. In contrast, pairing success of Wood Thrushes did not vary among woodlots ranging from 3 to 12 ha in a highly fragmented agricultural landscape (Friesen et al. 1999).

A considerable body of literature is now available dealing with predation on avian nests, which are a resource to nest predators, in relation to landscape pattern and type (e.g., see reviews by Paton 1994; Yahner 2000). For instance, predation rates on nests in small (<100 ha) woodlots surrounded by agriculture and suburbia were at least 25% compared with much lower rates in large forested tracts (Wilcove 1985). In forest-dominated landscapes, predation rates on nests were 50% in areas containing 50% fragmentation resulting from clearcutting, but predation rates were only 19% in areas containing 25% fragmentation (Yahner and Scott 1988). This suggests that predators, such as American Crows (*Corvus brachyrhynchos*), Blue Jays (*Cyanocitta cristata*), and raccoons are able to forage more efficiently in smaller than in larger patches, or in more-fragmented than in less-fragmented landscapes. Greater amounts of fragmentation generally increase the amount of edge habitat, which would be beneficial to predators that forage along edges (Yahner and Mahan 1996).

An understanding of the effects of landscape fragmentation on animal behavior may influence the future establishment of harvest quotas and seasons by wildlife agencies. Recall that the size of deer social units can vary with the degree of landscape "openness" (Figure 15.2). In Illinois, deer were more susceptible to hunters in highly fragmented agricultural landscapes with relatively small amounts of cover for deer than they were in landscapes with less fragmentation and more cover (Foster et al. 1997).

15.3.3 Policy Applications

Because roads fragment habitats and may affect the dispersal and abundance of some animal species, natural resource agencies and organizations have formulated policies that call for decommissioning of old roads and preserving roadless areas for the benefit of animal species requiring large undisturbed habitat for survival (e.g., The Wildlife Society 1998; D. DeCalesta, personal communication). Roadless areas also would have a major impact on other aspects of the behavior (e.g., migration, foraging) of these organisms.

15.4 Principles for Applying Landscape Ecology

Except for applications of concepts and principles of animal behavior (e.g., dispersal) in corridor design, resource managers have seldom applied other concepts or principles (e.g., size of social units, frequency of pair bonds, song rates) in relation to landscape fragmentation in on-the-ground conservation efforts. Consequently, the application principles that we suggest here are based on principles of wildlife ecology and behavior, not on demonstrated effects of field applications of behavior and landscape concepts. To apply behavioral and landscape concepts, managers must have a solid understanding about documented trends in animal responses to landscape fragmentation (see Table 15.1), and the home range size, habitat requirements, and natural history of target species.

15.4.1 Know About Home Range Size

Managers involved in design of corridors and wildlife reserves (e.g., protection areas) in fragmented landscapes must develop knowledge of home range sizes for target species. Some species, especially large vertebrates, have large home range sizes (Meffe and Carroll 1997). In fragmented landscapes, it may be impossible for resource managers to acquire a large enough continuous parcel of land to incorporate the home range of a large carnivore. However, knowing that a species has a large home range will make it clear that resource managers will have to acquire several smaller reserves located close to each other or connected with corridors to accommodate the home range of these large vertebrates.

15.4.2 Understand Habitat Requirements

To mitigate detrimental effects (e.g., population isolation, reduced dispersal) that landscape fragmentation has on a species, resource managers must develop detailed knowledge of the species' habitat requirements. For example, the habitat

TABLE 15.1. Trends in animal behavior that have been documented or inferred from fragmented landscapes, examples of organisms and habitats in which the trends have been studied, and pertinent references.

Behavior or behavioral trend	Examples of organisms studied	Habitat studied	Reference
Similar aggression levels in habitat patches and continuous habitats	White-footed mouse *Peromyscus leucopus*	Woodlots in agricultural landscapes	Mossman and Srivastava 1999
	White-footed mouse	Old-field patches in agricultural landscape	Diffendorfer et al. 1995
	Cotton rat *Sigmodon hispidus*		
	Prairie vole *Microtus orchrogaster*		
Reduced dispersal; inability to cross habitat gaps	Yellow-necked mouse *Apodemus flavicollis*	Forested habitat transected by roads	Mader 1984
	Amazonian mammals (e.g., black and white saki monkeys, *Pithecia pithecia;* three-toed sloth, *Bradypus variegates*)	Tropical rainforest	Dale et al. 1994
	Northern Spotted Owl *Strix occidentalis caurina*	Forested habitat managed by clearcutting	Miller et al. 1997
	Gray-tailed vole *Microtus canicaudus*	Experimental field plots	Wolff et al. 1997
	Mountain pygmy possum	Talus slopes fragmented by commercial development	Mansergh and Scotts 1989

Increased foraging efficiency in habitat patches	Blue Jay	Woodlots in agricultural and suburban landscapes	Keyser et al. 1998
	American Crow Gray fox *Urocyon cinereargenteus* Raccoon Virginia opossum *Didelphis virginiana*		
Increased foraging efficiency along edges	Blue Jay	Forested habitats	Marini et al. 1995
	American crow	Woodlots in agricultural landscape	Hannon and Cotterill 1998
Increased foraging efficiency in the interior of habitat patches	Small mammals Deer mouse *Peromyscus maniculatus* White-footed mouse	Woodlots in agricultural landscape	Hannon and Cotterill 1998
Increase in antipredatory behavior in fragmented landscapes	Eastern chipmunk	Forested habitat managed by clearcutting	Mahan 1996
Decreased pairing success in fragmented landscapes	Ovenbird	Woodlots in agricultural landscape	Villard et al. 1993

requirements of the New England cottontail (*Sylvilagus transitionalis*), a candidate for federally threatened or endangered status, include isolated stands of early-successional forest habitat amid an otherwise uncut, contiguous forest (Litivatis and Villafuerte 1996). By knowing the habitat requirements for this species, researchers have recommended a landscape management strategy that requires maintaining a series of early-successional habitats, each 15–75 ha in size and located less than 1 km from similar habitats (Litivatis and Villafuerte 1996). This strategy will help ensure successful dispersal and adequate amounts of habitat to sustain viable populations in the long term.

15.4.3 Know the Natural History of the Target Species

Territorial behavior, for example, may be affected by landscape fragmentation. In contiguous boreal forests, the red squirrel individually defends a territory surrounding a concentrated cache of conifer seeds, its principal food (Smith 1968). In contrast, in more fragmented agricultural landscapes where conifer trees exist as patches and conifer seeds are not as abundant, red squirrels do not cache food and subsist on a variety of food resources. Furthermore, in these patchy habitats, territoriality may not be exhibited (Yahner 1980).

15.5 Knowledge Gaps

We found few studies relating landscape fragmentation directly to animal behavior. Therefore, theoretical and empirical knowledge gaps need to be filled for the effective application of principles of animal behavior and landscape ecology to conservation. In this section, we identify some theoretical and empirical knowledge voids that impede the application of landscape ecology to animal conservation in fragmented landscapes.

15.5.1 Theoretical Voids

Two theoretical concepts, island biogeography and optimal foraging, have potential applications to understanding animal behavior in fragmented landscapes. Models developed from the theory of island biogeography have enabled conservation biologists to predict distributional patterns of flora and fauna and to understand better relationships between landscape fragmentation and species richness (MacArthur and Wilson 1967). To our knowledge, the potential effects of animal behavior on predictions generated by island biogeography models have not been examined. However, research on the behavior of animals occupying oceanic islands ("true" islands) may provide indications of how habitat islands created by landscape fragmentation could influence aggression in animals. For example, Garten (1976) found that mainland oldfield mice (*Peromyscus polionotus*) from Florida were more aggressive than were conspecifics from islands located off the Florida coast. In addition, Halpin (1981) discovered that mainland deer mice from British Columbia were very aggressive toward young unrelated mice. In the same study, adult mice from islands off the coast of British Columbia exhibited no aggression toward young unrelated mice. This trend of less-aggressive behavior on oceanic islands may not necessarily hold for habitat islands created by landscape fragmentation. Mossman and Srivastava (1999) found no difference in levels of aggression between adult male white-footed mice occupying isolated woodlots and those using continuous mature forest, whereas Linzey (1989) found that white-footed mice in clear cut areas were less-aggressive than were those in occupied undisturbed woodland areas.

Potential differences in aggression among animals that occupy continuous, undisturbed habitats versus isolated habitat islands may affect immigration and emigration rates. For instance, decreased levels of aggressive behavior among animals occupying habitat islands could promote colonization of these patches. Island-biogeographic models that depict colonization rates of islands located "near" versus "far" from mainland habitats (Wilcox 1980) have not been developed to incorporate differential levels of aggression as related to "near" versus "far" islands. Differences in aggressive behavior between animals that occupy isolated habitat islands may affect how corridors are used by dispersing animals. For example, aggressive animals may be more likely to use corridors, whereas less-aggressive individuals may remain more sedentary.

Optimal-foraging models were developed to provide theoretical predictions of how animals make choices about where and when to hunt for food and which kinds of prey to eat. These models provide important predictive information for managers interested in predator-prey interactions and the effects of landscape fragmentation on these interactions. Generally, predators prefer to forage in habitat patches with the highest density of prey in the landscape (Hassell and May 1974). Small mammals are an important prey item for many predators, and small mammals often reach high densities in remnant habitat patches in fragmented landscapes (Nupp and Swihart 1996; Wolff et al. 1997). Based on optimal-foraging models, predators should concentrate their foraging in these patches of high prey density, which should result in predation rates on small mammals that are higher in fragmented than in contiguous landscapes. Although many studies have focused on the susceptibility of forest songbirds to predation in isolated habitat patches (e.g., Gates and Gysel 1978; Robinson et al. 1995), there is relatively little information on predation effects on small mammals in relation to fragmented landscapes. Populations of eastern chipmunks occupying isolated forest patches in an agricultural landscape, however, have been shown to experience higher rates of predation and local extinction (Hendersen et al. 1985). In addition, New England cottontails were more likely to be preyed on by predators in small ($\leq$3 ha) habitat patches with high perimeter-area ratios (increased edge) than they were in larger patches with less edge (Brown and Litvaitis 1995; Villafuerte et al. 1997).

Theory related to plant-herbivore and plant-granivore interactions in isolated habitat patches also should be given more attention. Two studies of plant-granivore interactions illustrate the significance of such theory for conservation. White-footed mice feed on tree seeds and may cause an appreciable reduction in rates of oak

(*Quercus*) germination in managed forest stands (DeLong and Yahner 1996). In fragmented landscapes, mature seed-producing trees are often restricted to remnant managed patches; small mammals may concentrate their foraging activities in these profitable patches and, hence, deplete seed stocks needed for forest regeneration. As landscapes become increasingly fragmented, regeneration of certain tree species may be at greater risk than other species in isolated patches. Indeed, Hulme and Hunt (1999) documented that wood mice (*Apodemus sylvaticus*) virtually eliminated elm (*Ulmus*) seeds in isolated forested patches and could potentially cause localized extinctions of seed populations.

Furthermore, optimal-foraging theory predicts that as travel time between patches increases, predators will forage in a patch for a greater period of time. Thus, prey items (e.g., incubating birds, nestlings, bird eggs) in isolated patches can be exposed to higher predation pressures than when patches are located close together. Most research on predation in habitat patches versus that in continuous unfragmented landscapes has focused on the target prey, primarily nesting songbirds (Robinson et al. 1995). However, the differential foraging intensity by various species of mammalian and avian predators in habitat patches versus contiguous unfragmented landscapes deserves more study to provide additional insight into the susceptibility of prey in fragmented landscapes (Hannon and Cotterill 1998).

15.5.2 Empirical Voids

Martin (1998) provided an excellent overview of the empirical knowledge gaps between the fields of animal behavior and wildlife management. In particular, she suggested a detailed examination of animal behavior in fragmented landscapes as it relates to natal dispersal and movements between habitat patches and across gaps, and why some animals can persist in fragmented landscapes and others cannot.

Landscape fragmentation creates gaps that vary in width, length, length of persistence, and other characteristics that can have profound effects on the likelihood that individuals will move across these gaps and, hence, ensure long-term persistence of populations (Durant 1998). Agricultural habitats adjacent to fragmented woodlots act as habitat barriers to white-footed mice (Yahner 1983), and 20-m-wide roads are dispersal barriers to wood mice (Mader 1984). Eastern chipmunks seldom cross agricultural fields to reach isolated woodlots (Hendersen et al. 1985). Northern Spotted Owls dispersing across habitat gaps (clear-cuts or agricultural areas) experience lower survival than do those dispersing across contiguous landscapes (Miller et al. 1997). The unwillingness of bird species to cross certain habitats or gaps may make them more prone to endangerment or extinction (Reed 1999). In the central Amazon, tropical fauna with large home ranges are less able to cross gaps than are species with small home ranges (Dale et al. 1994). Hence, more research is needed to examine the gap-crossing abilities of animals, especially area-sensitive animals that are susceptible to discontinuities in an otherwise contiguous landscape.

Although time-intensive to conduct, behavioral studies dealing with time budgets of individual animals are necessary to understand how animals respond to, and persist in, fragmented landscapes. Quantification of the time that animals spend avoiding predators in fragmented versus unfragmented landscapes, for example, can be of particular importance because time spent in vigilance reduces time available to engage in other important behaviors, such as foraging (Lima and Dill 1990). In addition, changes in vigilance behavior can be a way of assessing the risk of predation in various habitats (Metcalfe 1984; Lima et al. 1985).

Changes in the landscape resulting from fragmentation, such as an increase in vegetative understory in remnant patches, increased amounts of edge, and reductions in numbers of mature seed-producing trees, could have profound effects on animals occupying modified landscapes (Forman and Godron 1986). We need information about how the fine-grained details of fragmented landscapes influence time budgets of individual animals. Mahan (1996), for instance, found that eastern chipmunks spent significantly more time in alert behavior (pause) in a fragmented forested landscape, presumably reconnoitering the area for potential predators, than they did in an unfragmented landscape. We encourage researchers to relate differences in behavioral patterns to individual survival and fitness, which would provide insight into why or how certain species persist in fragmented habitats.

Studies that examine reproductive and pairing success of individual animals occupying fragmented versus unfragmented landscapes should be expanded. Effects of landscape fragmentation on pairing success may be an important issue for forest birds that are declining, such as Golden-winged (*Vermivora chrysoptera*) and Cerulean Warblers (*Dendroica cerulea*) (Rosenberg and Wells 1995).

We need to move beyond examining predation rates in unfragmented versus fragmented landscapes by focusing on how foraging patterns vary among different predators in fragmented landscapes. Hannan and Cotterill (1998), for instance, noted that corvids foraged mainly at the edges of forest patches; in contrast, they found that small mammals (seed predators) were more efficient at foraging in the interior of large woodlots (>100 ha) away from the edges. In another study, Marini et al. (1995) found more species of avian nest predators near forest edges than in forest interiors, although this same trend in habitat use was not documented for mammalian nest predators.

Although researchers have documented increased predation rates on songbird nests in fragmented landscapes, the behavioral cues that predators use to locate and capture prey are not well-understood. Predators may use parental behavior and vocal begging by nestlings to locate bird nests (Ratti and Reese 1988), but the evidence is equivocal (Halupka 1998).

15.6 Research Approaches

In this section, we describe fundamental research approaches that can be used to help determine how fragmented landscapes influence animal behavior.

A study that focuses on the effects of landscape fragmentation on animal behavior should give careful consideration to experimental design. First, the design should carefully select study sites and landscapes that are tailored to the specific research question. For example, to elucidate the effect of landscape patterns on animal behavior, study sites should be selected that represent habitat remnants that vary in size, shape, and degree of isolation (Figure 15.1). Researchers may consider using landscapes that differ in the type of fragmentation (e.g., openings caused by agriculture vs. silviculture). Results obtained from animal behavior studies in one type of landscape, such as those for animals occupying woodlots amid agricultural land, may not necessarily apply to another landscape, such as results for animals inhabiting uncut forested stands in an extensively harvested forested landscape (see Yahner 2000).

Second, behavioral studies involve watching animals from close range (e.g., with field glasses; Yahner 1978) or remotely (e.g., with global positioning technology; Merrill et al. 1998). Studies that examine spacing mechanisms or time budgets of individual animals in relation to landscape fragmentation will require some means of uniquely marking individuals, which can be time- and labor-demanding (e.g., see protocols in Smith 1968; Altmann 1974; Yahner 1978; Hatchwell et al. 1996; Mahan 1996; McGregor and Peake 1998). Statistical comparisons of time budgets based on duration and rates of key behavior of individual animals in fragmented landscapes will help determine how fragmentation affects predator avoidance (e.g., based on measures of alert behavior), foraging (e.g., eating and food acquisition), and defense of limited resources (e.g., intraspecific agonistic interactions). Based on previous research, animals likely modify those behaviors in response to the effects of landscape fragmentation (e.g., Mahan 1996). In addition, researchers should determine the effects of other environmental factors, which could be affected by fragmentation, on animal behavior. These factors may include incident light, temperature, amount of downed woody debris, distance to a natural edge, canopy closure, and distance to human modified features (suburban development, paved roads, etc.)

Some behavioral studies may not require investigators to mark animals for individual recognition. Representative studies that fall into this category are those dealing with size and composition of social units (Hirth 1977), pairing success (Gibbs and Faaborg 1990), singing rates (McShea and Rappole 1997), and dispersal and movement patterns (Gibbs 1998) in relation to fragmented landscapes.

The behavior of animals should be addressed when designing or managing corridors (Clemmons and Buchholz 1997). The effects of width, length, and vegetative features (e.g., canopy coverage) of corridors should be examined in relation to animal behavior (e.g., probability of dispersal, time budgets). Better information on these aspects of corridor design and their effects on animal behavior will be vital to the conservation of species with reduced gap-crossing abilities that must move across fragmented landscapes (Mansergh and Scotts 1989; Machtans et al. 1996).

Nest predation studies have provided information on foraging patterns of predators in relation to patch size and edges in fragmented landscapes. Comparative studies of predation rates on artificial versus natural nests (e.g., Götmark et al. 1990; Butler and Rotella 1998) in the same fragmented habitats will help us understand how the behavior of nesting birds attracts different types of predators.

These types of studies in fragmented landscapes also will provide valuable information for managers interested in the effects of patch size and heterogeneity on predator activity in remnant patches.

Finally, an examination of the behavior of animals in response to population densities should be an integral component of environmental impact studies in fragmented landscapes (Goss-Custard and Durrell 1990). For example, feeding rates of individuals that occupy patches with different population densities could be compared to determine how the proportion of individuals failing to obtain adequate food varies with increases in population density.

In conclusion, studies designed to address behavioral responses to fragmented landscapes will be an important future area of research. From a conservation perspective, a future challenge will be to link animal behavior and landscape-level ecological processes and thereby better understand causal relationships between behavioral phenomena and landscape patterns (Lima and Zollner 1996; Yahner and Mahan 1997a).

Acknowledgments

Development of our ideas in this manuscript has been funded by various institutions, including The Pennsylvania State University, University of Minnesota, Ohio University, the U.S. Forest Service, and the Max McGraw Wildlife Foundation. We thank Dr. Michael Gannon and Dr. Jeffrey Kurland for reviewing this manuscript. We also thank S. Shawver for her assistance in preparing this manuscript.

References

Altmann, J. 1974. Observational study of behavior: sampling methods. *Behaviour* 49:227–267.

Ambrose, J.R., and Bratton, S.R. 1990. Trends in landscape heterogeneity along the borders of Great Smoky Mountains National Park. *Conserv. Biol.* 4:135–143.

Bendel, P.R., and Gates, J.E. 1987. Home range and microhabitat partitioning of the southern flying squirrel (*Glaucomys volans*). *J. Mammal.* 68:243–255.

Brower, J.E., and Zar, J.H. 1984. Field and laboratory methods for general ecology. Second Edition. Dubuque, Iowa: Wm. C. Brown Publishers.

Brown, A.L., and Litvaitis, J.A. 1995. Habitat features associated with predation of New England cottontails: what scale is appropriate? *Can. J. Zool.* 73:1005–1010.

Butler, M.A., and Rotella, J.J. 1998. Validity of using artificial nests to assess duck-nest success. *J. Wildl. Manage.* 62:163–171.

Clemmons, J.R., and Buchholz, R. 1997. Linking conservation and behavior. In *Behavioral Approaches to Conservation in the Wild,* eds. J.R. Clemmons and R. Buchholz, pp. 1–22. New York: Cambridge University Press.

Crook, J.H., Ellis, J.E., and Goss-Custard, J.D. 1976. Mammalian social systems: structure and function. *Anim. Behav.* 24:261–274.

Dale, V.H., Pearson, S.M., Offerman, H.L., and O'Neill, R.V. 1994. Relating patterns of land-use change to faunal biodiversity in the central Amazon. *Conserv. Biol.* 8:1027–1036.

DeLong, C., and Yahner, R.H. 1996. Predation on planted acorns in managed forested stands of central Pennsylvania. *Northeast Wildl.* 53:61–68.

deMaynadier, P.G., and Hunter, M.L., Jr. 1999. Forest canopy closure and juvenile emigration by pool-breeding amphibians in Maine. *J. Wildl. Manage.* 63:441–450.

Diffendorfer, J.E., Gaines, M.S., and Holt, R.D. 1995. Habitat fragmentation and movements of three small mammals (*Sigmodon, Microtus,* and *Peromyscus*). *Ecology* 76:827–839.

Durant, S. 1998. A minimum intervention approach to conservation: the influence of social structure. In *Behavioral Ecology and Conservation Biology,* ed. T. Caro, pp. 105–129. New York: Oxford University Press.

Eisenberg, J.F. 1966. The social organization of mammals. *Handbuch der Zoologie* 10:1–92.

Forman, R.T.T., and Godron, M. 1986. *Landscape Ecology.* New York: John Wiley and Sons.

Foster, J.R., Roseberry, J.L., and Woolf, A. 1997. Factors influencing efficiency of white-tailed deer harvest in Illinois. *J. Wildl. Manage.* 61:1091–1097.

Foster, M.L., and Humphrey, S.R. 1995. Use of highway underpasses by Florida panthers and other wildlife. *Wildl. Soc. Bull.* 23:95–100.

Friesen, L.E., Eagles, P.F.J., and MacKay, R.J. 1995. Effects of residential development on forest-dwelling Neotropical migrant songbirds. *Conserv. Biol.* 9:1408–1414.

Friesen, L.E., Wyatt, V.E., and Cadman, M.D. 1999. Pairing success of Wood Thrushes in a fragmented agricultural landscape. *Wilson Bull.* 111:279–281.

Garten, C. 1976. Relationships between aggressive behavior and genic heterozygosity in the oldfield mouse, *Peromyscus polionotus. Evolution* 30:59–72.

Gates, J.E., and Gysel, L.W. 1978. Avian nest dispersion and fledging success in field-forest ecotones. *Ecology* 58:871–883.

Gibbs, J.P. 1998. Amphibian movements in response to forest edges, roads, and streambeds in southern New England. *J. Wildl. Manage.* 62:584–589.

Gibbs, J.P., and Faaborg, J. 1990. Estimating the viability of Ovenbird and Kentucky Warbler populations in forest fragments. *Conserv. Biol.* 4:193–196.

Gill, F.B. 1990. *Ornithology.* New York: W.H. Freeman and Company.

Goss-Custard, J.D., and Durrell, S.E.A. Le V. Dit. 1990. Bird behaviour and environmental planning: approaches in the study of wader populations. *Ibis* 132:273–289.

Götmark, F., Neergaard, R., and Åhlund, M. 1990. Predation of artificial and real Arctic Loon nests in Sweden. *J. Wildl. Manage.* 54:429–432.

Haas, C.A. 1995. Dispersal and use of corridors by birds in wooded patches on an agricultural landscape. *Conserv. Biol.* 9:845–854.

Haddad, N.M. 1999. Corridor use predicted from behaviors at habitat boundaries. *Am. Nat.* 153:215–227.

Halpin, Z. 1981. Adult-young interactions in island and mainland populations of the deer mouse, *Peromyscus maniculatus. Oecologia* 51:419–425.

Halupka, K. 1998. Vocal begging by nestlings and vulnerability to nest predation in meadow pipits *Anthus pratensis:* to what extent do predation costs of begging exist? *Ibis* 140:144–149.

Hannon, S.J., and Cotterill, S.E. 1998. Nest predation in aspen woodlots in an agricultural area in Alberta: the enemy from within. *Auk* 115:16–25.

Harms, B., and Opdam, P. 1990. Woods as habitat patches for birds: application in landscape planning in The Netherlands. In *Changing Landscapes: An Ecological Perspective,* eds. I. Zonneveld and R.T.T. Forman, pp. 73–97. New York: Springer-Verlag.

Harris, L.D., and Silva-Lopez, G. 1992. Forest fragmentation and the conservation of biological diversity. In *Conservation Biology: The Theory and Practice of Nature Conservation, Preservation, and Management,* eds. P.L. Fiedler and S.K. Jain, pp. 197–237. New York: Chapman and Hall.

Harrison, R.L. 1992. Toward a theory of inter-refuge corridor design. *Conserv. Biol.* 6:293–295.

Hassell, M.P., and May, R.M. 1974. Aggregation of predators and insect parasites and its effect on stability. *J. Anim. Ecol.* 43:567–594.

Hatchwell, B.J., Chamberlain, D.E., and Perrins, C.M. 1996. The demography of Blackbirds *Turdus merula* in rural habitats: is farmland a sub-optimal habitat? *J. Appl. Ecol.* 33:1114–1124.

Hendersen, M.T., Merriam, G., and Wegner, G. 1985. Patchy environment and species survival: chipmunks in an agricultural mosaic. *Biol. Conserv.* 31:95–105.

Henein, K., and Merriam, G. 1990. The elements of connectivity where corridor quality is variable. *Landsc. Ecol.* 4:157–170.

Hesselton, W.T., and Hesselton, R.M. 1982. White-tailed deer. In *Wild Mammals of North America,* eds. J.A. Chapman and G.A. Feldhamer, pp. 878–901. Baltimore, Maryland: The Johns Hopkins University Press.

Hirth, D.H. 1977. Social behavior of white-tailed deer in relation to habitat. *Wildl. Monogr.* 53:1–55.

Hulme, P.E., and Hunt, M.K. 1999. Rodent post-dispersal seed predation in deciduous woodland: predator response to absolute and relative abundance of prey. *J. Anim. Ecol.* 68:407–428.

Hunter, M.L., Jr. 1996. *Fundamentals of Conservation Biology.* Cambridge, Massachusetts: Blackwell.

Keyser, A.J., Hill, G.E., and Soehren, E.C. 1998. Effects of forest fragment size, nest density, and proximity to edge on the risk of predation to ground-nesting passerine birds. *Conserv. Biol.* 12:986–994.

Knaapen, J.P., Scheffer, M., and Harms, B. 1992. Estimating habitat isolation in landscape planning. *Landsc. Urban Plan.* 23:1–16.

Lima, S.L., and Dill, L.M. 1990. Behavioral decisions made under the risk of predation: a review and prospectus. *Can. J. Zool.* 68:619–640.

Lima, S.L., Valone, T.J., and Caraco, T. 1985. Foraging-efficiency-predation-risk trade-off in gray squirrels. *Anim. Behav.* 33:155–165.

Lima, S.L., and Zollner, P.A. 1996. Towards a behavioral ecology of ecological landscapes. *Trends Ecol. Evol.* 11:131–134.

Linzey, A. 1989. Response of the white-footed mouse (*Peromyscus leucopus*) to the transition between disturbed and undisturbed habitats. *Can. J. Zool.* 67:505–512.

Litvaitis, J.A., and Villafuerte, R. 1996. Factors affecting the persistence of New England cottontail metapopulations: the role of habitat management. *Wildl. Soc. Bull.* 24:686–693.

MacArthur, R.H., and Wilson, E.O. 1967. *The Theory of Island Biogeography.* Princeton: Princeton University Press.

Machtans, C.S., Villard, M.-A., and Hannon, S.J. 1996. Use of riparian buffer strips as movement corridors by forest birds. *Conserv. Biol.* 10:1366–1379.

Mader, H.-J. 1984. Animal habitat isolation by roads and agricultural fields. *Biol. Conserv.* 29:81–96.

Mahan, C.G. 1996. *The Ecology of Eastern Chipmunks (Tamias striatus) in a Fragmented Forest.* Ph.D. dissertation. University Park: The Pennsylvania State University.

Mahan, C.G., and Yahner, R.H. 1998. Lack of population response by eastern chipmunks (*Tamias striatus*) to forest fragmentation. *Am. Midl. Nat.* 140:382–386.

Mansergh, I.M., and Scotts, D.J. 1989. Habitat continuity and social organization of the mountain pygmy possum restored by tunnel. *J. Wildl. Manage.* 53:701–707.

Marini, M.A., Robinson, S.K., and Heske, E.J. 1995. Edge effects on nest predation in the Shawnee National Forest, southern Illinois. *Biol. Conserv.* 74:203–213.

Martin, K. 1998. The role of animal behavior studies in wildlife science and management. *Wildl. Soc. Bull.* 26:911–920.

McGregor, P., and Peake, T. 1998. The role of individual identification in conservation biology. In *Behavioral Ecology and Conservation Biology,* ed. T. Caro, pp. 31–55. New York: Oxford University Press.

McLean, I.G. 1997. Conservation and the ontogeny of behavior. In *Behavioral Approaches to Conservation in the Wild,* eds. J.R. Clemmons and R. Buchholz, pp. 132–156. New York: Cambridge University Press.

McLellan, B.N., and Shackleton, D.M. 1988. Grizzly bears and resource-extraction industries: effects of roads on behavior, habitat use, and demography. *J. Appl. Ecol.* 25:451–460.

McShea, W.J., and Rappole, J.H. 1997. Variable song rates in three species of passerines and implications for estimating bird populations. *J. Field Ornithol.* 68:367–375.

Mech, L.D. 1970. *The Wolf: The Ecology and Behavior of an Endangered Species.* New York: Natural History Press.

Meffe, G.K., and Carroll, C.R., eds. 1997. *Principles of Conservation Biology.* Second Edition. Sunderland, Massachusetts: Sinauer Associates.

Menges, E.S. 1991. Seed germination percentage increases with population size in a fragmented prairie species. *Conserv. Biol.* 5:158–164.

Merrill, S.B., Adams, L.G., Nelson, M.E., and Mech, L.D. 1998. Testing releasable GPS radio collars on wolves and white-tailed deer. *Wildl. Soc. Bull.* 26:830–835.

Metcalfe, N.B. 1984. The effects of habitat on the vigilance of shorebirds: is visibility important? *Anim. Behav.* 32:981–985.

Miller, G.S., Small, R.J., and Meslow, E.C. 1997. Habitat selection by Spotted Owls during natal dispersal in western Oregon. *J. Wildl. Manage.* 61:140–150.

Mossman, C.A., and Srivastava, N.P. 1999. Does aggressive behavior of *Peromyscus leucopus* influence isolation of habitat islands? *Am. Midl. Nat.* 141:366–372.

Noss, R.F. 1991. Landscape connectivity: different functions at different scales. In *Landscape Linkages and Biodiversity,* ed. W.E. Hudson, pp. 27–39. Washington, DC: Island Press.

Nupp, T., and Swihart, R. 1996. Effect of forest patch area on population attributes of white-footed mice (*Peromyscus leucopus*) in fragmented landscapes. *Can. J. Zool.* 67:505–512.

Paton, P.W.C. 1994. The effect of edge on avian nest success: how strong is the evidence? *Conserv. Biol.* 8:17–26.

Ratti, J.T., and Reese, K.P. 1988. Preliminary test of the ecological trap hypothesis. *J. Wildl. Manage.* 52:484–491.

Reed, J.M. 1999. The role of behavior in recent avian extinctions and endangerments. *Conserv. Biol.* 13:232–241.

Robbins, C.S., Dawson, D.K., and Dowell, B.A. 1989. Habitat area requirements of breeding forest birds of the Middle Atlantic states. *Wildl. Monogr.* 103:1–34.

Robinson, S.K., Thompson, F.R., III, Donovan, T.M., Whitehead, D.R., and Faaborg, J. 1995. Regional forest fragmentation and the nesting success of migratory birds. *Science* 267:1987–1990.

Rosenberg, K.V., and Wells, J.V. 1995. *Importance of Geographic Areas to Neotropical Migrant Birds in the Northeast.* Final Report, Region 5. Hadley, Massachusetts: U.S. Fish and Wildlife Service.

Rudzinski, D.R., Graves, H.B., Sargeant, A.B., and Storm, G.L. 1982. Behavioral interactions of penned red and arctic foxes. *J. Wildl. Manage.* 46:877–884.

Saunders, D.A., Hobbs, R.J., and Margules, C.R. 1991. Biological consequences of ecosystem fragmentation: a review. *Conserv. Biol.* 5:18–32.

Smith, C.C. 1968. The adaptive nature of social organization in the genus of tree squirrels *Tamiasciurus. Ecol. Monogr.* 38:30–63.

Swihart, R.K., and Yahner, R.H. 1982. Eastern cottontail use of fragmented farmland habitat. *Acta Theriol.* 19:257–273.

The Wildlife Society. 1998. Road policy proposed for national forest system. *The Wildlifer* 287:1.

Trombulak, S.D., and C.A. Frissell. 2000. Review of ecological effects of roads on terrestrial and aquatic communities. *Conserv. Biol.* 14:18–30.

Villafuerte, R., Litvaitis, J.A., and Smith, D.F. 1997. Physiological responses by lagomorphs to resource limitation imposed by habitat fragmentation: implications for condition-sensitive predation. *Can. J. Zool.* 75:148–151.

Villard, M.-A., Martin, P.R., and Drummond, C.G. 1993. Habitat fragmentation and pairing success in the Ovenbird (*Seiurus aurocapillus*). *Auk* 110:759–768.

Wauters, L., Casale, P., and Dhondt, A. 1994. Space use and dispersal of red squirrels in fragmented habitats. *Oikos* 69:140–146.

Wilcove, D.S. 1985. Nest predation in forest tracts and the decline of migratory songbirds. *Ecology* 66:1211–1214.

Wilcox, B.A. 1980. Insular ecology and conservation. In *Conservation Biology: An Ecological-Evolutionary Perspective,* eds. M.E. Soulé and B.A. Wilcox, pp. 95–117. Sunderland, Massachusetts: Sinauer Associates.

Wolff, J.O., Schauber, E.M., and Edge, W.D. 1997. Effects of habitat loss and fragmentation on the behavior and demography of gray-tailed voles. *Conserv. Biol.* 11:945–956.

Yahner, R.H. 1978. The adaptive nature of the social system and behavior in the eastern chipmunk, *Tamias striatus. Behav. Ecol. Sociobiol.* 3:397–427.

Yahner, R.H. 1980. Burrow use by red squirrels. *Am. Midl. Nat.* 103:409–411.

Yahner, R.H. 1983. Population dynamics of small mammals in farmstead shelterbelts. *J. Mammal.* 64:380–386.

Yahner, R.H. 1996. Butterfly and skipper communities in a managed forest landscape. *Northeast Wildl.* 53:1–9.

Yahner, R.H. 1997. Long-term dynamics of bird communities in a managed forested landscape. *Wilson Bull.* 109:595–613.

Yahner, R.H. 1998. Butterfly and skipper use of nectar sources in forested and agricultural landscapes of Pennsylvania. *J. Pennsyl. Acad. Sci.* 71:104–108.

Yahner, R.H. 2000. *Eastern Deciduous Forest: Ecology and Wildlife Conservation.* Second Edition. Minneapolis: University of Minnesota Press.

Yahner, R.H., and Mahan, C.G. 1996. Depredation of artificial ground nests in a managed, forested landscape. *Conserv. Biol.* 10:285–288.

Yahner, R.H., and Mahan, C.G. 1997a. Behavioral considerations in fragmented landscapes. *Conserv. Biol.* 11:569–570.

Yahner, R.H., and Mahan, C.G. 1997b. Effects of logging roads on depredation of artificial ground nests in a forested landscape. *Wildl. Soc. Bull.* 25:158–162.

Yahner, R.H., and Scott, D.P. 1988. Effects of forest fragmentation on depredation of artificial nests. *J. Wildl. Manage.* 52:158–161.

16

Effects of Landscape Change on the Physical and Chemical Components of Aquatic Ecosystems

CAROLYN T. HUNSAKER AND ROBERT M. HUGHES

16.1 Introduction

This chapter introduces the importance of landscape structure in determining the characteristics and functioning of aquatic ecosystems. The focus is on rivers and streams with some discussion on lakes. Despite the underrepresentation of aquatic issues in the early landscape literature, landscape ecology is very relevant to several aquatic subject areas. Aquatic ecosystems manifest the properties of their water columns, channels or basins, riparian zones, and upland watersheds. This chapter addresses the effects of landscape change on the physical components (e.g., channel or lake shape and bank condition, substrate characteristics, water temperature, and stream discharge) and chemical components (e.g., water conductivity, nutrient concentrations, acid neutralizing capacity) of aquatic ecosystems, and it complements Chapter 17 of this volume, which addresses biointegrity issues of aquatic ecosystems.

The objectives of this chapter are to address landscape-related concepts, principles, and emerging ideas; recent management applications of these concepts and principles; principles for applying landscape ecology in aquatic conservation; theoretical and empirical knowledge gaps that prevent such applications; and research approaches for filling these gaps. We discuss physical and chemical issues (this chapter) separately from biointegrity issues (Chapter 17), but it is important to recognize that many of the same processes, patterns, principles, and knowledge gaps underlie both sets of issues. Although we only address freshwater systems, landscape ecology also is relevant to marine and coastal ecosystems.

16.2 Concepts, Principles, and Emerging Ideas

Landscape ecologists seek to understand better the relationships between landscape structure (including land cover) and ecosystem processes at various spatial scales (Turner 1989). *Land cover* refers to the entity covering the surface of the Earth. Sometimes land cover is expressed as vegetation classes only, and sometimes it is a mixture of vegetation classes and use classes such as forest, urban, agriculture, and so on. For the purposes of this chapter, both land cover and land use are components of landscape change. The word *structure* refers to the spatial relationships of ecosystem characteristics such as vegetation, animal distributions, and soil types. *Processes* or *functions* refer to the interactions—that is, the flow of energy, materials, and organisms—among the spatial elements. Because landscapes are spatially heterogeneous areas, or environmental mosaics, the structure and function of landscapes are scale-dependent. Landscape-ecology concepts and principles are important to the processes within the water column, the adjacent riparian zone (sometimes dry and sometimes flooded), and the terrestrial upland areas of the watershed. Frequently, terrestrial ecologists ignore the influence of aquatic systems on the ecosystems they study; however, aquatic ecologists can seldom ignore the influence of terrestrial areas (uplands) on aquatic ecosystems.

The discipline of landscape ecology has had less influence on aquatic ecology than it has had on terrestrial ecology. Few chapters in landscape-ecology books and few presentations at annual meetings of the U.S. Chapter of the International Association for Landscape Ecology have addressed aquatic issues. This difference is likely due to a combination of several factors. Scientists working on watershed issues have often taken a "landscape approach" because they had to characterize soil, geology, topography, vegetation, and climate to understand aquatic responses such as stream flow, water chemistry, and biotic communities. Perhaps scientists working on watershed issues have not felt the need for a new discipline such as landscape ecology. Another factor could be that models of aquatic systems require three spatial dimensions, plus time, so the importance of space and scale have been obvious (see Chapter 17). Freshwater systems are either spatially constrained (e.g., streams and rivers) or patchy (e.g., lakes and wetlands); thus, large amounts of continuous digital data are not necessarily available for aquatic systems, as they are for terrestrial ecosystems (soil maps, geology maps, vegetation cover, and land use). Noncontinuous data help explain why the application of landscape metrics has been much less frequent for aquatic systems than it has been for terrestrial systems.

16.2.1 General Concepts From Hydrology and Aquatic Ecology

There is a long history of research relating landscape characteristics, including land cover, to water chemistry and habitat quality (e.g., Omernik 1977; Reckhow et al. 1980; Lowrance et al. 1984; Peterjohn and Correll 1984; Hunsaker et al. 1986; Sivertun et al. 1988; Johnston et al. 1990; Hunsaker and Levine 1995; Allan et al. 1997; Johnson et al. 1997). Land-cover change affects aquatic ecosystems both directly and indirectly. Possible direct effects include changes in sediment loads from soil disturbance, in flow amount and timing from vegetation alteration or channel modification, in nutrient concentrations from soil disturbance or vegetation alteration, and in physical habitat structure such as woody debris, streambed characteristics, and channel characteristics. Effects on aquatic biota can occur directly, but in most cases effects on aquatic biota occur indirectly via a change in physical conditions such as streamflow or the chemical quality of water. For example, increased nutrient levels from increased agriculture or urbanization can cause algae blooms (a direct effect) that lower dissolved oxygen when they respire at night or decompose. The lower dissolved oxygen can cause animals to leave or die (an indirect effect). Increased nutrient levels are seldom high enough to have a direct toxic effect on fish or invertebrates.

Consideration of spatial scale (grain and extent) is critical for the use of landscape-ecology concepts in biological conservation. As issues being addressed by ecological research and resource management have become more complex and integrative, an interesting question has surfaced. "What is the appropriate spatial construct to characterize regions?" Before landscape ecologists, geographers struggled for a long time with spatial interrelationships and questions of geographic characterization. A regional characterization scheme is especially difficult for aquatic ecosystems because two popular approaches exist, ecoregions and watersheds, and both can be hierarchically constructed (Hunsaker et al. 1997). An *ecoregion* is an area (region) of relative homogeneity based on multiple landscape attributes. A *watershed* is an area of land where surface water drains to a specific point on a stream or to a lake or wetland; watersheds are based on topography and the fact that water flows downslope because of gravity. Watersheds are a logical analysis unit if chemical pollutants from point sources are the primary concern because the distribution and concentrations of these pollutants are primarily governed by the hydrologic network and dilution. Groundwater conforms to watershed boundaries less frequently than we would like to believe because aquifers have yet another set of boundaries. However, if diffuse or nonpoint-source pollution (which is governed by land-use pattern, geology, and soils) and ecosystem conditions as measured by aquatic organisms are of primary interest, then ecoregions (Gallant et al. 1989) are a useful analysis unit. Ecoregions are defined by physiography, land form, climate, soil, and land cover, and compared with watersheds, better characterize ecosystem homogeneity. For example, fish species found in headwater streams across many watersheds in the same ecoregion will be more similar to each other than they will be to fish species in the large rivers of the same watershed (e.g., river continuum concept, Figure 16.1).

Watersheds have been used by ecologists for a long time as an organizing paradigm for spatial scale. Biogeochemical cycling research is usually carried out on small experimental watersheds, and many of the National Science Foundation's Long Term Ecological Research sites are based around watersheds such as Hubbard Brook (Likens and Bormann 1974; Bormann and Likens 1981), Coweeta (Swank and Crossley 1987), and other long-term study sites (Correll 1986; Johnson and Van Hook 1989). Naiman (1992) presents current knowledge and future trends in watershed management and research with a focus on watershed and regional models, indicators of environmental change, and new techniques for integrated resource management.

Ecoregions have received more attention recently (Bailey 1976; Wiken 1986). Omernik (1987) developed a set of ecoregions (1:7,500,000-scale map) for the con-

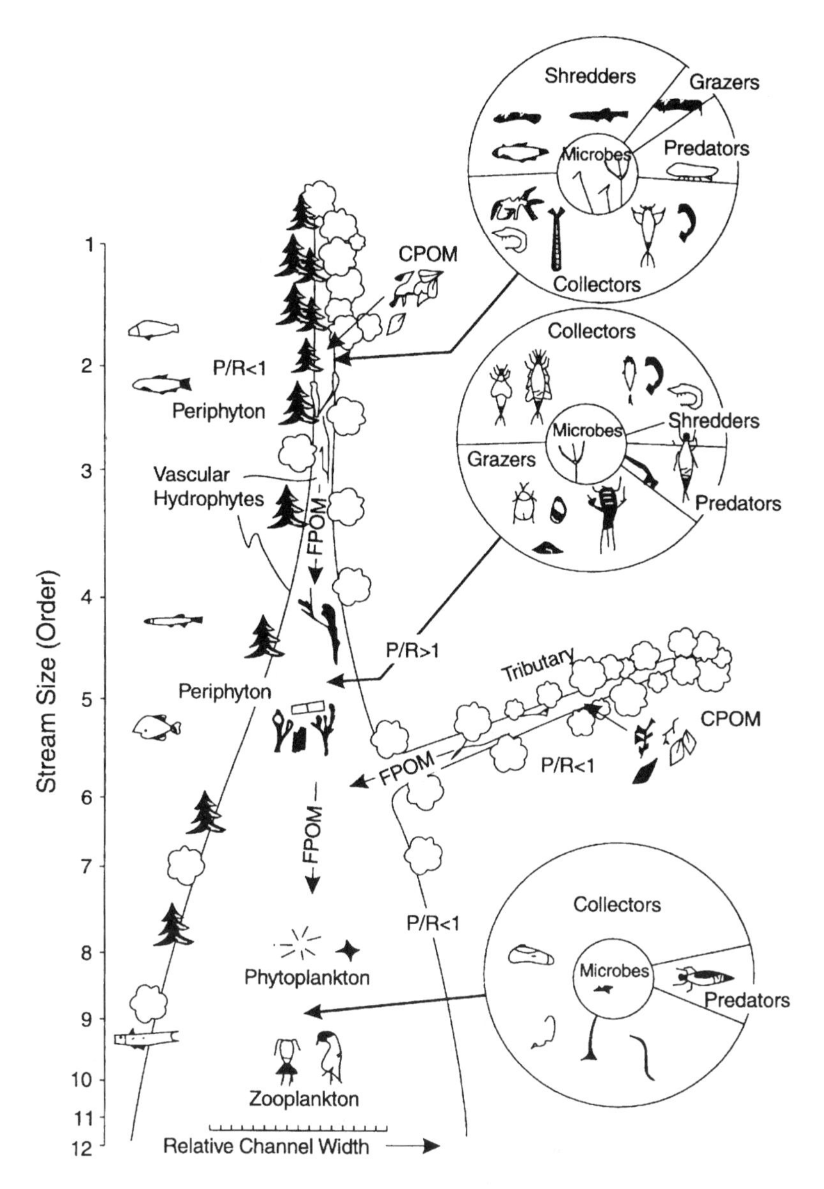

FIGURE 16.1. River continuum concept (Vannote et al. 1980) as illustrated by the relationship between stream size and the progressive shift in structural and functional attributes of lotic communities (from Hunsaker et al. 1997). P/R is the ratio of gross primary productivity to community respiration. CPOM stands for coarse particulate organic matter, and FPOM stands for fine particulate organic matter. The relative dominance (as biomass) of macroinvertebrates is shown on the right side of the figure. In the headwaters (orders 1–3), CPOM is the food or energy source and can be used by both shredders and collectors; grazers and collectors are important in mid-sized streams where FPOM is high; collectors such as phytoplankton and zooplankton dominate large rivers.

terminous United States that he believes are better suited to regional assessments for aquatic ecosystems than are watersheds such as the U.S. Geological Survey's (USGS) Hydrologic Units (half of which are not true watersheds). Several states (e.g., Ohio, Arkansas, and Minnesota) are using ecoregions for establishing variable water-quality criteria, including biocriteria (Omernik 1995). Ongoing ecoregion work to refine ecoregions, define subregions, and locate sets of reference sites is occurring in Iowa, Florida, Massachusetts, and parts of Alabama, Mississippi, Virginia, West Virginia, Maryland, Pennsylvania, Oregon, and Washington (Southerland and Stribling 1995), as well as in Idaho, Montana, Utah, Colorado, Kansas, Nebraska, South and North Dakota, Missouri, Kentucky, Indiana, Ohio, Wisconsin, Tennessee, and South and North Carolina (J. Omernik, personal communication).

Several ecological paradigms or concepts about rivers and streams emerged during the two decades since Hynes (1975) proposed that the nature of streams is tightly coupled with catchment or watershed characteristics (Cummins et al. 1984). The *river continuum* concept provided a template for examining how biotic attributes of rivers change within the longitudinal gradient from headwaters to outlets at the sea (Vannote et al. 1980; Minshall et al. 1985). Vannote et al. (1980) defined the network of streams in a river drainage system as a continuum of physical gradients and associated biotic adjustments (Figure 16.1). They classified streams as headwaters (orders 1–3), medium-sized streams (orders 4–6), and large rivers (orders >6); see Mosley and McKerchar (1993) for a discussion of stream order. Comparison of organic matter budgets in streams in different biomes provided the basis for the *riparian control* concept and demonstrated the importance of wood and leaves in lotic systems (Harmon et al. 1986; Gregory et al. 1991). The type and structure of plants along a stream control the energy inputs to the stream. For example, streams associated with old-growth forests typically have less primary productivity than do those in open areas because of low levels of sunlight, but carbon inputs are typically higher in old-growth due to allochthonous inputs. Also, dead tree structures may perform terrestrial and aquatic functions for many centuries because of their slow rates of decay. The *nutrient spiraling* concept (Webster and Patten 1979; Newbold et al. 1983) led to an understanding of how the nutrient cycling process is really a spiral process because of stream transport from upstream to downstream reaches. An idealized cycle is complete when a nutrient atom has been taken up by an organism from a dissolved available state, passed through the food chain, and returned to a dissolved available state for reutilization. The concept of *patch dynamics* in streams was introduced by Pringle et al. (1988). Both the catchment and the lotic environment can be considered to be a mosaic of patches or habitats in which materials and energy are transferred (connected) through food webs. For example, the run-riffle-pool sequence of patches influences the distribution and abundance of biota. The *ecotone* concept (Naiman and Décamps 1990; Holland et al. 1991) increased interest in the importance of transformations and fluxes of materials that occur across boundaries between functionally interconnected patches that form the riverine landscape. An ecotone is a transition zone or environmental gradient between two more homogeneous zones. For example, a riparian area is an ecotone at one scale, but at a finer scale the riparian area has ecotones at its upland and water interfaces. Stanford and Ward (1992) believe that the ecotone concept integrates the other concepts mentioned in this paragraph by emphasizing the functional connectivity inherent in all ecosystems.

The river continuum concept is worth expanding on because it illustrates the importance of using knowledge about spatial scale in the planning and execution of research, assessments, and monitoring. Forested headwater streams are influenced strongly by riparian vegetation, which reduces in-stream primary production by shading and contributes large amounts of organic detritus. As stream size increases, the in-stream primary production and organic transport from upstream becomes more important compared with terrestrial organic input. This change is thought to be reflected in the ratio of gross primary productivity to community respiration being less than one in headwaters, equal to then greater than one in medium-sized streams, then less than one again in large rivers with single channels. This conceptual model allows useful generalizations concerning the magnitude and variation through time and space of the organic matter supply, the structure of the invertebrate community, and resource partitioning along the length of the river. The river continuum concept views streams as longitudinally linked systems in which system-level processes (cycling of organic matter and nutrients, ecosystem metabolism, net metabolism) in downstream areas are linked to in-stream processes in upstream areas. Thus, streams can be viewed as continuous, spatially heterogeneous systems.

For more detail on the processes important to streams and rivers, we suggest chapters in Naiman and Bilby (1998) that address channel processes and classification, hydrology, stream quality, dynamic landscape systems, large woody debris, and nutrient cycles. They review the state of the knowledge and give examples from the Pacific Coastal Ecoregion.

16.2.2 Riparian and Ecotone Concepts

Concepts and results from riparian (Malanson 1993) and upland-aquatic ecotone research (Holland et al. 1991) are very relevant. A *riparian buffer* is a strip of relatively undisturbed vegetation positioned along a surface water body and downhill from a source of material release (Weller et al. 1998).

With effective retention, riparian buffers can reduce land discharges of nutrients, sediments, and associated chemicals; however, the importance of the riparian zone within the whole watershed remains poorly understood (Johnson et al. 1997). Some statistical comparisons among watershed landscapes have found little effect of streamside vegetation on water quality (Omernik et al. 1981; Hunsaker and Levine 1995; Johnson et al. 1997). Other statistical analyses (Osborne and Wiley 1988) and spatial models (Levine and Jones 1990; Hunsaker and Levine 1995; Soranno et al. 1996) suggest that near-stream areas do have a positive influence on water quality.

Understanding how scale, both data resolution and geographic extent, influences landscape characterization, and how terrestrial processes affect water quality, are critically important for model development and translation of research results from experimental watersheds to management of large drainage basins

(Hunsaker and Levine 1995). Weller et al. (1998) developed and analyzed models predicting landscape discharge based on material release by an uphill source area, the spatial distribution of riparian buffer along a stream, and retention within the buffer. They explore one of the basic themes of landscape ecology, the effects of pattern on process, and their work follows others that have used simple mathematical models to develop the body of theory and general principles underlying landscape ecology (Gardner et al. 1987; Milne 1992; O'Neill et al. 1992). Weller et al. (1998) provide simple models for material transport to the body of general theory describing the ecological effects of spatial patterning.

16.2.3 Landscape Principles for Physical and Chemical Characteristics of Surface Waters

We suggest two basic principles that are especially relevant to the physical and chemical characteristics of surface waters.

1. *Watershed characteristics govern the physical, chemical, and hydrologic properties of streams and lakes.*
2. *The riparian zone can mitigate (or alter) effects on physical and chemical properties of streams and lakes that arise from land-use change in the watershed.*

Watershed Characteristics Govern the Physical, Chemical, and Hydrologic Properties of Streams and Lakes.

The physical and chemical characteristics of streams and lakes are primarily a function of their watershed characteristics: geology, soils, vegetation, and topography. Climate and groundwater also influence the properties of streams and lakes, but these factors are a function of the larger landscape beyond the immediate watershed. Many studies have shown that the proportion of different land uses within a watershed can account for some of the variability in stream-water quality; several publications provide comprehensive reviews of the literature (Levine et al. 1993; Barling and Moore 1994; Castelle et al. 1994; Weller et al. 1998). For example, the water quality in a watershed with 50% agricultural land use and an intact forest riparian zone may be expected to be better (e.g., lower turbidity and nutrients) than that in a similar watershed without any riparian buffer.

A significant amount of literature relates watershed characteristics to lake chemistry. A good example of this research is the effort to separate natural from anthropogenic causes of lake acidification. Hunsaker et al. (1986) illustrated that wet deposition, lake elevation, and forest cover are the principal variables associated with variance in the data for headwater lake pH and acid-neutralizing capacity in the Adirondack Mountains in the eastern United States. For the Upper Midwest of the United States, statistical analyses using cluster and discriminant techniques showed that lake hydrologic type (e.g., drainage and seepage) and lake-surface area are characteristically associated with the variability in lake alkalinity (Eilers et al. 1983; Schnoor et al. 1986). Although Eilers et al. (1983) did not find surficial geology or soil type to be related to lake alkalinity in Wisconsin, Schnoor et al. (1986) found that bedrock geology, soil pH, and runoff helped discriminate between seepage or inflow lakes and drainage or spring lakes in Michigan. The shoreline development factor, percentage of watershed not in forest, and bedrock geology helped discriminate between seepage and inflow lakes. In Minnesota, Rapp et al. (1985) found that bedrock type, forest type, lake morphology, lake hydrologic type, and deposition were associated with the variability in lake alkalinity. A review of recent research illustrates that widespread changes in landscape cover, whether from changing land use, logging, fire, or broad-scale storm damage, can impact the chemistry of drainage waters in forested ecosystems (Sullivan et al. 1996). Modeling studies and ion-budget calculations performed for selected watersheds in Europe have suggested that acidic deposition and landscape processes are of approximately equal importance as regulators of surface water acid-base chemistry.

The Riparian Zone Can Mitigate (or Alter) Effects on Physical and Chemical Properties of Streams and Lakes That Arise from Land-Use Change in the Watershed.

Weller et al. (1998) model a hypothetical landscape containing two ecosystems: an uphill source ecosystem that releases waterborne materials, and a downhill riparian buffer that can take up those materials before they reach a stream. The spatial aspect is made of a grid of buffer cells grouped along a stream, with water transporting materials downhill through the buffer and into the stream (Figure 16.2a). Material uptake (surface and groundwater) is a simple first-order process, and material flux decreases exponentially with the width of buffer traversed. All buffer cells have identical retention capabilities for this model. Figure 16.2b indicates that landscape buffer transmission (the fraction of any source ecosystem discharge that would reach the stream), T, is less for uniform-width buffers than it is for buffers with Poisson-width distributions. The Poisson-width distribution gives variable buffer width along the stream channel, including some gaps (areas with zero widths). The difference between the uniform-width and Poisson buffers is greatest for narrow but highly retentive (cell transmission, t, near 0) buffers (Figure 16.2c). Weller et al. (1998) also found that as buffer retentiveness increases, gaps in the buffer are increasingly the sites of material delivery to the stream (Figure 16.2d). In landscapes with high retention potential, buffer retention rises steeply as the fraction of the landscape occupied by streamside buffer increases; however, buffer retention peaks and approaches a declining curve of slope $= -1$ (in this region further increases in buffer width merely raise source elimination while reducing buffer retention). This effect is illustrated in Figure 16.2e for Poisson-width distributions and uniform-width buffers.

The strength of correlation between T and various predictors varied with t (Figure 16.2f). For retentive buffers ($t = 0.1$), T was most strongly correlated with the frequency of gaps in the buffer. For unretentive buffers ($t = 0.9$), mean buffer width was the best predictor of T, whereas evenness of width was best at predicting T for moderately retentive buffers ($t = 0.5$). Because each predictor works

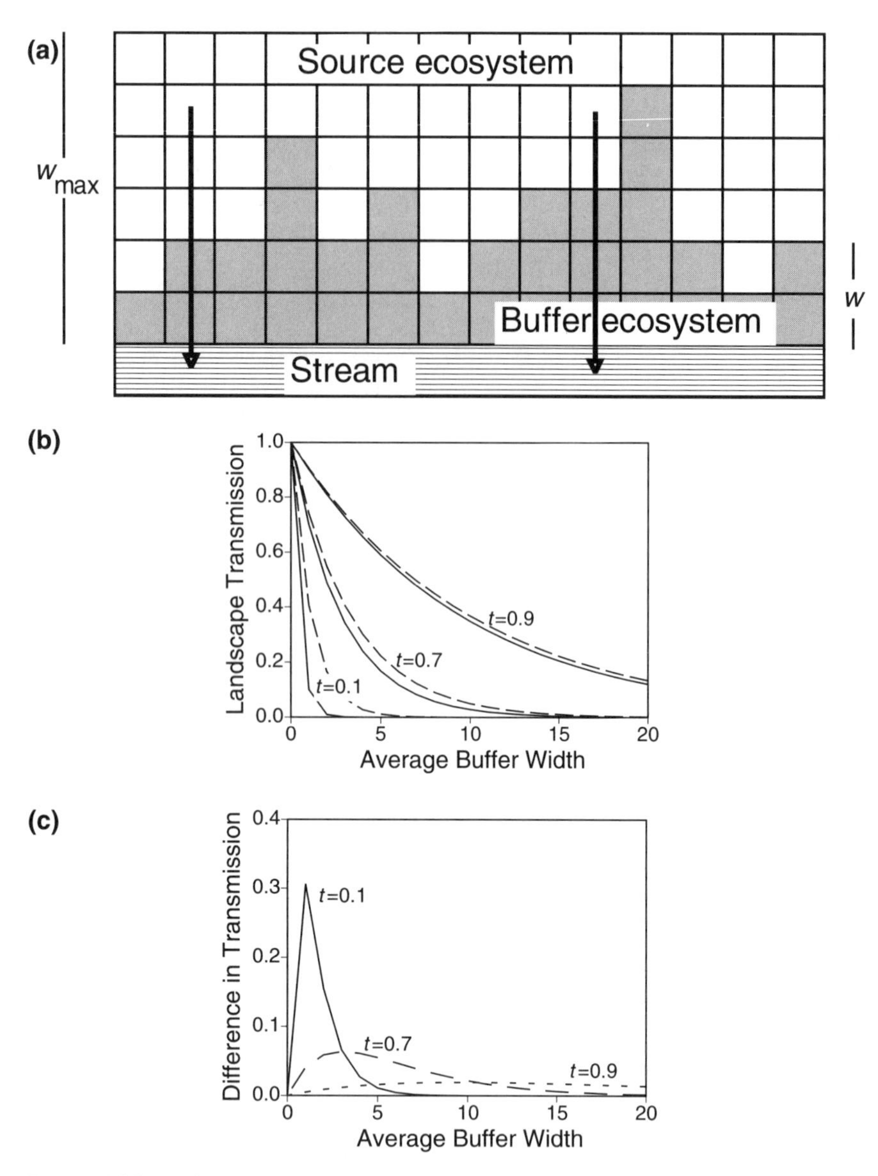

FIGURE 16.2. (a) Conceptual model of a landscape with a riparian buffer. The landscape is divided into cells that make up either the source ecosystem or the stream-buffer ecosystem. Water and materials flow downhill from the source ecosystem, through the buffer, and to the stream. (b) Landscape buffer transmission, T, versus average buffer width for three values of cell transmission, *t*. T is a fraction of the total material export from the source ecosystem to the buffer ecosystem. The solid curves are for uniform-width buffers, and the dashed curves are for buffers with Poisson-width distributions. (c) The difference between Poisson landscape buffer transmission and uniform-width landscape buffer transmission for three values of cell transmission, *t*. (d) Fraction of material discharge through gaps in buffers with Poisson-width distributions. The solid lines are the fraction of discharge through gaps for different values of cell transmission, *t*. The dashed line is the frequency of gaps. (e) Graphical comparison of buffer retention in buffers with Poisson-width distribu-

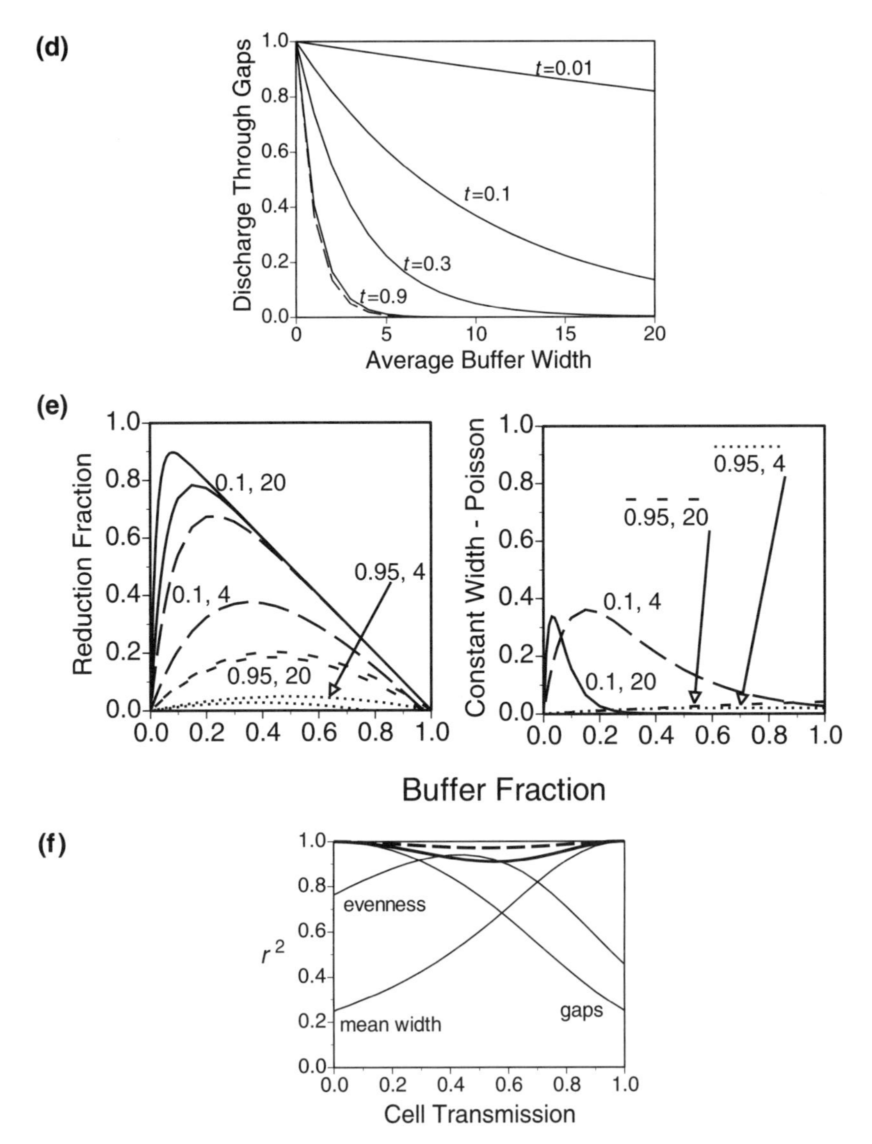

tions and uniform-width buffers. (Left) Buffer retention for the four combinations of cell transmission t and maximum width w_{max} indicated by the pairs of numbers on the figure. The vertical axis shows fractions relative to maximum discharge from a landscape completely covered by source ecosystem (no buffer). The horizontal axis is the fraction of the landscape occupied by streamside buffer. The upper curve in each pair (same line pattern) is for a uniform-width buffer, and the lower is for a Poisson buffer. (Right) The difference between uniform-width buffering and Poisson-width buffering for each pair of curves in the left panel. (f) Coefficients of determination (r^2) for predicting landscape buffer transmission, T, from statistics of the buffer-width distribution. Light solid lines are for the indicated univariate relationships. The heavy solid line is for models using two independent variables, mean buffer width, and gap frequency. The dashed line is for a three-variable model that adds evenness of buffer width (Weller et al. 1998).

best for a unique range of t values, combining predictors in multiple regression models gave extremely good estimates of T for any value of t. Multiple regression models predicting T from mean buffer width and the frequency of gaps always gave $r^2 > 0.91$ for any value of t. Adding evenness of width further raised the minimum r^2 for any t to 0.97 (Figure 16.2f). The results of the statistical analysis suggest that the mathematical model may be replaced by simpler statistical relationships for some applications (Weller et al. 1998).

16.3 Recent Applications

During the past five years, scientists and managers in state and federal agencies have become more familiar with landscape-ecology concepts and principles, and this knowledge is slowly being translated into policy and guidelines. People working on water quality or aquatic biodiversity issues have known about watershed processes and the importance of riparian zones for a long time; however, new empirical data and new tools such as landscape pattern metrics and geographic information systems make analyses of alternative policies and implementation of the principles and concepts more feasible. Landscape-ecology principles are rarely applied by local governments. For example, counties have trouble enforcing even straightforward grading ordinances and floodplain building restrictions, let alone riparian buffer zones for agricultural and silvicultural activities. Nevertheless, the increasing number of community-based watershed groups is an encouraging sign.

In this section, we provide some examples of recent state and federal applications of landscape ecology for conservation of the physical and chemical characteristics of aquatic ecosystems. We contacted scientists and managers in the U.S. Environmental Protection Agency (USEPA); the U.S. Department of Agriculture, Forest Service; and the State of California. In addition, a computerized keyword search of aquatic and environmental journals was conducted, numerous books were reviewed, and Web sites were searched for applications.

At the state level, water quality assessments have sometimes been designed using information from analyses based on ecoregions. As mentioned earlier, relationships between ecological regions and water quality have been examined for several states (Larsen et al. 1986; Rohm et al. 1987; Hughes and Larsen 1988; Larsen et al. 1988; Whittier et al. 1988; Lyons 1989). For example, Minnesota uses ecoregions to classify lakes and develops different expectations for the lakes within different ecoregions. This has enabled the state pollution control agency to communicate estimates of reasonable trophic state values and variability more effectively, and it has helped improve model predictions (Hughes et al. 1994). Using nutrient richness and ionic strength for Ohio, Larsen et al. (1988) concluded that land-classification systems such as Omernik's ecoregions can be used to characterize attainable stream water quality. A similar study with fish assemblages also showed correspondence to ecoregional pattern (Larsen et al. 1986).

Ohio is one of three states that has incorporated numeric biological criteria into water quality regulations.

At the federal level, a good example of incorporation of landscape-ecology concepts into policy occurred with the Northwest Forest Plan (NFP). Several federal agencies worked together to design a new approach to federal land management in the Pacific Northwest (Forest Ecosystem Management Assessment Team 1993). An aquatic conservation strategy for lands administered by the U.S. Department of Agriculture, Forest Service, and Bureau of Land Management (within the range of the Northern Spotted Owl, *Strix occidentalis caurina*) is outlined in the standards and guidelines of the NFP (U.S. Department of Agriculture and U.S. Department of Interior 1994), and we briefly summarize some of its objectives and components here. Four components make up the aquatic conservation strategy.

- Riparian reserves: lands along streams and unstable and potentially unstable areas where special standards and guidelines direct land use.
- Key watersheds: a system of large refugia comprising watersheds that are crucial to at-risk fish species and stocks and provide high-quality water.
- Watershed analysis: procedures for conducting analysis that evaluates geomorphic and ecological processes operating in specific watersheds. Watershed analysis provides the basis for monitoring and restoration programs and the foundation from which riparian reserves can be delineated.
- Watershed restoration: a comprehensive, long-term program of watershed restoration to restore watershed health and aquatic ecosystems, including the habitats supporting fish and other aquatic and riparian-dependent organisms.

Two of the aquatic conservation strategy objectives of the NFP are good examples of the influence of landscape ecology. Objective 1 is to "Maintain and restore the distribution, diversity, and complexity of watershed and landscape-scale features to ensure protection of the aquatic systems to which species, populations and communities are uniquely adapted." Objective 2 is to "Maintain and restore spatial and temporal connectivity within and between watersheds. These network connections must provide chemically and physically unobstructed routes to areas critical for fulfilling life history requirements of aquatic and riparian-dependent species."

The conservation strategy employs several tactics to approach the goal of maintaining the "natural" disturbance regime. Land-use activities need to be limited or excluded in those parts of the watershed prone to instability. The distribution of land-use activities, such as timber harvest and associated roads, must minimize increases in peak streamflows. Headwater riparian areas need to be protected, so that when debris slides and flows occur they contain coarse woody debris and boulders necessary for creating habitat farther downstream. Riparian areas along larger channels need protection to limit bank erosion, ensure an adequate and continuous supply of coarse woody debris to channels, and provide shade and microclimate protection. Watersheds currently containing the best habitat or those with the greatest potential for recovery should receive increased protection and receive highest priority for restoration programs (U.S. Department of Agriculture and U.S. Department of Interior 1994).

Under the NFP, 10 Adaptive Management Areas were established to encourage development and evaluation of new approaches; one of these, the Blue River Landscape Study (23,100 ha), occurs within the Central Cascades Adaptive Management Area. Here, scientists and managers are using models and empirical data to evaluate the difference between two approaches for achieving the objectives of the NFP. One approach, the "Interim Plan," is based on standard management activities under the NFP, in which a network of reserves is imbedded in a matrix of fairly intensive timber harvest with short rotations. The other approach under the Blue River Landscape Study, called the "Landscape Plan," uses historical disturbance regimes as a guide for management activities (Cissel et al. 1999). The first step in the Landscape Plan was to set aside special-area, aquatic, and corridor reserves. For the remaining area, timber harvest frequency and rotation age (100–260 years) were based on historical fire frequency; timber harvest intensity (15% to 50% overstory canopy retention) was based on historical fire severity; and the spatial patterns of timber harvest were based on the spatial patterns of historical fires. Fire prescriptions were integrated with the timber harvest prescriptions. Watershed restoration needs also were identified. Compared with the Interim Plan, the Landscape Plan is expected to produce more late-successional habitat in a less-fragmented landscape characterized by larger patches, more interior habitat, and less edge between old and young forests. A long-term, multiscale monitoring plan is being implemented to evaluate the effectiveness of the Landscape Plan, which, relative to the Interim Plan, is more customized to the watershed and more heterogeneous (i.e., less uniform). The monitoring framework is organized along a hierarchy of spatial scales as listed below.

- Watershed scale: landscape pattern, Northern Spotted Owls, and economics.
- Subwatershed scale: stand and landscape structure, stream discharge, and social acceptability.
- Small-stream scale: stream-breeding amphibians and stream temperature.
- Site scale: erosion, forest regeneration, stand development, and nonvascular plants.

A similar blending of science knowledge and management needs is occurring under the Sierra Nevada Conservation Framework in California. Management alternatives are being evaluated at the bioregional scale for the 11 national forests in the Sierra Nevada. A companion monitoring program is being designed that is very cognizant of landscape ecology, scale, and the interrelated nature of aquatic, riparian, and meadow ecosystems (Appendix E in U.S. Department of Agriculture 2000).

At the state level, the application of ecoregional classification has proven effective. For the large federal programs, such as the Northwest Forest Plan or the Sierra Nevada Conservation Framework, it is too early in the process to determine the effectiveness of landscape-ecology concepts for conserving the physical and chemical quality of aquatic ecosystems.

16.4 Principles for Applying Landscape Ecology

In a landscape, the flows of energy, matter, and species are determined to some extent by the spatial configuration of the elements. Forman and Godron (1986) identified seven principles of landscape ecology that directly address these relationships. Two of these are relevant to this chapter: "Nutrient flows in the landscape increase with disturbance, and landscapes will develop either physical system stability, resilience, or resistance to disturbance." Our principles for applying landscape ecology to surface waters are based on current ecological knowledge (1 and 2 below), modeling exercises (3 below), and managers' experience as much as scientifically documented effects (4 below).

Principle 1: Use Characteristics of the Adjacent Landscape to Predict and Control Water Chemistry.

Proportion of land cover is related to stream and lake chemistry; spatial pattern or arrangement also is related, but empirical studies show not as strongly as proportion. Other landscape characteristics (e.g., soil type, geology, and topography) can be of similar importance.

Principle 2: Use Knowledge About Physical Landscape Properties to Direct Aquatic Conservation.

Stream flow and sediment transport are functions of physical properties such as rainfall, topography, surface roughness, and soil properties that create a complex landscape mosaic. Flow and sediment properties influence water chemistry and the physical properties of habitat condition. Thus, all of these properties affect the condition of aquatic plants and animals as discussed in Chapter 17.

Principle 3: Stream Buffers Can Be Used to Protect Water Quality.

The average width of stream buffers (riparian areas) is the best predictor of landscape discharge for unrententive buffers. The frequency of gaps in stream buffers is the best predictor of landscape discharge for narrow, retentive buffers. Modeling exercises should be used to determine location-specific buffers unless empirical studies are available.

Principle 4: Resource Conservation Must Recognize and Incorporate Multiple Scales.

Both resource conservation and the scientific analyses that support management needs must recognize and incorporate multiple scales. Hydrologic classifications such as stream order and the river continuum concept are a reflection of scale. We must realize that geographic extent, data resolution, and type and number of categories considered are all interrelated and affect conclusions and decisions. The literature on environmental assessments recognizes this and suggests that one look at the scale below and the scale above the geographic area of interest (Hunsaker 1993; Hunsaker et al. 1993).

16.5 Knowledge Gaps

In this section, we identify the knowledge gaps that we believe are preventing applications of landscape ecology in the protection of physical and chemical conditions of aquatic ecosystems. By its nature, landscape ecology is an integrative science, and it can play an important role in defining conservation biology approaches for aquatic ecosystems. This need is obvious when we list theoretical and empirical knowledge gaps. The knowledge gaps for the physical and chemical components (this chapter) and the biotic components (Chapter 17) of aquatic ecosystems are very similar, but we emphasize somewhat different aspects as appropriate to each chapter. Several of these gaps are interrelated.

16.5.1 Theoretical Voids

We need to develop techniques to distinguish the cumulative effect of multiple stressors from that of natural covariates. It is essential to be able to separate the variables that affect ecosystem condition as much as possible to make management decisions. We need to know the shape of the curve between the response variable and the stressor variable(s) to be able to reduce or eliminate effects.

We need much more knowledge about how ecosystems change across multiple spatial scales, and which variables to measure at what resolution for a given scale. Pastor and Johnston (1992) pose two questions they consider central to unifying ecosystem and landscape ecology into a common theoretical framework: (1) "What processes at the landscape, regional, or global level control ecosystem processes at local scales and what are their spatial distributions?" and (2) "How do ecosystem properties percolate across the landscape to affect entire watersheds?" Discussions in this chapter on riparian buffers and the importance of land-cover amount versus spatial pattern all highlight this need. For practical reasons, our heavily instrumented watersheds and controlled experiments usually occur in first- or second-order watersheds. How then do we incorporate this knowledge into the management of larger river basins that are often the focus of the public and resource managers? Landscape ecologists are notorious for saying we can "scale up," but a significant amount of interdisciplinary collaboration on theory and modeling is needed before we can even come close to doing this. Various types of multivariate analysis are typically applied to examine relationships, but how do we know these are both statistically and ecologically appropriate?

We need to establish a clearer understanding of how terrestrial conditions determine conditions in aquatic ecosystems (Rathert et al. 1999). We can model and predict with more certainty how landscape changes affect stream flow and water quality than how they affect aquatic organisms, and much more effort has gone into analysis of effects of agricultural changes compared with analysis of effects of other land-use changes. Few theories and approaches of terrestrial and aquatic ecologists are very well-integrated. We need to improve both theoretical linkages and operational studies between terrestrial and aquatic ecologists. Unfortunately, the reward system for researchers discourages rather than promotes interdisciplinary work; such work also takes more time because of the high degree of coordination. Struggles within national research and monitoring efforts, and often misguided outside reviews, have highlighted our continuing need in this area (e.g., National Acid Deposition Program assessments and the USEPA's Environmental Monitoring and Assessment Program).

16.5.2 *Empirical Voids*

More ecological research funding should support holistic studies. Much of conservation biology is focused on single species rather than on ecosystems. More information is needed from research incorporating physical, chemical, and multi-assemblage variables examined at multiple geographic scales. This need is juxtapositioned against short-term, 2- or 3-year grants, and reductions in core research budgets at agencies that maintain long-term research sites and expensive research facilities.

Improved quantitative assessment techniques are necessary to implement conservation biology effectively (i.e., are we achieving the desired goal(s)?) Conservation biology is an applied science that requires solid tools to complement theoretical and empirical knowledge. A risk-assessment framework is recommended (Suter 1993; U.S. Environmental Protection Agency 1998), but techniques to quantify uncertainty in spatial data required for landscape analyses are limited (Hunsaker et al. 2001). Uncertainty comes from many sources, including data development, specifications such as resolution and categories, geographic information system manipulations, and data representation.

16.6 Research Approaches

This chapter and its companion (Chapter 17) highlight some of the different and similar challenges for landscape ecologists who address aquatic and terrestrial ecosystems. Scale issues and cumulative effects are challenges for both terrestrial and aquatic understanding, but the need for holistic or integrated research is especially critical for aquatic ecologists. The research approaches discussed here are directed toward helping to improve our knowledge about information gaps listed in the previous section.

16.6.1 *Approaches for Theoretical Research*

Closer collaboration among ecologists, assessment scientists, spatial statisticians, and quantitative geographers is needed to develop theory and methods to distinguish the cumulative effect of multiple stressors from that of natural covariates, to improve theoretical linkages between terrestrial and aquatic systems, and to foster operational studies between terrestrial and aquatic ecologists. Reid (1998) provides a good overview of watershed effects and watershed analysis for physical properties. Quantitative regional assessments, designed field studies, and laboratory experiments are needed to understand multiple stressors. Several recent efforts recognize this need (Water Environment Research Foundation 1998; Colodey and McDevitt 1999; Foran and Ferenc 1999).

Researchers should give a significant amount of thought to scale (grain or data resolution, geographic extent, number of classes in categorical data) and multiple lines of evidence when designing a study. These elements of scale are all interrelated, and we need more studies that help us to understand how scale affects results (some examples are Bartell and Brenkert 1991; Hunsaker et al. 1994; and O'Neill et al. 1996). We also need to realize that understanding scale requires research at both the regional and local scales (one or more small watersheds). For example, synoptic regional monitoring and assessment programs should be coordinated with focused, cause-and-effect research. They can share questions and answers about stressors or affecters and monitoring attributes (indicators or end points). The research manipulations may have continuous or daily measurements to address processes, whereas work on large geographic areas will often measure structure or surrogates for processes and have seasonal, annual, or longer time intervals of measurements. Sometimes coordinated laboratory or mesocosm research and modeling efforts also may be needed. Laboratory and mesocosm work can be more tightly controlled than can landscape manipulations. Modeling helps us better understand processes and variable or stressor sensitivity; often, it is the only way to evaluate landscape change for large geographic regions. Such integrated or holistic efforts are likely to be the ones that significantly advance our knowledge of how to address cumulative effects and to separate the effects of multiple stressors on aquatic ecosystems.

16.6.2 Approaches for Empirical Research

The research approaches in this section address the need for holistic studies and improved quantitative assessment techniques. A combination of surveys or monitoring programs, gradient studies, and ecological experiments at multiple scales and locations are needed. Activities such as the Blue River Landscape Study described in Section 16.3 are necessary. To date, most studies have shown that the proportion of vegetation, land use, or soil properties are most powerful in explaining the variance in water quality and sometimes the types of aquatic organisms that are present. Although knowing spatial pattern is useful, it is usually not the dominant factor. The extent to which results to date have been confounded by the elements of scale is not clear. Watershed analyses could be designed to investigate specifically when knowing spatial pattern (amount of edge between land uses, contagion, fractal dimension, etc.) is most important.

We can now easily calculate many landscape pattern metrics (Riitters et al. 1995). Many of these metrics are highly correlated or related. If one plans to address spatial pattern, the pattern metrics that are used should be selected based on logic and the hypotheses to be tested. Patch size and contagion are appropriate metrics for evaluating fragmentation of wetlands, whereas the amount of edge between wetlands and agriculture or urban areas would be more useful for evaluating the risk of increased downstream flooding. Pattern metrics that are useful for terrestrial research questions may not be useful for aquatic studies. For example, patch size of coniferous forest may be important to the survival of interior forest birds but has little relevance to trout; the proportion of continuous coniferous forest near streams is more likely to be useful for explaining trout presence or absence.

It is important to know the history of a watershed or region prior to designing new research. As this chapter explains, the characteristics of surface water are a result of the combined influences of the associated drainage area. Landscape change should include the many uses that European man and native Americans may have made of the land. Spatial data on logging, grazing, mining, channel alterations, landslides, fire, cultivation, and geology are all important. Historic information is especially important if one is trying to identify appropriate control or reference sites or determine cumulative effects. Such information is not likely to be in digital form and thus may need to be digitized from old maps or aerial photographs. The resolution and confidence in such data may often be lower than are those for current data. Sometimes historical information may be simply qualitative (e.g., about 200,000 sheep were grazed in the Sierra Nevada between location A and location B in 1890).

Well-designed empirical work is needed on uncertainty in spatial data. Hunsaker et al. (2001) review why ecologists need this information, discuss the many facets of the issue, and present examples of approaches. Much more work is needed to understand the issues and to make approaches more user-friendly. Quantification of uncertainty should be a standard component of spatial analysis for conservation biology.

Acknowledgments

We thank Kevin Gutzwiller for recognizing the need to bring more aquatic ecology into the landscape ecology literature. We also are grateful to the managers and scientists who responded to our inquiries about landscape-ecology applications. Reviews by Sean Eagan and John Cissell, U.S. Department of Agriculture, Forest Service, improved the manuscript. The ideas we present result from many years of work on regional and national projects and programs and are products of our interactions with many colleagues whom we cannot thank individually here. However, our viewpoints are personal and do not represent policies of the institutions with which we are affiliated.

References

Allan, J.D., Erickson, D.L., and Fay, J. 1997. The influence of catchment land use on stream integrity across multiple scales. *Freshwater Biol.* 37:149–161.

Bailey, R.G. 1976. *Ecoregions of the United States* (map 1:7,500,000). Ogden, Utah: U.S. Department of Agriculture, Forest Service, Intermountain Region.

Barling, R.D., and Moore, I.D. 1994. Role of buffer strips in management of waterway pollution: a review. *Environ. Manage.* 18:543–558.

Bartell, S.M., and Brenkert, A.L. 1991. A spatial-temporal model of nitrogen dynamics in a deciduous forest watershed. In *Quantitative Methods in Landscape Ecology,* eds. M.G. Turner and R.H. Gardner, pp. 379–398. New York: Springer-Verlag.

Bormann, F.H., and Likens, G.E. 1981. *Pattern and Process in a Forested Ecosystem.* New York: Springer-Verlag.

Castelle, A.J., Johnson, A.W., and Conolly, C. 1994. Wetland and stream buffer size requirements—a review. *J. Environ. Quality* 23:878–882.

Cissel, J.H., Swanson, F.J., and Weisberg, P.J. 1999. Landscape management using historical fire regimes: Blue River, Oregon. *Ecol. Appl.* 9:1217–1231.

Colodey, A.G., and McDevitt, C.A., eds. 1999. *On-Site Bioassay and Mesocosm Workshop Proceedings.* Vancouver, British Columbia, Canada: Environment Canada and BC Research Inc.

Correll, D.L., ed. 1986. *Watershed Research Perspectives.* Washington, DC: Smithsonian Institution Press.

Cummins, K.W., Minshall, G.W., Sedell, J.R., Cushing, C.E., and Petersen, R.C. 1984. Stream ecosystem theory. *Verhandlungen Internationale Vereinigung fur Theoretische und Angewandte Limnologie* 22:1818–1827.

Eilers, J.M., Glass, G.E., and Webster, K.E. 1983. Relationships between susceptibility of lakes to acidification and factors controlling lake water quality: hydrology as a key factor. *Can. J. Fish. Aquat. Sci.* 40:1896–1904.

Foran, J.A., and Ferenc, S.A. 1999. *Multiple Stressors in Ecological Risk and Impact Assessment.* Pensacola, Florida: Society of Toxicology and Chemistry.

Forest Ecosystem Management Assessment Team. 1993. *Forest Ecosystem Management: An Ecological, Economic, and Social Assessment.* 1993-793-071. Washington, DC: U.S. Government Printing Office.

Forman, R.T.T., and Godron, M. 1986. *Landscape Ecology.* New York: John Wiley and Sons.

Gallant, A.L., Whittier, T.R., Larsen, D.P., Omernik, J.M., and Hughes, R.M. 1989. *Regionalization as a Tool for Managing Environmental Resources.* EPA/600/3-89/060. Corvallis, Oregon: U.S. Environmental Protection Agency.

Gardner, R.H., Milne, B.T., Turner, M.G., and O'Neill, R.V. 1987. Neutral models for the analysis of broad-scale landscape pattern. *Landsc. Ecol.* 1:19–28.

Gregory, S.V., Swanson, F.J., McKee, W.A., and Cummins, K.W. 1991. An ecosystem perspective of riparian zones. *BioScience* 41:540–551.

Harmon, M.E., Franklin, F.J., Swanson, F.J., Sollins, P., Gregory, S.V., Lattin, J.D., Anderson, N.H., Cline, S.P., Aumen, N.G., Sedell, J.R., Lienkaemper, G.W., Cromack, K. Jr., and Cummins, K.W. 1986. Ecology of coarse woody debris in temperate ecosystems. *Advances Ecol. Res.* 15:133–302.

Holland, M.M., Risser, P.G., and Naiman, R.J., eds. 1991. *The Role of Landscape Boundaries in the Management and Restoration of Changing Environments.* New York: Chapman and Hall.

Hughes, R.M., Heiskary, S.A., Matthews, W.J., and Yoder, C.O. 1994. Use of ecoregions in biological monitoring. In *Biological Monitoring of Freshwater Ecosystems,* eds. S.L. Loeb and A. Spacies, pp. 125–151. Boca Raton, Florida: CRC Press.

Hughes, R.M., and Larsen, D.P. 1988. Ecoregions: an approach to surface water protection. *J. Water Pollut. Contr. Fed.* 60:486–493.

Hunsaker, C.T. 1993. Ecosystem assessment methods for cumulative effects at the regional scale. In *Environmental Analysis,* eds. S.G. Hildebrand and J.B. Cannon, pp. 480–493. Boca Raton, Florida: Lewis Publishers.

Hunsaker, C.T., Dickson, K., Waller, W., and Morgan, E. 1997. Watershed/regional assessment and in-stream monitoring. In *Biomonitoring in the Water Environment,* pp. 61–103. Alexandria, Virginia: Water Environment Federation.

Hunsaker, C., Goodchild, M., Friedl, M., and Case, T. 2001. *Spatial Uncertainty for Ecology.* New York: Springer-Verlag.

Hunsaker, C., Graham, R., Turner, R.S., Ringold, P.L., Holdren, G.R. Jr., and Strickland, T.C. 1993. A national critical loads framework for atmospheric deposition effects assessment: II. Defining assessment end points, indicators, and functional subregions. *Environ. Manage.* 17:335–341.

Hunsaker, C.T., and Levine, D.A. 1995. Hierarchical approaches to the study of water quality in rivers. *BioScience* 45:193–203.

Hunsaker, C.T., Malanchuk, J.L., Olson, R.J., Christensen, S.W., and Turner, R.S. 1986. Adirondack headwater lake chemistry relationships with watershed characteristics. *Water Air Soil Pollut.* 31:79–88.

Hunsaker, C.T., O'Neill, R.V., Timmins, S.P., Jackson, B.L., Levine, D.A., and Norton, D.J. 1994. Sampling to characterize landscape pattern. *Landsc. Ecol.* 9:207–226.

Hynes, H.B.N. 1975. The stream and its valley. *Verhandlungen Internationale Vereinigung fur Theoretische und Angewandte Limnologie* 19:1–15.

Johnson, D.W., and Van Hook, R.I., eds. 1989. *Analysis of Biogeochemical Cycling Processes in Walker Branch Watershed.* New York: Springer-Verlag.

Johnson, L.B., Richards, C., Host, C.E., and Arthur, J.W. 1997. Landscape influences on water chemistry in Mid-western stream ecosystems. *Freshwater Biol.* 37:193–208.

Johnston, C.A., Detenbeck, N.E., and Niemi, G.J. 1990. The cumulative effect of wetlands on stream water quality and quantity: a landscape approach. *Biogeochemistry* 10:105–141.

Larsen, D.P., Dudley, D.R., and Hughes, R.M. 1988. A regional approach for assessing attainable surface water quality: an Ohio case study. *J. Soil Water Conserv.* 43:171–176.

Larsen, D.P., Omernik, J.M., Hughes, R.M., Rohm, C.M., Whittier, T.R., Kinney, A.J., Gallant, A.L., and Dudley, D.R. 1986. Correspondence between spatial patterns in fish assemblages in Ohio streams and aquatic ecoregions. *Environ. Manage.* 10:815–828.

Levine, D.A., Hunsaker, C.T., Timmins, S.P., and Beauchamp, J.J. 1993. *A Geographic Information System Approach to Modeling Nutrient and Sediment Transport.* Technical Report ORNL-6736. Oak Ridge, Tennessee: Oak Ridge National Laboratory.

Levine, D.A., and Jones, W.W. 1990. Modeling phosphorus loading to three Indiana reservoirs: a geographic information system approach. *Lake Reserv. Manage.* 6:81–91.

Likens, G.E., and Bormann, F.H. 1974. Linkages between terrestrial and aquatic ecosystems. *BioScience* 24:447–456.

Lowrance, R., Todd, R., Fair, J. Jr., Hendrickson, O. Jr., Leonard, R., and Asmussen, L. 1984. Riparian forests as nutrient filters in agricultural watersheds. *BioScience* 34:374–377.

Lyons, J. 1989. Correspondence between the distribution of fish assemblages in Wisconsin streams and Omernik's ecoregions. *Am. Midl. Nat.* 122:163–182.

Malanson, G.P. 1993. *Riparian Landscapes.* Cambridge, United Kingdom: Cambridge University Press.

Milne, B.T. 1992. Spatial aggregation and neutral models in fractal landscapes. *Am. Nat.* 139:32–57.

Minshall, G.W., Cummins, K.W., Peterson, R.C., Cushing, C.E., Burns, D.A., Sedell, J.R., and Vannote, R.L. 1985. Developments in stream ecosystem theory. *Can. J. Fish. Aquat. Sci.* 42:1045–1055.

Mosley, M.P., and McKerchar, A.I. 1993. Streamflow. In *Handbook of Hydrology,* ed. D.R. Maidment, pp. 8.1–8.39. New York: McGraw-Hill Inc.

Naiman, R.J., ed. 1992. *Watershed Management: Balancing Sustainability and Environmental Change.* New York: Springer-Verlag.

Naiman, R.J., and Bilby, R.E., eds. 1998. *River Ecology and Management.* New York: Springer-Verlag.

Naiman, R.J., and Décamps, H., eds. 1990. *The Ecology and Management of Aquatic-Terrestrial Ecotones.* Paris: United Nations Educational, Scientific and Cultural Organization, and Carnforth, United Kingdom: Parthenon Publishing Group.

Newbold, J.D., Elwood, J.W., O'Neill, R.V., and Sheldon, A.L. 1983. Phosphorus dynamics in a woodland stream ecosystem: a study of nutrient spiralling. *Ecology* 64:1249–1265.

Omernik, J.M. 1977. *Nonpoint Source-Stream Nutrient Level Relationships: A Nationwide Study.* EPA-600/3-77-105. Corvallis, Oregon: U.S. Environmental Protection Agency.

Omernik, J.M. 1987. Ecoregions of the conterminous United States. *Ann. Assoc. Am. Geog.* 77:118–125.

Omernik, J.M. 1995. Ecoregions: a spatial framework for environmental management. In *Biological Assessment and Criteria, Tools for Water Resource Planning and Decision Making,* eds. W.S. Davis and T.P. Simon, pp. 49–62. Boca Raton, Florida: Lewis Publishers.

Omernik, J.M., Abernathy, A.R., and Male, L.M. 1981. Stream nutrient levels and proximity of agricultural and forest land to streams: some relationships. *J. Soil Water Conserv.* 36:227–231.

O'Neill, R.V., Gardner, R.H., and Turner, M.G. 1992. A hierarchical neutral model for landscape analysis. *Landsc. Ecol.* 7:55–61.

O'Neill, R.V., Hunsaker, C.T., Timmins, S.P., Jackson, B.L., Jones, K.B., Riitters, K.H., and Wickham, J.D. 1996. Scale problems in reporting landscape pattern at the regional scale. *Landsc. Ecol.* 11:169–180.

Osborne, L.L., and Wiley, M.J. 1988. Empirical relationships between land use/cover and stream water quality in an agricultural watershed. *J. Environ. Manage.* 26:9–27.

Pastor, J., and Johnston, C.A. 1992. Using simulation models and geographic information systems to integrate ecosystem and landscape ecology. In *Watershed Management: Balancing Sustainability and Environmental Change,* ed. R.J. Naiman, pp. 324–346. New York: Springer-Verlag.

Peterjohn, W.T., and Correll, D.L. 1984. Nutrient dynamics in an agricultural watershed: observations on the role of a riparian forest. *Ecology* 65:1466–1475.

Pringle, C.M., Naiman, R.J., Bretschko, G., Karr, J.R., Oswood, M.W., Webster, J.R., Welcomme, R.L., and Winterbourn, M.J. 1988. Patch dynamics in lotic systems: the stream as a mosaic. *J. N. Am. Benthol. Soc.* 7:503–524.

Rapp, G., Allert, J.D., Liukkonen, B.W., Ilse, J.A., Loucks, O.L., and Glass, G.E. 1985. Acid deposition and watershed characteristics in relation to lake chemistry in northeastern Minnesota. *Environ. Internat.* 11:425–440.

Rathert, D., White, D., Sifneos, J.C., and Hughes, R.M. 1999. Environmental correlates of species richness for native freshwater fish in Oregon, USA. *J. Biogeog.* 26:257–273.

Reckhow, K.H., Beaulac, M.N., and Simpson, J.T. 1980. *Modeling Phosphorus Loading and Lake Response under Uncertainty: A Manual and Compilation of Export Coefficients.* EPA 440/5-80-011. Washington, DC: U.S. Environmental Protection Agency.

Reid, L.M. 1998. Cumulative watershed effects and watershed analysis. In *River Ecology and Management,* eds. R.J. Naiman and R.E. Bilby, pp. 476–501. New York: Springer-Verlag.

Riitters, K.H., O'Neill, R.V., Hunsaker, C.T., Wickham, J.D., Yankee, D.H., Timmins, S.P., Jones, K.B., and Jackson, B.L. 1995. A factor analysis of landscape pattern and structure metrics. *Landsc. Ecol.* 10:23–39.

Rohm, C.M., Giese, J.W., and Bennett, C.C. 1987. Evaluation of an aquatic ecoregion classification of streams in Arkansas. *J. Freshwater Ecol.* 4:127–140.

Schnoor, J.L., Nikolaidis, N.P., and Glass, G.E. 1986. Lake resources at risk to acidic deposition in the upper Midwest. *J. Water Pollut. Contr. Fed.* 58:139–148.

Sivertun, A., Reinelt, L.E., and Castensson, R. 1988. A GIS method to aid in non-point source critical area analysis. *Internat. J. Geog. Infor. Sys.* 2:365–378.

Soranno, P.A., Hubler, S.L., Carpenter, S.R., and Lathrop, R.C. 1996. Phosphorus loads to surface waters: a simple model to account for spatial patterns of land use. *Ecol. Appl.* 6:865–878.

Southerland, M.T., and Stribling, J.B. 1995. Status of biological criteria development and implementation. In *Biological Assessment and Criteria, Tools for Water Resource Planning and Decision Making,* eds. W.S. Davis and T.P. Simon, pp. 81–96. Boca Raton, Florida: Lewis Publishers.

Stanford, J.A., and Ward, J.V. 1992. Aquatic resources in large catchments: recognizing interactions between ecosystem connectivity and environmental disturbance. In *Watershed Management: Balancing Sustainability and Environmental Change,* ed. R.J. Naiman, pp. 91–124. New York: Springer-Verlag.

Sullivan, T.J., McMartin, B., and Charles, D.F. 1996. Re-examination of the role of landscape change in the acidification of lakes in the Adirondack Mountains, New York. *Sci. Total Environ.* 183:231–248.

Suter, G.W. 1993. *Ecological Risk Assessment.* Boca Raton, Florida: Lewis Publishers.

Swank, W.T., and Crossley, D.A., Jr. 1987. *Forest Hydrology and Ecology at Coweeta.* New York: Springer-Verlag.

Turner, M.G. 1989. Landscape ecology: the effect of pattern on process. *Ann. Rev. Ecol. Syst.* 20:171–197.

U.S. Department of Agriculture. 2000. *Sierra Nevada Forest Plan Amendment, Draft Environmental Impact Statement, Appendices, Vol. 3.* Sacramento, California: USDA Forest Service, Pacific Southwest Region.

U.S. Department of Agriculture and U.S. Department of the Interior. 1994. *Record of Decision for Amendments to Forest Service and Bureau of Land Management Planning Documents Within the Range of the Northern Spotted Owl; Standards and Guidelines for Management of Habitat for Late-Successional and Old-Growth Forest Related Species Within the Range of the Northern Spotted Owl.* Portland, Oregon: U.S. Government Printing Office.

U.S. Environmental Protection Agency. 1998. *Guidelines for Ecological Risk Assessment.* EPA/630/R-95/002F. Washington, DC: Risk Assessment Forum.

Vannote, R.L., Minshall, G.W., Cummins, K.W., Sedell, J.R., and Cushing, C.E. 1980. The river continuum concept. *Can. J. Fish. Aquat. Sci.* 37:130–137.

Water Environment Research Foundation. 1998. *Critical Research Needs for Understanding Ecosystem Risks from Multiple Stressors.* Project 97-IRM-3. Alexandria, Virginia: Water Environment Research Foundation.

Webster, J.R., and Patten, B.C. 1979. Effects of watershed perturbation on stream potassium and calcium dynamics. *Ecol. Monogr.* 49:51–72.

Weller, D.E., Jordan, T.E., and Correll, D.L. 1998. Heuristic models for material discharge from landscapes with riparian buffers. *Ecol. Appl.* 8:1156–1169.

Whittier, T.R., Hughes, R.M., and Larsen, D.P. 1988. Correspondence between ecoregions and spatial patterns in stream ecosystems in Oregon. *Can. J. Fish. Aquat. Sci.* 45:1264–1278.

Wiken, E. 1986. *Terrestrial Ecozones of Canada.* Ecological Land Classification Series Number 19. Ottawa, Ontario, Canada: Environment Canada.

17

Effects of Landscape Change on Aquatic Biodiversity and Biointegrity

ROBERT M. HUGHES AND CAROLYN T. HUNSAKER

17.1 Introduction

Linkages between land cover and aquatic biointegrity were reported by 19th-century scientists (Grove 1995). So why should we still be concerned about relating land use and land cover to biodiversity in aquatic ecosystems? World-wide, Moyle and Leidy (1992) estimated that 1,800 species (20%) of freshwater fish have recently become extinct or are seriously declining. Williams et al. (1989) classified 364 fish taxa (31%) in North America as endangered, threatened, or of special concern, and Miller et al. (1989) listed 43 taxa as recently extinct. Master (1990) reported that 65% of crayfish species and 73% of unionid mussel species in North America are recently extinct or at risk. Compared with terrestrial animals, aquatic species are at higher risk. Ricciardi and Rasmussen (1999) estimated that extinction rates for the freshwater fauna of North America are five times those of terrestrial animals and comparable to rates in tropical rainforests. This concern extends to North American marine fish, among which 33 taxa are listed as vulnerable to extirpation as a result of habitat degradation (Musick et al. 2000).

Given these recent declines in aquatic biodiversity and integrity, we wrote this chapter to better link recent landscape-ecology concepts with biodiversity conservation in lakes and streams. We begin with a discussion of natural ecological concepts pertinent to landscape ecology and biodiversity (Section 17.2.1), and we follow with a discussion of anthropogenic factors that commonly limit biodiversity at the landscape scale (Section 17.2.2). Section 17.3 describes examples of how resource-management agencies have applied key landscape-ecology concepts to biodiversity conservation in aquatic systems. We then list a set of principles for applying landscape ecology in aquatic-biodiversity conservation (Section 17.4). We conclude with sections on critical knowledge gaps (Section 17.5) and general research approaches for filling them (Section 17.6). We feel the concepts and principles we discuss are applicable to all aquatic systems, but most examples pertain to streams.

17.2 Concepts, Principles, and Emerging Ideas

To understand aspects of landscape ecology that are useful in the conservation of aquatic ecosystems, one must understand the major landscape-scale factors associated with aquatic biodiversity and biointegrity. For convenience, we classify those factors into two groups, natural and human, recognizing that they often covary (Omernik 1987; Allen et al. 1999). Biodiversity and biointegrity are controlled by natural factors that act at different spatial and temporal scales and affect species and assemblages differently (Figure 17.1). Because aquatic systems are islands in terrestrial landscapes, and because those islands are periodically disturbed, island biogeography and metapopulations are important concepts. It is also necessary to realize that water bodies extend across and under the landscape in three dimensions and naturally vary dramatically through time. Natural disturbances are typically short-term *pulse disturbances* (e.g., floods, droughts, fires; Yount and Niemi 1990) that often affect biointegrity in a positive manner, as opposed to continuous anthropogenic *press disturbances* (e.g., nutrient enrichment, channel modification; Figure 17.1). Keeping these concepts in mind will facilitate biodiversity conservation and restoration.

17.2.1 Natural Factors

Among the natural factors, we consider six basic ecological concepts (Figure 17.1). (1) Different assemblages and species are affected by their environments at

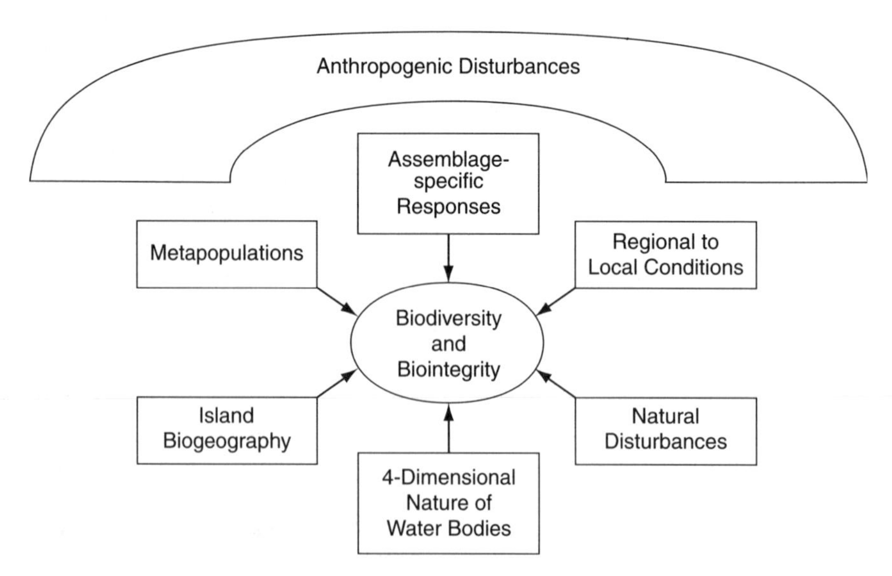

FIGURE 17.1. Key natural factors (rectangles) associated with biodiversity and biointegrity, and the press (continuous, chronic) nature of anthropogenic disturbances that also limit biodiversity and biointegrity.

different scales. In northeast United States lakes, for example, large-bodied species such as fish and macroinvertebrates were affected more by landscape-scale variables than were small-bodied species such as diatoms and zooplankton (Allen et al. 1999). This indicates the fallacy of a common notion that protecting one group, rather than entire systems, will preserve the others.

(2) Local assemblages are influenced by their immediate environments and by the landscape. Based on analyses of fish, macroinvertebrate, and algae data from Australia, Sweden, and the United States for lakes and streams, Hawkins et al. (2000) concluded that basin and ecoregion classifications accounted for relatively little biological variation. Tonn (1990; Figure 17.2) demonstrated how a lake fish assemblage is a product of continental, regional, lake-type, and local environmental filters. Because landscape ecology is a multiscale science, we will discuss its applications to conservation biology at multiple scales.

(3) Natural disturbances, or their absence, affect biodiversity. For example, salmonid stocks in streams are adapted to variable environments, with floods and debris flows critical for creating a mosaic of habitats (Reeves et al. 1995). Floods are so important to fishes and wetlands that Bayley (1995) argued that the absence of floods, not the floods themselves, is the disturbance. It is critical that managers allow for natural disturbances while seeking to correct or mitigate human disturbances.

(4) Because of the fluidity of water, aquatic ecosystems have four dimensions (vertical, lateral, longitudinal, and temporal; Ward 1989). First, there is a vertical dimension involving connections with the water table or *hyporheic zone*. Ignoring this dimension results in the disappearance of streams, lakes, and wetlands where ground water pumping or runoff is excessive.

The second, or lateral dimension, is equally important. Floodplains, riparian zones, and other wetlands are critical elements of lakes and streams, supplying habitat complexity, nutrients, structure, shade, flow continuity, and microclimate modification. Altered riparian zones are commonplace, producing warmer water, simplified habitat structure, lower base flows, and increased flood severity (Gregory et al. 1991).

Third, the longitudinal dimension is most obvious in rivers, and it was described as a river continuum (Vannote et al. 1980; Figure 16.1 in this volume). Recognition of this dimension typically means being aware of the effects of upstream conditions, but downstream hydrological and biological conditions have equally important effects on upstream sites (Pringle 1997). Lakes and wetlands, especially those that are connected by streams, also have a surficial flow component. In addition, larger lakes often have finer bottom substrates and different water quality and biota in protected areas compared with shores exposed to prevailing winds. Ignoring the longitudinal dimension in southern Florida wetlands threatens the integrity of the Everglades (Toth et al. 1998).

The fourth dimension is time. Natural and anthropogenic changes in lakes and wetlands are recorded in bottom sediments (Smol 1992), and changes in historical river channels are evident from land surveys (Benner and Sedell 1997) and historical data (McIntosh et al. 1994). Failure to understand past conditions of

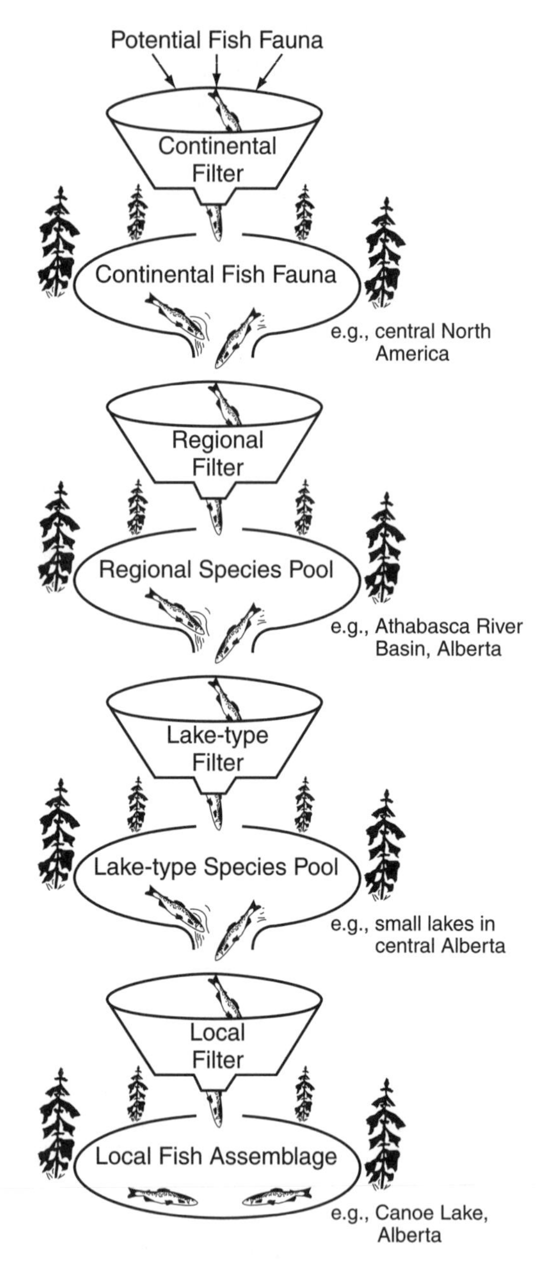

FIGURE 17.2. Various scales of landscape characteristics affecting aquatic biota (from Tonn 1990, with permission from the American Fisheries Society).

water bodies and natural succession severely limits our conservation and restoration efforts.

(5) Although the concept of *island biogeography* is most often associated with lands in water bodies, it also is relevant for water bodies in land masses. Large water bodies and those with connections to other lakes, rivers, or wetlands typically support more species than do small or isolated systems. Waters that have long been isolated from others contain endemic species. Water-body size is a key variable affecting species richness at global (Oberdorff et al. 1995), regional (Hugueny 1989), and local scales (Fausch et al. 1984). Thus, to understand the effects of land use on species richness in aquatic ecosystems, one must adjust expectations for water body sizes and their proximity to refugia.

(6) Levins (1969) explained that *metapopulations* consist of temporarily isolated populations of a species. Although populations in some areas may be extirpated by local disturbances, those areas are recolonized from neighboring populations. Habitat size and isolation govern local extinction and colonization of populations, just as they govern species richness on islands. Li et al. (1995) suggested how metapopulation dynamics may be applicable to regional salmonid management at basin scales, and Taylor (1997) applied metapopulation dynamics to a single endangered species in a set of stream pools. The metapopulation perspective is useful in interpreting how landscape changes may affect species by eliminating key populations.

17.2.2 Anthropogenic Factors

Given the ability to respond to natural disturbances, how do aquatic ecosystems respond to human disturbances? Working in three eastern Ontario rivers, Corkum (1990) determined that macroinvertebrate assemblages were more strongly associated with land use than with site location, drainage basin, or season. Catchment agriculture and urbanization negatively influenced stream substrate and channel morphology for macroinvertebrates in Minnesota streams (Richards and Host 1994). Anderson and Vondracek (1999) also observed a negative effect with increased agriculture, but land use was associated with less variance in insect metrics than were temperature, season, and ecoregion in North Dakota prairie potholes. In Pennsylvania, Croonquist and Brooks (1993) found that sensitive bird species did not occur without a riparian corridor ≥50 m wide, whereas Fischer et al. (2000) concluded from a number of studies that >100 m is needed. Livestock grazing has deleterious effects on riparian areas and surface waters, especially in the western United States, where it is the predominant land use (Platts 1991; Fleischner 1994; Kauffman et al. 1997). Macroinvertebrates required multiple years for recovery following clear cut logging of catchments in North Carolina (Tebo 1955; Gurtz and Wallace 1984), and sensitive fish in Michigan had not recovered from the effects of clear-cut logging after 50 years (Richards 1976). Strip mining in Tennessee and Virginia also had decades-long effects on macroinvertebrates and fish (Matter et al. 1978; Vaughan et al. 1978). Channel modifications by dams (Carlson and Muth 1989; Ebel et al. 1989) and channelization (Tarplee et al. 1971; Hortle and Lake 1983) have even longer-term effects on biota, especially fish and mussels, which do not recover until channels return to natural flow and channel patterns.

In other words, although aquatic systems can recover from pulse disturbances like floods and droughts, they are degraded by press (continuous, chronic) disturbances (Yount and Niemi 1990) in their catchments. Consequently, the leading stressors associated with aquatic species imperilment are chronic disturbances from agriculture, urbanization, hydroelectricity generation, and alien species (Richter et al. 1997). The first three lead directly to habitat degradation and pollution, which Allan and Flecker (1993) and Wilcove et al. (1998) considered leading causes of imperilment along with alien species and overexploitation. Generally though, any landscape alteration that impairs water quality, physical habitat structure, flow regime, food source, or biotic interactions can alter the biological integrity of aquatic ecosystems (Karr and Dudley 1981). Frequently, stressors are cumulative (Figure 17.3).

Recent studies have attempted to separate effects of catchment and riparian disturbances. Steedman (1988) reported that the best predictor of mean fish assemblage integrity for multiple sites in Ontario catchments was percent catchment forest cover. Wang et al. (1997), in a sample of 103 Wisconsin streams, reported stronger relationships between fish assemblage integrity and catchment land use than riparian land use. They found that declining integrity scores generally occurred when agriculture and urban land uses exceeded 50% and 10%, respectively. For macroinvertebrates and fish in southern Appalachian streams, Harding et al. (1998) reported that conserving riparian forest fragments was less protective than was conserving entire catchments, and that land use in the 1950s was a better predictor of biointegrity than was current land use. In Michigan, Richards et al. (1996) concluded that riparian and catchment land uses were comparable in their effects on stream habitat, but riparian conditions more strongly influenced macroinvertebrates (Richard et al. 1997). Studying 38 Wisconsin catchments, Stewart et al. (2000) found that fish and macroinvertebrate assemblages were significantly correlated with land use within a 30-m riparian buffer, but not with catchment land use outside of that buffer. A series of studies in the Raisin River Basin in Michigan helps clarify some of these contradictions (Roth et al. 1995; Allan et al. 1997; Lammert and Allan 1999). When seven widely distributed catchments were sampled, percent agricultural land use at the catchment and riparian scales was associated with 50% and <7%, respectively, of the variability in fish assemblage integrity. When only three catchments were sampled more intensively, catchment agriculture was insignificant but percent riparian agriculture was associated with 22% of the variability in fish integrity. Riparian conditions also were more important for macroinvertebrate integrity. As Hunsaker and Levine (1995) reported, the number and distribution of catchments sampled, together with the number of sites sampled per catchment, governed the scale of effects detected. We focus on a number of catchments across multiple regions in this chapter.

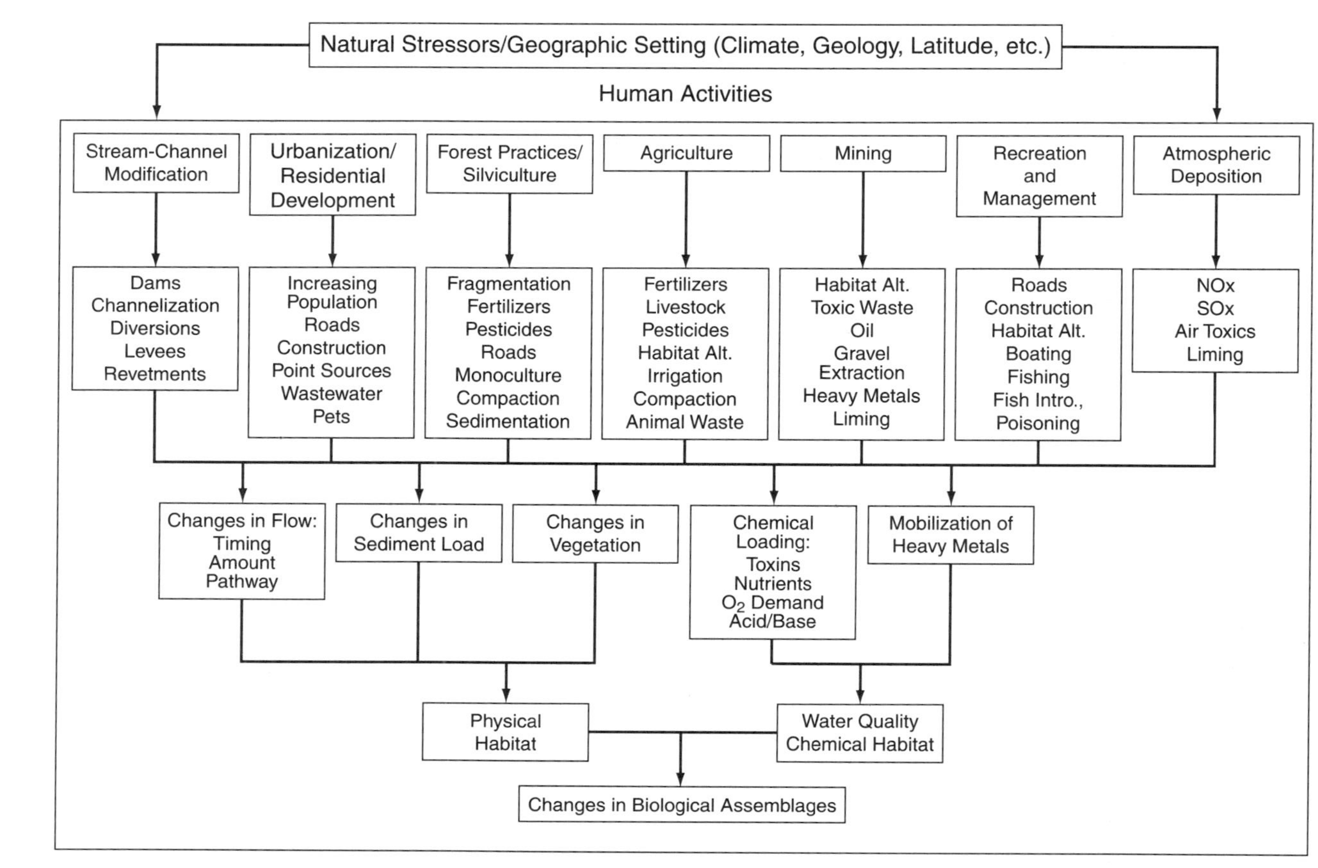

FIGURE 17.3. Cumulative effects of natural conditions and human activities on aquatic biota (from Bryce et al. 1999, with permission from the American Water Resources Association).

17.3 Recent Applications

We contacted offices of the following institutions by telephone or e-mail for examples of management applications of the six ecological concepts listed above: Columbia River Intertribal Fish Commission, International Joint Commission, National Marine Fisheries Service (NMFS Northwest Office), Oak Ridge National Laboratory, Ohio River Valley Water Sanitation Commission (ORSANCO), South Florida Water Management District, U.S. Bureau of Land Management (USBLM Prineville District), U.S. Natural Resources Conservation Service (USNRCS Western Region, Oregon and Coos County), U.S. Environmental Protection Agency (USEPA Office of Water, Regions 3 and 10, Corvallis Laboratory), U.S. Fish and Wildlife Service (USFWS Southwest Region, Indiana and Oregon Offices), U.S. Forest Service (USFS Intermountain, Pacific Northwest, and Southwest Regions and Research Stations), and U.S. Geological Survey (USGS Upper Mississippi Research Group). In addition, six independent researchers were contacted by e-mail or telephone, World Wide Web sites of federal agencies were searched for leads, the contents of the journal *Conservation Ecology* were read on the Web, and a computerized keyword search of aquatic and environmental management journals was conducted. Examples of applications of each of the key concepts are summarized below.

17.3.1 Different Assemblages Are Affected at Different Scales

To protect salmonids and owls, the Forest Ecosystem Management and Assessment Team (1993) restricts federal land management agencies from intensively logging key watersheds and riparian zones throughout western Oregon and Washington and northern California. However, recognizing that protecting these species is insufficient for other taxa at risk, these agencies also are required to "survey and manage" for less-charismatic assemblages, including algae, macroinvertebrates, and amphibians. In 1999, several timber sales were halted by court order because the USFS and USBLM failed to conduct those surveys.

17.3.2 Local and Widespread Influences on Aquatic Systems

The Forest Ecosystem Management and Assessment Team (1993) prescription to protect riparian habitats and key catchments also reflects awareness that both local and distant conditions affect stream biota. However, Espinosa et al. (1997) reported that best management plans and forest plans for catchments and riparian zones were unsuccessful in the Clearwater National Forest. Major reasons for failure included continued overexploitation of timber, simplistic modeling (improper resolution and incorrect assumptions for natural recovery rates, harvest levels, and harvest impact), and inadequate monitoring (incomplete, qualitative indicators). Espinosa et al. (1997) concluded that these failures are common to national forests managed for "multiple uses." Local actions by watershed councils were found ineffective in changing urbanization, water diversions, and industrial forestry practices (Huntington and Sommarstrom 2000).

17.3.3 Importance of Natural Disturbances

As with periodic fire in terrestrial landscapes, there is increasing interest in the use of natural disturbances to help restore aquatic ecosystems. Among the best examples are intentional floods on the upper Mississippi, Colorado, Rio Grande, and Kissimmee Rivers (Molles et al. 1998; Schmidt et al. 1998; Sparks et al. 1998; Toth et al. 1998). In all cases, native wetland vegetation responded positively. See Poff et al. (1997) for a discussion of how a natural flow regime promotes river conservation and restoration.

17.3.4 Four-Dimensional Nature of Water Bodies

The Forest Ecosystem Management and Assessment Team (1993) requires land management agencies to consider riparian and upstream conditions and to plan for temporal changes, but there is little recognition of hyporheic and downstream dimensions. The USBLM and USFS, by restricting grazing from riparian zones along small rangeland streams, have observed rapid recovery of riparian woody plants, channel narrowing and deepening, and increased salmonid populations in the resulting cooler waters (Platts 1991; Gutzwiller et al. 1997; Federal Interagency Stream Restoration Working Group 1998). Iowa State University and cooperating farmers (Isenhart et al. 1997) installed riparian buffer strips along 3.8 km of Bear Creek, Iowa. The USFS in Alaska, demonstrating the importance of stream-ocean connectivity, added 60 salmon carcasses to a stream to evaluate their effects on stream productivity (Wipfli et al. 1999). They observed increased biofilm, chlorophyll a, and macroinvertebrate density as a result. Thirteen culverts were repaired or replaced with salmon-friendly ones in the Coquille Basin, Oregon. Together with riparian fencing and bank stabilization, these changes increased beaver density and fish diversity (Hudson and Heikkila 1997). Both culvert improvements and carcass placements are increasingly common in the Pacific Northwest, where they are employed by land-management agencies, fishery agencies, and watershed-restoration groups.

17.3.5 Island Biogeography and Metapopulations

The concepts of island biogeography and metapopulation dynamics are often applied to conservation and recovery of endangered and threatened species. For example, delisting the endangered Oregon chub (*Oregonichthys crameri*) requires a total of at least 20 populations, each with >500 individuals. The populations must be stable or increasing for at least five years, and >three populations must occur in each of three major subbasins of the Willamette River (Scheerer et al. 1999). To protect multiple salmonid species, the USFS selected 137 key catchments dispersed across western Oregon and Washington, where timber harvest will be highly restricted; each catchment was >15 km^2, with current or potential high-quality physical and chemical habitat (Reeves and Sedell 1992). Wherever possible, these key catchments were selected near other key catchments to facilitate colonization following natural disturbances. Streams with riparian forest buffers 100 m wide will also link the key catchments. The NMFS (Waples 1995) has employed evolutionarily significant units, which incorporate metapopulation and island biogeography concepts, for listing and managing salmon species under the Endangered Species Act. Mundy et al. (1995) argued that the units were too small to reflect the life-history diversity of Columbia River salmon metapopulations.

It is too early to determine successes or failures from applying landscape-ecology theory to conservation biology. The above examples are simply too recent; final judgment must be preceded by several decades that include extremes in weather and vegetation recovery. It is important to recognize, however, that these applications do *not* involve the errors made in earlier restoration efforts that focused on single species, single assemblages, or surrogates like water quality and physical habitat (National Research Council 1992; Cooke et al. 1993). In addition, applications with the greatest potential are those that work with nature because they allow natural catchment and riverine processes occurring at regional scales, rather than site-specific technological fixes, to restore aquatic ecosystems (Bradbury et al. 1995; Williams et al. 1997; Winter and Hughes 1997). Nonetheless, without comprehensive monitoring programs, we will never be able to evaluate rigorously the success of attempted restorations (National Research Council 1992; Spence et al. 1996; Paulsen et al. 1998).

Given the above limits, the applications in this section indicate that some positive effects on biota have occurred as a result of implementing landscape-ecology principles. Allowing natural flows in rivers has helped restore native hydrophytic vegetation. Improved road crossings (culverts) have increased anadromous fish passage. Protecting riparian zones from livestock grazing has increased densities and mean sizes of salmonids. Adding salmon carcasses to streams increased the density and productivity of benthic organisms. Establishing multiple populations of Oregon chub resulted in survival of several populations despite a 50-year flood that decimated other populations. In the former three cases (restoring flows, correcting culverts, and protecting riparian zones), the human controls were temporarily removed and the waters were allowed to recover naturally. In the latter two cases (carcass addition and establishing multiple populations), there were insufficient populations for natural recovery, so humans transported fishes, placed them in appropriate habitats, and let natural factors operate.

17.4 Principles for Applying Landscape Ecology

We found few examples of field applications in which quantitative biological results were tested and recorded. Therefore, the five principles that follow are based on current ecological knowledge and managers' observations, instead of on scientifically documented effects. We therefore offer them as testable hypotheses as much as principles.

Conserve Multiple Assemblages and Ecosystems.

Focusing on a single species or assemblage instead of entire systems tends to create problems with other biota.

Apply Conservation Efforts at Multiple Scales.

Although it is difficult for scientists and managers, let alone interested citizens, to employ a broad perspective, this is necessary even when the concern is a specific site. At the same time, most individual actions occur at the site level, so a site perspective cannot be sacrificed for a regional or basin perspective. However, if regional and catchment processes are not allowed to function naturally, site-specific actions are unlikely to be successful. Healthy catchments, like healthy humans, can typically heal quickly from an injury, whereas sick ones with many injuries recover much more slowly.

Promote Natural Pulse Disturbances.

Ecosystems naturally experience disturbances and recover from them. Often, the preferred mechanism for recovery is simply to remove the anthropogenic press disturbance and allow the catchment and the aquatic ecosystem to respond to the natural disturbance regime (especially flow dynamics) and recover. This typically results in systems that look messy (debris jams and snags in rivers and lakes, wetlands that are periodically flooded or muddy, and multiple and shifting channels in rivers). These physical processes, addressed in Chapter 16, are important for sustaining biological integrity and natural levels of biodiversity.

Include Knowledge of Four Dimensions in Management Efforts.

Aquatic ecosystems are more than what we see. They typically connect laterally under the ground, vertically to the water table, upstream and downstream, or upwind and downwind in the case of lakes. They also change through time. Actions that alter water, sediment, wood, and biotic movements across these dimensions typically have cascading effects. Such actions should no longer be considered as having unforeseen impacts. To conserve or restore aquatic systems, it is necessary to have a historical perspective—that is, to know how and to what degree systems have changed.

Treat Freshwaters as Islands.

Communities, assemblages, species, and stocks are periodically eliminated from these islands, especially those that are small and distant from others. It is extremely difficult to conserve and restore biota in an increasingly fragmented landscape. Like classic, isolated marine islands, freshwater ecosystems, because of lower colonization rates and the resultant vacant niches, are prone to substantial disruption by alien species. Lakes, wetlands, and rivers of all types must be set aside as refugia and linked by streams with healthy riparian systems. It also is wise to protect as many systems as possible from human disturbances, rather than investing scarce resources on highly altered ones. This is analogous to human societies, in which it is most cost-effective to invest in public health and disease prevention as opposed to terminal treatments.

17.5 Knowledge Gaps

Many knowledge gaps hinder application of landscape ecology in biological conservation; we believe the ones listed below are the most important. Each gap needs filling if landscape ecology is to be used effectively to conserve aquatic biodiversity and biointegrity.

17.5.1 Theoretical Voids

1. We must learn how to distinguish the cumulative effects of multiple stressors from the effects of natural covariates, both of which affect ecosystem condition. It is essential to be able to tease apart the effects of these variables to make management decisions (Moyle and Light 1996; Allan et al. 1997). Closer collaboration among ecologists, spatial statisticians, and quantitative geographers is needed to develop theory and methods for such efforts. In addition, we must foster more scientific collaboration between scientists and managers because scientists lack the funds to experiment at relevant landscape scales, and resource managers rarely design their actions as experiments and monitor the results. This is the intent of adaptive management (Forest Ecosystem Management and Assessment Team 1993), but the scientific study designs and quantitative ecological monitoring are typically missing.
2. We need better theory on the types of stressors and natural disturbances that are most important at regional or basin versus local scales. What do regional stressor and response variables have in common as opposed to local variables? Intuitively, one would think that mobile stressors or those that limit material flows would be most likely to have regional effects. For example, what do the movements (or restricted movements) of water, sediments, large woody debris, disease and alien organisms, and bioaccumulating toxins have in common? What are the management implications of focusing at regional versus site scales? Interdisciplinary collaboration and theory development are necessary here too.
3. We need to understand more clearly how terrestrial factors determine conditions in aquatic ecosystems (Rathert et al. 1999). Unfortunately, few theories and approaches of terrestrial and aquatic ecologists are well-integrated. If we are to improve aquatic ecosystem integrity by improving terrestrial ecosystem integrity, the theoretical linkages must be improved. This requires increased collaborative theoretical work by terrestrial and aquatic ecologists, especially for large geographic areas.

4. We need to learn how to develop sustainable human societies. Managers increasingly talk about ecosystem management and sustainable development, but we have yet to demonstrate successful and sustainable human management (Soulé and Terborgh 1999). If we cannot sustainably manage our species, what basis for optimism is there in managing anything as complex as ecosystems? Perhaps the same conditions that lead to healthy ecosystems lead to healthy human populations—or maybe not. These linkages need to be developed by natural scientists working with social scientists and politicians to produce common theories and predictive models (Healy 1997; Priester and Kent 1997), and with catchment residents (who provide a reality check on the applicability of those theories and models). Certainly, landscape ecology's European roots include human society as a major factor, but in the United States there are few examples in which the natural and social sciences are linked to arrive at a theory of resource management.

17.5.2 Empirical Voids

1. Despite considerable monitoring, there is amazingly little quantitative knowledge of the actual status and trends of aquatic biological communities or of their major stressors. Consequently, management actions are frequently ill-conceived and ineffective. For example, Paulsen et al. (1998) and Yoder and Rankin (1998) demonstrated that the conditions of fish and macroinvertebrate assemblages are inaccurately assessed if surrogate indicators and biased sampling designs are employed. National Research Council (1992), Adler et al. (1993), and Public Employees for Environmental Responsibility (1999), in their reviews of the Nation's water body restoration programs, called for more comprehensive and more rigorous monitoring.
2. We need to acquire much more knowledge about how ecosystems vary across multiple spatial scales and how to better integrate the lessons learned (Allan et al. 1997; Allen et al. 1999; Stewart et al. 2000). Controlling variables that are important at one scale are often less so at another. Typically, managers manipulate at the site level, but vagile species respond to conditions across a much greater area. In addition, the fluid nature and connectivity of water mean that local actions are translated great distances, especially in lotic ecosystems. To protect biota effectively, managers need to know which environmental attributes are most important to conserve or restore and at which scales.
3. A tiny proportion of ecological research support goes to holistic studies. More information is needed from research incorporating physical, chemical, and multiassemblage data examined at multiple geographic scales. These should incorporate different types of aquatic ecosystems in different parts of the country (Messer et al. 1991; Hughes et al. 2000).
4. We cannot yet accurately model how land uses affect aquatic biota (Espinosa et al. 1997; Chapter 24 in this volume). Existing models are data poor and

rarely or incompletely verified. Filling both the theoretical and empirical gaps listed above should aid model development, but the models will still require considerable field verification.

17.6 Research Approaches

17.6.1 Approaches for Theoretical Research

A combination of multivariate analyses and residuals analyses of survey data and gradient studies seem appropriate for closing the four theoretical gaps (see Allen et al. 1999). There is considerable potential for collaborative work between Environmental Monitoring and Assessment Program (EMAP) aquatic and terrestrial ecologists and EMAP design and analytical statisticians. In fact, the early designers of EMAP stressed the need for such integration (Messer et al. 1991), but changes in USEPA administrators and budgets halted collaboration. Similar potential exists for the same four disciplines at the research laboratories of the USFS. In both cases, we are unaware of joint research projects or papers that have integrated all four disciplines working together at theory development and testing. A rational way to start is by finding common concepts, like those listed for aquatic systems in Section 17.2, developing predictions for both terrestrial and aquatic ecosystems, and testing them.

A greater research gap than that between statisticians and ecologists exists among natural scientists, social scientists, managers, politicians, and local residents (theoretical gap 4). Perhaps this is why so little of what ecologists know and value is translated into public policies to conserve ecosystems. Huntington and Sommarstrom (2000) offer one example of merging ecological and sociological research. Another admittedly simplistic example of a potential approach is suggested by Bradbury et al. (1995), in which the President of the Oregon Senate worked closely with a group of scientists to answer a specific salmon restoration question. These people had different backgrounds but shared a common interest to produce a product with minimal funding in six months. A third example is that of a civil engineer who developed a model for predicting the behavior of human systems (Bella 1997). He used this model to explain how the aggregation of individual human actions resulted in outcomes ranging from salmon extirpation to the Challenger explosion. In all three cases, people were forced to think from the perspective of a different discipline.

17.6.2 Approaches for Empirical Research

A combination of surveys, gradient studies, experimental restoration projects, and ecological experiments at multiple scales and locations are needed. In addition, management prescriptions should be directed toward learning and hypothesis testing, not just toward meeting an exploitation level or technological fix. The United States has begun two federal research programs (EMAP, National Water

Quality Assessment [NAWQA]) with the potential to fill the first three theoretical and empirical gaps listed above. Both employ intensive, quantitative sampling of biotic and abiotic variables at sites that are selected to represent basins and regions. Given the scale and complexity of the gaps, a federal approach is most likely to succeed, but close collaboration with academic and state institutions also is essential for success.

Previous and planned EMAP studies of New England lakes, and streams in Appalachia, the Central Plains, and the 12 western states, should provide data for the first three empirical gaps. Design and field sampling protocols exist for these surveys (Paulsen et al. 1991; Baker et al. 1997; Peck et al. 2000a,b). EMAP employs a statistical study design that allows inference to entire mapped populations of waters with known confidence intervals. EMAP indicators are based on quantitative sampling of physical and chemical habitat, as well as of multiple assemblages (vertebrates, invertebrates, algae). Several analytical approaches have been applied (Whittier et al. 1997; Allen et al. 1999; O'Connor et al. 2000), but these need further testing. Analyses have included principal components analysis, factor analysis, canonical correspondence analysis, and multivariate scatter plots. In addition to journal publications, results have been used in draft regional assessment reports prepared by environmental managers for selected regions (U.S. Environmental Protection Agency 2000).

The NAWQA Program conducts many gradient studies throughout the United States to develop predictive models that link general land use (% urban, % agriculture) with the responses of multiple aquatic assemblages (empirical gaps 2–4; e.g., Cuffney et al. 1997; Wentz et al. 1998; Williamson et al. 1998). To produce models that are more predictive of the effects of land-use changes, and to better evaluate restoration projects, restoration studies with design and sampling protocols similar to NAWQA's are needed. Further research involving land-use effects on aquatic ecosystems, similar to that of Allan et al. (1997), Richards et al. (1997), and Wang et al. (1997), but in other regions of the country and with additional assemblages, is also recommended.

Huntington and Sommarstrom (2000) offer a protocol for evaluating both the ecological and sociological components of watershed councils. Their protocol provides ecological criteria for determining what councils do well and what they do poorly based on such factors as ecological goals, conservation plans or strategies, restoration priorities, and monitoring and evaluation. Sociological criteria include council organization, staffing, decision-making process, conflict resolution, funding, and communication. Scores for the mean ecological restoration values of projects were directly related to scores for sociological processes. Ecological/sociological research such as this is essential for bridging the gap between understanding how humans behave and the ecological effects of those actions.

The latter may be a fitting thought for concluding and summarizing this chapter. We have tried to challenge the reader to think differently and beyond familiar spaces, times, and disciplines. Clearly, we cannot continue to apply the same social and scientific approaches that have led to widespread species and ecosystem extirpation if we want to reverse those conditions and trends.

Acknowledgments

We thank Kevin Gutzwiller for inviting us to prepare this chapter. We also are grateful to the many managers and scientists who responded to our phone calls and e-mails requesting information about ecological applications, especially D. McCullough, T. Nelson, H. Regier, and B. Streif. Timely and critical reviews by Phil Kaufmann, Denis White, and Kevin Gutzwiller much improved the manuscript. Suzanne Pierson produced the final figures. Many of the thoughts presented here are products of interactions with colleagues at the USEPA Laboratory, USFS Laboratory, and Oregon State University, all in Corvallis, Oregon. However, the viewpoints are ours and do not represent policies of the institutions with which we are affiliated. The manuscript was prepared outside of normal work hours, and it was not subjected to our institutional review and clearance processes.

References

Adler, R.W., Landman, J.C., and Cameron, D.M. 1993. *The Clean Water Act: 20 Years Later.* Covelo, California: Island Press.

Allan, J.D., Erickson, D.L., and Fay, J. 1997. The influence of catchment land use on stream integrity across multiple scales. *Freshwater Biol.* 37:149–161.

Allan, J.D., and Flecker, A.S. 1993. Biodiversity conservation in running waters. *BioScience* 43:32–43.

Allen, A.P., Whittier, T.R., Larsen, D.P., Kaufmann, P.R., O'Connor, R.J., Hughes, R.M., Stemberger, R.S., Dixit, S.S., Brinkhurst, R.O., Herlihy, A.T., and Paulsen, S.G. 1999. Concordance of taxonomic composition patterns across multiple lake assemblages: effects of scale, body size, and land use. *Can. J. Fish. Aquat. Sci.* 56:2029–2040.

Anderson, D.J., and Vondracek, B. 1999. Insects as indicators of land use in three ecoregions in the Prairie Pothole Region. *Wetlands* 19:648–664.

Baker, J.R., Peck, D.V., and Sutton, D.W. 1997. *Field Operations Manual for Lakes.* EPA/620/R-97/001. Corvallis, Oregon: U.S. Environmental Protection Agency.

Bayley, P.B. 1995. Understanding large river-floodplain ecosystems. *BioScience* 45:153–158.

Bella, D.A. 1997. Organizational systems and the burden of proof. In *Pacific Salmon and Their Ecosystems: Status and Future Options,* eds. D.J. Stouder, P.A. Bisson, and R.J. Naiman, pp. 617–638. New York: Chapman and Hall.

Benner, P.A., and Sedell, J.R. 1997. Upper Willamette River landscape: a historic perspective. In *River Quality: Dynamics and Restoration,* eds. A. Laenen and D.A. Dunnette, pp. 23–47. Boca Raton, Florida: Lewis Publishers.

Bradbury, B., Nehlsen, W., Nickelson, T.E., Moore, K.M.S., Hughes, R.M., Heller, D., Nicholas, J., Bottom, D.L., Weaver, W.E., and Beschta, R.L. 1995. *Handbook for Prioritizing Watershed Protection and Restoration to Aid Recovery of Native Salmon.* Eugene, Oregon: Pacific Rivers Council.

Bryce, S.A., Larsen, D.P., Hughes, R.M., and Kaufmann, P.R. 1999. Assessing relative risks to aquatic ecosystems: a mid-Appalachian case study. *J. Am. Water Resour. Assoc.* 35:23–36.

Carlson, C.A., and Muth, R. 1989. The Colorado River: lifeline of the American Southwest. In *Proceedings of the International Large River Symposium,* ed. D.P. Dodge, pp. 220–239. Canadian Special Publication in Fisheries and Aquatic Sciences 106. Ottawa, Ontario, Canada: Department of Fisheries and Oceans.

Cooke, G.D., Welch, E.B., Peterson, S.A., and Newroth, P.R. 1993. *Restoration and Management of Lakes and Reservoirs.* Boca Raton, Florida: Lewis Publishers.

Corkum, L.D. 1990. Intrabiome distributional patterns of lotic macroinvertebrate assemblages. *Can. J. Fish. Aquat. Sci.* 47:2147–2157.

Croonquist, M.J., and Brooks, R.P. 1993. Effects of habitat disturbance on bird communities in riparian corridors. *J. Soil Water Conserv.* 48:65–70.

Cuffney, T.F., Meador, M.R., Porter, S.D., and Gurtz, M.E. 1997. *Distribution of Fish, Benthic Invertebrate, and Algal Communities in Relation to Physical and Chemical Conditions, Yakima River Basin, Washington, 1990.* U.S. Geological Survey Water Resources Investigations Report 96-4280. Denver, Colorado: U.S. Geological Survey.

Ebel, W.J., Becker, C.D., Mullan, J.W., and Raymond, H.L. 1989. The Columbia River—toward a holistic understanding. In *Proceedings of the International Large River Symposium,* ed. D.P. Dodge, pp. 205–219. Canadian Special Publication in Fisheries and Aquatic Sciences 106. Ottawa, Ontario, Canada: Department of Fisheries and Oceans.

Espinosa, F.A., Jr., Rhodes, J.J., and McCullough, D.A. 1997. The failure of existing plans to protect salmon habitat in the Clearwater National Forest in Idaho. *J. Environ. Manage.* 49:205–230.

Fausch, K.D., Karr, J.R., and Yant, P.R. 1984. Regional application of an index of biotic integrity based on stream fish communities. *Trans. Am. Fish. Soc.* 113:39–55.

Federal Interagency Stream Restoration Working Group. 1998. *Stream Corridor Restoration: Principles, Processes, and Practices.* Springfield, Virginia: National Technical Information Service.

Fischer, R.A., Martin, C.O., and Fischenich, J.C. 2000. Improving riparian buffer strips and corridors for water quality and wildlife. In *Riparian Ecology and Management in Multi-Land Use Watersheds,* eds. P.J. Wigington and R.L. Beschta, pp. 457–462. Middleburg, Virginia: American Water Resources Association.

Fleischner, T.L. 1994. Ecological costs of livestock grazing in western North America. *Conserv. Biol.* 8:629–645.

Forest Ecosystem Management and Assessment Team. 1993. *Forest Ecosystem Management: An Ecological, Economic, and Social Assessment.* 1993-793-071. Washington, DC: U.S. Government Printing Office.

Gregory, S.V., Swanson, F.J., McKee, W.A., and Cummins, K.W. 1991. An ecosystem perspective of riparian zones: focus on links between land and water. *BioScience* 41:540–551.

Grove, R.H. 1995. *Green Imperialism: Colonial Expansion, Tropical Island Edens and the Origins of Environmentalism, 1600–1860.* Cambridge, United Kingdom: Cambridge University Press.

Gurtz, M.E., and Wallace, J.B. 1984. Substrate-mediated response of stream invertebrates to disturbance. *Ecology* 65:1556–1569.

Gutzwiller, L.A., McNatt, R.M., and Price, R.D. 1997. Watershed restoration and grazing practices in the Great Basin: Mary's River of Nevada. In *Watershed Restoration: Principles and Practices,* eds. J.E. Williams, C.A. Wood, and M.P. Dombeck, pp. 360–380. Bethesda, Maryland: American Fisheries Society.

Harding, J.S., Benfield, E.F., Bolstad, P.V., Helfman, G.S., and Jones, E.B.D., III. 1998. Stream biodiversity: the ghost of land use past. *Proc. Natl. Acad. Sci.* 95:14843–14847.

Hawkins, C.P., Norris, R.H., Gerritsen, J., Hughes, R.M., Jackson, S.K., Johnson, R.K., and Stevenson, R.J. 2000. Evaluations of the use of landscape classifications for the prediction of freshwater biota: synthesis and recommendations. *J. N. Am. Benthol. Soc.* 19:541–556.

Healy, M.C. 1997. Paradigms, policies, and prognostications about the management of watershed ecosystems. In *River Ecology and Management: Lessons from the Pacific Coastal Ecoregion,* eds. R.J. Naiman and R.E. Bilby, pp. 662–682. New York: Springer-Verlag.

Hortle, K.G., and Lake, P.S. 1983. Fish of channelized and unchannelized sections of the Bunyip River, Victoria. *Australian J. Marine Freshwater Res.* 34:441–450.

Hudson, W.F., and Heikkila, P.A. 1997. Integrating public and private restoration strategies: Coquille River of Oregon. In *Watershed Restoration: Principles and Practices,* eds. J.E. Williams, C.A. Wood, and M.P. Dombeck, pp. 235–252. Bethesda, Maryland: American Fisheries Society.

Hughes, R.M., Paulsen, S.G., and Stoddard, J.L. 2000. EMAP-Surface Waters: a multiassemblage, probability survey of ecological integrity in the U.S.A. *Hydrobiologia* 423:429–443.

Hugueny, B. 1989. West African rivers as biogeographic islands: species richness of fish communities. *Oecologia* 79:235–243.

Hunsaker, C.T., and Levine, D.A. 1995. Hierarchical approaches to the study of water quality in rivers. *BioScience* 45:193–203.

Huntington, C.W., and Sommarstrom, S. 2000. *An Evaluation of Selected Watershed Councils in the Pacific Northwest and Northern California.* Eugene, Oregon: Pacific Rivers Council.

Isenhart, T.M., Schultz, R.C., and Colletti, J.P. 1997. Watershed restoration and agricultural practices in the Midwest: Bear Creek of Iowa. In *Watershed Restoration: Principles and Practices,* eds. J.E. Williams, C.A. Wood, and M.P. Dombeck, pp. 318–334. Bethesda, Maryland: American Fisheries Society.

Karr, J.R., and Dudley, D.R. 1981. Ecological perspective on water quality goals. *Environ. Manage.* 5:55–68.

Kauffman, J.B., Beschta, R.L., Otting, N., and Lytjen, D. 1997. An ecological perspective of riparian and stream restoration in the western United States. *Fisheries* 22(5):12–24.

Lammert, M., and Allan, J.D. 1999. Assessing biotic integrity of streams: effects of scale in measuring the influence of land use/cover and habitat structure on fish and macroinvertebrates. *Environ. Manage.* 23:257–270.

Levins, R. 1969. Some demographic and genetic consequences of environmental heterogeneity for biological control. *Bull. Entomol. Soc. Am.* 15:237–240.

Li, H.W., Currens, K., Bottom, D., Clarke, S., Dambacher, J., Frissell, C., Harris, P., Hughes, R.M., McCullough, D., McGie, A., Moore, K., Nawa, R., and Thiele, S. 1995. Safe havens: refuges and evolutionarily significant units. *Am. Fish. Soc. Symp.* 17:371–380.

Master, L. 1990. The imperiled status of North American aquatic animals. *Biodiversity Network News* 3(3):1–8.

Matter, W.J., Ney, J.J., and Maughan, O.E. 1978. Sustained impact of abandoned surface mines on fish and benthic invertebrate populations in headwater streams of southwestern Virginia. In *Surface Mining and Fish/Wildlife Needs in the Eastern United States,* eds. D.E. Samuel, J.R. Stauffer, C.H. Hocutt, and W.T. Mason, pp. 203–215. FWS/OBS-78/81. Washington, DC: U.S. Fish and Wildlife Service.

McIntosh, B.A., Sedell, J.R., Smith, J.E., Wissmar, R.C., Clarke, S.E., Reeves, G.H., and Brown, L.A. 1994. Historical changes in fish habitat for selected river basins of eastern Oregon and Washington. *Northwest Sci.* 68:36–53.

Messer, J.J., Linthurst, R.A., and Overton, W.S. 1991. An EPA program for monitoring ecological status and trends. *Environ. Monitor. Assess.* 17:67–78.

Miller, R.R., Williams, J.D., and Williams, J.E. 1989. Extinctions of North American fishes during the past century. *Fisheries* 14(6):22–38.

Molles, M.C., Jr., Crawford, C.S., Ellis, L.M., Valett, H.M., and Dahm, C.N. 1998. Managed flooding for riparian ecosystem restoration. *BioScience* 48:749–756.

Moyle, P.B., and Leidy, R.A. 1992. Loss of biodiversity in aquatic ecosystems: evidence from fish faunas. In *Conservation Biology: The Theory and Practice of Nature Conservation, Preservation, and Management,* eds. P.L. Fiedler and S.K. Jain, pp. 127–169. New York: Chapman and Hall.

Moyle, P.B., and Light, T. 1996. Biological invasions of fresh water: empirical rules and assembly theory. *Biol. Conserv.* 78:149–161.

Mundy, P.R., Backman, T.W.H., and Berkson, J.M. 1995. Selection of conservation units for Pacific salmon: lessons learned from the Columbia River. *Am. Fish. Soc. Symp.* 17:28–38.

Musick, J.A., Harbin, M.M., Berkeley, S.A., Burgess, G.H., Eklund, A.M., Findley, L., Gilmore, R.G., Golden, J.T., Ha, D.S., Huntsman, G.R., McGovern, J.C., Parker, S.J., Poss, S.G., Sala, E., Schmidt, T.W., Sedberry, G.R., Weeks, H., and Wright, S.G. 2000. Marine, estuarine, and diadromous fish stocks at risk of extinction in North America (exclusive of Pacific salmonids). *Fisheries* 25(11):6–30.

National Research Council. 1992. *Restoration of Aquatic Ecosystems: Science, Technology, and Public Policy.* Washington, DC: National Academy Press.

Oberdorff, T., Guegan, J.F., and Hugueny, B. 1995. Global scale patterns of fish species richness in rivers. *Ecography* 18:345–352.

O'Connor, R.J., Walls, T.E., and Hughes, R.M. 2000. Using multiple taxonomic groups to index the ecological condition of lakes. *Environ. Monitor. Assess.* 61:207–228.

Omernik, J.M. 1987. Ecoregions of the conterminous United States. *Ann. Assoc. Am. Geog.* 77:118–125.

Paulsen, S.G., Hughes, R.M., and Larsen, D.P. 1998. Critical elements in describing and understanding our nation's aquatic resources. *J. Am. Water Resour. Assoc.* 34:995–1005.

Paulsen, S.G., Larsen, D.P., Kaufmann, P.R., Whittier, T.R., Baker, J.R., Peck, D.V., McGue, J., Hughes, R.M., McMullen, D., Stevens, D., Stoddard, J.L., Lazorchak, J., Kinney, W., Selle, A.R., and Hjort, R. 1991. *EMAP-Surface Waters Monitoring and Research Strategy: Fiscal Year 1991.* EPA/600/3-91/022. Corvallis, Oregon: U.S. Environmental Protection Agency.

Peck, D.V., Averill, D.K., Lazorchak, J.M., and Klemm D.J. 2000a. *Field Operations Manual for Non-Wadeable Rivers and Streams.* Corvallis, Oregon: U.S. Environmental Protection Agency.

Peck, D.V., Lazorchak, J.M., and Klemm D.J. 2000b. *Field Operations Manual for Wadeable Streams.* Corvallis, Oregon: U.S. Environmental Protection Agency.

Platts, W.S. 1991. Livestock grazing. In *Influences of Forest and Rangeland Management on Salmonid Fishes and Their Habitats,* ed. W.R. Meehan, pp. 389–424. Special Publication 19. Bethesda, Maryland: American Fisheries Society.

Poff, N.L., Allan, J.D., Bain, M.B., Karr, J.R., Prestegaard, K.L., Richter, B.D., Sparks, R.E., and Stromberg, J.C. 1997. The natural flow regime: a paradigm for river conservation and restoration. *BioScience* 47:769–784.

Preister, K., and Kent, J.A. 1997. Social ecology: a new pathway to watershed restoration. In *Watershed Restoration: Principles and Practices,* eds. J.E. Williams, C.A. Wood, and M.P. Dombeck, pp. 28–48. Bethesda, Maryland: American Fisheries Society.

Pringle, C.M. 1997. Exploring how disturbance is transmitted upstream: going against the flow. *J. N. Am. Benthol. Soc.* 16:425–438.

Public Employees for Environmental Responsibility. 1999. *Murky Waters.* Washington, DC: Public Employees for Environmental Responsibility.

Rathert, D., White, D., Sifneos, J.C., and Hughes, R.M. 1999. Environmental correlates of species richness for native freshwater fish in Oregon, USA. *J. Biogeog.* 26:257–273.

Reeves, G.H., Benda, L.E., Burnett, K.M., Bisson, P.A., and Sedell, J.R. 1995. A disturbance-based ecosystem approach to maintaining and restoring freshwater habitats of evolutionarily significant units of anadromous salmonids in the Pacific Northwest. *Am. Fish. Soc. Symp.* 17:334–349.

Reeves, G.H., and Sedell, J.R. 1992. An ecosystem approach to the conservation and management of freshwater habitat for anadromous salmonids in the Pacific Northwest. In *Biological Diversity in Aquatic Management,* eds. J.E. Williams and R.J. Neves, pp. 408–415. Washington, DC: Wildlife Management Institute.

Ricciardi, A., and Rasmussen, J.B. 1999. Extinction rates of North American freshwater fauna. *Conserv. Biol.* 13:1220–1222.

Richards, C., Haro, R.J., Johnson, L.B., and Host, G.E. 1997. Catchment and reach-scale properties as indicators of macroinvertebrate species traits. *Freshwater Biol.* 37:219–230.

Richards, C., and Host, G.E. 1994. Examining land use influences on stream habitats and macroinvertebrates: a GIS approach. *Water Resour. Bull.* 30:729–738.

Richards, C., Johnson, L.B., and Host, G.E. 1996. Landscape-scale influences on stream habitats and biota. *Can. J. Fish. Aquat. Sci.* 53:295–311.

Richards, J.S. 1976. Changes in fish species composition in the Au Sable River, Michigan from the 1920's to 1972. *Trans. Am. Fish. Soc.* 105:32–40.

Richter, B.D., Braun, D.P., Mendelson, M.A., and Master, L.L. 1997. Threats to imperiled freshwater fauna. *Conserv. Biol.* 11:1081–1093.

Roth, N.E., Allan, J.D., and Erickson, D.L. 1995. Landscape influences on stream biotic integrity assessed at multiple spatial scales. *Landsc. Ecol.* 11:141–156.

Scheerer, P.D., Apke, G.D., and McDonald, P.J. 1999. *Oregon Chub Research: Middle Fork Willamette and Santiam River Drainages.* Contract Number E96970022. Corvallis, Oregon: Oregon Department of Fish and Wildlife.

Schmidt, J.C., Webb, R.H., Valdez, R.A., Marzolf, G.R., and Stevens, L.E. 1998. Science and values in river restoration in the Grand Canyon. *BioScience* 48:735–748.

Smol, J.P. 1992. Paleolimnology: an important tool for effective ecosystem management. *J. Aquat. Ecosystem Health* 1:49–58.

Soulé, M.E., and Terborgh, J. 1999. Conserving nature at regional and continental scales—a scientific program for North America. *BioScience* 49:809–817.

Sparks, R.E., Nelson, J.C., and Yin, Y. 1998. Naturalization of the flood regime in regulated rivers. *BioScience* 48:706–720.

Spence, B.C., Lomnicky, G.A., Hughes, R.M., and Novitzki, R.P. 1996. *An Ecosystem Approach to Salmonid Conservation.* TR-4501-96-6057. Portland, Oregon: National Marine Fisheries Service.

Steedman, R.J. 1988. Modification and assessment of an index of biotic integrity to quantify stream quality in southern Ontario. *Can. J. Fish. Aquat. Sci.* 45:492–501.

Stewart, J.S., Downes, D.M., Wang, L., Wierl, J.A., and Bannerman, R. 2000. Influences of riparian corridors and hydrology on aquatic biota in agricultural watersheds. In *Riparian Ecology and Management in Multi-Land Use Watersheds,* eds. P.J. Wigington and R.L. Beschta, pp. 209–214. Middleburg, Virginia: American Water Resources Association.

Tarplee, W.H., Jr., Louder, D.E., and Weber, A.J. 1971. *Evaluation of the Effects of Channelization on Fish Populations in North Carolina's Coastal Plain Streams.* Raleigh, North Carolina: North Carolina Wildlife Resources Commission.

Taylor, C. 1997. Fish species richness and incidence patterns in isolated and connected stream pools: effects of pool volume and spatial position. *Oecologia* 110:560–566.

Tebo, L.B. 1955. Effects of siltation, resulting from improper logging, on the bottom fauna of a small trout stream in the southern Appalachians. *Progr. Fish Cult.* 17:64–70.

Tonn, W.M. 1990. Climate change and fish communities: a conceptual framework. *Trans. Am. Fish. Soc.* 119:337–352.

Toth, L.A., Melvin, S.L., Arrington, D.A., and Chamberlain, J. 1998. Hydrologic manipulations of the channelized Kissimmee River. *BioScience* 48:757–764.

U.S. Environmental Protection Agency. 2000. *Mid-Atlantic Highlands Stream Assessment.* EPA/903/R-00/015. Philadelphia, Pennsylvania: U.S. Environmental Protection Agency.

Vannote, R.L., Minshall, G.W., Cummins, K.W., Sedell, J.R., and Cushing, C.E. 1980. The river continuum concept. *Can. J. Fish. Aquat. Sci.* 37:130–137.

Vaughan, G.L., Talak, A., and Anderson, R.J. 1978. The chronology and character of recovery of aquatic communities from the effects of strip mining for coal in east Tennessee. In *Proceedings of the Symposium on Surface Mining and Fish and Wildlife Needs in the Eastern US,* ed. E.E. Samuel, pp. 119–125. Knoxville, Tennessee: University of Tennessee.

Wang, L., Lyons, J., Kanehl, P., and Gatti, R. 1997. Influences of watershed land use on habitat quality and biotic integrity in Wisconsin streams. *Fisheries* 22(6):6–12.

Waples, R.S. 1995. Evolutionarily significant units and the conservation of biological diversity under the Endangered Species Act. *Am. Fish. Soc. Symp.* 17:8–27.

Ward, J.V. 1989. The four-dimensional nature of lotic ecosystems. *J. N. Am. Benthol. Soc.* 8:2–8.

Wentz, D.A., Bonn, B.A., Carpenter, K.D., Hinkle, S.R., Janet, M.L., Rinella, F.A., Uhrich, M.A., Waite, I.R., Laenen, A., and Bencala, K.E. 1998. *Water Quality in the Willamette Basin, Oregon, 1991–95.* U.S. Geological Survey Circular 1161. Denver, Colorado: U.S. Geological Survey.

Whittier, T.R., Halliwell, D.B., and Paulsen, S.G. 1997. Cyprinid distributions in northeast U.S.A. lakes: evidence of regional-scale minnow biodiversity losses. *Can. J. Fish. Aquat. Sci.* 54:1593–1607.

Wilcove, D.S., Rothstein, D., Dubow, J., Phillips, A., and Losos, E. 1998. Quantifying threats to imperiled species in the United States. *BioScience* 48:607–615.

Williams, J.E., Johnson, J.E., Hendrickson, D.A., Contreras-Balderas, S., Williams, J.D., Navarro-Mendoza, M., McAllister, D.E., and Deacon, J.E. 1989. Fishes of North America endangered, threatened, or of special concern: 1989. *Fisheries* 14(6):2–20.

Williams, J.E., Wood, C.A., and Dombeck, M.P., eds. 1997. *Watershed Restoration: Principles and Practices.* Bethesda, Maryland: American Fisheries Society.

Williamson, A.K., Munn, M.D., Ryker, S.J., Wagner, R.J., Ebbert, J.C., and Vanderpool, A.M. 1998. *Water Quality in the Central Columbia Plateau, Washington and Idaho, 1992–95.* U.S. Geological Survey Circular 1144. Denver, Colorado: U.S. Geological Survey.

Winter, B.D., and Hughes, R.M. 1997. Biodiversity. *Fisheries* 22(1):22–29.

Wipfli, M.S., Hudson, J.P., Chaloner, D.T., and Caouette, J.P. 1999. Influence of salmon spawner densities on stream productivity in southeast Alaska. *Can. J. Fish. Aquat. Sci.* 56:1600–1611.

Yoder, C.O., and Rankin, E.T. 1998. The role of biological indicators in a state water quality management process. *Environ. Monitor. Assess.* 51:61–88.

Yount, J.D., and Niemi, G.J. 1990. Recovery of lotic communities and ecosystems from disturbance—a narrative review of case studies. *Environ. Manage.* 14:547–569.

18

Time Lags in Metapopulation Responses to Landscape Change

KEES (C.J.) NAGELKERKE, JANA VERBOOM, FRANK VAN DEN BOSCH, AND KAREN VAN DE WOLFSHAAR

18.1 Introduction

Landscape ecologists and conservation biologists usually assume a causal relation between the observed current landscape pattern and the species distribution pattern. This assumption may not always be correct because species distribution patterns can reflect past as well as present landscape conditions (Figure 18.1).

Time delays in the reaction of biodiversity to changed circumstances are of major importance because they may determine how much extinction or other ecosystem change is awaiting us. Habitat destruction is generally seen as the dominant threat to biodiversity (e.g., Pimm et al. 1995; Pimm 1998). Recorded species extinction, however, is much less than expected theoretically (Heywood and Stuart 1992; Whitmore 1997). Some authors therefore reason that the predictions are too pessimistic (e.g., Budiansky 1994). Others argue that extinction takes place with a delay and that many extant species are doomed (e.g., Heywood et al. 1994; Whitmore 1997; Pimm 1998). Hence, it is extremely important to investigate the extent and mechanisms of time-delayed biodiversity changes.

In this chapter, we focus on the role of metapopulation dynamics in delayed responses to landscape changes. Note, however, that other processes, like succession, range expansion, and evolution, also can be responsible for time lags. We first introduce the principles of metapopulation lags, explain relevant concepts, and discuss recent work. We continue with principles for application, identify major theoretical and empirical knowledge voids, and end by suggesting research approaches for filling those voids.

18.2 Concepts, Principles, and Emerging Ideas

18.2.1 Principles of Time Lags in Metapopulation Dynamics

A *metapopulation* is a system of discrete subpopulations, with some turnover, linked by dispersal. Of N habitable patches per unit area, only n are occupied at any given moment. *Metapopulation dynamics* (the course of *occupancy* [n] over time) is determined by the turnover processes of local extinction and coloniza-

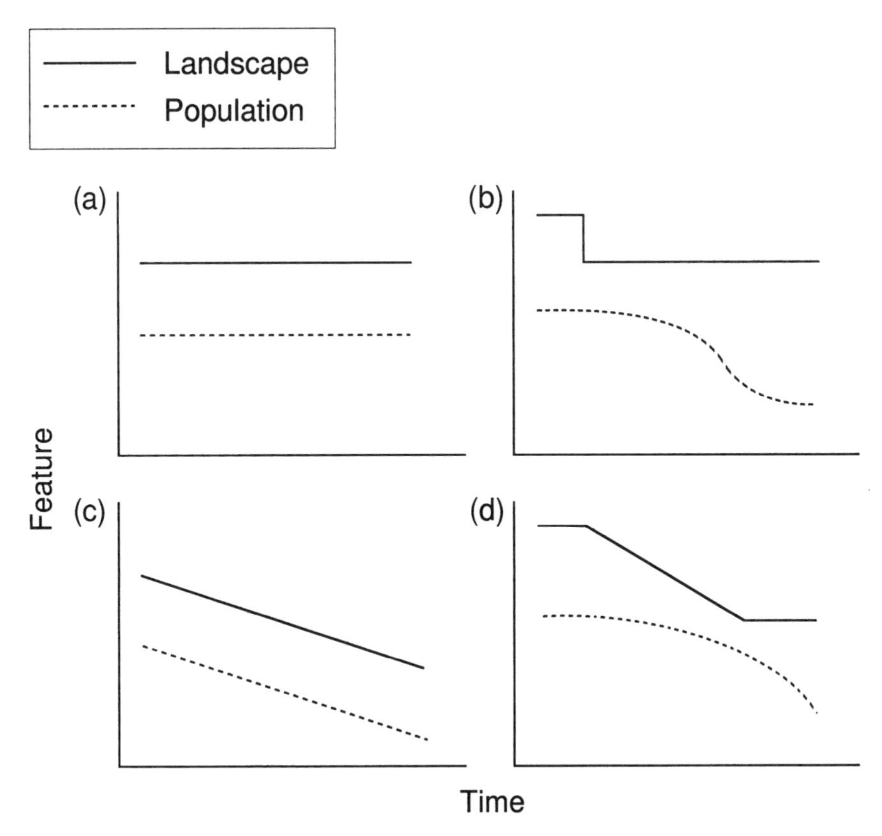

FIGURE 18.1. Illustration of the time-lag phenomenon. The landscape can be constant over time (a, right part of b) or gradually changing (c, d). The (meta)population may appear to be in equilibrium with the landscape (a, c) or lagging behind. In (b), the population responds to an earlier sudden landscape change; in (d), the population response is lagging behind a gradual landscape change. The landscape feature may be, for example, area or percentage of habitat, or number or average size of habitat patches. The (meta)population feature may be, for example, population size, range, or patch occupation.

tion. The metapopulation equilibrium (n^*) is achieved when these turnover processes balance. This equilibrium is decreased by a lower colonization rate or a higher local extinction rate. Metapopulation dynamics may take place on a long time scale, simply because of the probabilistic nature of the basic processes. Local extinction and (re)colonization rates, expressed as probability per year, often are on the order of 10^{-2} to 10^{-1} (e.g., Verboom et al. 1991). For an extinction rate of $e = 0.01$, the mean time to local population extinction equals $1/e =$ 100 years. A subpopulation may persist for tens or even hundreds of years without alarming conservation managers. Similar reasoning applies to colonization. Because the dynamics of the whole metapopulation are driven by colonization and extinction, effects of landscape change can go unnoticed for long periods.

Time lags in metapopulation response may be encountered in the following cases: (1) following a relatively fast landscape change that decreases n^*, the metapopulation may linger near the old equilibrium for a while (Figure 18.1b); (2) following a landscape change that increases the metapopulation equilibrium, the metapopulation may expand slowly. Intuitively, the first case should be due to a low local extinction rate, and the second case to a low colonization rate. We show below, however, that what really matters is the size of the *difference* between colonization and extinction. A special case of (1) occurs when the landscape has deteriorated so much that the new equilibrium is metapopulation extinction. Cases related to (2) include the invasion of an alien species, and metapopulation recovery after a catastrophe or cessation of persecution.

18.2.2 Concepts Relevant to Time Lags in Metapopulation Response

Delays in the response of biodiversity to landscape degradation have been discussed by Tilman et al. (1994). For a community of species that hierarchically competed for the same habitat, they found an *extinction debt*—the loss of species following a delay after landscape degradation. Time delays, however, are a robust metapopulation phenomenon not dependent on the competitive interactions assumed by Tilman et al. (1994). The other side of the coin is a *colonization credit,* which is the slow reappearance of species after landscape restoration. The *return time* is the time the metapopulation needs to return to equilibrium after perturbation. It is used as a measure of the time lag.

Relaxation applies to communities of species in isolated areas like islands, mountain tops, and forest fragments. It is the progressive local loss of species caused by an excess of extinction over colonization, after a decline in area due to factors like sea level rise, climate change, or habitat destruction (reviewed by Diamond 1984).

Ghosts of the landscape past appear when the distribution of a species is better explained by a prior landscape configuration than it is by the current one. *Living dead* are (meta)populations that will certainly go extinct because they have become nonviable. An obvious example is the last individual of a Galápagos giant turtle subspecies (Caccone et al. 1999), but more subtle cases may escape our attention. Hanski and Kuussaari (1995) estimate that in Finland, 10 of 94 resident butterflies are represented by nonequilibrium metapopulations heading for extinction.

18.3 Recent Applications

18.3.1 The Lack of On-the-Ground Applications

We failed to find a single explicit field application of the time-lag concept in relation to metapopulation dynamics, although we consulted many colleagues and systematically searched the Web and all relevant literature databases we could think of. The extinction debt concept has been used to predict future species loss from current landscapes (e.g., Brooks et al. 1997; Cowlishaw 1999), but not in an explicit metapopulation context. This absence of applications did not come as a surprise, because the concepts, principles, and emerging ideas are all quite recent. For example, the general discussion about extinction debts just began in 1994. Moreover, no general principles for applying the concepts have been published, and the theory is still far from complete. The focus of both conservation managers and researchers tends to be on short-term processes rather than on long-term transient dynamics. Rather than discussing actual on-the-ground applications of the concepts of Section 18.2, we discuss how the concepts have been involved in research contexts. We present recent evidence for time lags in metapopulation responses to landscape change (Section 18.3.2) as well as theoretical models of these phenomena. The latter will address time lags in the Levins metapopulation model and mainland-island model (Section 18.3.3), time lags in a model assuming evolved dispersal (Section 18.3.4), the role of stochasticity in small networks (Section 18.3.5), time lags in spatially explicit models (Section 18.3.6), and time lags in mosaic landscapes with multispecies assemblages (Section 18.3.7).

18.3.2 Evidence for Time Lags in Metapopulation Responses to Landscape Change

There are several examples of metapopulations expanding with a time lag. The number of patches occupied by the butterfly *Hesperia comma,* which depends on heavily grazed habitat, expanded slowly in Britain after rabbit mortality by myxomatosis subsided (Thomas and Jones 1993). The butterfly *Proclossiana eunomia* slowly increased its occupation of a habitat network after reintroduction in central France (Neve et al. 1996). The range of the sea otter (*Enhydra lutis*) in North America is expanding steadily since the cessation of persecution (Estes 1990). After decimation due to extreme conditions in wintering habitat, marshland birds in The Netherlands recovered quickly in less-fragmented habitat, but recovery in more-fragmented habitat had not yet taken place after eight years (Foppen et al. 1999). There is a body of literature about sluggish expansion following introduction of alien species. For example, the zebra mussel *Dreissena polymorpha* is only slowly spreading into isolated inland waters in the United States, whereas it quickly expanded through interconnected waters (Johnson and Padilla 1996). The other side of the time-lag coin, metapopulations declining with a lag after landscape degradation, has been documented for carabid beetles (De Vries 1996; Petit and Burel 1998).

Whether declining fragmented populations still are metapopulations, that is, whether (re)colonization of patches still occurs, is often unclear. Declining species will often mimic a metapopulation without being one (Simberloff 1997). For example, on isolated Spanish dunes, Obeso and Aedo (1992) found only extinctions of plant species and no colonizations.

Metapopulation dynamics arise from colonization and extinction. The evidence for lags in these basic processes, which will translate into metapopulation delays, is much more abundant than is evidence for lags in metapopulations.

There is much evidence (reviewed by Diamond 1984 and Pimm 1991) of relaxation on islands or mountain tops. Relaxation in habitats fragmented by humans is exemplified by the extinction of mammals in Tanzanian and North American reserves (Newmark 1995, 1996), small mammals in Californian shrub fragments (Bolger et al. 1997), and many organisms in tropical forest fragments (Corlett and Turner 1997). Other possible evidence for a time lag in extinction is that Java and other long-settled islands have a remarkably poor lowland bird fauna (Van Balen 1999), whereas recently deforested islands nearby have lost few species as yet (Brooks et al. 1997). Low colonization rates also are often reported. There are many examples of plants colonizing newly available areas at very low rates (Peterken and Game 1984; Eriksson 1996; Grashof-Bokdam 1997). For many European plant species, modern human landscapes and practices seem to be especially detrimental to dispersal (Poschlod and Bonn 1998). Tropical inner-forest trees can be very slow in colonizing adjacent secondary forests, especially when their seed dispersers have disappeared (Corlett and Turner 1997). Saproxylic insects characteristic of old forests also colonize very slowly (Warren and Key 1991).

Examples of ghosts of the landscape past are given for carabid beetles by Petit and Burel (1998), and for forest plants by Van Ruremonde and Kalkhoven (1991) and Grashof-Bokdam and Geertsema (1998). Grashof-Bokdam and Geertsema (1998) showed that present occurrence of forest plant species was significantly related to the amount of currently occupied forest that was already present nearby about 150 years ago, but not to the contemporary amount of occupied adjacent forest.

18.3.3 Time Lags in Simple Metapopulation and Mainland-Island Models

In real landscapes, patches generally differ in size and connectivity. However, for creating general insights, it is useful to study simple models that ignore those differences. We consider the Levins metapopulation model (Levins 1969) and the related mainland-island model (MacArthur and Wilson 1967). In both models, all patches are equal, but the latter model differs from the former in that the colonization rate for an "island" is constant, whereas in the Levins model, patch colonization rate is proportional to the number of patches occupied. Real-life metapopulations may lie in between these two extreme cases, because colonization may occur both from within and outside of the metapopulation.

The Levins model has the form

$$\frac{dn(t)}{dt} = cn(t)(N - n(t)) - en(t), \tag{1}$$

where occupancy $n(t)$ is the density (or number) of occupied patches at time t, N is the total density (or number) of patches, c is the colonization rate per occupied patch and per empty patch, and e is the extinction rate per occupied patch. The equilibrium densities are

$$n^* = N - \frac{e}{c} \quad (cN > e) \tag{2a}$$

and

$$n^* = 0 \quad (cN \leq e). \tag{2b}$$

Hence, when $cN > e$, the metapopulation is viable.

We define T_{50} as the time it takes to reduce a deviation from an equilibrium value by 50%. For this measure of return time, we find that for a viable metapopulation (F. van den Bosch, unpublished results),

$$T_{50} = \frac{\ln 2}{cN - e} \qquad \left(n^* = N - \frac{e}{c}\right) \tag{3a}$$

and for a nonviable metapopulation,

$$T_{50} = \frac{\ln 2}{e - cN} \qquad (n^* = 0). \tag{3b}$$

Strictly, these equations are only true for small deviations, but they give a reasonable approximation. Clearly, when a metapopulation is viable, the return time increases with increasing patch extinction rate (e), decreasing colonization rate (c), and decreasing number of patches (N). For nonviable metapopulations, the dependence of return time on the parameters is reversed. Figure 18.2a shows that return times are especially large for parameter combinations in which the equilibrium is close to the viability boundary, $p = 0$, where p is the proportion of occupied patches ($= n^*/N$) at equilibrium (see also Figure 18.3).

When most colonizations of several smaller patches take place from a large "mainland," a more appropriate model is the mainland-island model,

$$\frac{dn(t)}{dt} = c'(N - n(t)) - en(t), \tag{4}$$

where c' is the number of colonizations per empty patch per time unit. The equilibrium for this model is $n^* = c'N/(c' + e)$, and the metapopulation is now always viable, because there will always be colonization from the mainland. The return time becomes

$$T_{50} = \frac{\ln 2}{c' + e}. \tag{5}$$

From Figure 18.2b, we see that return times are small when the colonization and extinction rates are large. Speed is determined by the sum of the colonization and extinction parameters. This contrasts with Levins metapopulations for which it is the difference between the parameters e and cN that determines return time. For comparable rates of colonization (per empty patch; hence, c' vs. cn) and extinction, the mainland-island model always reacts faster than does the Levins model.

Landscape degradation can occur by a decrease in the amount of habitat N, an increase in the local extinction rate e (e.g., an increase in disturbance rate), or an increase in landscape resistance to dispersal that decreases c. To compare the effects of different kinds of degradation, we investigate return times for impacts that result in equal changes in n^* and hence are of equal size. For the Levins case,

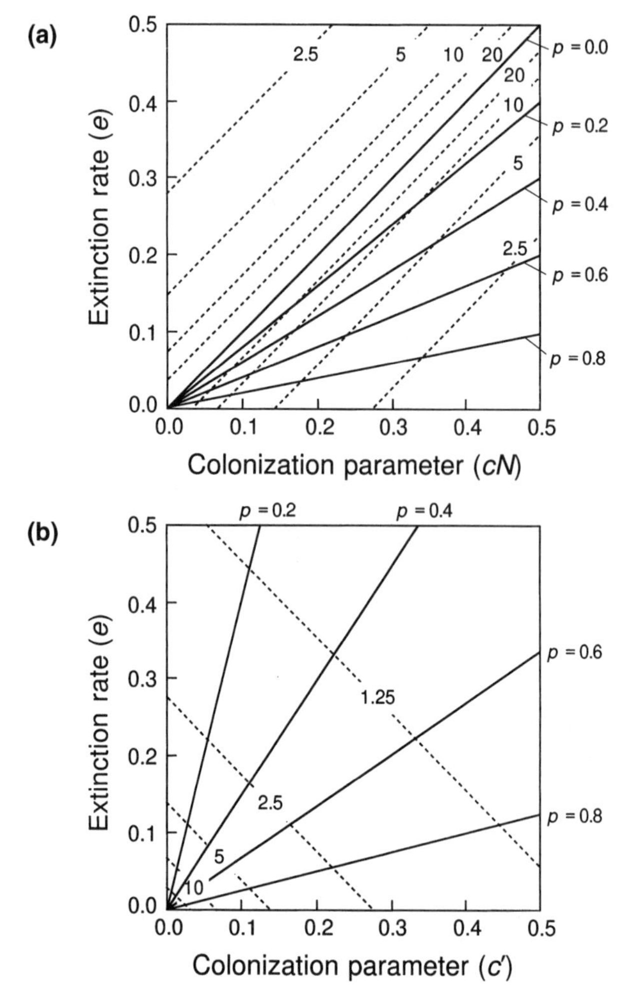

FIGURE 18.2. The return time in years (T_{50}, dotted lines) and the equilibrium fraction of occupied patches ($p = n^*/N$, solid lines) as functions of the colonization parameters (cN or c') (year^{-1}) and extinction rate (e) (year^{-1}) for the Levins model (a) and the mainland-island model (b).

habitat destruction and an increase in e do not differ in speed of reaction, but reaction to an increase in dispersal resistance is much slower (Figure 18.3). Note, however, that increases in e have, compared with decreases in N or in c, no obvious natural maximum; hence, reaction to increased disturbance can become very fast. In all cases, return times are highest near the viability border, as predicted by Eqs. 3a and 3b. In the mainland-island model, an increase in dispersal resistance also produces a slower reaction than does a change in e, but changes in N never produce out-of-equilibrium situations.

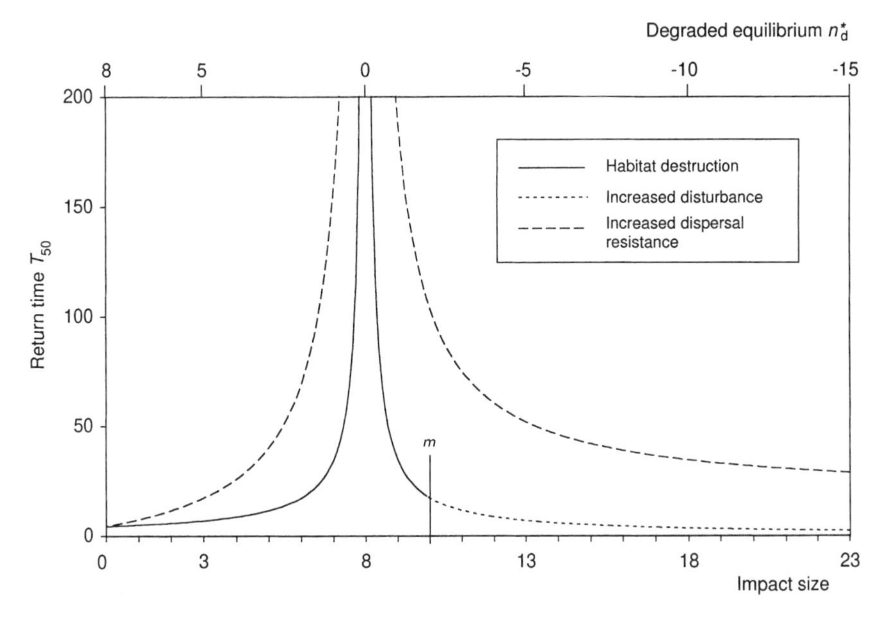

FIGURE 18.3. Comparison of the influences of three types of landscape degradation on the return time of a metapopulation according to the Levins model. The size of the impact is defined as the reduction in the equilibrium number of occupied patches, n^* (i.e., $n^*_p - n^*_d$). Negative values of n^*_d are virtual results obtained by using Eq. 2a for nonviable metapopulations, necessary to assess the size of impacts resulting in extinction. The parameter values for the pristine landscape are $N = 10$, $e = 0.04$, and $c = 0.02$, resulting in $n^*_p = 8$ (Eq. 3a). Impact size for habitat destruction is bounded (vertical line m). Left of m, the curves for habitat destruction and increased disturbance coincide.

18.3.4 Time Lags Assuming Evolved Dispersal

From Eqs. 3a and 3b, we can predict return time from the parameters and, for example, investigate the influence of differences in e on delay length, while keeping c constant. However, we often do not know both e and c. Fortunately, e and c may not be independent. Data show that, generally, species from unstable habitats are also good dispersers (Hanski 1999). Supporting this, evolutionary theory predicts that dispersal will be molded by natural selection in response to the local extinction rate (Comins et al. 1980; Ferriere et al. In press). To incorporate this, we use the simple evolutionary relationship that the optimal rate of dispersal d^* is equal to the extinction rate e (Van Valen 1971). Further, we translate dispersal rate d into c with a scaling factor α; hence, $c = \alpha e$. The value of α will depend on such factors as fecundity and dispersal ability. An assumption is that evolution is slow compared with landscape change, and that therefore the dispersal rate that evolved in the pristine landscape does not change during landscape deterioration. Hence, to be clear, we do *not* investigate evolutionary reactions to degradation, but study the consequences of earlier parameter evolution.

We can now investigate how this evolutionary dependency between e and c influences the reaction to landscape change. Return time and equilibrium for a viable metapopulation are now given by

$$T_{50} = \frac{\ln 2}{e(\alpha N - 1)}, \qquad n* = N - \frac{1}{\alpha}, \tag{6a}$$

and for a doomed metapopulation by

$$T_{50} = \frac{\ln 2}{e(1 - \alpha N)}, \qquad n* = 0. \tag{6b}$$

We put e equal to the habitat disturbance rate and compare species from habitats that differ in stability. Other causes of local extinction, for example small population size, have the same effect as high habitat disturbance.

First, we investigate habitat destruction (decreasing N). Table 18.1 and Eq. 3b show that when disregarding evolved dispersal, we would conclude that after a fatal decline in the amount of habitat, species from habitats with a high disturbance rate go extinct faster (shorter return time) than do species of more stable habitats. However, when the end point is not extinction but a new positive equilibrium (Eq. 3a), the return time increases with local extinction rate. Including evolved dispersal changes our predictions markedly. As long as the amount of habitat does not differ between species that differ in e, return times will always *decrease* with increasing e (Eqs. 6a and 6b). Species from habitats with a high (low) disturbance rate will react rapidly (slowly) to changes in the amount of habitat, irrespective of the end point (viable or nonviable metapopulation). These outcomes are intuitively more appealing than are the results using independent rates. For example, we expect weeds to react speedily to increased opportunities. Clearly, there are "slow" species and "fast" species, and evolution ensures that reaction speed correlates positively with the disturbance rate of the habitat. Slow species live in relatively stable habitats and react with a long lag to changes in the amount of habitat, whereas fast species live in often-disturbed habitat and react with a short lag.

We now consider effects of increasing the disturbance rate e above the pristine rate. The model without evolved dispersal predicts that after a given increase in disturbance rate, a species facing extinction (Eq. 3b) fades out faster (has a shorter return time) when the pristine disturbance rate was high compared with when it was low. Incorporating evolved dispersal again brings new insights. The velocity with which doomed species approach extinction now *decreases* with increasing pristine disturbance rate (Table 18.1), contrary to the effect of habitat destruction. Note that a fatal increase in disturbance of a metapopulation results in a denominator of Eq. 6b given by $e_{\mathrm{p}} (1 - \alpha N) + \Delta_e$, where Δ_e is the increase in the disturbance. Because the metapopulation was viable in the pristine state, $(1 - \alpha N)$ is negative and the return time increases with pristine disturbance rate e_{p}. This implies that species from originally relatively stable habitats (low pristine e) go extinct sooner. The reason is that their evolved low colonization rate is easily

TABLE 18.1. The influence of evolved dispersal on predictions about reaction speeds of metapopulations under the Levins model, as influenced by the pristine disturbance frequency e_p (assumed to be equal to the local extinction rate). The dispersal rate has evolved in the pristine landscape and does not change after landscape degradation. The effects of several landscape degradation types are shown. n^*: equilibrium occupancy, c: colonization rate, α: evolutionary relationship between c and e_p; subscripts, p: pristine state, d: degraded state; + and – : effect of larger e_p on reaction speed; a "+" means that species with high e_p have a faster reaction (shorter time lag) compared with species with low e_p; a "–" means that species with high e_p have a slower reaction (longer time lag) compared with species with low e_p. Results (C. J. Nagelkerke, unpublished) are given for both viable ($n^* > 0$) and nonviable ($n^* = 0$) metapopulations.

Degradation type		Without evolved dispersal (c independent of e)	With evolved dispersal ($c_p = \alpha e_p$)
Pristine landscape	$(n^* > 0)$	–	+
	$(n^* = 0)$	+	+
Habitat destruction	$(n^*_d > 0)$	–	+
	$(n^*_d = 0)$	+	+
Increased disturbance	$(n^*_d > 0)$	–	+
	$(n^*_d = 0)$	+	–
Increased dispersal resistance	$(n^*_d > 0)$	–	+
	$(n^*_d = 0)$	+	+

swamped by an increase in local extinction rate. The added disturbance causes the originally slow species to become the faster ones.

For increased resistance to dispersal, the results are comparable to those of habitat destruction.

18.3.5 Time Lags in Simple Stochastic Models

The Levins model is deterministic, but in reality metapopulations will show stochastic fluctuations in occupancy n, which means that small metapopulations can go extinct by chance, hence after a variable delay, even when their deterministic n^* is positive. Gurney and Nisbet (1978) gave the following approximation for the expected time to metapopulation extinction (T_M) in small networks:

$$T_M = \frac{1}{e}\exp\left[\frac{n*^2}{2(N - n*)}\right]. \tag{7}$$

Realized time to extinction can, however, be extremely variable (Lehman and Tilman 1997). Eq. 7 means that for a given n^*, the delay to stochastic extinction T_M will be shorter when N is larger and hence p is smaller and e and cN nearly balance. This is because fluctuations are more violent when only a small proportion of a network is occupied. See Hanski et al. (1996) and Hanski (1999) for a further discussion of stochastic extinction in small networks. One point raised by these authors is that T_M will be shorter when there is spatially correlated environmental stochasticity.

18.3.6 Time Lags in Spatially Explicit Models

In spatially explicit models, patches can differ in size and connectivity. The equilibrium occupancy, n^*, now depends on the exact configuration of patches and the dispersal structure of the metapopulation (Hanski 1999). Time lags may become longer as the spatial structure becomes more heterogeneous. We discuss two case studies.

Hanski's incidence function model (Hanski 1994, 1997, 1998) includes both patch size and configuration. Hanski and coworkers (Hanski et al. 1996; Hanski 1998) found that after habitat destruction, the lag in this model was much longer when the end point was metapopulation extinction than it was when the end point was a lower positive equilibrium. Hanski et al. (1996) suggested that this occurred because when extinction was the end point, local extinction occurs last in the largest patches with the lowest extinction rates, and their demise may take a long (and very unpredictable!) time. Metapopulation shrinkage to a lower equilibrium is caused mainly by local extinction in small patches with a high e, and should go faster. It is clear that in habitat networks with unequal patch size, mean extinction and colonization rates may change during the shrinkage process, and such rate changes may have important consequences, including a long delay to final metapopulation extinction.

In the second case study, Van de Wolfshaar (1999) analyzed landscapes for time lags with a spatially explicit deterministic model. Landscape structure and interpatch dispersal rate estimates were taken to apply to the butterfly *Melitaea cinxia* in Finland or to the European badger (*Meles meles*) in The Netherlands. Eliminating connectivity variance decreased the time lags by a factor of 2–100. No variance corresponds to the assumption used in Sections 18.3.3 and 18.3.4 that all patches are equally accessible from all other patches. Hence, simple models may underestimate the time lag.

See Hanski et al. (1996) and Hanski (1998, 1999) for a discussion of the expected time to stochastic extinction in small, spatially realistic networks.

18.3.7 Time Lags in Mosaic Landscapes With Multispecies Assemblages

When we understand the reactions of the individual species, it is possible to predict the reaction of a whole assembly of species when a landscape degrades or is allowed to recover. As an example, we present simulations using the models with evolved dispersal from Section 18.3.4 and including some immigration from outside of the landscape. Landscapes can be seen as a mosaic of different habitats that vary in disturbance rate (Figure 18.4). Each species faces the patch extinction rate specific for the habitat it uses.

We first investigate the effect of changes in the relative amounts of the different habitats. A pristine landscape will be dominated by relatively stable habitats, whereas a degraded landscape will have more habitat that is often disturbed (Figure 18.4). After degradation, species from more stable habitats will decline or go

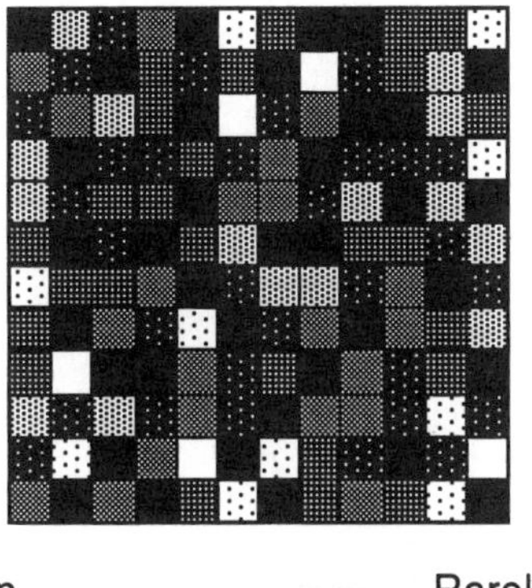

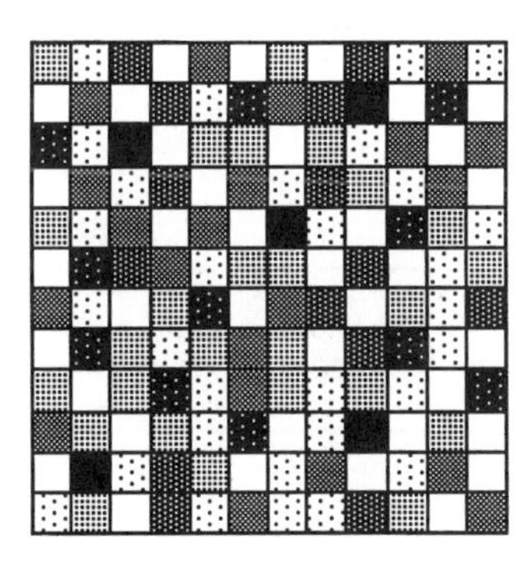

FIGURE 18.4. The landscape as a mosaic of habitats that vary in disturbance rate. (a) Pristine landscape and (b) degraded landscape.

extinct because their habitat diminishes, whereas species from often-disturbed habitats profit. However, the increase of the fast, disturbance-dependent species will occur at a higher pace than the decline of the slow species from the more stable habitats. This results in a temporary increase in diversity (Figure 18.5a), by immigration, before the number of species settles at its new lower equilibrium. Because many species from more stable habitats will disappear only after a delay,

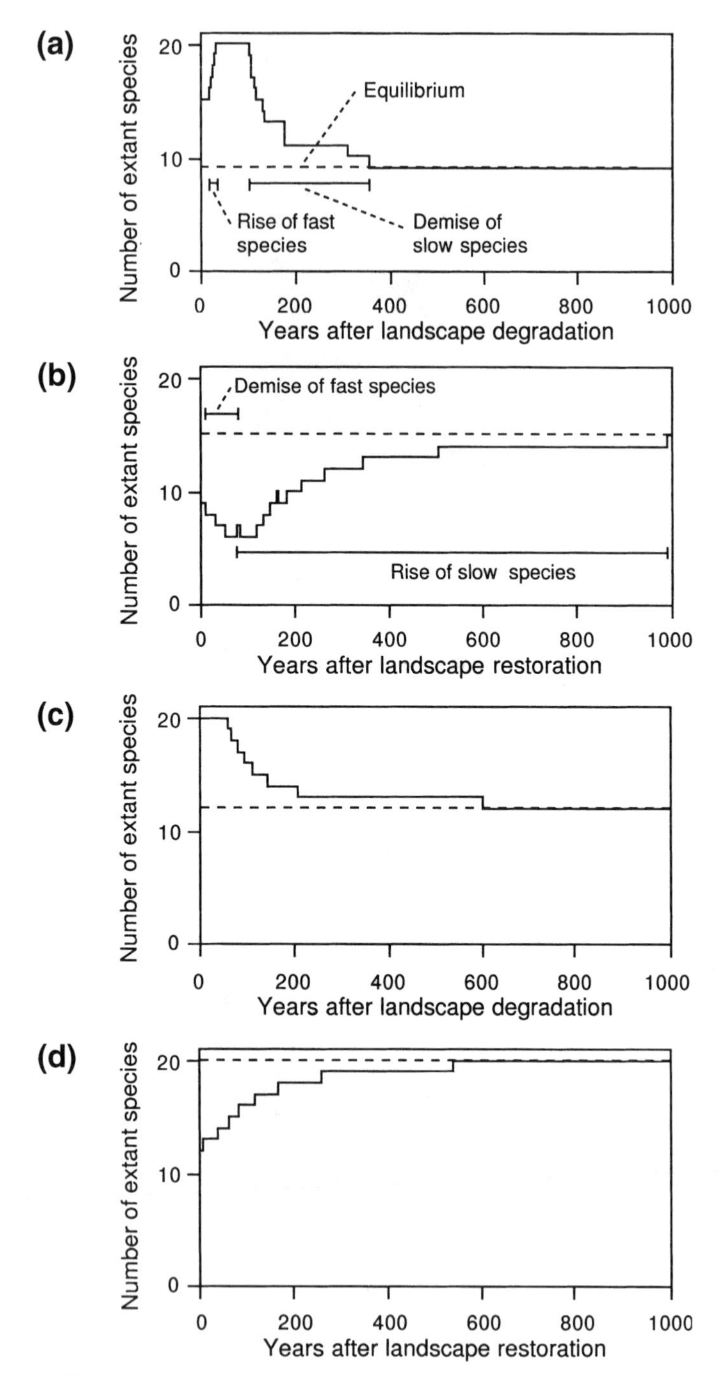

FIGURE 18.5. Examples of simulation results for total number of species over time in a mosaic landscape after different types of landscape degradation and restoration. There are 20 habitats with pristine disturbance rates ranging from 0.005 to 0.1 per year. Every species has a metapopulation in its habitat network. There is only one species for every

there is an extinction debt. In contrast, after restoration, a colonization credit appears (Figure 18.5b). Biodiversity then decreases temporarily before rebounding to its pristine level—the fast species quickly react to the decline in their habitat, whereas the slow species only return after a delay. Clearly, there is a strong asymmetry between degradation and recovery; that is, recovery is not just degradation in reverse.

A general increase in the disturbance rate causes declines in occupancy for all species. Species from the originally most stable habitats suffer the most, and contrary to the effect of habitat destruction, they also are the ones that decline the fastest. There are now no temporary increases or decreases in the number of species (Figures 18.5c and 18.5d), and there is little asymmetry between degradation and recovery. Whereas (formerly) slow species are now the first to disappear, they also are the last to reappear when the disturbance rate returns to the pristine value.

18.4 Principles for Applying Landscape Ecology

The most important message is that time lags matter. Many metapopulations may be living dead, and protection of the current landscape may not guarantee long-term survival. In Sections 18.4.1–18.4.3, we provide general, tentative application principles based on the simple mathematical models presented in Section 18.3 and on empirical findings.

18.4.1 When to Expect Time Lags and Why They Are Important

Delays can occur both during metapopulation decline and expansion. In response to degradation, a slow metapopulation decline, or even a temporary increase in biodiversity, may lead to unjustified optimism when monitoring short-term change. Such delays in decline may mask the fate of a large part of biodiversity. On the positive side, delays can provide a buffer. They may enable species to persist during a period of pollution or habitat destruction until better times return (Kellman 1996). For example, longevity may have helped the Golden Eagle (*Aquila chrysaetos*) in western Scotland to maintain its breeding density through the 1960s despite reproductive failure due to organochlorine pollution (Newton 1998). When properly identified, a time lag also gives us time for action before it is too late. Slow recovery after restoration or after a disaster may lead to unjustified pessimism, especially when a temporary biodiversity decline occurs. Delays in recovery also increase the possibility of stochastic

habitat, and species do not interact. Landscape changes are applied suddenly, and some immigration from outside is allowed. (a) Changing the relative amounts of the different habitats toward habitats with a high disturbance rate (e.g., from Figure 18.4a to Figure 18.4b). (b) Restoring the original habitat distribution. (c) Applying a general increase (of 0.06 per year) in disturbance rates. (d) Restoring the original disturbance rates. After C.J. Nagelkerke (unpublished manuscript).

metapopulation extinction. Furthermore, slow expansion of a nonindigenous species can lead to underestimation of its potential impact but also provides time for countermeasures.

Time lags other than those caused by pure metapopulation dynamics may also be important. Species may change over time due to evolution or behavioral adjustment, allowing them to cope with the environmental change. Also, communities may take time to reorganize (Kellman et al. 1996), after which they may be more resistant to stresses of the new situation and be able to maintain many species, albeit in new structures and combinations (Kellman 1996). For example, the deterioration of new tropical forest fragments may not be representative of the longer-term condition (Brokaw 1998).

18.4.2 Time Lags Depend on Habitat, Type of Degradation, Landscape, Species, and Ecological Setting

Being able to estimate time-lag length helps to interpret the present, to predict the future, and to devise appropriate actions. When lags are long, it may be difficult, or at least take a long time, to observe the effects of harmful environmental changes or of conservation measures taken to improve persistence. Establishing causality will often be nearly impossible. Remedial action is less urgent, but slow declines do not necessarily indicate that drastic measures will not ultimately be needed. When lags are short, conservation measures have extra urgency. Below, we discuss factors influencing time-lag length.

1. Pristine habitat stability and type of degradation determine which species are fast or slow.

In Section 18.3, we showed that pristine local extinction rate e (and hence disturbance frequency) and type of degradation are important determinants of reaction speed. Species from stable habitat such as old-growth forests will react slower to changes in the amount of habitat or in dispersal resistance than will species from often-disturbed habitat, such as river banks. One consequence is that after habitat destruction, we expect in the more stable habitats a higher prevalence of ghosts of the landscape past. When the disturbance rate is increased, however, species from originally stable habitats disappear fast, but recover slowly after eventual restoration. In this case, there can then be large differences in time scales between decline and recovery.

2. A balance of colonization and extinction parameters curbs speed.

In the Levins model, speed is set by the difference between the colonization and extinction parameters cN and e. When parameters are nearly equal, and hence metapopulations are on the brink of viability, metapopulation dynamics may give rise to long lags between landscape change and biotic response. In small networks, stochastic fluctuations may then become important and cause fast extinction.

3. Reaction speed differs between types of degradation.

Reaction to impacts of similar size (see Section 18.3.3) will be equally fast for habitat destruction and increased disturbance, but slower for increases in dispersal resistance. Reaction speed to increases in disturbance can be very fast however.

4. Landscape configuration is important.

Lags will be long, assuming disturbance remains the same, if patches are large and isolated (Hanski 1997) because then *e* and *c* will both be low. The effects of heterogeneity between patches in size or connectivity still need to be elucidated fully (see Section 18.3.5).

Principles 1–4 are based on the classic Levins model. When there is a mainland-island configuration, speed depends on the sum of the parameters, and reactions are faster, possibly much faster. Species from stable habitats are always slower in a mainland-island situation. Habitat destruction (of the islands) produces no time lags, but reaction to increased dispersal resistance is slower than is reaction to increased disturbance, as in the Levins model. Degradation of the mainland source has the same effect as increased dispersal resistance.

5. Lag times will differ among species, taxonomic groups, and ecological settings.

Habitat stability is not the only factor in local extinction. Small local populations may also die out from, for example, demographic stochasticity. Hence, attributes like density, longevity, and other life-history characteristics are important (reviewed by Pimm 1991). Therefore, lag times will differ among species that share the same habitat. Fragments of temperate old forests seem to lose their characteristic carabid beetles much faster (<100 years; Gruttke 1997) than they lose their plants (hundreds of years; Peterken and Game 1984; Dzwonko and Loster 1989). In temperate areas, butterflies decline much faster than do their food plants (Erhardt and Thomas 1991; Erhardt 1995). In tropical forest remnants, plants also respond slower than do animals (Corlett and Turner 1997). This apparently general, slower reaction by plants may be due to their long life time or seed banks (Eriksson 1996), small individual area requirements, or (in the case of expansion) low rate of colonization (Peterken and Game 1984; Eriksson 1996; Grashof-Bokdam 1997). Among plants in tropical forest fragments, trees and ferns have a lower rate of loss than do orchids (Corlett and Turner 1997). In all of these cases, however, it is not known whether the differences concern only the speed of decline or also the end point. More specific are Brooks et al. (1999), who claim for birds in tropical forest fragments a return time, T_{50}, of approximately 50 years. The survival of much tropical tree diversity during repeated glacial forest contractions points to persistence time scales longer than 10,000–20,000 years (Kellman et al. 1996).

Lags will be most prominent when both (1) species have low *e* and *c,* and hence are slow, and (2) dispersal resistance is increased. In that case, short-term persistence will be the most misleading. Forest plants provide an example—they have low *e* and *c,* and their dispersal is easily impaired. Forest plants indeed seem to be especially slow. This also seems true for saproxylic insects from old forests, as 18th-century parks where the habitat seems good still have an impoverished saproxylic fauna (Warren and Key 1991).

In general, low local extinction rates occur in species with large or stable local population size, due to small individual area requirements, long individual life span, low sensitivity to environmental stochasticity, low patch disturbance rate, large patch size, or large within-patch heterogeneity. Low colonization rate occurs in species that have low dispersal rate (high site fidelity), short dispersal distances, inefficient dispersal behavior, or small establishment chance, combined with large interpatch distance, large between-patch resistance, or both.

18.4.3 Manage Time Lags in Preferred Directions

Perhaps the only universal principle for preserving species in the face of time lags is that we should decrease local extinction and enhance colonization. When temporary, such measures slow the speed of decline or speed recovery. In the first case, they increase the time window for more fundamental measures (or for an unmanaged return of better times). Speeding recovery decreases the possibility of stochastic metapopulation extinction and other deleterious effects of small population size. The fastest way to accelerate recovery may be by boosting colonization. When more permanent, such measures also increase the equilibrium, but their effect on return time can then be counterintuitive. For example, in declining but viable metapopulations, countermeasures *reduce* the time needed to reach an equilibrium. Local extinction risk can be decreased by (1) increasing patch carrying capacity by increasing patch size, or managing habitat quality; (2) increasing population growth rate by managing habitat quality or decreasing human-induced mortality such as traffic mortality, hunting, and poaching; and (3) interfering in cases of extreme downward fluctuations in numbers, by feeding when food is scarce, vaccination, predator control, and special conservation measures such as temporarily closing roads (Goodman 1987). Measures for promoting colonization include translocating individuals, sowing, reintroduction programs, creating corridors, and situating habitat restoration projects adjacent to occupied existing habitat.

When communities need reorganization, for example the formation of protective edges around fragments, we can facilitate this, especially when degradation has proceeded too rapidly for spontaneous restructuring (Kellman et al. 1998).

However, we must be vigilant for possible unwanted side effects of management measures, such as pollution of the gene pool; introduction of predators, pests, diseases, or parasites; and effects on other species. Furthermore, we must be prepared for surprises as stochastic events take place, species adapt to their new environment, or communities reorganize.

18.5 Knowledge Gaps

Here, we identify the major gaps preventing application of the concept of time lags in metapopulation responses to landscape change. Theoretical knowledge is still far from complete (cf., Simberloff 1992), and empirical knowledge is almost nonexistent. Prediction is made difficult by (1) lack of knowledge about basic metapopulation processes and the evolutionary plasticity of species, and (2) the nascent state of a multifaceted theory that needs to incorporate models that are spatially more realistic. The lack of empirical knowledge about time lags makes it difficult to test theory.

18.5.1 Theoretical Voids

Models can be an important help for interpreting empirical results, searching for generalities, and (especially important for time-lag research) extrapolating over time. There is a need for a multifaceted theory. We need simple, general rules for recognizing situations in which time lags are expected and for guessing their order of magnitude, and we need complex models that yield specific estimates of the expected time length of the lags. Most work has been done on simple, deterministic, spatially implicit models, and the most important void concerns spatially explicit, realistic, stochastic models, because these potentially can produce markedly different predictions. Major factors influencing time lags that should be investigated are heterogeneity in patch size, quality, and connectivity; stochastic fluctuations in small networks; environmental correlation among patches; influence of dispersal on local population dynamics; source-sink dynamics; and life-history characteristics such as individual life span. Further voids include more comprehensive evolutionary models that take, for example, habitat amount and dispersal distance into account; and how landscape dynamics, such as ecological succession, and evolutionary dynamics add their time lags to those associated with metapopulation dynamics.

18.5.2 Empirical Voids

There is a need for records of time lags classified according to species characteristics (e.g., longevity, dispersal characteristics) and landscape characteristics (e.g., pace, type, and severity of degradation). Accurate records of changes over time in the landscape as well as in species distributions are needed. Information about long lags is particularly important, as is the untangling of natural and anthropogenic causes of nonequilibrium situations.

Other empirical voids concern the basic processes of local extinction and, especially, colonization (and hence dispersal; cf., Opdam 1990). Knowledge about the rates of the basic processes is essential to predict time lags, and deterioration of these rates can be an important component of environmental degradation. How do species-specific characteristics (e.g., longevity, social structure), landscape characteristics (e.g., reserve size, heterogeneity, habitat dynamics), and interspecific interactions (e.g., competition, predation) influence local extinction rate? Is there still a functioning metapopulation with local extinction accompanied by recolonization, or will absence of colonization result in regional extinction? Especially relevant questions concerning dispersal are (1) what is the occurrence of rare long-distance dispersal events, often crucial for colonization (Cain et al. 1998)? (2) what is the role of interactions such as zoochorous seed dispersal? and (3) to what extent is dispersal compromised in modern landscapes?

The last void concerns evolution. How and how fast do characteristics relevant to metapopulation dynamics adapt to the landscape? For example, species of traditionally stable habitats such as late-successional stages are thought to have evolved into poor colonizers, but this needs more testing. Possible tradeoffs between different aspects of colonization ability and persistence ability also deserve significant study. Many species seem to have adapted to traditional cultural landscapes (e.g., Thomas and Morris 1995). But how widespread and important is this? Are generation time and genetic heterogeneity crucial factors? Can evolutionary changes also compromise metapopulation survival ("evolutionary suicide"; e.g., Leimar and Norberg 1997)?

18.6 Research Approaches

First, a general point applicable to both theoretical and empirical research should be made. Species-specific characteristics *and* landscape features together determine time lags. Therefore, the concept of ecologically scaled landscape indices developed by Vos et al. (2001) can be of particular use. This approach involves looking at landscapes "from a species' perspective" and reducing parameter space from landscape statistics and life-history characteristics to scaled indices such as "local carrying capacity" and "connectivity," which capture the essentials of both species and landscape.

18.6.1 Approaches for Theoretical Research

The effects of the major factors mentioned in Section 18.5.1 should be investigated using a series of models, ranging from simple, strategic models (May 1973) to spatially explicit, stochastic or individual-based models. See Durrett and Levin (1994), Hanski and Gilpin (1997), Hanski (1999), and Holyoak and Ray (1999) for guidance on general spatial models. A promising middle ground between cumbersome and untransparent spatially explicit or individual-based models on the one hand, and unrealistic simple models on the other hand, is the "mesoscale" approach advocated by Casagrandi and Gatto (1999), which uses statistical distributions of abundance per patch. For prediction of the fate of specific metapopulations, population viability analysis (PVA) software such as ALEX, VORTEX, and RAMAS/metapop is available. For references and a comparison of five PVA software packages, see Brook et al. (1997). Such packages enable the user to add demographic and environmental stochasticity, inbreeding (in some), and other details. The more realistic models can be used for impact assessment, and the simpler ones, such as the Levins model, can be used to gain a better general understanding. Simple models provide no concrete answers for specific problems, but detailed PVA software provides no general understanding, results are often very sensitive to the parameter values chosen, and there is a severe risk of introducing errors. For basic modeling strategy and guidance for how to deal with these pitfalls of modeling, we refer researchers to Burgman et al. (1993). See Ferriere et al. (In press) for approaches to modeling evolutionary dynamics, and see Remmert (1991) and Stelter et al. (1997) for methods to model landscape dynamics.

18.6.2 Approaches for Empirical Research

Specific protocols are not easy to supply because as yet there is no experience with empirical research on time lags, and many empirical voids require application of general ecological methods. However, some guidance can be given.

To obtain records of time lags, we can learn a lot from natural and human-induced catastrophes such as hurricanes and oil spills, and from gradual landscape changes involving habitat loss, habitat fragmentation, and habitat-quality degradation; restoration projects also provide opportunities. We must monitor both landscape changes and effects on biodiversity closely. To have controls in both space and time, the best research protocol is BACI (Before, After, Control, Impact) (Green 1979). This will usually be possible only in cases of planned impacts. We should especially pay attention to differences among taxa, habitats, landscapes, ecosystems, and impacts. The populations of a selected group of indicator species should be monitored on the basis of habitat patches that cover, if possible, a complete habitat network or at least a comprehensive part of the landscape (Opdam et al. 1993). Gaining insight about the turnover processes is an important aim. Monitoring for numbers or densities is preferable, but determining presence and absence also can provide insight (cf., Kareiva et al. 1997). Until now, research has concentrated on species with "fast" metapopulations, like birds and butterflies, and on habitat destruction. Researchers need to pay more attention to "slow" species and to dispersal impairment, in which metapopulation decline may be less apparent but in the long term more severe.

Just as the impact of ongoing change of landscapes and land use may be visible only in the future, the patterns in species distributions witnessed today may reflect landscapes and land use in the past. We can study the latter by "ghost hunting" (searching for ghosts of the landscape past) with historical maps and records to look for bygone landscape features that correlate with current biodiversity patterns (e.g., Grashof-Bokdam and Geertsema 1998). This also is an approach to obtain information about long time lags. Other ways to study long delays are trying to find comparable systems in which the history of the focal impact is much longer, or searching for natural analogs of impacted systems (e.g., Kellman et al. 1996). Such study of long-term historical effects can help to predict the robustness of present systems. We also stress the importance of developing long-term monitoring programs. The difference in time scale between time lags and the opportunities to study them is a major constraint.

Dispersal is difficult to investigate because for many species, effective dispersal from one patch to another is rare. As a consequence, deterioration of dispersal is much more difficult to measure than is habitat destruction or increased disturbance. Techniques and specific research protocols for studying dispersal can be found in Turchin (1998) and the references therein.

When colonization and extinction processes cannot be observed directly, species distribution data and their turnover in time can be analyzed using logistic regression; an example of this approach is the application of the "incidence function model" (Verboom et al. 1991; Hanski 1994, 1997, 1999). Such methods, however, imply deriving processes from patterns, and a major drawback is that by disregarding time lags, one may get seriously overoptimistic predictions for non-equilibrium (declining) metapopulations (Ter Braak et al. 1998).

See Dieckmann et al. (1999) and Ferriere et al. (In press) for the emerging field of research in evolutionary conservation. Research should concentrate on how habitat use (e.g., Van Balen 1999) and dispersal-related traits (e.g., Hill et al. 1999) evolve in response to landscape change.

Acknowledgments

Paul Opdam reviewed the scientific content of this chapter and gave suggestions for improving chapter structure. Kevin Gutzwiller gave many suggestions for improving the manuscript. Jan Bruin made improvements to the writing style. Kees Nagelkerke was funded by the Priority Programme "Biodiversity in Disturbed Ecosystems" of The Netherlands Organization for Scientific Research.

References

Bolger, D.T., Alberts, A.C., Sauvajot, R.M., Potenza, P., McCalvin, C., Tran, D., Mazzoni, S., and Soulé, M.E. 1997. Response of rodents to habitat fragmentation in coastal southern California. *Ecol. Appl.* 7:552–563.

Brokaw, N. 1998. Fragments past, present and future. *Trends Ecol. Evol.* 13:382–383.

Brook, B.W., Lim, L. Harden, R., and Frankham, R. 1997. Does population viability analysis software predict the behaviour of real populations? A retrospective study on the Lord Howe Island Woodhen *Tricholimnas sylvestris* (Sclater). *Biol. Conserv.* 82:119–128.

Brooks, T.M., Pimm, S.L., and Collar, N.J. 1997. Deforestation predicts the number of threatened birds in insular southeast Asia. *Conserv. Biol.* 11:382–394.

Brooks, T.M., Pimm, S.L., and Oyugi, J.O. 1999. Time lag between deforestation and bird extinction in tropical forest fragments. *Conserv. Biol.* 13:1140–1150.

Budiansky, S. 1994. Extinction or miscalculation? *Nature* 370:105.

Burgman, M.A., Ferson, S., and Akçakaya, H.R. 1993. *Risk Assessment in Conservation Biology.* New York: Chapman and Hall.

Caccone, A., Gibbs, J.P., Ketmaier, V., Suatoni, E., and Powell, J.R. 1999. Origin and evolutionary relationships of giant Galápagos tortoises. *Proc. Natl. Acad. Sci. USA* 96:13223–13228.

Cain, M.L., Damman, H. and Muir, A. 1998. Seed dispersal and the Holocene migration of woodland herbs. *Ecol. Monogr.* 68:325–347.

Casagrandi, R., and Gatto, M. 1999. A mesoscale approach to extinction risk in fragmented habitats. *Nature* 400:560–562.

Comins, H.N., Hamilton, W.D., and May, R.M. 1980. Evolutionarily stable dispersal strategies. *J. Theor. Biol.* 82:205–230.

Corlett, R.T., and Turner, I.M. 1997. Long-term survival in tropical forest remnants in Singapore and Hong Kong. In *Tropical Forest Remnants; Ecology, Management and Conservation of Fragmented Communities,* eds. W.F. Laurance and R.O. Bierregaard, pp. 333–345. Chicago: University of Chicago Press.

Cowlishaw, G. 1999. Predicting the pattern of decline of African primate diversity: an extinction debt from historical deforestation. *Conserv. Biol.* 13:1183–1193.

De Vries, H. 1996. *Viability of Ground Beetle Populations in Fragmented Heathlands.* Ph.D. dissertation. Wageningen, The Netherlands: Agricultural University.

Diamond, J.M. 1984. "Normal" extinctions of isolated populations. In *Extinctions,* ed. M.H. Nitecki, pp. 191–246. Chicago: University of Chicago Press.

Dieckmann, U., O'Hara, B., and Weisser, W. 1999. The evolutionary ecology of dispersal. *Trends Ecol. Evol.* 14:88–90.

Durrett, R., and Levin, S.A. 1994. Stochastic spatial models: a user's guide to ecological applications. *Phil. Trans. R. Soc. Lond. B Biol. Sci.* 343:329–350.

Dzwonko, Z., and Loster, S. 1989. Distribution of vascular plant species in small woodlands on the western Carpathian foothills. *Oikos* 56:77–86.

Erhardt, A. 1995. Ecology and conservation of alpine Lepidoptera. In *Ecology and Conservation of Butterflies,* ed. A.S. Pullin, pp. 258–276. London: Chapman and Hall.

Erhardt, A., and Thomas, J.A. 1991. Lepidoptera as indicators of change in the seminatural grasslands of lowland and upland Europe. In *The Conservation of Insects and their Habitats,* eds. N.M. Collins and J.A. Thomas, pp. 213–236. London: Academic Press.

Eriksson, O. 1996. Regional dynamics of plants—a review of evidence for remnant, source-sink and metapopulations. *Oikos* 77:248–258.

Estes, J.A. 1990. Growth and equilibrium in sea otter populations. *J. Anim. Ecol.* 59:385–401.

Ferriere, R., Dieckmann, U., and Couvet, D., eds. In press. *Evolutionary Conservation Biology.* Berlin: Springer-Verlag.

Foppen, R., Ter Braak, C.J.F., Verboom, J., and Reijnen, R. 1999. Dutch Sedge Warblers *Acrocephalus schoenobaenus* and West-African rainfall: empirical data and simulation modelling show low population resilience in fragmented marshlands. *Ardea* 87:113–127.

Goodman, D. 1987. Consideration of stochastic demography in the design and management of biological reserves. *Nat. Resour. Model.* 1:205–234.

Grashof-Bokdam, C.J. 1997. Forest species in an agricultural landscape in The Netherlands: effects of habitat fragmentation. *J. Veget. Sci.* 8:21–28.

Grashof-Bokdam, C.J., and Geertsema, W. 1998. The effect of isolation and history on colonization patterns of plant species in secondary woodland. *J. Biogeog.* 25:837–846.

Green, R.H. 1979. *Sampling Design and Statistical Methods for Environmental Biologists.* New York: John Wiley and Sons.

Gruttke, H. 1997. Impact of landscape changes on the ground beetle fauna (Carabidae) of an agricultural landscape. In *Habitat Fragmentation and Infrastructure,* ed. K. Canters, pp. 149–159. Delft, The Netherlands: Ministry of Transport, Public Works and Water Management.

Gurney, W.S.C., and Nisbet, R.M. 1978. Single-species population fluctuations in patchy environments. *Am. Nat.* 112:1075–1090.

Hanski, I. 1994. A practical model of metapopulation dynamics. *J. Anim. Ecol.* 63:151–162.

Hanski, I. 1997. Metapopulation dynamics: from concepts and observations to predictive models. In *Metapopulation Biology: Ecology, Genetics, and Evolution,* eds. I. Hanski and M.E. Gilpin, pp. 69–91. San Diego: Academic Press.

Hanski, I. 1998. Metapopulation dynamics. *Nature* 396:41–49.

Hanski, I. 1999. *Metapopulation Ecology.* Oxford, United Kingdom: Oxford University Press.

Hanski, I., and Gilpin, M.E., eds. 1997. *Metapopulation Biology: Ecology, Genetics, and Evolution.* San Diego: Academic Press.

Hanski, I., and Kuussaari, M. 1995. Butterfly metapopulation dynamics. In *Population Dynamics,* eds. N. Cappuccino and P.W. Price, pp. 149–171. San Diego: Academic Press.

Hanski, I., Moilanen, A., and Gyllenberg, M. 1996. Minimum viable metapopulation size. *Am. Nat.* 147:527–541.

Heywood, V.H., Mace, G.M., May, R.M., and Stuart, S.N. 1994. Uncertainties in extinction rates. *Nature* 368:105.

Heywood, V.H., and Stuart, S.N. 1992. Species extinctions in tropical forests. In *Tropical Deforestation and Species Extinction,* eds. T.C. Whitmore and J.A. Sayer, pp. 91–117. London: Chapman and Hall.

Hill, J.K., Thomas, C.D., and Lewis, O.T. 1999. Flight morphology in fragmented populations of a rare British butterfly, *Hesperia comma. Biol. Conserv.* 87:277–283.

Holyoak, M., and Ray, C. 1999. A roadmap for metapopulation research. *Ecol. Letters* 2:273–275.

Johnson, L.E., and Padilla, D.K. 1996. Geographic spread of exotic species: ecological lessons and opportunities from the invasion of the zebra mussel *Dreissena polymorpha. Biol. Conserv.* 78:23–33.

Kareiva, P., Skelly, D., and Ruckelshaus, M. 1997. Reevaluating the use of models to predict the consequences of habitat loss and fragmentation. In *The Ecological Basis of Conservation: Heterogeneity, Ecosystems, and Biodiversity,* eds. S.T.A. Pickett, R.S. Ostfeld, M. Shachak, and G.E. Likens, pp. 156–166. New York: Chapman and Hall.

Kellman, M. 1996. Redefining roles: plant community reorganization and species preservation in fragmented systems. *Global Ecol. Biogeog. Letters* 5:111–116.

Kellman, M., Tackaberry, R., and Meave, J. 1996. The consequences of prolonged fragmentation: lessons from tropical gallery forests. In *Forest Patches in Tropical Landscapes,* eds. J. Schelhas and R. Greenberg, pp. 37–58. Washington, DC: Island Press.

Kellman, M., Tackaberry, R., and Rigg, L. 1998. Structure and function in two tropical gallery forest communities: implications for forest conservation in fragmented systems. *J. Appl. Ecol.* 35:195–206.

Lehman, C.L., and Tilman, D. 1997. Competition in spatial habitats. In *Spatial Ecology: The Role of Space in Population Dynamics and Interspecific Interactions,* eds. D. Tilman and P. Kareiva, pp. 185–203. Princeton: Princeton University Press.

Leimar, O., and Norberg, U. 1997. Metapopulation extinction and genetic variation in dispersal-related traits. *Oikos* 80:448–458.

Levins, R. 1969. Some demographic and genetic consequences of environmental heterogeneity for biological control. *Bull. Entomol. Soc. Am.* 15:237–240.

MacArthur, R.H., and Wilson, E.O. 1967. *The Theory of Island Biogeography.* Princeton: Princeton University Press.

May, R.M. 1973. *Stability and Complexity in Model Ecosystems.* Princeton: Princeton University Press.

Neve, G., Barascud, B., Hughes, R., Aubert, J., Descimon, H., Lebrun, P., and Baguette, M. 1996. Dispersal, colonization power and metapopulation structure in the vulnerable butterfly *Proclossiana eunomia* (Lepidoptera, Nymphalidae). *J. Appl. Ecol.* 33:14–22.

Newmark, W.D. 1995. Extinction of mammal populations in western North American national parks. *Conserv. Biol.* 9:512–526.

Newmark, W.D. 1996. Insularization of Tanzanian parks and the local extinction of large mammals. *Conserv. Biol.* 10:1549–1556.

Newton, I. 1998. Pollutants and pesticides. In *Conservation Science and Action,* ed. W.J. Sutherland, pp. 66–89. Oxford, United Kingdom: Blackwell.

Obeso, J.R., and Aedo, C. 1992. Plant-species richness and extinction on isolated dunes along the rocky coast of northwestern Spain. *J. Veget. Sci.* 3:129–132.

Opdam, P. 1990. Dispersal in fragmented populations: the key to survival. In *Species Dispersal in Agricultural Habitats,* eds. R.G.H. Bunce and D.C. Howard, pp. 3–17. London: Belhaven Press.

Opdam, P., van Apeldoorn, R., Schotman, A., and Kalkhoven, J. 1993. Population responses to landscape fragmentation. In *Landscape Ecology of a Stressed Environment,* eds. C.C. Vos and P. Opdam, pp. 147–171. London: Chapman and Hall.

Peterken, G.F., and Game, M. 1984. Historical factors affecting the number and distribution of vascular plant species in the woodlands of central Lincolnshire. *J. Ecol.* 72:155–182.

Petit, S., and Burel, F. 1998. Effects of landscape dynamics on the metapopulation of a ground beetle (Coleoptera, Carabidae) in a hedgerow network. *Agric. Ecosyst. Environ.* 69:243–252.

Pimm, S.L. 1991. *The Balance of Nature? Ecological Issues in the Conservation of Species and Communities.* Chicago: University of Chicago Press.

Pimm, S.L. 1998. Extinction. In *Conservation Science and Action,* ed. W.J. Sutherland, pp. 20–38. Oxford, United Kingdom: Blackwell.

Pimm, S.L., Russell, G.J., Gittleman, J.L., and Brooks, T.M. 1995. The future of biodiversity. *Science* 269:347–350.

Poschlod, P., and Bonn, S. 1998. Changing dispersal processes in the central European landscape since the last ice age: an explanation for the actual decrease of plant species richness in different habitats? *Acta Bot. Neerland.* 47:27–44.

Remmert, H., ed. 1991. *The Mosaic-Cycle Concept of Ecosystems.* Berlin: Springer-Verlag.

Simberloff, D. 1992. Do species-area curves predict extinction in fragmented forests? In *Tropical Deforestation and Species Extinction,* eds. T.C. Whitmore and J.A. Sayer, pp. 75–89. London: Chapman and Hall.

Simberloff, D. 1997. Biogeographic approaches and the new conservation biology. In *The Ecological Basis of Conservation: Heterogeneity, Ecosystems, and Biodiversity,* eds. S.T.A. Pickett, R.S. Ostfeld, M. Shachak, and G.E. Likens, pp. 274–284. New York: Chapman and Hall.

Stelter, C., Reich, M., Grimm, V., and Wissel, C. 1997. Modelling persistence in dynamic landscapes: lessons from a metapopulation of the grasshopper *Bryodema tuberculata. J. Anim. Ecol.* 66:508–518.

Ter Braak, C.J.F., Hanski, I., and Verboom, J. 1998. The incidence function approach to modeling of metapopulation dynamics. In *Modeling Spatiotemporal Dynamics in Ecology,* eds. J. Bascompte and R.V. Solé, pp. 167–188. Berlin: Springer-Verlag and Landes Bioscience.

Thomas, C.D., and Jones, T.M. 1993. Partial recovery of a skipper butterfly (*Hesperia comma*) from population refuges: lessons for conservation in a fragmented landscape. *J. Anim. Ecol.* 62:472–481.

Thomas, J.A., and Morris, M.G. 1995. Rates and patterns of extinction among British invertebrates. In *Extinction Rates,* eds. J.H. Lawton and R.M. May, pp. 111–130. Oxford, United Kingdom: Oxford University Press.

Tilman, D., May, R.M., Lehman, C.L., and Nowak, M.A. 1994. Habitat destruction and the extinction debt. *Nature* 371:65–66.

Turchin, P., ed. 1998. *Quantitative Analysis of Movement: Measuring and Modeling Population Redistribution in Plants and Animals.* Sunderland, Massachusetts: Sinauer Associates.

Van Balen, B. 1999. *Birds on Fragmented Islands: Persistence in the Forests of Java and Bali.* Ph.D. dissertation. Wageningen, The Netherlands: University of Wageningen.

Van de Wolfshaar, K. 1999. *Metapopulaties in een Veranderend Landschap. Een Onderzoek naar de Terugkeertijd van Metapopulaties.* Wageningen, The Netherlands: Institute for Forestry and Nature Research.

Van Ruremonde, R.H.A.C., and Kalkhoven, J.T.R. 1991. Effects of woodlot isolation on the dispersion of plants with fleshy fruits. *J. Veget. Sci.* 2:377–384.

VanValen, L. 1971. Group selection and the evolution of dispersal. *Evolution* 25:591–598.

Verboom, J., Schotman, A., Opdam, P., and Metz, J.A.J. 1991. European Nuthatch metapopulations in a fragmented agricultural landscape. *Oikos* 61:149–156.

Vos, C.C., Verboom, J., Opdam, P.F.M., and Ter Braak, C.J.F. 2001. Towards ecologically scaled landscape indices. *Am. Nat.* 157:24–41.

Warren, M.S., and Key, R.S. 1991. Woodlands: past, present and potential for insects. In *The Conservation of Insects and Their Habitats,* eds. N.M. Collins and J.A. Thomas, pp. 155–211. London: Academic Press.

Whitmore, T.C. 1997. Tropical forest disturbance, disappearance, and species loss. In *Tropical Forest Remnants: Ecology, Management and Conservation of Fragmented Communities,* eds. W.F. Laurance and R.O. Bierregaard, pp. 3–12. Chicago: University of Chicago Press.

Section IV

Conservation Planning

19

Using Broad-Scale Ecological Information in Conservation Planning: Introduction to Section IV

KEVIN J. GUTZWILLER

Planning is the critical operation that links knowledge of landscape ecology to its successful application on the ground. If we are to make conservation headway, the crucial step of planning must make use of broad-scale spatial relations and predictive principles (e.g., Dramstad et al. 1996). Isolated, local-scale efforts to solve a conservation problem are often rendered ineffective by ecological forces that operate at broad spatial scales and across jurisdictional boundaries. Under these conditions, conservation success depends significantly on two factors: the degree to which information about broad-scale ecological influences is applied, and the extent to which stakeholders act in concert (spatially and temporally) to solve their common problem (see Lambeck 1999). To be effective, these efforts require collaborative planning.

The biological ramifications of broad-scale conservation projects demand that conservation planning be based whenever possible on supporting analyses. For example, the effectiveness of existing reserve systems, the potential for expanding them, and species' sensitivity to and persistence under current and anticipated environmental conditions, should be assessed. The potential value of habitat networks, viability projections from metapopulation models, and different prescriptions of habitat layouts warrant examination. For aquatic systems, predictions of the ecological potential of specific waters should be weighed. Section IV addresses these key issues and related methods in the context of incorporating broad-scale ecological knowledge into conservation planning.

19.1 Planning Approaches

Many broad-scale planning approaches have been developed (see Haber 1990; Naveh and Lieberman 1990; Ruzicka and Miklos 1990), and each has a unique scope of geographic relevance and degree of conservation emphasis. Chapter 20 considers two types of planning in which biological conservation is paramount. One type emphasizes representation of biodiversity through systems of reserves, an approach that has been used extensively throughout the world. A second type focuses on maintaining ecological processes and conditions necessary for the long-term persistence of biodiversity. Chapter 20 elaborates on an emerging approach to the latter method that involves managing threatening processes (e.g., habitat loss and isolation) so that those species that are most sensitive to them are protected; this approach assumes that protection is simultaneously afforded to species that are less-sensitive to these processes.

19.2 Acquiring Important Planning Data

Ideally, conservationists should implement projects that ensure protection of biological processes and associated species within large areas over long time periods (Scott et al. 1999). The ability to plan (hence execute) these projects depends importantly on the availability of several kinds of information. For example, because naturally connected landscapes are usually better able to support native biodiversity than are those whose habitats are disconnected, knowledge about the conservation potential of habitat networks is invaluable. Chapter 21 examines approaches for acquiring such data. Many natural habitats have undergone fragmentation, and many naturally occurring metapopulations are being impacted by human activities. As a result, the use of metapopulation models to supply data for conservation planning, explained in Chapter 22, has now become essential in numerous situations.

Habitat amount and arrangement in a landscape are important determinants of the occurrence and persistence of many species. Depending on the spatial distribution of existing habitat, plans for landscape development, and the feasibility of restoring destroyed habitats, various spatial configurations of habitat may be possible. In circumstances such as these, the optimal location of habitats for meeting the needs of one or more organisms is usually not obvious, and planners need efficient methods for determining the spatial distributions of habitat that would best satisfy specific conservation objectives. Chapter 23 describes approaches for assessing optimal placement of habitats for conservation planning.

Broad-scale conservation planning for aquatic organisms requires knowledge of the ecological potential of an area's water bodies. A variety of approaches have been used to predict ecological potential for aquatic species, but statistical-modeling approaches that use map-based information about the landscape context of water bodies (landscape-context modeling) have distinct advantages over traditional site-based and regional classification methods. Landscape-context modeling is emerging as a sound and practical means of generating important planning data for aquatic systems. Chapter 24 discusses these issues and provides essential information for all who are responsible for developing plans to conserve aquatic biodiversity.

References

Dramstad, W.E., Olson, J.D., and Forman, R.T.T. 1996. *Landscape Ecology Principles in Landscape Architecture and Land-Use Planning.* Washington, DC: Island Press.

Haber, W. 1990. Using landscape ecology in planning and management. In *Changing Landscapes: An Ecological Perspective,* eds. I.S. Zonneveld and R.T.T. Forman, pp. 217–232. New York: Springer-Verlag.

Lambeck, R.J. 1999. *Landscape Planning for Biodiversity Conservation in Agricultural Regions: A Case Study From the Wheatbelt of Western Australia.* Biodiversity Technical Paper Number 2. Canberra, Australia: Department of the Environment and Heritage.

Naveh, Z., and Lieberman, A.S. 1990. *Landscape Ecology: Theory and Application.* Student Edition. New York: Springer-Verlag.

Ruzicka, M., and Miklos, L. 1990. Basic premises and methods in landscape ecological planning and optimization. In *Changing Landscapes: An Ecological Perspective,* eds. I.S. Zonneveld and R.T.T. Forman, pp. 233–260. New York: Springer-Verlag.

Scott, J.M., Norse, E.A., Arita, H., Dobson, A., Estes, J.A., Foster, M., Gilbert, B., Jensen, D.B., Knight, R.L., Mattson, D., and Soulé, M.E. 1999. The issue of scale in selecting and designing biological reserves. In *Continental Conservation: Scientific Foundations of Regional Reserve Networks,* eds. M.E. Soulé and J. Terborgh, pp. 19–37. Washington, DC: Island Press.

20

Landscape and Regional Planning for Conservation: Issues and Practicalities

ROBERT J. LAMBECK AND RICHARD J. HOBBS

20.1 Introduction

Conservation planning attempts to address conservation problems by drawing on our knowledge of natural ecosystems and our understanding of human impacts on these systems. However, attempts to use existing theory to counteract these impacts have revealed significant limitations to our ecological knowledge. Although existing theory can describe how the components of ecosystems interact, very little of this theory can be used to specify the details of ecosystem modification required to ameliorate or reverse the detrimental impacts of human land use on biological diversity. Our capacity to describe problems in great detail does not translate easily into an ability to prescribe solutions.

This chapter examines these limitations as well as our current capabilities in conservation planning. We describe some of the existing concepts and principles that have emerged from landscape ecology and assess their usefulness for conservation planning, and we discuss a new approach that is being developed in the agricultural regions of Western Australia. We then describe some recent applications of these approaches and extract from them a range of principles that will help land managers to develop conservation plans. We identify knowledge gaps that currently constrain our ability to achieve meaningful conservation outcomes, and we describe research approaches that will help close these gaps.

20.2 Concepts, Principles, and Emerging Ideas

20.2.1 Landscape and Regional Planning: Rationale and Terminology

Biophysical processes and human impacts operate at scales ranging from centimeters to hundreds of square kilometers (Fahrig and Merriam 1994). Conservation planning must therefore consider the requirements for maintaining species and processes that operate at local, landscape, and regional scales. We use the term *landscape* to describe areas that contain a mix of local ecosystems or land uses repeated over the land surface (Forman 1995). *Regions* are considered to be broad geographical areas with a common macroclimate and sphere of human activity and interest (Forman 1995). The term *bioregion* is used to delineate environmentally homogeneous areas that are independent of social boundaries (e.g., Thackway and Creswell 1992).

A limitation of these definitions is that regions, as defined by Forman, often encompass too much environmental variation to be useful for conservation planning, and bioregions fail to take into account human dimensions of the landscape, which are often major drivers of biotic impoverishment. We therefore introduce the term *Conservation Management Zone* to describe areas that are not only biophysically uniform, in the sense of a bioregion, but also have similar patterns of land use. This planning unit is described in greater detail below.

For the purpose of this chapter, we consider *planning* to be the specification of the type of landscape elements required, the spatial extent of those elements, and their positioning, to meet clearly stated management objectives. To be effective, conservation planning must ensure first, that biological diversity is adequately represented and second, that the biota are able to persist through time. The following sections describe some planning approaches for representing and maintaining biological diversity.

20.2.2 Planning to Represent Biodiversity

The establishment of reservations has underpinned much conservation planning throughout the world. Unfortunately, selection of reserves has usually been ad hoc, with criteria such as scenic beauty or unsuitability for resource extraction determining where reserves are located (Pressey et al. 1993). Clearly, these criteria fail to ensure that the full range of biological diversity is protected. To address this problem, more systematic approaches to reserve selection have been developed, primarily in Australia, South Africa, and North America. In Australia and South Africa, identification of areas best suited for reservation has focused on reserve selection algorithms that identify the minimum set of areas required to best represent the variation in the region being assessed (for examples, see McKenzie et al. 1989; Pressey and Nicholls 1989; Margules et al. 1994; Lombard et al. 1997).

These algorithms can be used to sample directly the range of biological diversity in a region if we know how that diversity is distributed. Alternatively, where information about the distribution of species is incomplete, surrogate data, such as environmental variables, can be sampled (Austin et al. 1984; Nix 1992). Because these surrogates are believed to be correlated with the distribution of components of biodiversity, it is assumed that representation of the range of environmental variation should also capture the range of biotic variation.

Representation of biodiversity requires not only that biological *hotspots* (areas of high species richness, levels of endemism, or numbers of rare or threatened species) are protected (Curnutt et al. 1994; Reid 1998), but also that areas with low levels of biodiversity containing unique elements are conserved. This is the notion of *complementarity* (Margules et al. 1988; Pressey et al. 1993), which seeks to ensure that selected areas collectively capture the greatest proportion of diversity in a region. Related to complementarity is the concept of *irreplaceability* (Pressey et al. 1994), which is a measure of a site's uniqueness and represents the options that may be lost if a site is not considered. Another critical feature of rational reserve selection is *flexibility,* the capacity to identify sites that can be substituted for each other without compromising the conservation goal (Pressey et al. 1994). These three elements—complementarity, irreplaceability, and flexibility—provide the foundations for much of the algorithm-based reserve selection literature.

Comprehensiveness, adequacy, and representativeness also are critical factors to consider when selecting areas for reservation (Pressey 1998). *Comprehensiveness* reflects the need to represent the full variety of vegetation types in a system of reserves. *Adequacy* specifies the amount of each vegetation type to be represented and in what configuration, and *representativeness* covers the need to represent the range of biological diversity in each vegetation type.

In the United States, *gap analysis* (Scott et al. 1993; Kiester et al. 1996) is used to assess how well various components of biological diversity are represented in existing reserve systems. This approach examines the distribution of species, vegetation types, ecosystems, or hotspots of species richness and assesses the extent to which each is represented in the conservation estate. Gaps in the representation of these units are then identified as priorities for conservation.

20.2.3 Planning for Persistence

Procedures for representing biological diversity generally fail to address the issue of long-term persistence of species that are reserved. To retain biota over time, planning must take into account the maintenance of ecosystem processes at rates and via pathways appropriate for meeting the needs of the biota to be protected. Conservation planning must therefore address the resource requirements of plants and animals, as well as processes responsible for water, nutrient, and energy cycling.

Two broad approaches can be taken when planning for persistence: general enhancement and strategic planning. *General enhancement* aims to improve on the current situation by increasing the probability of species persisting without specifying the magnitude of the desired change. *Strategic planning,* on the other hand, aims to address a specified outcome such as conserving populations of a species, protecting groups of species, retaining all species and their associated functions, or reintroducing species that have disappeared from an area. The outcomes of adopting one or other of these approaches will be quite different. Therefore, it is worth exploring each in greater detail.

General enhancement approaches largely employ ecological principles to develop conservation plans. These principles are derived from observations accumulated from different locations at different times. Planners attempt to interpret this accumulated knowledge in the context of new problems in new locations.

These principles, drawn from island biogeography (MacArthur and Wilson 1963), species-area relationships (Connor and McCoy 1979; Boecklen and Gotelli 1984), niche theory (Cody 1968; Connell 1975), and metapopulation theory (Hanski 1991; Harrison 1994), have obvious implications for planners. They tell us that larger remnants generally support more species than do smaller ones (Opdam 1991; Collinge 1998), that species are more likely to persist in more-connected landscapes than they are in fragmented ones (Fahrig and Merriam 1985), that wider corridors are more likely to facilitate movement than are narrow corridors (La Polla and Barrett 1993), and that habitats with vertical and horizontal heterogeneity are better than are structurally simple monocultures (Boecklen 1986).

A limitation of these principles is that when managing a specific problem in a particular landscape, they cannot tell us how big a remnant should be, how many habitat patches are required, or what the appropriate width of a corridor should be. General principles can identify directions in which to proceed but fail to specify the magnitude of the response required. In the absence of better information and in the face of urgent ecological problems, these principles may be all that is available and should be used to guide conservation planning. However, it is important to recognize that landowners in production landscapes will wish to allocate the minimum possible area to conservation. They will therefore expect to have more precise recommendations than those provided by general principles.

If we wish to meet a more strategic goal, such as retaining particular species, a biological community, or all species in a region, it will be necessary to meet the resource needs of the species that we wish to protect. Strategic conservation planning must therefore have clearly stated goals, and it must specify the type, number, and placement of landscape elements, or the types and rates of processes required to meet those goals. To date, most strategic conservation plans have focused on single species. These studies generally examine the demographic attributes of the species of interest and only consider environmental variables when they affect birth and death rates and, hence, persistence probabilities. Management plans then target the limited set of environmental factors that affect these demographics.

In an attempt to increase the benefit derived from species-based management, attention has been directed toward keystone and umbrella species. *Keystone species* are those that have an impact on the dynamics of an ecosystem that is disproportionate to their numbers. *Umbrella species* are those whose resource requirements encompass the needs of other species. Interest in these types of species stems from the assumption that their management will confer benefits to a range of additional species (Bond 1993; Launer and Murphy 1994; Simberloff 1998). Although there is considerable merit to the argument that the needs of some species encompass the requirements of others, or that management of a keystone species will benefit other components of the system, there are still no widely accepted criteria for identifying such species (but see Lambeck 1997) or for predicting the extent to which other components of an ecosystem will benefit.

In recognition that single-species studies are not sufficient for dealing with widespread biotic impoverishment, multispecies approaches to conservation planning are being developed (Lambeck 1997; Noon et al. 1997). However, multispecies plans are often little more than aggregated single-species plans. Although such plans may generate greater biodiversity benefit than would a single-species plan, they will fail to prevent species loss if they do not address the needs of all species at risk. In the absence of a conceptual framework for selecting the different species that should be the focus of attention, aggregated multispecies plans may fail to account for important interactions between species and their environment and fail to consider ecosystem processes that are essential for maintaining community dynamics.

Conservation plans also must consider the spatial scales required to maintain processes such as nutrient, water, and energy cycling, and to absorb the effects of disturbances such as fire, floods, and storms. When these processes are considered, the focus tends to move away from individual species toward management of natural communities and whole ecosystems (Franklin 1993; Risser 1995) in recognition that important interactions and functions may be overlooked when only one or a few species are considered (Wiens 1997). Because these interactions may occur at broad spatial scales, planning has expanded to regional scales (see, for example, the Florida and California habitat conservation plans described in Noss et al. 1997). The evolution of conservation plans through a hierarchy of scale and complexity reflects a hope that by protecting larger areas and more complex systems, the probability of protecting the biota in an area will increase. However, apart from the perception that bigger areas encompassing more heterogeneity support more species, there is no theoretical framework for specifying the spatial scale over which such projects should be applied or the relative effort that should be directed toward representing and managing different landscape elements within these ecosystems.

The remaining paragraphs of this section outline a new conservation planning framework that aims to prevent the further loss of species from fragmented agricultural landscapes by linking species-based and process-based approaches to conservation planning. To meet this goal, it is necessary to ensure that the resource requirements of the constituent species are met and that ecosystem processes function at rates and via pathways appropriate for meeting the needs of those species. These requirements present planners with a dilemma—how does one manage resources and processes to meet the needs of all species without considering each species individually?

A solution to this problem can be found by linking species' requirements with ecological processes. The loss of species from a landscape is clearly attributable to the presence of some limiting or threatening processes. It is the management of these processes that is required to protect species. To protect all species threatened by any given process, it will be necessary to manage the process at a level that protects the most sensitive species. If species that are most sensitive to a threat are protected, then less-sensitive species also should be protected from that threat. Lambeck (1997) describes such species as *focal species,* species toward which we direct our primary management efforts. Where there are multiple threats, there will be multiple focal species, and where these threats affect multiple habitat types, there will be a focal species for each threat in each habitat type. The result of this approach is a multispecies umbrella, a limited set of sensitive species whose requirements, if met, should meet the needs of all other less-sensitive species.

Any comprehensive approach to conservation management would consider all threatening processes in a landscape. However, because this chapter focuses on spatial aspects of landscape planning, we consider here only those threats that require specification of the type, amount, and positioning of habitat. In this context, the primary threats are habitat loss and habitat isolation.

Species considered threatened by each of these processes are grouped and ranked in terms of their sensitivity. Presence-absence surveys of habitat patches indicate the distribution of species considered area-limited or isolation-limited. Analysis of spatial attributes of vegetation is then undertaken to determine the characteristics of habitat patches in which species do and do not occur. This enables us to specify the minimum habitat area and the maximum interpatch distance (e.g., Figure 20.1) required for these species to occur. Remnants that do not meet these criteria are identified, and the amount of reconstruction required to create adequate habitat is specified. Similarly, it is possible to identify remnants that are too isolated for the most dispersal-limited species and to identify positions in the landscape that need construction of intermediate habitat patches. An example of the application of this approach is presented in Section 20.3.2.

Because individual land managers or local groups of managers are unlikely to be able to ensure the persistence of all species in the area they manage, it is necessary to provide a regional planning framework that identifies the contribution that they can make to achieving that goal at a regional scale. Unfortunately,

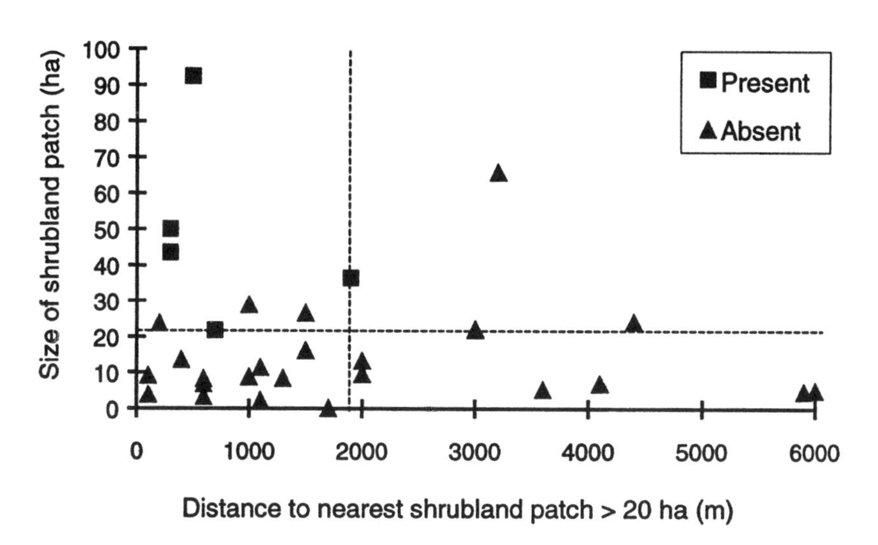

FIGURE 20.1. Pattern of patch occupancy by Western Yellow Robins (*Eopsaltria griseogularis*). This species was not found in patches of shrubland smaller than 20 ha or in patches more than 2 km from the nearest occupied patch, suggesting that habitat patches need to exceed 20 ha in area and be less than 2 km apart to be occupied.

the requirements for meeting such a goal will differ from location to location as environmental conditions, patterns of land use, and species complement vary. If the same recommendations are not appropriate for all locations, it is necessary to identify areas that are homogeneous enough that design parameters derived within an area can be legitimately extrapolated across the remainder of that area. These are areas that are biophysically homogeneous and have similar human land-use patterns, the Conservation Management Zones defined above. These zones are identified by first partitioning the region of interest into bioregions—in this case, areas that have equivalent geomorphology and climate (Thackway and Cresswell 1992)—and then further subdividing each bioregion into areas having similar land-use patterns. The identification of landscapes with similar patterns can be achieved by using an array of landscape metrics that are available in geographic information system (GIS) packages. Measures such as percent cover of vegetation, proportion of different vegetation types, number of patches, mean remnant size, contagion, and isolation are commonly used to characterize landscapes (Hulshoff 1995; O'Neill et al. 1996).

By undertaking a focal-species analysis within a Conservation Management Zone, it is possible to develop planning recommendations that apply to the entire zone. This approach produces guidelines for action that address strategic conservation goals in a spatially explicit manner and that have relevance over relatively large areas. It is important to recognize that this methodology has been developed for reconstructing highly fragmented agricultural landscapes, and that its suitability as an approach for guiding habitat removal has not been ascertained.

20.3 Recent Applications

Acquiring information about the application of conservation planning principles is not a trivial task, owing to the fact that many practitioners do not publish the results of their efforts in the primary scientific literature. Much of the information that is published resides in technical reports and semipopular magazine articles that are not widely accessible. The applications described below were selected following a search of the formal scientific literature, combined with requests for technical reports from colleagues known by us to be attempting on-ground application of conservation planning principles. In this section, we provide examples of the application of procedures for representing biological diversity in reserve systems, and for designing conservation systems to ensure species persistence at landscape and regional scales.

20.3.1 Planning to Represent Biodiversity: Reserve Selection Algorithms and Gap Analysis

In Australia, formal reserve-selection methods have been applied in a research context to wetlands (Margules et al. 1988), semiarid shrublands (McKenzie et al.

1989), mallee (multistemmed eucalypt) woodlands (Margules and Nicholls 1987), and temperate forests (Kirkpatrick 1983; Bedward et al. 1992; Margules and Nicholls 1993; Pressey 1998). However, it is only recently that these planning approaches have found their way into the decision-making arena in Australia. The Australian National Forest Policy (Anonymous 1992) has required each of the state and territory governments to partition the forest estate between categories of extraction and protection using the concepts of comprehensiveness, adequacy, and representativeness (Pressey 1998). In New South Wales, estimates of irreplaceability (Pressey 1998) also were used to set conservation priorities. This planning process culminated in the identification of nine new national parks; over 800,000 ha of forest from which timber extraction was deferred subject to further investigation of its conservation value; extensive new wilderness areas; and agreements on the amount of timber to be extracted over the following five years (Resource and Conservation Assessment Council 1996; Pressey 1998).

Although most reserve selection exercises in Australia have been applied at regional scales, Brooker and Margules (1996) prioritized conservation areas at both regional and local scales in agricultural regions of Western Australia. Not surprisingly, the two analyses provided different results, with remnants considered important in the local analysis being less important in the regional analysis. These differences are important for different land managers in the region. The regional analysis provides guidance for government authorities responsible for selecting and managing reserves, whereas the local analysis assists private landholders in managing smaller remnants. The results of this analysis are currently being used to guide the local community in the development of a nature conservation plan.

The results of formal reserve selection procedures also are beginning to be applied in South Africa, with new reserves being implemented or planned in the Agulhas Plain Region (Lombard et al. 1997), the Succulent Karoo biome (Desmet et al. 1999), the Table Mountain area (Trinder-Smith et al. 1996), the Cape Floristic Region (Cowling et al. 1998), and the Western Cape coastal lowlands (Heijnis et al. 1997).

In North America, reserve selection has focused on closing the gaps in existing conservation networks. Of particular note is the development of a state-wide reserve network in Florida, based on a detailed assessment of the distribution and dynamics of a set of focal species, which included mammals, birds, reptiles, and amphibians (Cox et al. 1994; Noss and Cooperrider 1994). This approach sought to produce planning recommendations that would meet a set of minimum conservation goals for the region. Using a GIS, it assessed the degree of security provided to selected species by the current reserve system and identified Strategic Habitat Conservation Areas requiring additional protection. The combination of setting clear conservation goals, using available data coupled with population viability modeling, and applying an overtly regional approach resulted in the production of clear management options for the region.

20.3.2 Planning for Persistence at Landscape Scales: Focal-Species Planning in the Wheatbelt of Western Australia

The focal-species planning approach has been applied to four watersheds in the wheatbelt of Western Australia, each covering an area of approximately 20,000–30,000 ha (Lambeck 1999). Survey results for area-limited and dispersal-limited species (e.g., Figure 20.1) were used to calculate the minimum habitat area and the maximum interpatch distance that were required for the most demanding species to have a 60% probability of occurrence in a remnant. GIS routines were used to identify all remnants that had insufficient habitat or were too isolated to meet the needs of these species. Maps were produced that indicated the extent to which each patch needed to be expanded to have an equivalent probability of being occupied by the most area-limited species (Figure 20.2). These

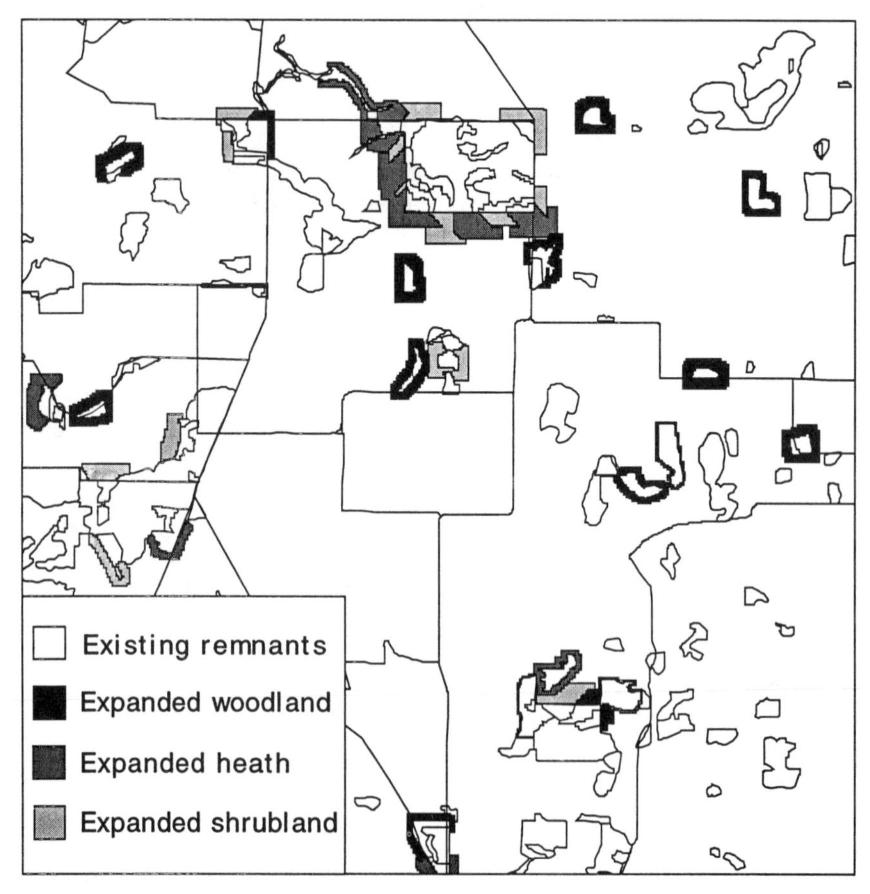

FIGURE 20.2. Map of a wheatbelt watershed indicating the extent to which limiting habitat types need to be expanded to meet the needs of area-limited focal species.

outcomes are currently being used to guide the development of conservation plans for each subcatchment.

Although this focal-species planning approach identifies the minimum patch sizes and maximum interpatch distances, it does not specify the number of patches required to support viable populations of the focal species. Future work aims to explore this question and to test whether landscapes that support viable populations of focal species will also support viable populations of nonfocal taxa.

20.3.3 Combining Representation and Persistence at Regional and National Scales: The Wildlands Project in North America

The Wildlands Project aims to protect and restore the natural heritage of North America through the establishment of a connected system of wildlands (Soulé and Terborgh 1999). Although still in its planning stages, the project emphasizes the design, implementation, and management of regional networks of protected areas at a continental scale. The primary elements of such a network include large core areas that are primarily dedicated to conservation (Noss et al. 1999), corridors that link these core areas (Dobson et al. 1999), and multiple-use buffer zones that surround and protect the core areas and corridors (Groom et al. 1999).

To achieve national- and regional-scale outcomes, the Wildlands Project recommends linking multiple smaller-scale efforts via a regionally uniform planning framework (Simberloff et al. 1999). This framework should include components addressing both reserve selection and reserve design, employing formal procedures for reserve selection, such as those described in Section 20.2.2, and species-based approaches to reserve design (Noss et al. 1999). The use of focal species to derive design parameters is recommended, with top carnivores being identified as key species for determining the appropriate size of core areas. This species-based approach has many similarities to the focal-species approach presented in Section 20.2.3, in that the requirement for using multiple focal species is acknowledged. It differs, however, in that there is no formal procedure for selecting focal species and there is no link between the choice of focal species and threatening processes. By not explicitly considering the full array of threats, and selecting focal species to represent each threat, the risk remains that numerous species may still be vulnerable in spite of meeting the needs of top carnivores.

20.4 Principles for Applying Landscape Ecology

Because land managers are rarely expert in the various disciplines that contribute to effective resource management, it is necessary to identify management principles that can be readily applied to particular situations. It is important to recognize, however, that many of these principles may provide guidance about the type of actions that are required, but they may not provide details about the magnitude of those actions. In this section, we draw together a range of principles from landscape ecology that can assist natural resource managers. These are largely drawn from theory because formal land-use planning for nature conservation has not been widely applied, and in cases in which it has, there has been insufficient time to assess the consequences.

These principles are presented in three groups: general principles, which are likely to be appropriate for most conservation planning scenarios; strategic planning principles, which apply to situations in which there is sufficient biological information to provide spatially explicit recommendations about the type, amount, and placement of habitat; and general enhancement principles, which apply to situations in which there is limited biological information and little opportunity to gather such information. These sets of principles are not mutually exclusive, and there will often be situations in which they can be combined.

20.4.1 General Principles

General Principle 1: Conservation goals must be clearly articulated and agreed to by all parties. If this is not done, it will not be possible to determine the appropriateness of alternative actions or the adequacy of the outcome.

General Principle 2: Conservation goals must be appropriate for the spatial extent of the area being managed. If a conservation goal cannot be realistically met in a given area, either the extent of the planning area must be changed or the conservation goal must be modified to reflect what is achievable at the scale of management.

General Principle 3: The scale of management must match the spatial scale of processes that are being managed. For example, management of hydrological processes should at least cover local watersheds, and management of species should cover an area that ensures first, that the habitat requirements of individuals are met and second, that population processes are maintained.

General Principle 4: Regional conservation planning should aim to ensure that land-use practices in a region do not result in the loss of species from that region. Although the distribution and abundance of species may change as a result of the land uses applied, they should not change to the extent that species are eliminated from the region.

20.4.2 Strategic Planning Principles

Strategic Planning Principle 1: Planning at regional scales must aim to represent the range of biological diversity in the region and ensure that all components of that diversity are able to persist in the region in the long term.

Strategic Planning Principle 2: Criteria of comprehensiveness, representativeness, and adequacy should underpin the selection of areas for nature conservation. Selection procedures also should be able to identify the relative importance of different sites (irreplaceability), enable choices between alternative sites (flexibility), and ensure that species-poor sites with unique species are included in addition to species-rich sites (complementarity).

Strategic Planning Principle 3: Planning at a local scale should ensure that patches of habitat can support breeding units of the most habitat-demanding species, and that these patches are within reach of neighboring patches for the most dispersal-limited species. Linking vegetation should provide habitat for species with low dispersal ability.

Strategic Planning Principle 4: Planning at the individual remnant scale should focus on the type, number, and configuration of habitat patches within a remnant, and on threats that affect the condition of those patches. Local landholders managing small portions of a landscape cannot solve problems that arise from impacts over a much larger area. Therefore, they can only seek to ensure that their management actions contribute to goals set at larger spatial scales.

Strategic Planning Principle 5: Regional conservation planning should be applied to areas that are relatively homogeneous in terms of their biophysical and anthropogenic characteristics. Guidelines developed within such areas can then be legitimately extrapolated to the remainder of the area.

Strategic Planning Principle 6: Local actions should be clearly linked to regional strategies. If conservation agencies are unable to achieve conservation goals within a designated conservation estate, it will be necessary to achieve that goal in partnership with private landholders. For these landholders to feel that their contribution is of value, it will be necessary to identify the actions that they can take, and to demonstrate that these actions will contribute to the regional goal.

Strategic Planning Principle 7: Actions that attempt to ensure no further species loss from a region must ensure that the needs of constituent species are met. This requires some degree of species focus in regional conservation plans.

Strategic Planning Principle 8: Species-based approaches must be able to demonstrate that the benefits extend beyond the target species. If keystone or umbrella species are employed, criteria for their selection must be clearly articulated.

Strategic Planning Principle 9: Ecosystem processes must function at rates and via pathways that ensure that the needs of the biota (and of the human population) are met.

Strategic Planning Principle 10: Threatening processes must be managed at levels that protect the most sensitive species. If species most sensitive to a given threat are protected, other species less sensitive to that threat also should benefit.

20.4.3 General Enhancement Principles

General Enhancement Principle 1: Larger areas of habitat are generally better than are smaller areas. Larger areas usually encompass more environmental variation and hence contain a greater array of habitats that, in turn, support a wider variety of species. Larger areas also are more likely to encompass ecosystem processes that are essential for the maintenance of biota.

General Enhancement Principle 2: Diverse habitat will generally support more species than will uniform habitat. This diversity can be achieved by having a range of patch types and a diversity of species within each patch type. However, if patches are too small, they may not provide the resources required by sedentary habitat specialists.

General Enhancement Principle 3: More-connected landscapes are more likely to maintain population processes than are more-fragmented landscapes. If individual reserves or remnants cannot support viable populations of the constituent biota, it will be necessary to have a connected landscape that can support numerous subpopulations of some species. A metapopulation structure will increase the probability of recolonization following local extinction.

General Enhancement Principle 4: Wider corridors are likely to facilitate movement for a wider array of species than are narrow corridors. For dispersal-limited species that do not have viable populations in individual habitat patches, these corridors will have to provide habitat.

The application of these general enhancement principles will result in a landscape similar to the 'spatial solution' proposed by Forman and Collinge (1996): a landscape with a few large patches connected by stepping stones; linear vegetation, preferentially along streams and rivers; and additional "bits of nature" scattered throughout the landscape.

20.5 Knowledge Gaps

Although substantial progress has been made in developing planning procedures for nature conservation, a number of significant knowledge gaps still remain. In some cases, these gaps reflect inadequate theoretical frameworks. More often, however, the frameworks are relatively sound but lack empirical support. In addition, our limited capacity to move effectively between theoretical concepts and practical application presents a major hurdle. The following sections outline some of the more significant issues that must be addressed if we are to progress beyond the general, unparameterized principles that are currently available, to more explicit recommendations that managers can implement.

20.5.1 Theoretical Voids

There is no robust theoretical framework to guide planners when dealing with a range of processes and species that operate at different spatial scales. Although we have identified the need to link regional planning to local outcomes, the science of scaling up and down among local, landscape, and regional scales is still poorly developed. Can solutions derived from local case studies be applied to larger areas simply by repeating the recommendations over the whole area, or do we need to change the management response as the spatial extent of the management unit changes? For example, the maintenance of a viable population of a particular species over a small area may require a significant portion of that area to be allocated to conservation, thereby jeopardizing the viability of affected landholders. Can the same outcome be achieved by addressing the problem over a larger area with less effort per unit area? For some species, increasing the management area may enable a shift from emphasis on single populations confined to large patches to more dispersed metapopulation structures that may require less conservation effort per unit area.

The theoretical basis for predicting the effects of regional habitat changes on local habitat requirements of species also is limited. In a landscape that has only a small amount of vegetation, particular species may only be able to persist in larger blocks of habitat, but as the amount of habitat is increased at a regional scale, the same species may use smaller blocks, possibly because they are able to exploit a number of adjoining patches. These types of relationships have important implications for planners.

Further clarification of the role of single species in conservation planning is required. The foundations for using particular species as surrogates for other components of an ecosystem remains problematic. Although the concepts of keystone and umbrella species have been widely accepted, their use in planning is not straightforward (Simberloff 1998). There are no clear criteria for selecting such species and no rationale for predicting the extent of the benefit that would accrue to the ecosystem in which they occur. Most species will respond to changes in their environment and, hence, could be considered an indicator of something. Similarly, many species have requirements that encompass some needs of other species, allowing virtually any taxa to be nominated as an umbrella species. Although keystone species may significantly influence a system, they are unlikely to provide a solution for managing all aspects of complex ecosystems. The identification of keystone species is difficult, and it is possible that some ecosystems may not have such species.

Frameworks for simultaneously considering biophysical and human determinants of species distributions and persistence probabilities also require further development. Although considerable effort has been invested in understanding biotic responses to environmental variability, and similarly, the effects of human land use on plant and animal populations, these two approaches have rarely been drawn together in a form useful to planners. The Conservation Management Zones, described above, represent an attempt to identify regions that are homogeneous with respect to both biophysical and anthropogenic patterns. However, the categorization of continuous variables will always be somewhat arbitrary, and the biological significance of these categories is rarely known. Attempts to classify landscapes will only be of value to planners if the classification is linked to known biotic responses. Consequently, it will be essential to link the theory of landscape description with an understanding of biotic responses to landscape variability.

The theoretical foundations of reserve selection procedures are relatively robust, and recent developments in this area have focused on minor refinements of the selection criteria. These selection procedures are typically based on the concepts of comprehensiveness, representativeness, and adequacy. Although it is relatively easy to assess how comprehensive or representative a reserve system is, it is more difficult to determine how adequate it is. Commonly, multiple representations of taxa or vegetation types, or specified proportions of original distributions, are used to define adequacy, but these are invariably arbitrary. Adequate conservation estates will ultimately be those that retain their biotic values and ecosystem processes over some specified time period. The theoretical framework for determining the requirements for achieving this is poorly developed.

20.5.2 Empirical Voids

A significant constraint to effective conservation planning is the limited empirical support for many of the theories and procedures that are proposed. In the face of enormous ecological complexity, urgent need to act, and limited resources, scientists and managers seek shortcuts, such as surrogates for detailed ecological data, and general principles derived from experience. In many instances, the empirical support for these shortcuts is limited.

A wide range of environmental variables or selected taxa have been proposed as surrogates for quantitative knowledge about the distribution of species, the role of particular species in ecosystem processes, and the relative value of different species for management. Although the notions of indicator, umbrella, and keystone species are intuitively appealing, their value to planners is limited by the paucity of empirical support. Empirical support for focal species planning is limited to a few taxa in a few locations (Lambeck, unpublished data). Quantitative data are required to test the hypothesis that landscapes designed and managed for a limited suite of sensitive species will protect other, less-sensitive species.

Another assumption that has not been tested is that we have the technical capacity to implement the required landscape treatments. Our understanding of how to recreate habitat is still in its infancy. For instance, we will rarely know which elements of habitat (e.g., structure, composition) are likely to be most important for many species that use that habitat. A large area will not suffice if critical resources are missing. There also is likely to be a relatively long lead time before essential habitat features reappear in recreated habitat, especially in long-lived perennial plant communities.

20.6 Research Approaches

20.6.1 Approaches for Theoretical Research

Understanding the responses of species to alternative management scenarios at landscape scales is not readily amenable to experimental manipulation. Consequently, it will be necessary to develop a theoretical understanding of the effect of landscape changes at different scales by developing population models that predict the likely responses of different species to alternative landscape configurations. By building predictive population models for a suite of species representing a range of taxonomic groups with different life-history traits, it may be possible to compare alternative landscape designs to determine those that confer the greatest benefit to the widest array of species types. Such models could be used to develop an understanding of the relative benefits of attempting to address conservation goals with intensive action at local scales compared with more-dispersed activities linked over a larger area.

Predicting the effect of regional habitat changes on local habitat requirements also is likely to be best addressed by population modeling. The development of population viability models for real organisms, for which data are available or, if data are lacking, for hypothetical organisms representing a range of life-history attributes, could provide some insights into the changing probability of particular habitat patches being occupied by different types of species as landscape configurations are altered.

Population modeling also could be used to explore the role of single species, such as focal species, in conservation planning. By explicitly linking vulnerable species to threatening processes, the focal-species approach (Lambeck 1997) attempts to provide a theoretical justification for deriving multispecies umbrellas that should, in theory, protect the full spectrum of biological diversity in a region. However, the underlying assumption that nonfocal species will be adequately catered for in landscapes designed for focal species has not been tested. By developing population models for focal species and a selected suite of nonfocal taxa that represent a range of different taxonomic groups, it may be possible to determine the extent to which landscapes designed and managed for a limited suite of focal taxa are likely to benefit other species.

The development of a theoretical framework for understanding the combined influence of biophysical and human influences on biodiversity will depend largely on the acquisition of empirical data that demonstrate consistent relationships between the distribution and abundance of species, and various environmental and anthropogenic variables. Because there is no *a priori* basis for partitioning continuous landscape variables into discrete classes, it may be necessary to initially partition regions into a manageable number of relatively uniform zones and then compare biotic responses within and between zones. If no differences can be detected between particular zones, these could be combined into single management units to which consistent management regimes can be applied. Where there is considerable biotic variation within zones, it will be necessary to partition them further to identify areas that are sufficiently homogeneous that consistent management guidelines apply. For issues such as these, the development of new theory is unlikely to make significant progress in the absence of accompanying empirical data.

Reserve-selection exercises must not only represent the biota in a region, but also ensure that the resultant reserves are adequate. To achieve this, it will be necessary to combine algorithm-based approaches or gap analyses with species-based approaches that specify requirements for persistence. The combination of these approaches should provide reserve-design parameters that will ensure that the selected reserves continue to meet the needs of the species that are represented.

20.6.2 Approaches for Empirical Research

To test the effectiveness of focal species, such as indicator, umbrella, and keystone species, for indicating changes in an ecosystem of for protecting a wider array of biodiversity, it is essential to state clearly the objective being addressed. Clear criteria also must be specified for selecting such species, and a logical argument supporting the purported benefits must be presented. The collection of presence-absence data from throughout the region of interest will be the minimum requirement for determining whether the probability of occurrence of nonfocal taxa exceeds the probability of occurrence of focal species. However, information on presence or absence is not sufficient to ascertain probabilities of persistence. Relatively abundant species undergoing rapid decline may be more vulnerable than may be rarer species with stable populations. Detailed autecological studies of carefully selected taxa will therefore be required to determine whether the persistence of umbrella species also will ensure the persistence of other species. An array of nonfocal species, representing different taxonomic groups and different life-history attributes, should be selected for these detailed studies.

Because we have very limited experience in recreating particular types of habitat to meet specified conservation goals, it will be necessary to ensure that conservation activities, particularly those involving landscape restoration, are long-term and are conducted within an adaptive framework (Lessard 1998). Management decisions often need to be made with insufficient information, so it is essential to formulate management suggestions as testable hypotheses. This will enable the focused acquisition of relevant data to assess the adequacy of proposed actions. For example, the application of a range of habitat reconstruction treatments with different degrees of initial complexity could be used to determine whether the achievement of a conservation goal requires the complete recreation of an ecosystem, or whether the establishment of a limited suite of early succession species will result, in time, in the development of more complex communities as additional species find their way into the treated area.

The previous sections have selectively outlined some procedures for addressing theoretical and empirical knowledge gaps. However, it is important to recognize that theory and data alone are insufficient for land managers confronted with the need to solve real problems in particular locations. It also is essential to provide pathways from general principles to particular actions. Although ecological planning principles are important as a means of organizing knowledge and conveying important relationships to managers, it is necessary to develop procedures that enable managers to derive appropriate parameters for these relationships. It is not sufficient to tell a manager that larger remnants are better than are smaller ones, or that a landscape must contain "a few large patches." Protocols must also be developed that enable managers to determine the appropriate configuration of landscape elements to meet the specific needs of the area they are managing. The focal-species approach described in Section 20.2.3 represents an attempt to provide such a procedure. Similar approaches will be required if land managers are to deal effectively with the array of additional environmental issues with which they are confronted.

Acknowledgments

The ideas presented in this paper are the outcome of ongoing discussions with numerous colleagues. Dennis Saunders and Kevin Gutzwiller provided valuable comments on earlier drafts of the manuscript. The focal-species planning approach was developed with funding from the Biodiversity Section of Environment Australia.

References

Anonymous 1992. *National Forest Policy Statement: A New Focus for Australia's Forests.* Canberra: Australian Government Publishing Service.

Austin, M.P., Cunningham, R.B., and Flemming, P.M. 1984. New approaches to direct gradient analysis using environmental scalars and statistical curve-fitting procedures. *Vegetatio* 55:11–27.

Bedward, M., Pressey, R.L., and Keith, D.A. 1992. A new approach for selecting fully representative reserve networks: addressing efficiency, reserve design and land suitability with an iterative analysis. *Biol. Conserv.* 62:115–125.

Boecklen, W.J. 1986. Effects of habitat heterogeneity on the species-area relationships of forest birds. *J. Biogeog.* 13:59–68.

Boecklen, W.J., and Gotelli, N.J. 1984. Island biogeographic theory and conservation practice: species-area or specious-area relationships? *Biol. Conserv.* 29:63–80.

Bond, W.J. 1993. Keystone species. In *Ecosystem Function of Biodiversity,* eds. E.D. Schulze and H.A. Mooney, pp. 237–253. Berlin: Springer-Verlag.

Brooker, M.G., and Margules, C.R. 1996. The relative conservation value of remnant patches of native vegetation in the wheatbelt of Western Australia: I. Plant diversity. *Pac. Conserv. Biol.* 2:268–278.

Cody, M.L. 1968. On the methods of resource division in grassland bird communities. *Am. Nat.* 102:107–147.

Collinge, S.K. 1998. Spatial arrangement of habitat patches and corridors: clues from ecological field experiments. *Landsc. Urban Plan.* 42:157–168.

Connell, J.H. 1975. Some mechanisms producing structure in natural communities: a model and evidence from field experiments. In *Ecology and Evolution of Communities,* eds. M.L. Cody and J.M. Diamond, pp. 460–486. Cambridge, Massachusetts: Harvard University Press.

Connor, E.F., and McCoy, E.D. 1979. The statistics and biology of the species-area relationship. *Am. Nat.* 113:791–833.

Cowling, R.M., Heijnis, C., Lombard, A.T., Pressey, R.L., and Richardson, D.M. 1998. *Systematic Conservation Planning for the CAPE Project. Conceptual Approach and Protocol for the Terrestrial Biodiversity Component.* Report on project number IPC 9803 CAPE. Cape Town, South Africa: World Wide Fund for Nature.

Cox, J., Kautz, R., MacLaughlin, M., and Gilbert, T. 1994. *Closing the Gaps in Florida's Wildlife Habitat Conservation System.* Tallahassee: Florida Game and Freshwater Fish Commission.

Curnutt, J., Lockwood, J., Luh, H.K., Nott, P., and Russell, G. 1994. Hotspots and species diversity. *Nature* 367:326–327.

Desmet, P.G., Barret, T., Cowling, R.M., Ellis, A.G., Heijnis, C., le Roux, A., Lombard, A.T., and Pressey, R.L. 1999. *A Systematic Plan for a Protected Area System in the Knersvlakte Region of Namaqualand.* Report of project IPC 9901, Leslie Hill Succulent Karoo Trust. Cape Town, South Africa: World Wide Fund for Nature.

Dobson, A., Ralls, K., Foster, M., Soulé, M.E., Simberloff, D., Doak, D., Estes, J.A., Mills, L.S., Mattson, D., Dirzo, R., Arita, H., Ryan, S., Norse, E.A., Noss, R.F., and Johns, D. 1999. Connectivity: maintaining flows in fragmented landscapes. In *Continental Conservation: Scientific Foundations of Regional Reserve Networks,* eds. M.E. Soulé and J. Terborgh, pp. 129–170. Washington, DC: Island Press.

Fahrig, L., and Merriam, H.G. 1985. Habitat patch connectivity and population survival. *Ecology* 66:1762–1768.

Fahrig, L., and Merriam, H.G. 1994. Conservation of fragmented populations. *Conserv. Biol.* 8:60–71.

Forman, R.T.T. 1995. Some general principles of landscape and regional ecology. *Landsc. Ecol.* 10:133–142.

Forman, R.T.T., and Collinge, S.K. 1996. The "spatial solution" to conserving biodiversity in landscapes and regions. In *Conservation of Faunal Diversity in Forested Landscapes,* eds. R.M. DeGraaf and R.I. Miller, pp. 537–568. London: Chapman and Hall.

Franklin, J.F. 1993. Preserving biodiversity: species, ecosystems or landscapes? *Ecol. Appl.* 3:202–205.

Groom, M., Jensen, D.B., Knight, R.L., Gatewood, S., Mills, L., Boyd-Heger, D., Mills, L.S., and Soulé, M.E. 1999. Buffer zones: benefits and dangers of compatible stewardship. In *Continental Conservation: Scientific Foundations of Regional Reserve Networks,* eds. M.E. Soulé and J. Terborgh, pp. 171–198. Washington, DC: Island Press.

Hanski, I.A. 1991. Single-species metapopulation dynamics: concepts, models and observations. *Biol. J. Linn. Soc.* 42:17–38.

Harrison, S. 1994. Metapopulations and conservation. In *Large-Scale Ecology and Conservation Biology,* eds. P.J. Edwards, R.M. May, and N.R. Webb, pp. 111–128. Cambridge, Massachusetts: Blackwell.

Heijnis, C., Lombard, A.T., and Cowling, R.M. 1997. *Identification of Priority Areas for a Proposed Biosphere Reserve in the Coastal Lowlands of the Western Cape.* Report on project number ZA 523. Cape Town, South Africa: World Wide Fund for Nature.

Hulshoff, R.M. 1995. Landscape indices describing a Dutch landscape. *Landsc. Ecol.* 10:101–111.

Kiester, A.R., Scott, J.M., Csuti, B., Noss, R.F., Butterfield, B., Sahr, K. and White, D. 1996. Conservation prioritization using GAP data. *Conserv. Biol.* 10:1332–1342.

Kirkpatrick, J.B. 1983. An iterative method for establishing priorities for the selection of nature reserves: an example from Tasmania. *Biol. Conserv.* 25:127–134.

Lambeck, R.J. 1997. Focal species: a multi-species umbrella for nature conservation. *Conserv. Biol.* 11:849–856.

Lambeck, R.J. 1999. *Landscape Planning for Biodiversity Conservation in Agricultural Regions.* Biodiversity Technical Paper Number 2. Canberra: Environment Australia.

La Polla, V.N., and Barrett, G.W. 1993. Effects of corridor width and presence on the population dynamics of the meadow vole (*Microtus pennsylvanicus*). *Landsc. Ecol.* 8:25–37.

Launer, A.E., and Murphy, D.D. 1994. Umbrella species and the conservation of habitat fragments: a case of a threatened butterfly and a vanishing grassland ecosystem. *Biol. Conserv.* 69:145–153.

Lessard, G. 1998. An adaptive approach to planning and decision-making. *Landsc. Urban Plan.* 40:81–87.

Lombard, A.T., Cowling, R.M., Pressey, R.L., and Mustart, P.J. 1997. Reserve selection in a species-rich and fragmented landscape on the Agulhas Plain, South Africa. *Conserv. Biol.* 11:1101–1116.

MacArthur, R.H., and Wilson, E.O. 1963. An equilibrium theory of insular zoogeography. *Evolution* 17:373–387.

Margules, C.R., Cresswell, I.D., and Nicholls, A.O. 1994. A scientific basis for establishing networks of protected areas. In *Systematics and Conservation Evaluation,* eds. P.L. Forey, C.J. Humphreys, and R.I. Vane-Wright, pp. 327–350. Oxford, United Kingdom: Clarendon Press.

Margules, C.R., and Nicholls, A.O. 1987. Assessing the conservation value of remnant habitat 'islands': mallee patches on the western Eyre Peninsula, South Australia. In *Nature Conservation: The Role of Remnants of Native Vegetation,* eds. D.A. Saunders, G.W. Arnold, A.A. Burbidge, and A.J.M. Hopkins, pp. 89–102. Chipping Norton, Australia: Surrey Beatty and Sons.

Margules, C.R., and Nicholls, A.O. 1993. Where should nature reserves be located? In *Conservation Biology in Australia and Oceania,* eds. C. Moritz, J. Kikkawa, and D. Doley, pp. 339–346. Chipping Norton, Australia: Surrey Beatty and Sons.

Margules, C.R., Nicholls, A.O., and Pressey, R.L. 1988. Selecting networks of reserves to maximize biological diversity. *Biol. Conserv.* 43:63–76.

McKenzie, N.L., Belbin, L., Margules, C.R., and Keighery, G.J. 1989. Selecting representative reserve systems in remote areas: a case study in the Nullarbor Region, Australia. *Biol. Conserv.* 50:239–261.

Nix, H.A. 1992. Environmental domain analysis. In *Environmental Regionalisation,* ed. R. Thackway, pp. 56–58. Canberra: Australian National Parks and Wildlife Service.

Noon, B., McKelvey, K., and Murphy, D. 1997. Developing an analytical context for multispecies conservation planning. In *The Ecological Basis of Conservation: Heterogeneity, Ecosystems, and Biodiversity,* eds. S.T.A. Pickett, R.S. Ostfeld, M. Shachak, and G.E. Likens, pp. 43–59. New York: Chapman and Hall.

Noss, R.F., and Cooperrider, A. 1994. *Saving Nature's Legacy: Protecting and Restoring Biodiversity.* Washington, DC: Defenders of Wildlife and Island Press.

Noss, R.F., Dinerstein, E., Gilbert, B., Gilpin, M., Miller, B.J., Terborgh, J. and Trombulak, S. 1999. Core areas: where nature reigns. In *Continental Conservation: Scientific Foundations of Regional Reserve Networks,* eds. M.E. Soulé and J. Terborgh, pp. 99–128. Washington, DC: Island Press.

Noss, R.F., O'Connell, M.A., and Murphy, D.D. 1997. *The Science of Conservation Planning: Habitat Conservation Under the Endangered Species Act.* Washington, DC: Island Press.

O'Neill, R.V., Hunsaker, C.T., Timmins, S.P., Jackson, B.L., Jones, K.B., Riitters, K.H., and Wickham, J.D. 1996. Scale problems in reporting landscape pattern at the regional scale. *Landsc. Ecol.* 11:169–180.

Opdam, P. 1991. Metapopulation theory and habitat fragmentation: a review of holarctic breeding bird studies. *Landsc. Ecol.* 5:93–106.

Pressey, R.L. 1998. Algorithms, politics and timber: an example of the role of science in a public, political negotiation process over new conservation areas in production forests. In *Ecology for Everyone: Communicating Ecology to Scientists, the Public and the Politicians,* eds. R. Wills and R. Hobbs, pp. 73–87. Chipping Norton, Australia: Surrey Beatty and Sons.

Pressey, R.L., Humphries, C.R., Margules, C.R., Vane-Wright, R.I., and Williams, P.H. 1993. Beyond opportunism: key principles for systematic reserve selection. *Trends Ecol. Evol.* 8:124–128.

Pressey, R.L., Johnson, I.R., and Wilson, P.D. 1994. Shades of irreplaceability: measuring the potential contribution of sites to a reservation goal. *Biodiver. Conserv.* 3:242–262.

Pressey, R.L., and Nicholls, A.O. 1989. Application of a numerical algorithm to the selection of reserves in semi-arid New South Wales. *Biol. Conserv.* 50:263–278.

Reid, W.V. 1998. Biodiversity hotspots. *Trends Ecol. Evol.* 13:275–280.

Resource and Conservation Assessment Council. 1996. *Draft Interim Forestry Assessment Report.* Sydney, Australia: Resource and Conservation Assessment Council.

Risser, P.G. 1995. Biodiversity and ecosystem function. *Conserv. Biol.* 9:742–746.

Scott, J.M., Davis, F., Csuti, B., Noss, R., Butterfield, B., Groves C., Anderson, H., Caicco, S., D'Erchia, F., Edwards, T.C. Jr., Ulliman, J., and Wright, R.G. 1993. Gap analysis: a geographic approach to protection of biological diversity. *Wildl. Monogr.* 123:1–41.

Simberloff, D.S. 1998. Flagships, umbrellas, and keystones: is single-species management passé in the landscape era. *Biol. Conserv.* 83:247–257.

Simberloff, D., Doak, D., Groom, M., Trombulak, S., Dobson, A., Gatewood, S., Soulé, M.E., Gilpin, M., Martínez del Rio, C. and Mills, L. 1999. Regional and continental restoration. In *Continental Conservation: Scientific Foundations of Regional Reserve Networks,* eds. M.E. Soulé and J. Terborgh, pp. 65–98. Washington, DC: Island Press.

Soulé, M.E., and Terborgh, J., eds. 1999. *Continental Conservation: Scientific Foundations of Regional Reserve Networks.* Washington, DC: Island Press.

Thackway, R., and Creswell, I.D. 1992. *Environmental Regionalisations of Australia.* Canberra: Environmental Resources Information Network.

Trinder-Smith, T.H., Lombard, A.T., and Picker, M.D. 1996. Reserve scenarios for the Cape Peninsula: high-, middle- and low-road options for conserving the remaining biodiversity. *Biodiver. Conserv.* 5:649–669.

Wiens, J.A. 1997. The emerging role of patchiness in conservation biology. In *The Ecological Basis of Conservation: Heterogeneity, Ecosystems, and Biodiversity,* eds. S.T.A. Pickett, R.S. Ostfeld, M. Shachak, and G.E. Likens, pp. 93–107. New York: Chapman and Hall.

21

Assessing the Conservation Potential of Habitat Networks

PAUL OPDAM

21.1 Introduction

Habitat networks are supposed to offer a solution for habitat fragmentation. The notion is that when natural habitat becomes fragmented during economic development of a landscape, individual areas are no longer large enough for persistent populations. Connected as a network, the habitat remnants may still offer conditions for long-term conservation. In landscapes where many land-use functions are combined, landscape and conservation planners need quantitative rules for developing sustainable habitat networks. Planners also need instruments to generate alternative options and scenarios in a search for the most effective, best accepted, and most economically stable network design. In this chapter, I discuss existing concepts of habitat networks and evaluate recent applications of network assessment. Then I propose a new prognostic method called *landscape cohesion assessment,* in which an actual landscape is compared with a reference database of landscapes that offer sustainable conditions for a selection of target species. I finish by discussing research needed to develop the method further and to underpin it with empirical knowledge.

21.2 Concepts, Principles, and Emerging Ideas

Literature abounds with terms that describe networks of landscape sites (see references in Table 21.1), but it is not always clear whether such descriptions are consistent with the network definition I propose. In this chapter, I define a *patch* (sensu Forman 1995) as the site where the habitat conditions of a species are realized. A *network* can be defined physically (as a collection of spatially distinct patches interconnected by linear elements) and functionally (as a collection of patches linked by a flow of, for instance, individuals). I emphasize functional networks in this chapter and elaborate below on the definition of this network type.

21.2.1 Network Definition

To a human observer, a series of patches of similar landscape units (with similar ecosystems) connected by linear elements, like forest patches connected by

TABLE 21.1. Examples of various types of bioecological networks found in the literature. The typology is based on whether there were explicit and operational conservation aims (e.g., the long-term conservation of one or several target species), and whether the network design criteria used were underpinned by knowledge of metapopulation processes.

Explicit and operational conservation aims?	Process-based design criteria for sustainability?	
	No	Yes
No	Europe (Ribaut 1995) Slovakia (Doms et al. 1995) Czech Republic (Kubeš 1996) Belgium (De Blust et al. 1995) Denmark (Brandt 1995) New York City (Yaro and Hiss 1996 in Flores et al. 1998)	Not applicable
Yes	The Netherlands (Van Zadelhoff and Lammers 1995) Germany (Burkhardt et al. 1995) Greece (Troumbis 1995)	Victoria, Australia (Bennett 1999) Rhine Valley (Foppen et al. 1999)

hedgerows, may appear as a landscape network. In most literature, habitat networks are conceived as "a more or less continuous structure of interconnected linear elements" (e.g., Forman 1995). Contrarily, in my definition, linkages between patches are established not through physical elements, but by a stream of individuals. If the linkage is possible without any visible connecting structure, the network appears through the human eye as a spatially independent collection of patches (Figures 21.1a, b). The stream of individuals may be enhanced or concentrated by physical interconnections in the form of corridor-like landscape elements (Figure 21.1c), or may even be impossible without corridors (Figure 21.1d).

A species-specific application of the network concept is somewhat impractical if one wants to develop design standards for landscapes to protect biodiversity. Often, conservation planning deals not with single species, but with biodiversity or a list of target species (Bennett 1999). Intuitively, it can be assumed that species differing in habitat and spatial requirements use different network patterns in the same landscape (Vos et al. 2001). Therefore, when developing a knowledge base for designing networks, the challenge is to integrate species-linked criteria into a multispecies approach. This is a key issue in this chapter.

In this chapter, I define a *habitat network* as an explicit collection of patches in which habitat conditions for a particular species are realized; these patches are embedded in a matrix of nonhabitat and linked by movements of individuals of that species. The key characteristic is the *functional cohesion* between the sites, defined as the overall result of successful dispersal events in the network per generation. Without that cohesion, the same total area of habitat has a lower conservation potential to the species. The dispersal movements link the patch populations to form a metapopulation (Merriam 1988; Opdam 1988, 1990; Opdam et al.

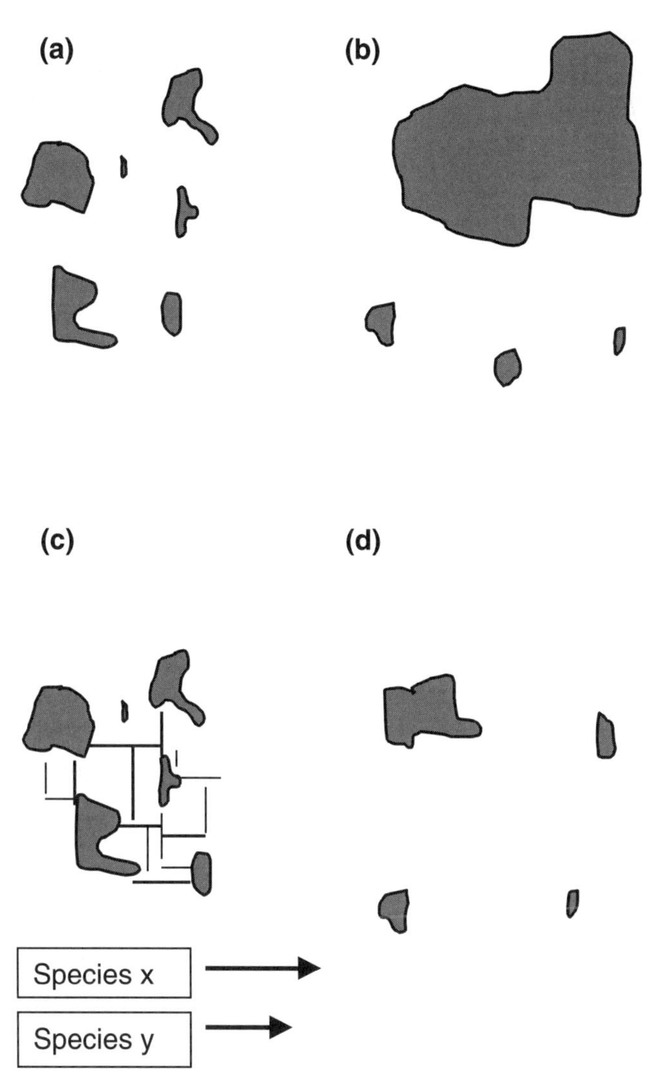

FIGURE 21.1. Four patterns of habitat for species x and y in landscapes of the same spatial scale. The gray patches represent habitat. Lines in the landscape matrix are elements (corridors) that enhance and direct movement of species y but are irrelevant to those of species x. The arrows indicate the distance covering 80% of the dispersal movements. Landscapes (a), (b), and (c) are habitat networks for species x because the dispersal flow can bridge the gap between the patches of habitat. However, species y is not able to link the patches in (a) and (b) because outside of habitat patches, it needs corridors for dispersal. Landscape (c) contains a habitat network for species y. Landscape (d) does not contain a habitat network for species x or y because interpatch dispersal is not possible.

1994; Hanski 1997). A *patch network* is the multispecies counterpart of the habitat network. In this case, the network is interconnected by movements of many species with similar habitat requirements. These movements encompass dispersal and other movement types, such as daily movements between food patches within a territory. The focus on networks becomes common practice wherever fragments of natural vegetation have remained amidst land intensively used by humans.

21.2.2 Other Network Concepts

Because of different geographic, natural, and social-economic conditions, different countries have developed different network approaches. Based on Brandt (1995) and Jongman and Kristiansen (1998), I distinguish three network types: greenways (mainly North-American and Australian and some European urban regions), geoecological networks (Eastern Europe), and bioecological networks (mainly Western Europe, Africa, Australia, and USA).

Predecessors of greenways were already known in the urban planning tradition of the 19th century. In several metropolitan areas, both in the United States and in Europe, *green belt* systems consisted of urban parks and forests designed to connect a city with surrounding nature areas. Their main purpose was to satisfy the recreational needs of inhabitants of crowded and polluted cities. Green belts provided some fresh air and open space (Kavaliauskas 1995; Jongman and Kristiansen 1998). The concept of the *greenway* is directly derived from the original concept of the green belt. Greenways are defined as linear systems of land that are planned, designed, and managed for a combination of purposes, including ecological, recreational, cultural, aesthetic, or other purposes that are compatible with the concept of sustainable land use (see Machado et al. 1995 and Flores et al. 1998 for examples in metropolitan areas).

The concept of *ecostabilization* emerged from a holistic approach in landscape ecology, based on an integration of geomorphological, hydrological, and climatological approaches. Because these networks are mainly designated on the basis of geoecological information, we will call them *geoecological networks.* According to Kavaliauskas (1995) and Jongman and Kristiansen (1998), the basic notion behind ecostabilization is that the landscape is stabilized by alternating strictly delimited zones of intensively used land with a more stable "nature frame," thereby "counterweighing" impacts on the landscape from economic pressure. The ecostabilization framework encompasses all areas that are supposed to contribute to regulation of water, nutrient, and organism flows, and may cover 40% to 70% of the land. I found it impossible to infer how the nature frame was delineated and which underlying theories were used. Mander et al. (1995) admitted that the weakness of the approach is that no optimization criteria have been developed and tested. They propose a set of criteria at three levels of scale. However, population-level criteria for long-term conservation of biodiversity are not included in their list.

Bioecological networks are based on distributions and processes in species populations only. The narrower scope makes it easier to underpin the network design with ecological theory, to use models, and to develop criteria for assessment of network functioning. In this context, the theory of island biogeography has been an important source of inspiration, but it has been replaced by the theory of metapopulations (Diamond 1975; Opdam et al. 1994; Hanski and Simberloff 1997). Whereas the geoecological approach is very strict in the separation of economic and nature zones in the landscape, the bioecological network can be an integrated part of urban and farm landscapes, and the area covered by the network may be well below 10%. Although never defined explicitly, bioecological networks are usually conceived in the sense of patch networks (the multispecies counterpart of the habitat network). They occur in two types: with and without physically connecting landscape elements that are designated as corridor zones.

Although all three network types reviewed here may include components of a habitat or patch network, the bioecological network comes closest to my definition. This network type is most commonly encountered in the literature; it is described in various forms differing in whether explicit and operational conservation aims and criteria for sustainability (based on metapopulation processes) have been formulated (Table 21.1).

21.2.3 Requirements for Assessment Methods

To assess the effectiveness of a habitat network, we need an explicit conservation aim as a reference point. Does the network add to that aim? Would the same amount of habitat without the cohesion of the patches enhance conservation effectiveness? Just monitoring or mapping a species does not enable us to answer these questions. For example, the presence of a species in a network does not prove its long-term persistence. And a monitoring data series showing a species' presence over 10 years does not prove that the persistence is attributable to network cohesion. Simply by observing dispersal between patches, for instance through a corridor, we do not learn much about the extinction risk of the population at the network level. An assessment ("Does the network give long-term protection?") should imply a judgment of the risk of extinction over a longer time frame and a broader spatial scale than we can encompass by direct observations. Therefore, we need a theoretical background on which to build the assessment, and an instrument with which we can extrapolate in time and space.

Basically, I claim here that in practice we only have three options. One option is to build a system of spatially explicit criteria for sustainable networks and assess whether a particular network meets the criteria. A second option is to build mechanistic simulation models of populations in a network structure and apply these models on particular networks to determine whether target species can persist due to network cohesion. The third option is to perform an experiment: redesign half of the landscape according to criteria for cohesion, and do not treat the other half; subsequently, measure and compare populations between the two landscapes. The percentage of occupied patches could be used as an indicator of performance. It could be difficult to keep habitat conditions equal and, at the same time, separate treated and untreated landscapes enough to prevent dispersing individuals from the untreated area from influencing the performance of the population in the treated area. To me, the third option seems impractical and only applicable to underpin theoretical assumptions about network functioning. Moreover, a *prognostic assessment,* which is completed before any landscape measure is taken, will help to make the best decisions today. We simply do not have the time to wait for the outcome of experiments or conclusions from long-term monitoring. So, I suggest that assessment for habitat networks should usually be done in the planning stage, long before the effectiveness can be measured in the field, and therefore based on the conservation *potential* of the landscape. That is why I elaborate below on the first and second options only.

21.3 Recent Applications

In this section, I focus on the bioecological network concept and describe various network-assessment approaches that have been applied to evaluate the conservation value of this type of network. I searched the literature thoroughly enough to distinguish the major types of applications, but I have not attempted to be complete in my citation of examples. Unfortunately, on this subject, considerable gray literature exists, and this material is difficult to obtain.

21.3.1 Theoretical Basis for Assessment

The habitat network is effective if the habitat quality, as well as the spatial arrangement of the patches and the resistance of the landscape matrix, allow persistent network populations. The basic notion is that the network population is viable, whereas local populations are not.

Completely isolated populations will go extinct if, on average, reproduction is lower than mortality plus emigration. This so-called deterministic extinction may happen during times of unsuitable environmental conditions, or after habitat deterioration. Additionally, in small populations, extinction may occur as a result of stochastic variation in the gain and loss of individuals (Shaffer 1981; Goodman 1987). Habitat networks must compensate for these sources of local extinction. Dispersing individuals establish new populations in deserted habitat patches, prevent local extinction, and support sink populations in low-quality habitat. If dispersal is the cohesive factor in a spatially structured population, then persistence depends on the degree of cohesion that the habitat network permits.

I assume with Thomas and Hanski (1997), in spite of little empirical evidence (Harrison and Taylor 1997), that spatially discontinuous populations in fragmented landscapes with small-to-moderate habitat coverage often tend to behave like metapopulations in the sense that regional extinction depends on the interaction of local populations (Vos et al. 2001). A classic metapopulation characteristically shows instability at the patch level, but not necessarily in all patches. Many suitable patches in a landscape may be unoccupied at some point in time. In spite of this, long-term population stability at the landscape level is attainable, provided that local extinctions are compensated by recolonizations (Verboom et al. 1993; Hanski 1997).

Two phenomena related to the distribution of a metapopulation across its habitat network argue for a process-based assessment method rather than a pattern-based approach. First, a metapopulation is often absent in a considerable part of the habitat network, and the metapopulation's spatial distribution changes over time. This implies that for a viable metapopulation, the habitat network must encompass a larger (perhaps much larger) number of patches than is occupied in any given year. Second, a metapopulation distribution pattern may not be in equilibrium with the actual habitat pattern, if this has been changed recently. Due to this time lag between landscape change and metapopulation response, the actual landscape may no longer have conservation potential while the species is still present in adequate numbers. This means that an analysis of a snap-shot distribution does not necessarily reveal information about conditions for long-term persistence. Note that the spatial distribution of habitat across the network determines which type of spatially structured population we are dealing with. More classic metapopulations occur in cases in which all patches are more or less equal in size and far enough apart to restrict dispersal rate. In cases in which one patch is extremely large relative to the other patches, the network population exhibits a mainland-island population structure (as a special case of a metapopulation; Hanski 1997).

21.3.2 Methods Based on Distribution Data of Species

Three types of approaches attempt to designate the number and location of reserves for conservation of biodiversity based on the interpretation of species distribution data. In "gap analysis" (Scott et al. 1993), regional maps of vegetation and animal distributions are compared to determine whether concentrations of target species fall outside of the boundaries of currently protected areas. Vegetation, vertebrate, and butterfly distributions are used as indicators of overall biodiversity. Vegetation maps are constructed from satellite images, and species distribution data are based either on empirical counts or on predictions from vegetation maps by habitat models. Flather et al. (1997) evaluated this approach and suggested that advances in landscape ecology be incorporated into it.

The second type is based on simplifications of the species-area relationship of island biogeographical theory. Margules et al. (1988), for instance, combined two algorithms, one of which aimed at maximizing species richness in a selection of sites, and the second of which ensured that all species and habitat types were included. They concluded that 50% of the total wetland area in an Australian floodplain must be preserved to maintain plant biodiversity. Alternatives to this approach have been proposed by Pressey and Nicholls (1989) and Lomolino (1994).

The third type is based on constructing multiple-regression models of species occurrence in landscape networks (e.g., Van Dorp and Opdam 1987; Opdam 1991 for a review of bird studies; Verboom et al. 1991; Vos and Chardon 1998). The spatial variation in presence-absence or density data is correlated with habitat and landscape features to build a multifactor regression model, which is used to develop rules of thumb for spatial configuration and matrix structure.

None of these methods is explicitly based on spatial processes in the population. Only the third type explicitly considers the relationship between the distribution and the spatial configuration of the landscape network. However, even if a regression model includes significant parameters for network shape, there is no guarantee that the networks necessarily support viability. The landscape network may change and no longer allow viability while the population is still there (Chapter 18). Another limit to the predictive power of regression models is that the correlations on which they are based, even if these reflect causal relationships, may not apply to other landscapes where the local configuration of habitat and the presence of other important landscape features differ from those used to derive the models. A more practical problem with all three approaches is their dependence on complete data sets of species' spatial distributions, which simply do not exist for many areas. Of course, the advantage of these methods is that algorithms are simple and based on empirical data.

21.3.3 Methods Based on Indices of Landscape Pattern

Landscape indices are popular among landscape ecologists (see Gustafson 1998 for a review). Usually, these measures are not explicitly related to any ecological process, and consequently, they cannot be used as descriptors of required habitat arrangements. Most of these indices relate to landscape connectivity only (e.g., Metzger and Décamps 1997) rather than to a combination of connectivity and patch-area distribution. From a comparative test of popular connectivity measures versus a dispersal model, Schumaker (1996) concluded that the predictive power of these measures is poor. The advantage of landscape indices is that they are simple, related to the landscape rather than to the species level, and can be applied irrespective of the availability of species distribution data. They require detailed geographic information system (GIS)-based landscape data to infer species-specific habitat patterns. However, to be ecologically relevant, landscape indices should always be differentiated and scaled according to species characteristics (Gustafson 1998; Verboom et al. 2001; Vos et al. 2001).

P. Opdam, J. Verboom, and R. Reijnen (unpublished manuscript) introduced the term *landscape cohesion* to describe a species-specific, integrated landscape indicator directly linked with metapopulation persistence. Landscape cohesion is determined by habitat quality and coverage, spatial arrangement of habitat, and landscape matrix permeability. Hence, landscape cohesion integrates how many individuals the landscape can maximally contain (landscape carrying capacity), the effect of the spatial distribution of the habitat on the extent to which these spatially distributed individuals constitute a coherent spatially structured population, and the effect of the matrix on the cohesion within that network population. Landscape cohesion is a species-specific concept based on spatial processes in network populations, and may therefore be well-suited for habitat network assessment. The concept of landscape cohesion resembles "connectivity" as defined by Fahrig and Merriam (1994), who focus on dispersal routes rather than on matrix permeability. Usually, connectivity is used as a pattern-based measure to describe the degree to which patches are structurally linked (Forman 1995). *Patch cohesion* (Schumaker 1996) was proposed to quantify connectivity as perceived by dispersing organisms, and hence relates to the individual rather than to the metapopulation level (see also Bennett 1999). Operational methods based on landscape cohesion have appeared in research reports (Reijnen and Koolstra 1998; Foppen et al. 1999). Verboom et al. (2001) proposed the *key patch* approach; this is an operational form of habitat network assessment based on the presence in the network of a large patch in which the local population has a very small probability of going extinct.

21.3.4 Methods Based on Mechanistic, Spatially Explicit Metapopulation Models

Metapopulation models are suitable tools for assessing the conservation potential of a landscape on the basis of population processes (Verboom et al. 1993). The weakness of metapopulation models is parameter uncertainty, due to little field evidence or wide variation in field estimates. Therefore, to attain high predictive power with metapopulation models, it is critical to use real distribution data to calibrate estimates of the most sensitive parameters. Good examples of calibrated metapopulation models are described by Lindenmayer and Lacy (1995), Lindenmayer and Possingham (1995), and Smith and Gilpin (1997). Thomas and Hanski (1997) and Vos et al. (2001) present examples of metapopulation models based on presence-absence patterns during one or several years. Figure 21.2 describes an assessment obtained by applying the model METAPHOR on possible future scenarios for the Rhine River. Different water management options named after rivers (Loire and Mississippi) were considered. In the Loire River option, the river dynamics are given more room within the floodplains, resulting in larger units of natural areas with larger distances in between. In the Mississippi scenario, the former flood basins in the Rhine Delta are connected to the river system to restore them as spillways, resulting in large macrophyte marshes. For a few small and large bird species typical of river forests and marshes, the potential for persistence in the expected habitat network was predicted and compared with the situation expected without a change in river management. The conclusions played a role in decisions about future river management.

The advantages of metapopulation-model approaches over methods mentioned in previous sections are predictive power and the mechanistic basis. Models that contain all of the key factors can predict the persistence of species for any landscape configuration or change. The main disadvantage (apart from model uncertainties) is that landscapes are usually not planned for a single species, but for an array of species, each with its unique critical thresholds. In theory, it is possible to do a metapopulation-based analysis for a number of target species separately, and then to find the best compromise by combining the results. But for most planning exercises, this procedure will be too laborious and

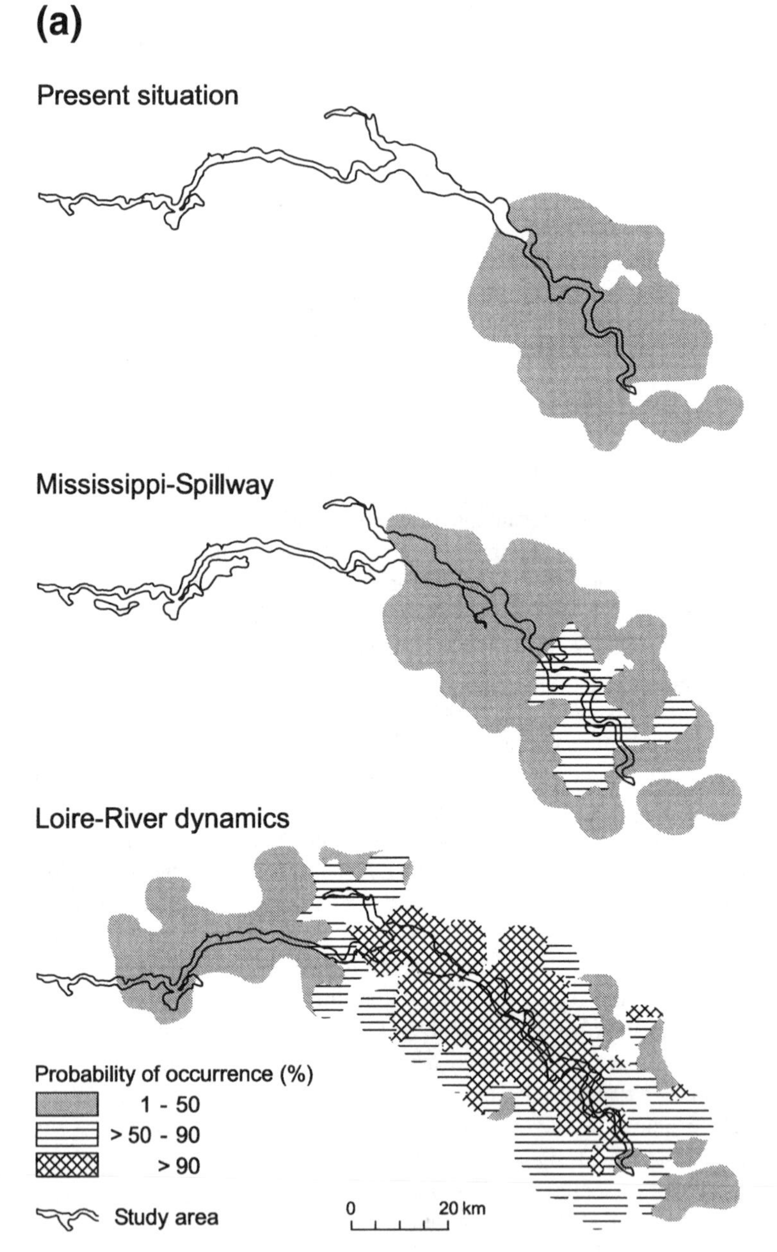

FIGURE 21.2. Example of an assessment of the conservation potential of habitat expected to develop in two possible scenarios for part of the valley of the Rhine River, The Netherlands and Germany. The model used (METAPHOR) is an individual-based stochastic metapopulation model system, for this case adapted and calibrated for the Middle Spotted Woodpecker (*Dendrocopos medius*), a riparian old-growth forest indicator species with small dispersal capacity. The two scenarios were conceptualized assuming different water

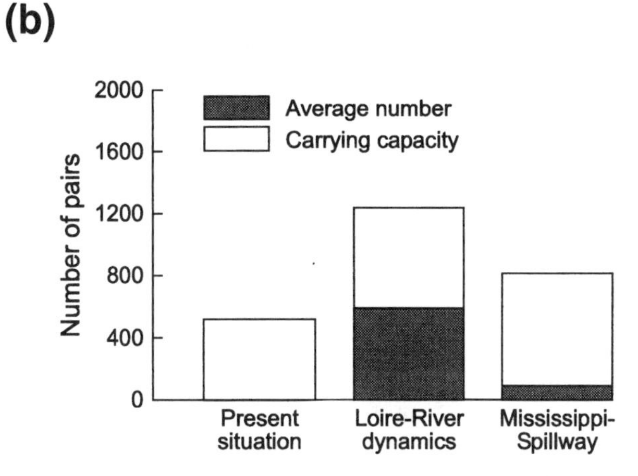

regimes in the river system; (a) depicts the simulated distribution patterns expressed as different levels of occurrence probabilities; (b) summarizes the calculated number of woodpecker pairs, both average and maximally possible carrying capacity. The metapopulation extinction chance is smaller than 5% in the Loire-River scenario only. From Reijnen et al. 1995, reproduced by permission of RIZA© 1995.

complicated. Metapopulation models can, however, be used to generate general rules of thumb about minimal area and interpatch distance for spatial conditions that allow persistence (Verboom et al. 2001). This promising route is explained further in the next section.

21.4 Principles for Applying Landscape Ecology

21.4.1 General Principles for Assessing the Conservation Potential of Patch Networks

Based on 20 years of experience in applying landscape ecology in planning and policy, I have developed the following principles for network assessment. Most importantly, the assessment method should be based on the relationship between spatial population processes and the landscape pattern. The method must be able to integrate requirements of species that differ widely in their response to landscape structure and in the spatial and temporal scales at which they relate to habitat amount and spatial arrangement. The method should be able to assess the conservation potential of a network independent of actual species distribution data. The output parameter must be directly related to the long-term persistence chance of the selected species (or a model representative of a group of species whose responses to landscape change are similar—a *species profile*). The structure of the method must be modular, so that it can be improved with new empirical knowledge. And, last but not least, the output should be simple for nonecologists to understand. In my view, landscape cohesion assessment (P. Opdam, J. Verboom, and R. Reijnen, unpublished manuscript) is the only method that meets these requirements. The method is based on the assumption that it is possible to tell from the configuration of the landscape network whether it allows persistence of a selected number of species or species profiles. Subsequently, the following steps are taken (Figure 21.3): define conservation aims; define target or indicator species; delineate habitat networks for selected species (profiles); for each species separately, determine the landscape cohesion of the network and compare this to a reference framework of potentially sustainable network patterns; assess the conservation potential of the network for each species; and integrate species-level information into biodiversity-level measures of conservation potential. This procedure can be performed electronically in a GIS environment. In the following subsections, I provide specific principles for applying landscape cohesion assessment and discuss techniques for implementing those principles. These principles and techniques, derived from frequent interactions with applied ecologists, plan-

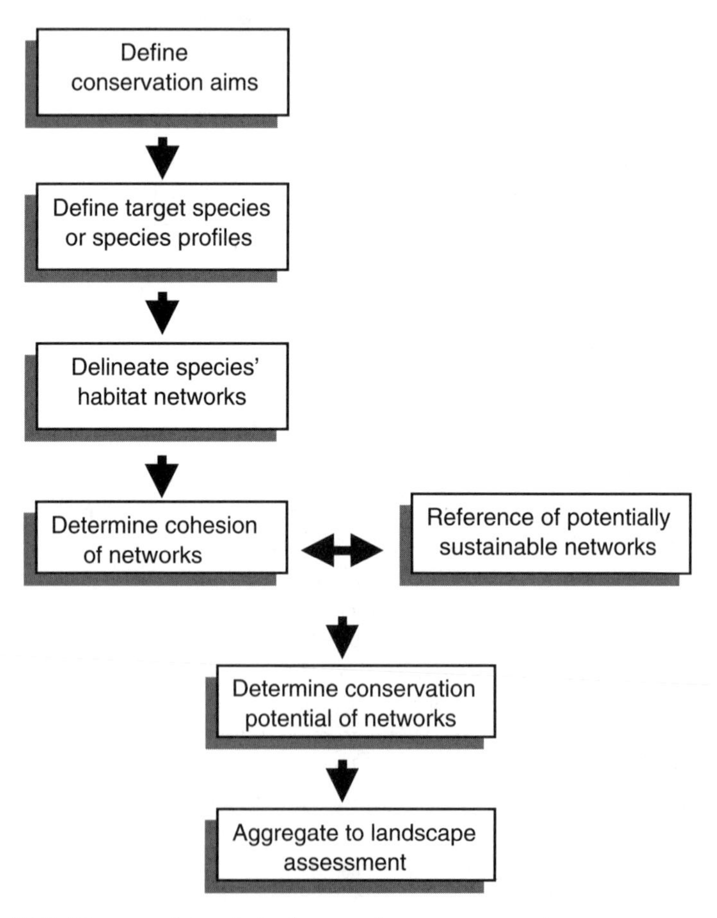

FIGURE 21.3. Diagram of steps in landscape cohesion assessment.

ners, and politicians, were developed during the last 10 years at my institution, the Department of Landscape Ecology of ALTERRA (former DLO-Institute for Forestry and Nature Research) in The Netherlands.

21.4.2 Principle 1: Determine Operational Goals of Biodiversity Conservation

To assess the potential of a network for conservation of biodiversity, we need a reference level for comparison. Determining the reference is not a scientist's decision. Reference points may be inferred from statements on conservation priorities, such as the species the human society of a region wishes to conserve, or a general statement about the conservation of biodiversity. In The Netherlands, for example, the government has developed a system of ecosystem types with target species, based on a nature conservation plan that had been approved in Parliament. Thus, one may decide to focus on ecosystem types for which the region is outstanding or unique in the country or the world, or to target rare or declining species. Ideally, this is a political or democratic process in which the ecologist has the role of facilitator in the discussion. Another way to put biodiversity conservation into practice is to reduce the variation among species to a more manageable system of species profiles. If the aim of the network is to guarantee long-term persistence of all species, it should be determined which species need the cohesion of the patch network. Vos et al. (2001) proposed a system of species profiles as indicators for aspects of biodiversity that depend on spatial characteristics of the network.

21.4.3 Principle 2: Determine the Acceptable Risk of Regional Extinction

Apart from defining the conservation goal, we need to determine the acceptable risk of extinction. Because population persistence is subject to stochastic processes, the risk of extinction is never zero. Close-to-zero risk levels for all species require excessively large areas, particularly for species that require large territories. So it may be more manageable to aim for reasonably small extinction chances per network, and spread the risk over several networks. In my experience in The Netherlands, a 5% level of extinction in a period of 100 years per network proved to be workable, provided that it is plausible that the networks are linked by exceptional long-distance dispersers.

21.4.4 Principle 3: Determine Habitat Networks

Maps of the distribution of habitat are inferred from spatial databases. The selected aims (Principle 1) determine whether this interpretation must be specific for species (in the case of target species), or more general at the level of ecosystem types (in the case of species profiles). This habitat map should be developed into a map of patches for local populations (P. Opdam, J. Verboom, and R. Reijnen, unpublished manuscript). Pieces of habitat close together may be interpreted as belonging to one patch. Subsequently, the patches close enough together to fall within the dispersal range of species are assumed to constitute networks.

21.4.5 Principle 4: Determine Conservation Potential of Networks

Based on a system of indices and standards for networks that allow persistent populations (Verboom et al. 2001), procedures should be developed to assess whether the networks obtained for the planning area provide the spatial conditions for persistence. For example, one can use computer-generated landscapes to simulate the domain of persistence for a range of variation in habitat configuration, either for specific species or for species profiles (Vos et al. 2001). The characteristics of the habitat networks can be conceptualized as axes of a hyperspace of all possible networks. With the help of metapopulation and dispersal-simulation models, one can determine in which part of the hyperspace the species can persist. Alternatively, one may construct a reference database with time series of species distributions in habitat networks.

The result of this step is an assessment of the conservation potential of existing or expected habitat networks in an area, based on metapopulation dynamics. Figure 21.4 depicts an application of the LARCH (Landscape Analysis and Rules for the Cohesion of Habitat) model, a rule-based GIS model based on these principles (Foppen et al. 1999).

21.5 Knowledge Gaps

Landscape cohesion assessment typically determines *a priori* the conservation potential of networks in a landscape by integrating process-based information about the spatial structure of the landscape into simple multispecies indicators. Landscape cohesion assessment is still far from being accepted and developed for regional application. We have developed the LARCH model (Foppen et al. 1999) to an operational stage, and it is being used to detect conservation bottlenecks in landscapes and to plan spatial conditions for biodiversity conservation. Nevertheless, this approach is in an early stage of development. I regard the following uncertainties and knowledge gaps as most important, either because they are crucial to metapopulation persistence but underrepresented in research to date, or because they are crucial for generating aggregated maps showing conservation potential for biodiversity instead of for just single species.

21.5.1 Theoretical Voids

Theory of Landscape Cohesion

Landscape cohesion is a unifying concept at the landscape scale that allows aggregation of species-level information. The concept is new (P. Opdam, J. Verboom, and R. Reijnen, unpublished manuscript), and practical methods to calculate landscape

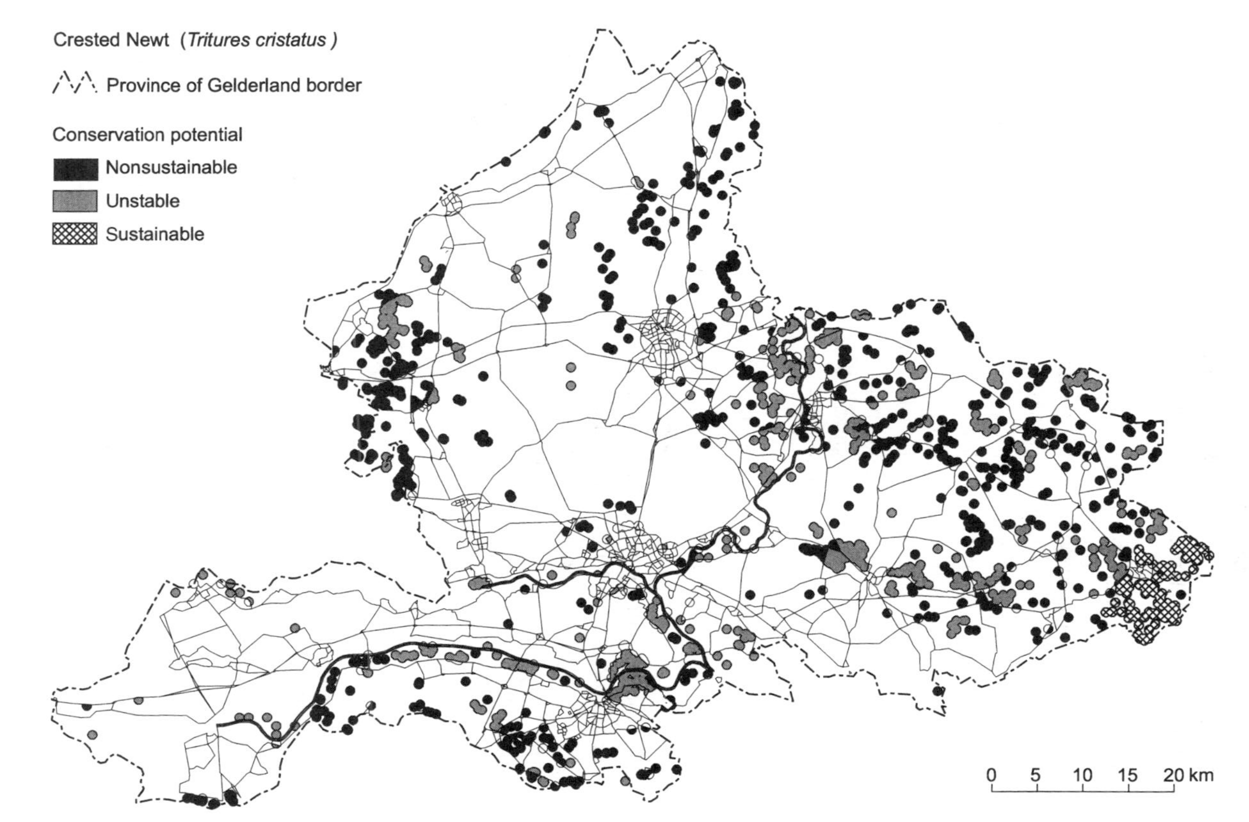

FIGURE 21.4. Example of results of an assessment of the conservation potential of landscape areas by applying the LARCH model. From Reijnen and Koolstra 1998, reproduced by permission of IBN-DLO ©1998.

cohesion are not readily available (but see Verboom et al. 2001 for a first example). The relationship between ecologically scaled landscape indices (Vos et al. 2001) and extinction thresholds for metapopulations needs further clarification.

Spatial Dynamics of Habitat Networks

In most metapopulation studies to date, the spatial structure of the habitat network was assumed to have been constant over time. Consequently, effects of changes in the habitat configuration on metapopulation viability, including the response time (Chapter 18), are not well-known. For slow-moving species, like forest plants, the response time can take more than a century (Grashof-Bokdam and Geertsema 1998). Clearly, compared with a network that is static over time, a network of dynamic habitat needs more cohesion and more total area to offer sustainability (Hanski 1999). Theory about effects of static and dynamic habitat networks on the persistence of slow- and fast-responding species needs development.

Generality of Metapopulation Theory for Habitat Networks

Landscape cohesion assessment depends very much on assumptions from metapopulation theory. But, of course, metapopulation theory does not adequately describe the distribution of every spatially structured population at the landscape scale. For example, a population may exist in a continuous network of narrow strips of habitat (e.g., roadside verges or hedgerows). In such a network, the loss of individuals to unsuitable adjacent areas may be an important cause of death (Vermeulen 1995; Van Dorp et al. 1997) and probably causes gaps in the population. We need to extend metapopulation theory into this domain of landscape networks.

Habitat Suitability Aspects at the Network Scale

Species, such as avian predators, that perceive the landscape patchwork as a heterogeneous habitat may use the patchwork during daily or seasonally searches for resources. For this situation, the assessment of the patch network must be based on the quality of food patches and the distance the individual has to cover daily. To extend our reference database, we need spatially explicit habitat suitability models that take into account the effect of landscape heterogeneity on individual performance. Note that in these cases, the measure for conservation potential will be in terms of habitat suitability instead of population persistence.

Uncertainties in the Predictive Tool

The reference system for landscape cohesion assessment is largely based on model simulations, and because models involve many uncertainties, it is necessary to estimate the uncertainty of conclusions obtained with landscape cohesion assessment models. Those uncertainties must be assessed in relation to the required confidence of the prediction. The divergence of landscape scenarios influences allowable uncertainty in model output; nature conservation potentials for scenarios that differ widely may still be distinguishable with enough confidence even at high degrees of uncertainty in the predictive tool. I do not know of any methodological study that considers the uncertainty of model output with reference to the difference between landscape scenarios.

21.5.2 Empirical Voids

Lack of Empirical Studies on Metapopulations at the Landscape Scale

In metapopulation studies, theoretical studies with simulation models are overemphasized. Good empirical cases at the landscape level, supported by modeling studies, are scarce (e.g., Verboom et al. 1991; Thomas and Hanski 1997; Vos et al. 2000). Such combinations are particularly needed to extend locally obtained field results beyond the limits of the study area and to interpret distribution patterns in terms of persistence chance. In this context, the lack of calibrated metapopulation models needs consideration.

Multispecies Interactions in Networks

Another shortcoming is the single-species focus of most present metapopulation approaches. Consequently, rules for sustainable landscapes calculated by means of metapopulation models exclude the role of interspecific relations. At best, if empirical data are used, the impact of the average predator or competitor is only implicit in birth and death parameters. We need empirical studies of multispecies systems in fragmented landscapes (e.g., Bengtsson 1991) to develop a basis for taking multispecies relations into account.

Dispersal and Colonization

Our knowledge of how dispersal is influenced by landscape pattern is really poor. Field studies linking dispersal distance and direction to the variation of landscape pattern are extremely scarce. Most dispersal studies are biased toward following an individual's daily movements through its home range, but dispersal may be influenced by other forces. The role of corridors has been illustrated by observational and experimental studies (see Bennett 1999 for a recent overview), but most evidence does not allow conclusions about the effectiveness of different kinds of corridors for a variety of species, or about how variation in the landscape matrix relates to dispersal success. Another void is the understanding of factors controlling settlement (e.g., the role of social attraction in territory establishment).

Parameters for Landscape Cohesion

If we aim to assess the conservation potential by landscape pattern alone, it is essential to have generally applicable landscape parameters that are ecologically relevant to spatial cohesion. Such parameters need to vary in proportion to the variation of the persistence probability of a species or to the number of species for which the landscape offers sustainability (Vos et al. 2001). Empirical studies that relate spatial distribution to landscape patterns use a wide array of measures of configuration, which complicates or even prevents any comparison between landscapes and regions. There is a need to develop a common basis for measures of landscape cohesion.

Knowledge Required to Identify Habitat Networks

Landscape cohesion assessment requires that we can recognize the habitat of species in the field and find the habitat characteristics in the spatial database of the region. For this purpose, we do not need sophisticated data on habitat suitability. Many vertebrate species in temperate ecosystems have been studied well enough to recognize, in the field, the pattern of their habitat and the landscape elements that may transmit their dispersal. Usually, the bottleneck lies in the quality of the land-cover map. Land-cover maps may be not detailed enough, or they may be incomplete. The legend of a land-cover map may be designed for land-use interpretation and lack essential ecological information, such as the structure of the vegetation or the presence of small landscape elements. For specific target species, we need maps that enable us to distinguish between habitat and nonhabitat, and between high- and low-quality habitat.

21.6 Research Approaches

Considering the number of uncertainties and knowledge gaps, one could argue that adequate prediction of the conservation potential of landscapes based on persistence rules is far beyond our scientific horizon. I disagree fundamentally with such a conservative point of view and instead make a plea to integrate all that we do know to help politicians and conservation and landscape planners as much as we can to improve the quality of their decision making. Below are suggestions for how to work on the voids.

In landscape ecology, it is essential to integrate theoretical and empirical studies. Empirical studies are indispensable for formulating and testing hypotheses, and for learning differences between species' responses to spatial scale and landscape change. Theoretical studies guide empirical work and help us understand complex multifactor relationships in a discipline in which full-scale field experiments are typically impractical. Also, models are indispensable for generalizing lessons from specific field studies to other landscape patterns and species. Finally, models and field results must be used in combination to generate generic rules for sustainable landscapes and to construct the reference database for landscape cohesion assessment. So, in my presentation of research approaches, theoretical and empirical work are intertwined.

Theory and Parameters for Landscape Cohesion

The essential step in landscape cohesion assessment is to develop the reference database of sustainable habitat networks. For this, a database that reflects the real-world landscape variation encountered in landscape planning must be developed. This can be done by means of computer-generated landscape patterns and simulation models that are specified for species profiles and calibrated for real-world representatives per profile (Vos et al. 2001). It is necessary to include the effect of environmental stochasticity, for which some profiles are more susceptible than are others, and the effect of matrix permeability (Chapter 6). Distribution patterns and other data from long-term monitoring schemes are very useful here for calibration, but researchers also can underpin the modeling approach through brief and local field sampling. Important measures for metapopulation performance (related to persistence) are percentage of occupied patches per year (Vos et al. 2001), and turnover rates in presence-absence patterns (Hanski 1997). Monitoring data expressed in terms of these measures are useful for testing landscape cohesion predictions. However, monitoring schemes must be more detailed than is the usual approach of recording presence or absence per large grid; they must contain information on the patch scale. Another fruitful approach may be to predict the decreasing conservation potential along a gradient of decreasing landscape cohesion and compare that with a sample of distribution data from real-world landscapes along that gradient.

Metapopulations in a Variety of Habitat Networks

I promote empirical case studies in spatially explicit patch networks to form the basis of our understanding of how species respond to landscape patterns. Comparing spatial patterns of a single species in different landscapes, and comparing spatial patterns of different species in a single landscape, will provide the necessary empirical basis for all of the modeling exercises. It will be necessary to simplify the immense variation among species with respect to (for example) life histories, habitat requirements, and dispersal behavior. Vos et al. (2001) propose a system of species profiles that may be useful to select representative life forms that matter in a landscape context. This provides a theoretical framework for selecting species in field studies. Considering the overemphasis of metapopulation studies on temperate forest birds and dry grassland butterflies, our empirical knowledge on responses of species to habitat patterns at the landscape level must be extended with representatives of all species profiles. Special emphasis is needed to reveal the effect of the spatial dynamics of habitat networks on landscape cohesion, taking into account differences in response time among species (see Chapter 18).

Important extensions to these metapopulation studies must focus on the impacts of multispecies and multitrophic interactions on predictions by landscape cohesion assessment tools. This is particularly important when target species are specified. Which species can only persist in the presence or absence of other species? Are there any thresholds in spatial cohesion that are critical for the coexistence of competitors or predator-prey relationships?

The generality of the metapopulation theory is a critical issue when dealing with continuous networks of linear landscape elements. Responses of populations to spatial configuration of such networks are best studied with a mix of empirical and modeling studies. I recommend testing the hypothesis by Opdam et al. (2000)

that area coverage of the network, its degree of continuity, and the density of intersections and nodes are the three most important spatial features of this network type for biodiversity. Comparative field surveys can offer initial underpinning for this theory. We also must develop a new type of spatially explicit model that is able to deal with interdependence of density fluctuations between nearby network units (e.g., intersections and nodes vs. adjacent narrow elements). Modeling studies involving different spatial configurations can teach us about the impact of losses in areas adjacent to networks. The probability that an individual crosses the edge of the network and its subsequent fate must be studied in experimental field studies with capture-recapture techniques in combination with modeling of movement patterns (e.g., Vermeulen 1995; Vos 1999).

Dispersal

In landscape cohesion assessment, the expected ability of individuals to move between patches is the criterion for clustering patches into networks. One challenge here is to distinguish, in field studies, between dispersal and home-range movements; individuals may have different motives and respond differently to landscape pattern during a daily feeding routine than when they are searching for a place to settle in an unfamiliar landscape. A second challenge is to develop general measures of landscape permeability for each species profile. According to Vos et al. (2001), species profiles representing the variety of responses to spatial pattern are based on spatial requirements and dispersal effectiveness, ecologically scaled for the relevant scale of the landscape. So we must incorporate habitat configuration features and, for profiles representing ground-moving species, matrix permeability into our understanding of how landscape patterns determine dispersal. The best approach may be to determine, for an array of species representing various species profiles, how individuals decide where to move from different positions in the landscape. For instance, one can measure the percentage of individuals deciding to pass an edge between a hedgerow and a field, or the probability that a gap in a road verge is crossed, and use that as an estimate of a model parameter in a dispersal model. Subsequently, these models can be calibrated on movement patterns measured by capture-mark-recapture or telemetric monitoring studies (see Chapter 6 for more details and literature). The dilemma here is how to go about getting rid of all the details we do not need when eventually we scale the local species studies up to the level appropriate for landscape cohesion assessment. Models can help to resolve this problem by enabling us to find the key factors.

Habitat Suitability

We need studies that relate landscape pattern to habitat quality. Most studies driven by metapopulation or island-biogeographic theory assume that habitat patches are homogeneous units, and they consider the landscape matrix to be either irrelevant or nonhabitat that affects dispersal movements (Wiens 1997). We need to extend to the landscape scale theories and models about time and energy constraints during the feeding and care of offspring. We need field studies to understand the effect of changing the spatial distribution of landscape elements within home ranges on these constraints and, consequently, on the performance of individuals. Can we identify thresholds or risk levels in densities of feeding patches where the suitability of landscapes as habitat for medium-sized birds and mammals changes? Studies across a range of landscape heterogeneity that link demographic parameters (e.g., reproductive performance) to the spatial characteristics of the landscape will support modeling studies and field studies on behavioral responses to landscape pattern. Through this approach, the results can be generalized and simplified for application in landscape cohesion assessment.

Analysis of Uncertainty in Habitat Maps and Assessment Tools

Predictions of the conservation potential of networks are a combination of predictions about the suitability of landscape units as habitat for a species, and predictions about the potential of the habitat pattern to maintain persistence of the network population. The accuracy of such predictions can be estimated and calibrated with distribution data obtained in monitoring schemes and other surveys. For example, the impact of different habitat suitability rules on the predicted habitat pattern can be compared with the actual distribution of a species. However, in doing this, we have to separate the accuracy of the predicted persistence probability from that of the underlying habitat predictions. An extra difficulty is that metapopulation dynamics exhibit a great deal of chance behavior. Because of that, the validation of a predicted extinction chance requires hundreds of metapopulations. We need statistics to calibrate metapopulation models on time series of metapopulation distribution patterns. Note that landscape cohesion assessment uses a reference database assuming a 5% extinction probability in 100 years, implying that one out of 20 target species will be extinct after 100 years. Studies testing the uncertainty of landscape cohesion assessment must account for this phenomenon by increasing the sample size. A good option is to set up monitoring schemes for a series of habitat networks and survey presence and absence over a long time series.

We need more insight into the accuracy of assessment tools in comparison to the type of prediction we make. When comparing possible alternative scenarios for landscape development, systematic deviations are leveled out. Hence, the prediction for scenarios ("This scenario is better than that one.") is more certain than is an absolute prediction of the effect of a landscape plan on a species' population, because in the latter, all uncertainties in the model are mirrored in the advice. We need studies that quantify the uncertainty of habitat cohesion models in relation to different types of application. We should compare the overlap in uncertainty ranges for pairs of landscape scenarios and determine how different the scenarios must be to provide discriminating advice about which scenario to choose.

Acknowledgments

The concept of landscape cohesion assessment was developed during a number of projects carried out at the Department of Landscape Ecology of ALTERRA, the former DLO-Institute for Forestry and Nature Research, The Netherlands. I thank my colleagues for their contributions during this process. Joost van Kuijk provided literature research on planned networks. The manuscript has benefited from constructive comments by Rogier Pouwels.

References

Bengtsson, J. 1991. Interspecific competition in metapopulations. *Biol. J. Linn. Soc.* 42:219–237.

Bennett, A.F. 1999. *Linkages in the Landscape: The Role of Corridors and Connectivity in Wildlife Conservation.* Gland, Switzerland and Cambridge, United Kingdom: The World Conservation Union (IUCN) Forest Conservation Programme.

Brandt, J. 1995. Ecological networks in Danish planning. *Landschap* 12(3):63–76.

Burkhardt, R., Jaeger, U., Mirbach, E., Rothenburger, A., and Schwab, G. 1995. Design of the habitat network of Rheinland-Pfalz State (Germany). *Landschap* 12(3):99–110.

De Blust, G., Paelinckx, D., and Kuijken, E. 1995. The green main structure for Flanders. *Landschap* 12(3):89–98.

Diamond, J.M. 1975. The island dilemma: lessons from modern biogeographic studies for the design of nature reserves. *Biol. Conserv.* 7:129–146.

Doms, M., Steffek, J., and Jancova, M. 1995. Ecological networks in Slovakia. *Landschap* 12(3):39–50.

Fahrig, L., and Merriam, G. 1994. Conservation of fragmented populations. *Conserv. Biol.* 8:50–59.

Flather, C.H., Wilson, K.R., Dean, D.J., and McComb, W.C. 1997. Identifying gaps in conservation networks: of indicators and uncertainty in geographic-based analyses. *Ecol. Appl.* 7:531–542.

Flores, A., Pickett, S.T.A., Zipperer, W.C., Pouyat, R.V., and Pirani, R. 1998. Adopting a modern ecological view of the metropolitan landscape: the case of a greenspace system for the New York City Region. *Landsc. Urban Plan.* 39:295–308.

Foppen, R., Geilen, N., and Van der Sluis, T. 1999. *Towards a Coherent Habitat Network for the Rhine.* Research Report 99/1. Wageningen, The Netherlands: Institute for Forestry and Nature Research.

Forman, R.T.T. 1995. *Land Mosaics: The Ecology of Landscapes and Regions.* Cambridge, United Kingdom: Cambridge University Press.

Goodman, D. 1987. Consideration of stochastic demography in the design and management of biological reserves. *Nat. Resour. Model.* 1:205–234.

Grashof-Bokdam, C.J., and Geertsema, W. 1998. The effect of isolation and history on colonization patterns of plant species in secondary woodland. *J. Biogeog.* 25:837–846.

Gustafson, E.J. 1998. Quantifying landscape spatial pattern: what is the state of the art? *Ecosystems* 1:143–156.

Hanski, I. 1997. Metapopulation dynamics, from concepts and observations to predictive models. In *Metapopulation Biology: Ecology, Genetics and Evolution,* eds. I. Hanski and M.E. Gilpin, pp. 69–91. London: Academic Press.

Hanski, I. 1999. Habitat connectivity, habitat continuity, and metapopulations in dynamic landscapes. *Oikos* 87:209–219.

Hanski, I., and Simberloff, D. 1997. The metapopulation approach, its history, conceptual domain, and application to conservation. In *Metapopulation Biology: Ecology, Genetics and Evolution,* eds. I. Hanski and M.E. Gilpin, pp. 5–26. London: Academic Press.

Harrison, S., and Taylor, A.D. 1997. Empirical evidence for metapopulation dynamics. In *Metapopulation Biology: Ecology, Genetics and Evolution,* eds. I. Hanski and M.E. Gilpin, pp. 27–42. London: Academic Press.

Jongman, R.H.G., and Kristiansen, I. 1998. *National and Regional Approaches for Ecological Networks in Europe.* Strasbourg: Council of Europe.

Kavaliauskas, P. 1995. The nature frame. Lithuanian experience. *Landschap* 12(3):17–26.

Kubeš, J. 1996. Biocentres and corridors in a cultural landscape. A critical assessment of the 'territorial system of ecological stability.' *Landsc. Urban Plan.* 35:231–240.

Lindenmayer, D.B., and Lacy, R.C. 1995. A simulation study of the impacts of population subdivision on the mountain brushtail possum *Trichosurus caninus* Ogilby (Phalangeridae: marsupialia) in south-eastern Australia. I. Demographic stability and population persistence. *Biol. Conserv.* 73:119–129.

Lindenmayer, D.B., and Possingham, H.P. 1995. Modelling the viability of metapopulations of the endangered Leadbeaters's possum in south-eastern Australia. *Biodiver. Conserv.* 4:984–1018.

Lomolino, M.V. 1994. An evaluation of alternative strategies for building networks of nature reserves. *Biol. Conserv.* 69:243–249.

Machado, J.R., Andresen, M.T., Tico, A.T., Ahern, J., and Fabos, J.G. 1995. Metropolitan landscape planning. A greenway vision for the Lisbon Metropolitan Area (AML). *Landschap* 12(3):111–121.

Mander, U., Palang, H., and Jagomägi, J. 1995. Ecological networks in Estonia. Impact of landscape change. *Landschap* 12(3):27–38.

Margules, C.R., Nicholls, A.O., and Pressey, R.L. 1988. Selecting networks of reserves to maximise biological diversity. *Biol. Conserv.* 43:63–76.

Merriam, G. 1988. Landscape dynamics in farmland. *Trends Ecol. Evol.* 3:16–20.

Metzger, J.-P., and Décamps, H. 1997. The structural connectivity threshold: an hypothesis in conservation biology at the landscape scale. *Acta Œcologia* 18:1–12.

Opdam, P. 1988. Populations in fragmented landscape. *Münstersche Geog. Arb.* 29:75–77.

Opdam, P. 1990. Dispersal in fragmented populations: the key to survival. In *Species Dispersal in Agricultural Habitats,* eds. R.G.H. Bunce and D.C. Howard, pp. 3–17. London: Belhaven Press.

Opdam, P. 1991. Metapopulation theory and habitat fragmentation: a review of Holarctic breeding bird studies. *Landsc. Ecol.* 5:93–106.

Opdam, P., Foppen, R., Reijnen, R., and Schotman, A. 1994. The landscape ecological approach in bird conservation: integrating the metapopulation concept into spatial planning. *Ibis* 137:139–146.

Opdam, P., Grashof, C., and Van Wingerden, W. 2000. Green veins: a spatial concept for combining land use functions and biodiversity in farmland. *Landschap* 17(1):45–51.

Pressey, R., and Nicholls, A.O. 1989. Application of a numerical algorithm to the selection of reserves in semi-arid New South Wales. *Biol. Conserv.* 50:263–278.

Reijnen, R., Harms, W.B., Foppen, R.P.B., De Visser, R., and Wolfert, H.P. 1995. *RHINE-ECONET. Ecological Networks in River Rehabilitation Scenarios: A Case Study for the Lower Rhine.* Lelystad, The Netherlands: Rijks Instituut voor de Zuivering van Afvalwater.

Reijnen, R., and Koolstra, B. 1998. *Evaluation of Corridor Effectiveness in the Gelderland Province.* Research Report 372. Wageningen, The Netherlands: Institute for Forestry and Nature Research.

Ribaut, J.P. 1995. Ecological networks. Experience gained by the Council of Europe. *Landschap* 12(3):11–16.

Schumaker, N. 1996. Using landscape indices to predict habitat connectivity. *Ecology* 77:1210–1235.

Scott, J.M., Davis, F., Csuti, B., Noss, R., Butterfield, B., Groves, C., Anderson, H., Caicco, S., D'Erchia, F., Edwards, T.C. Jr., Ulliman, J., and Wright, R.G. 1993. Gap analysis: a geographic approach to protection of biological diversity. *Wildl. Monogr.* 123:1–41.

Shaffer, M.L. 1981. Minimum population sizes for species conservation. *BioScience* 31:131–134.

Smith, A.T., and Gilpin, M.E. 1997. Spatially correlated dynamics in a pika population. In *Metapopulation Biology: Ecology, Genetics and Evolution,* eds. I. Hanski and M.E. Gilpin, pp. 407–428. London: Academic Press.

Thomas, C.D., and Hanski, I. 1997. Butterfly metapopulations. In *Metapopulation Biology: Ecology, Genetics and Evolution,* eds. I. Hanski and M.E. Gilpin, pp. 359–386. London: Academic Press.

Troumbis, A. 1995. Ecological networks in Greece. Current status, administration approaches and reserve initiatives. *Landschap* 12(3):51–62.

Van Dorp, D., and Opdam, P.F.M. 1987. Effects of patch size, isolation and regional abundance on forest bird communities. *Landsc. Ecol.* 1:59–73.

Van Dorp, D., Schippers, P., and Van Groenendael, J.M. 1997. Migration rates of grassland plants along corridors in fragmented landscapes assessed with a cellular automated model. *Landsc. Ecol.* 12:39–50.

Van Zadelhoff, E., and Lammers, W. 1995. The Dutch Ecological Network. *Landschap* 12(3):77–88.

Verboom, J., Foppen, R., Chardon, P., Opdam, P., and Luttinkhuizen, P. 2001. Introducing the key patch approach for habitat networks with persistent populations: an example of marshland birds. *Biol. Conserv.* 100:89–101.

Verboom, J., Metz, J.A.J., and Meelis, E. 1993. Metapopulation models for impact assessment of fragmentation. In *Landscape Ecology of a Stressed Environment,* eds. C.C. Vos and P. Opdam, pp. 172–191. London: Chapman and Hall.

Verboom, J., Schotman, A., Opdam, P., and Metz, J.A.J. 1991. European Nuthatch metapopulations in a fragmented agricultural landscape. *Oikos* 61:149–156.

Vermeulen, H.J.W. 1995. *Road-Side Verges: Habitat and Corridor for Carabid Beeties of Poor Sandy and Open Areas.* Ph.D. dissertation. Wageningen, The Netherlands: Wageningen Agricultural University.

Vos, C.C. 1999. *A Frog's-Eye View of the Landscape: Quantifying Connectivity for Fragmented Amphibian Populations.* IBN-Scientific Contribution 18. Wageningen, The Netherlands: DLO-Institute for Forestry and Nature Research.

Vos, C.C., and Chardon, J.P. 1998. Effects of habitat fragmentation and road density on the distribution pattern of the moor frog *Rana arvalis. J. Appl. Ecol.* 35:44–56.

Vos, C.C., Ter Braak, C.J.F, and Nieuwenhuizen, W. 2000. Empirical evidence of metapopulation dynamics; the case of the tree frog (*Hyla arborea*). *Ecol. Bull.* 48:165–180.

Vos, C.C., Verboom, J., Opdam, P.F.M., and Ter Braak, C.J.F. 2001. Towards ecologically scaled landscape indices. *Am. Nat.* 157:24–41.

Wiens, J.A., 1997. Metapopulation dynamics and landscape ecology. In *Metapopulation Biology: Ecology, Genetics and Evolution,* eds. I. Hanski and M.E. Gilpin, pp. 43–62. London: Academic Press.

22

Use of Metapopulation Models in Conservation Planning

DAVID R. BREININGER, MARK A. BURGMAN, H. RESIT AKÇAKAYA, AND MICHAEL A. O'CONNELL

22.1 Introduction

We focus on approaches to predict the effects of landscape change on population viability by first introducing concepts and emerging ideas. We then review recent applications of metapopulation models and summarize principles for applying metapopulation models in conservation planning. We end by identifying knowledge gaps that prevent broader applications of metapopulation models and by providing general research approaches for filling those gaps.

We limit our chapter to demographic and environmental considerations. Genetic consequences of fragmentation are usually a less-immediate threat (Lande 1988; Holsinger and Vitt 1997), and we focus on model analyses that emphasize management time frames rather than evolutionary time frames (Driscoll 1998). We have not addressed how genetics should be incorporated into population models for conservation planning (cf., VORTEX; Lacy 1993) and acknowledge that there are many gaps in understanding the movements of genes in metapopulations (Antonovics et al. 1997; Sork et al. 1998). We also exclude multispecies interactions but acknowledge that these too are serious problems facing the application of landscape ecology (Crooks and Soulé 1999).

22.2 Concepts, Principles, and Emerging Ideas

22.2.1 *Metapopulations: A Broad Definition*

We define a *metapopulation* as a set of populations that may exchange individuals through migration, dispersal, or human-mediated movement. This definition includes metapopulations with similar-sized populations and "mainland-island" systems, in which a few populations are larger than are others. It includes subpopulations that have similar dynamics and those with different dynamics (e.g., source-sink systems). It includes metapopulations with frequent local extinctions and those in which local extinctions are rare. Other, more restrictive definitions of metapopulations reflect particular approaches to modeling. Over the past decade, the trend in metapopulation concepts has moved from abstract models toward real-world applications. We view the metapopulation concept as a set of tools for analyzing species viability in fragmented landscapes.

22.2.2 *Why Are Metapopulation Models Used?*

Many practical questions related to landscape planning demand quantitative and predictive answers. These questions often focus on single species, either because legal concerns or management objectives target individual species or because the only tractable questions are those that relate to indicator or umbrella species. Metapopulation models are used because they are quantitative, their results can be replicated, they organize available information, they incorporate uncertainties and natural variability, they help set priorities for empirical research, their assumptions can be explicit and transparent, and they can provide a platform for achieving consensus on conservation strategies (Burgman et al. 1993).

Species are arranged in metapopulations because of landscape structure. Both natural heterogeneity and fragmentation by humans result in patchy landscapes (Wiens 1997). These landscapes are rarely static. In some cases, a patchy landscape is inhabited by populations that exchange individuals through a matrix of hostile environmental conditions. In other cases, many parts of the landscape are habitable, to varying degrees. When a species exists in multiple populations, a single-population model often gives inaccurate answers because it ignores among-population factors introduced by spatial structure (Wooton and Bell 1992). Because of spatial factors such as the correlation of fluctuations and dispersal of individuals among populations, there is no simple way of combining results of single-population models to predict metapopulation dynamics. The only way to model a set of populations correctly is to consider them simultaneously using a model.

Metapopulations can be impacted by factors that affect single populations (e.g., habitat loss) and factors that operate at the metapopulation level (e.g., decreased dispersal). Management options for metapopulations include those that apply to single populations, but with the added complication that the spatial distribution of such efforts must be specified or prioritized. Management questions related to translocation, reserve design, and habitat corridors are best addressed with a metapopulation model. In the face of habitat fragmentation, the use of other methods that ignore spatial structure may result in an overly optimistic picture of persistence.

22.2.3 *Theory*

Metapopulation theories began with the ideas that fragmented populations can occur as systems of empty and occupied patches that are maintained by stochastic processes of extinction and recolonization (Hanski 1999). These ideas do not apply to all species, so additional population structures and dynamics have been proposed (Harrison and Taylor 1997).

Source subpopulations have more births than deaths, and emigration exceeds immigration (Pulliam 1996). This definition is often refined to account for complications due to density dependence and environmental variability. *True sinks* have deaths that exceed births and immigration that exceeds emigration and therefore cannot persist by themselves at low or high densities. True sinks may only be distinguished at low densities where density dependence has little influence on survival, reproductive success, or immigration pressure (Watkinson and Sutherland 1995). *Pseudo-sinks* act as sources at low densities and as sinks at high densities.

The identification of source and sink populations is complicated by temporal and spatial variability in demography and dispersal. Habitat patches can have mortality that exceeds reproduction and temporarily have emigration that exceeds immigration (Breininger 1999). This results from differences in habitat quality among patches where individuals move from lower-quality to higher-quality patches or where individuals must move to find mates.

Other issues arise in the context of metapopulation dynamics. The location of a sink population (e.g., as a "stepping stone" between other populations) may enable it to contribute to the connectivity, hence the viability, of the metapopulation. Sink habitats are often valuable to the entire metapopulation because they provide for an overall larger population that is better able to withstand catastrophic events. Not all source and sink systems are metapopulations because source habitat patches can be contiguous with sink patches (Wiens and Rotenberry 1981; Howe et al. 1991).

Influences of density dependence can include not only the reduction of population growth rates at high densities, but also reduced population growth at low density; this latter influence is termed the *Allee effect.* Various density-dependence theories that have been applied in metapopulation models are reviewed elsewhere (Burgman et al. 1993; White 2000). Applications have begun to test alternatives to source-sink theory to explain dispersal among heterogeneous patches (e.g., McPeek and Holt 1992; Diffendorfer 1998).

22.2.4 Approaches to Metapopulation Modeling

The range of metapopulation models we describe here demonstrates tradeoffs between flexibility (realism) and practicality (data requirements) that relate to demographic structure and spatial structure. *Occupancy models* have the simplest demographic structure, describing each population as present (occupied) or absent (extinct) (Hanski 1999). Intermediate complexity is found in *structured* (or *frequency-based*) *models* that describe each population in terms of the abundances of age classes or life-history stages (Akçakaya 2000). These models incorporate spatial dynamics by modeling dispersal and temporal correlation among populations (Akçakaya and Baur 1996).

At the other extreme are *individual-based models,* which describe spatial structure of the locations of territories or the locations of individuals in the population (Lamberson et al. 1992). Some models use a regular grid where each cell can be modeled as a potential territory (Pulliam et al. 1992). Another approach uses a habitat suitability map to determine the spatial structure (Akçakaya et al. 1995).

All of these approaches have been applied to specific conservation management questions. The appropriate choice depends on the complexity of the problem at hand and the data available.

22.2.5 *Strengths and Weaknesses of the Metapopulation Approach*

A common criticism of the metapopulation approach for landscape planning is its inherent single-species focus. Recommendations resulting from the assessment of a single species may include strategies that do not benefit other elements of biodiversity. However, population dynamics are quantified and may be better understood than landscape and ecosystem processes. Metapopulation end points (e.g., risk of extinction) are well-defined, whereas end points such as ecosystem health or biotic integrity are vague. With its risk-based language, the metapopulation approach can address questions related to human impact and landscape management in terms of the viability of indicator and umbrella species. One way of dealing with the single-species limitation is to select target species that are representative of the natural community, that are sensitive to potential human impact, and whose conservation will protect other species (Noon et al. 1997). Another way to deal with this limitation is to combine results from different target species.

If the goal of a landscape analysis is to conserve species, then alternatives to metapopulation models also have limitations. Because of the rapid proliferation of data from remote sensing and the analytical abilities of geographic information systems (GIS), many landscape-ecology studies deal with metrics such as patch-size distribution, fractal dimension, shape index, and other descriptions of spatial structure. Reliance on landscape metrics has drawbacks. First, the patches that form the structure are often arbitrarily defined and may have no relationship with the habitat requirements of species. Second, the spatial scale is often arbitrarily selected and may be inappropriate for the metapopulation dynamics of many species. Third, there is often no strong relation between these metrics and end points that are relevant to successful conservation (Schumaker 1996). Metapopulation models are directly relevant to conservation because their currency is the viability of species. Some landscape studies concentrate on factors that are part of a metapopulation approach, such as connectivity and dispersal. But, the best way to make such measures relevant to conservation is to use them in metapopulation models. There is no reason for management to aim for increased dispersal, for example, unless it can be shown that dispersal limits the species of interest within a given landscape.

22.3 Recent Applications

We identified recent applications of metapopulation models by direct communications with 200 scientists from 20 countries, by reviewing 200 metapopulation model applications, by reviewing 30 natural resource programs, and by posting information requests on electronic bulletin boards. Early metapopulation models that we reviewed were often developed only to improve the conceptual understanding of population dynamics. Theoretical applications had considerable impact when they were motivated by the needs of managers, or when they informed managers about possibilities they faced when making decisions. Developers of early metapopulation models frequently made no attempts to construct them so that they could be directly applied for conservation planning. Recently, the focus of modeling has shifted to a broad array of real management problems. Approximately 80% of the scientists who responded to our surveys reported results relevant to management questions; these scientists attempted to make their results available to managers. Approximately 70% of these applications had been developed since 1995 and nearly all since 1990.

Only 3% of the natural resource programs we reviewed routinely used metapopulation models in 1995, but this increased to 17% in 2000. Even for programs in which metapopulation models were applied routinely, their use was usually restricted to supporting the management of one or a few high-profile species. This was due largely to development costs and the scarcity of skills and data necessary to build and maintain models. The percentage of programs that occasionally used the results of metapopulation models increased from 20% to 60% between 1995 and 2000.

22.3.1 Measures of Success

We did not find any applications among the 200 that we reviewed in which the evaluation of model success was based on achieving a population size or demographic state predicted by the model. In most cases, too little time had elapsed since the model predictions were made to evaluate them, or no attempts were made to measure success. Model results are seldom the only factor that leads to a decision. Respondents suggested that we should consider successful applications to be cases in which the biology of the organism was reasonably incorporated into the model and in which the model was used to assist in decision making. Models were not used in planning when managers lacked the ability or confidence to use them, or when scientists believed that the models were not ready for use by managers.

22.3.2 Uses of Metapopulation Thinking for Planning

Metapopulation models have been used to assist in developing goals regarding spatial configurations and habitat management schedules needed to minimize extinction. Some problems cannot be evaluated without the use of models because they involve complex, protracted situations. Evaluations of catastrophes or cumulative impacts from human activities that occur over decades or centuries are best handled by models (Dunning et al. 1995).

22.3.3 *Ranking Management Options and Advice*

Most models resulted in qualitative advice such as comparison of management alternatives and assessment of human impacts (Verboom 1996). These involved reporting comparative results that were considered to be robust under a range of assumptions about model structure and parameter uncertainty, and recommendations for data acquisition that followed from sensitivity analyses (Akçakaya and Atwood 1997). The use of relative results (e.g., difference in metapopulation viability with and without impact) rather than absolute results (e.g., risk of extinction) makes the models less-sensitive to uncertainties in input data and model structure (Akçakaya and Raphael 1998). Thus, modelers emphasize relative results, even when quantitative predictions are available and reported.

When models were used as evaluation tools, we found that they often contributed to an improved understanding of problems without relying on specific, quantitative predictions. Perhaps the most consistent use of metapopulation studies has been to identify the most important parameters in a model to guide future field research and to provide direction for management (Lindenmayer and Possingham 1996).

22.3.4 *Studies Linking Landscape Dynamics to Metapopulation Dynamics*

Some metapopulation applications incorporate habitat changes (Holthausen et al. 1995; Lindenmayer and Possingham 1995; DeAngelis et al. 1998; Drechsler et al. 1998; Menges and Dolan 1998). These changes can result from habitat management (Quintana-Ascencio et al. 1998) or catastrophes that must be considered in planning (e.g., Akçakaya and Atwood 1997; Bradstock et al. 1998; McCarthy and Lindenmayer 1999). Habitat changes often have greater effects on large populations over long periods of time than do stochastic factors, which are more severe for smaller populations (MacDonald et al. 1999).

23.3.5 *Approaches for Expanding the Role of Modeling*

Most landscape decisions involve choosing among uncertain strategies when time and money are limited (e.g., size versus connectivity). If an optimal solution is state-dependent, dynamic programming may be able to find a solution when the number of required simulations exceeds the ability to evaluate every possible solution. Management of successional habitat, translocation decisions, determining the most important patches, and incorporating economics are all examples of situations in which dynamic programming can assist landscape planning for metapopulation persistence (Possingham 1996, 1997).

Remote sensing and GIS provide powerful tools to describe landscape patterns and quantify changes over time, but maps are only models; theories to apply these technologies, for predicting population processes, have not kept pace (Wiens 1997). Metapopulation modeling provides a mechanism to link landscape ecology to theories of population dynamics, and a mechanism to link empirical studies and theory to conservation planning using GIS. Landscape features, road densities, and disturbance regimes have been analyzed using GIS to make inferences about population processes within metapopulations (Bushing 1997). Other modeling approaches such as hydrodynamic models have modeled larval dispersal and connectivity to evaluate potential source-sink relationships (James et al. 1997). Bayesian approaches have been useful for classifying risk (Taylor et al. 1996).

22.4 Principles for Applying Landscape Ecology

The following principles are not based on demonstrated effects of model applications that have achieved a population size or a demographic state predicted by models. Adequate data about such effects do not exist. Instead, these principles reflect our beliefs about how models are best used to assist in decision making and are based on information we derived from scientific literature and communications with our colleagues.

22.4.1 Importance of Spatial Structure

Species viability is likely to be affected by the rate and arrangement of habitat change that varies among land-use alternatives (Drechsler and Wissel 1998). Population viability analyses (PVA) should compare alternatives quantitatively and identify nonlinear responses between habitat change and species viability. Habitat destruction usually reduces the viability of species. Landscape planning should determine the significance of indirect habitat changes that result from fragmentation (e.g., road mortality, disruption of fire regimes) because these can greatly decrease population viability (Breininger et al. 1999).

Habitat fragmentation usually reduces viability because of edge effects and increased mortality during dispersal. It is often difficult to separate the effects attributed to habitat loss from those associated with population subdivision (Kareiva et al. 1997). It is important to identify thresholds of habitat destruction or degradation where extinction is the likely outcome because relationships between species viability and habitat change are often nonlinear (Lande 1988; Doak 1995). At high levels of fragmentation, population sizes become too low and isolation among populations becomes too high for individuals to find territories or mates.

Changes in habitat contiguity will often influence dispersal patterns and need evaluation. Dispersal among populations may lead to recolonization of empty patches by immigration from neighboring populations, improving overall metapopulation persistence (Haight et al. 1998). Decreases in habitat connectivity often decrease population viability. However, there are exceptions related to infectious diseases, when population subdivision decreases threats from catastrophes, and when species that have poor selective abilities disperse into sink habitats (Akçakaya and Baur 1996; Hess 1996). The effectiveness of dispersal in reducing extinction risks depends on the correlation of environmental fluctuations experienced by different populations. If the correlation is high, all populations decline simultaneously, reducing recolonization rates of empty patches. If the fluctuations are at least partially independent, some patches can act as sources of emigrants (Burgman et al. 1993). Extinction risks are often sensitive to spatial correlation in environmental fluctuation (LaHaye et al. 1994) and disturbance regime (McCarthy and Lindenmayer 1999).

22.4.2 Landscape Features Critical for Survival

Certain patches or habitat conditions are more critical than are other patches or conditions for species persistence, and these critical habitat features need identification (Moilanen et al. 1998). The degradation or destruction of certain areas can greatly increase extinction probabilities of subpopulations or entire metapopulations, especially if these areas are sources or are important for connecting populations (Beier 1993). Many studies show that there also is a threshold patch size, below which patches are too small to be of value, but these sizes vary among species (McKelvey et al. 1992; Litvaitis and Villafuerte 1996; Rushton et al. 1999). The best way to develop a sense of the relative importance of specific patches is through the interpretation of a spatially explicit model.

Landscape planners must consider potential source-sink relationships (Pulliam 1996). Undetected source-sink metapopulation structure can obscure signals that managers need for conservation action, because numbers in sinks are unreliable indicators of population status (Cooper and Mangel 1998). Disastrous consequences can result for source-sink metapopulations in which management jurisdictions do not coincide with metapopulation boundaries (Donovan et al. 1995; Hokit et al. 1999). Greater cooperation is needed among landowners and agencies to manage species with metapopulation dynamics that span different land ownerships.

Whereas efforts to protect species typically consider only occupied habitat, coherent strategies must consider unoccupied, potential habitat because unoccupied patches can be critical for population persistence (Menges 1990; Hanski et al. 1996). Particular concern for unoccupied habitat needs to occur for species associated with habitats subject to periodic disturbance, or for species that consist of small extinction-prone local populations that persist regionally because of a balance between local extinctions and colonizations (Hanski 1997). Some species with high site tenacity may take years to recolonize suitable, unoccupied patches, even though in the long run these unoccupied patches may be crucial to population persistence. Many individuals in a population may temporarily be in areas where their long-term persistence is always doubtful, so restoration of appropriate, unoccupied habitat may be the only chance for long-term persistence (Stith et al. 1996). Plants, such as many species of *Acacia,* may persist for years or decades in seed banks within apparently unoccupied patches long after all adults have died. Evaluating the importance of vacant habitat relies on understanding the distribution and ecology of a species. Integrating this knowledge so that management decisions are transparent and reliable requires a metapopulation model.

22.4.3 *Reserve Design*

Population modeling and island biogeography result in basic rules of reserve design (Noss et al. 1997). Examples are that patches need to be large and have high-quality habitat, and that interpatch distances should be small. System-specific conditions and objectives require that general principles about reserve design be replaced by recommendations developed from specific metapopulation models (Hanski 1997). The need for several large reserves versus many small reserves often depends on species and their dispersal abilities, environmental variation, characteristics of catastrophic events, and landscape opportunities that remain.

Similarly, general principles are hard to find for the design of linkages, so conservation planners must consider dynamics specific to species of concern and regional metapopulation dynamics. Achieving long-term population viability is not always best served by focusing on linkages across large regions, because the dynamics of some species can be more strongly influenced by local factors. Centrally located, large patches are often more important than are small, peripherally located patches. These tendencies must be balanced by concerns for representing the range of genetic variation, maintaining species across their existing ranges to provide for changing environments, and guarding against disease and catastrophic events.

Patch quality and connectivity influence population dynamics. In addition, the matrix or type of habitat adjacent to reserves must be considered in metapopulation models and landscape planning (Lawes et al. 2000). Some habitat types may result in negative edge effects because they supply predators or parasites. Some species may be poor discriminators of habitat quality, so it may be important to have distinct boundaries (McKelvey et al. 1992). In some cases, the arrangement of suitable habitat is considered to be more important than is the amount (Letcher et al. 1998).

22.4.4 *Limitations of Models*

Planners must consider that misuse of models can occur because few data are available to construct models, models provide insight for a small proportion of the biota, model outputs are difficult to validate, and alternative model structures can affect results (Conroy et al. 1995). Applications should present confidence limits in the results imposed by model uncertainty (Ludwig 1998). The answer is not to avoid modeling but to develop models that consider all data, to incorporate uncertainties, and then to judge whether the results are useful (Akçakaya 2000). A model must be simple to allow parameterization with available data, yet sufficiently complex to approximate the dynamics of the species and answer the questions posed.

Many types of models are available, and not all model types are appropriate for all species or planning activities. The simplest metapopulation models assume that patches are either occupied or not, ignoring local population dynamics, habitat size and quality, and trends in abundance. Where population dynamics operate on time scales that are fast with respect to the probabilities of extinction and colonization, these are reasonable assumptions (Hanski 1997). The advantage of these occupancy models is that they do not require detailed demographic data; the disadvantage is that simple occupancy models may require unreasonable assumptions for some applications (Kindvall 1999).

The most realistic models are spatially explicit and individual-based. The cost of realism often includes many model parameters that are difficult to estimate and many assumptions that are difficult to verify. Parameter estimation errors can reduce the reliability of spatially explicit models (Ruckelshaus et al. 1997), but these errors have been overestimated (Mooij and DeAngelis 1999). Individual-based models have utility when details of individual behavior are important, when genetic processes can be included in the model, and when modelers want to explore the fine details of the spatial arrangement of individuals (DeAngelis and Gross 1992; McCarthy 1996).

22.4.5 When Should Models Be Used?

We believe planners should use metapopulation models because they link landscape ecology to the population dynamics of species of conservation concern. Although data availability limits the realism and dependability of models, landscape management still suffers from pitfalls regarding limited data availability when models are not used. Without models, it is often difficult to attribute population responses to specific management actions because results are seldom measured and management decisions are made with no documentation of the assumptions that led to decisions. We argue that planning should use models to provide a structured process for summarizing knowledge, identifying assumptions, and directly linking decisions to predictions of population dynamics. Model structures and assumptions can become focal points where management decisions are evaluated. When data are poor, models should still be used to determine data needs and a structure for adaptive management (Boyce 1997). It is important that planning efforts couple monitoring efforts with modeling because models can predict extinction risks before the fate of the populations are realized by monitoring (Doak 1995). Planners should not judge the success of a model by its explanatory power or forecasting ability alone (Conroy et al. 1995). In planning, models also should be judged by how well they contribute to experimental design, surveying, monitoring, parameter estimation, and the development of management priorities.

22.4.6 Why Are Metapopulation Models Not Used More?

Natural resource programs must culture the ability to evaluate changes in population viability associated with human activities. Many human factors influence choices for using models, and misconceptions should be addressed to develop modeling cultures within conservation programs (Starfield 1997). Many model applications are not used because they ignore social or economic factors. Natural resource programs must incorporate economic factors into landscape and population models so that these programs have a greater impact on decisions (Liu 1993).

22.5 Knowledge Gaps

Modeling of metapopulations has proceeded ahead of field biology despite the eminence of metapopulation ideas (Doak and Mills 1994). Although we describe theoretical and empirical gaps separately below, successful applications require close linkages between modeling and field research. Below, we identify gaps that have repeatedly been stated as factors that limit the application of metapopulation models in landscape planning.

22.5.1 Theoretical Voids

Guidelines are needed to assist researchers and managers in selecting model structures and understanding how model structure influences planning applications. Guidance also is needed for selecting the spatial scale of metapopulation studies.

Explicit links between metapopulation models and landscape dynamic models are needed to predict temporal changes in a species' habitat due to vegetation dynamics, disturbance regimes, human activities, and climatic change (Burgman et al. 1993). Few applications have linked population models to the processes of landscape change, especially where landscape change varies in uncertain spatial and temporal patterns. Methods are needed to subdivide landscapes into relevant units that can be used to assign demographic rates to individual patches, because degrees of habitat suitability often do not occur with stable or discrete boundaries that coincide with boundaries used by individual animals (Breininger et al. 1998).

The accuracy of PVA is often compromised by the lack of understanding of statistical methods for (1) estimating model parameters from limited data, and (2) partitioning observed temporal variance of demographic rates into environmental variation, demographic stochasticity, density dependence, Allee effects, and sampling error.

The appropriate level of uncertainty needs to be integrated into landscape applications of metapopulation models using sensitivity analyses. Incorporating processes of landscape change adds new uncertainties into metapopulation analyses. Management alternatives and measures of habitat suitability involve different types of information, including professional judgments, vague policies, unknown statistical distributions, and statistical dependencies among variables (Burgman et al. 2001). Exploring all possibilities using Monte Carlo simulations is exceedingly difficult, and Monte Carlo analysis with worst-case assumptions can underestimate or overestimate risks (Ferson 1996).

Greater guidance is needed for presenting quantitative risk estimates and developing adaptive strategies to improve estimates because biologists are increasingly being expected to quantify risks. PVA is best used for the comparison of management objectives (Beissinger and Westphal 1998), but the dichotomies between relative and quantitative predictions are not always clear, especially because conservation actions entail costs (Brook et al. 2000b).

There are gaps in the identification of data characteristics needed for testing theories and specific models. Frequently, the assumptions and structure of theoretical work need to be deciphered to determine the data needed for model testing. Assistance also is needed in prioritizing data to test theoretical models, recognizing that prioritization will be specific to particular models or species.

Methods to present populations and attributes of their viability in the form of maps are needed because maps are a fundamental part of planning. Most habitat maps involve only polygons enclosing land use or vegetation features and do not realistically represent populations and their viability. Even maps that involve accurate ground truthing of species' abundances represent spatial patterns that could have resulted from different processes and by themselves may offer little understanding (Hanski 1999).

There must be greater integration of metapopulation modeling and economics for landscape ecology to improve decision making (Liu 1993). The results of metapopulation studies often are not incorporated into management efforts because they ignore economic constraints. Economic constraints are often poorly understood, and there is a need to develop theoretical foundations for formulating economic dimensions of metapopulation models.

22.5.2 *Empirical Voids*

There are few experimental tests of spatial models, yet we know that fragmentation and related spatial effects threaten populations of conservation concern (Harrison and Taylor 1997). Research is needed to evaluate how well maps represent real populations and how changes in mapped features are related to population dynamics measured in the field.

A greater understanding is needed of how habitats outside conservation reserves influence population dynamics within conservation reserves, because land-use planners frequently emphasize zoning requirements and buffers around reserves. It is generally assumed that buffers and zoning have positive effects on population viability, but these areas can have negative effects and their significance can vary across time. For example, hard edges are probably better in some reserves (McKelvey et al. 1992). Species of conservation concern that temporarily occur outside of potential reserves may be important colonists once reserves are acquired and restored (Breininger et al. 1999).

Applications of metapopulation models to landscape conservation and management issues are limited most by demographic and dispersal data. Nearly all demographic and dispersal data come from time series that are too short and too limited in geographic scope to provide reliable estimates of environmental variability, density dependence, Allee effects, successional changes, seed banks, catastrophic events, temporal and spatial autocorrelations, and responses to habitat change. Demographic data for local conditions are critical for applying metapopulation models to landscape management. Despite widely cited concerns expressed almost two decades ago (Van Horne 1983), measures of habitat suitability still remain focused on habitat use or abundance rather than on demography (Garshelis 2000). Data are needed to quantify how patch size, quality, and arrangement affect reproductive success, survival, and dispersal of various life-history stages. Data are needed on how organisms partition heterogeneous landscapes into units of varied demographic success.

Quantitative data on dispersal are needed because they define the rates of exchange among subpopulations and the scale at which species perceive their environment. Data are needed on the parameters and shapes of dispersal-distance functions, and on how spatial and temporal variations in patch quality, boundary effects, landscape connectivity, and patch context influence exchanges among subpopulations. Behavioral-ecology data are needed to understand how dispersing individuals respond to landscape variation, how they move to find new habitat, how dispersal rates differ between the sexes, and whether conspecific attraction is important. There are gaps in our understanding about how environmental variation, habitat quality, population density, and habitat changes influence dispersal rates.

22.6 Research Approaches

To assist in evaluating how land use will influence population viability, theoreticians and empiricists should collaborate to bridge specific gaps between the modeling data needed and the standard data routinely collected for particular taxa.

22.6.1 Approaches for Theoretical Research

To assist empiricists and managers in selecting and applying metapopulation models, theoreticians should make products accessible and offer assistance in designing field studies and adaptive-management strategies (Kareiva 1990). Models should be user-friendly, portable, operate on spatial and temporal scales appropriate to management, use input and output measures that can be measured affordably, and have visualizations to enable managers to understand the effects of alternative actions (Turner et al. 1995). Guidance on matching species biology and available data to appropriate model structures and software packages is available for general PVA (e.g., Akçakaya and Sjogren-Gulve 2000; Brook 2000; Brook et al. 2000a; Matsinos et al. 2000). Model comparisons need to be extended to include relationships between population viability and landscape change.

Metapopulation modeling of land-use plans should consider modeling at different scales where populations are distributed across large geographic areas. An advantage for limiting spatial extent is that fewer data are generally required, which may be appropriate when there are many uncertainties associated with landscape change that need to be considered for populations dominated by local birth and death processes. Limiting spatial extent can be a disadvantage when broad-scale population processes influence the dynamics of local populations and where applications have little relevance to the viability of the species in general.

Evaluating land-use management alternatives requires modeling population responses to changes over time (e.g., Holthausen et al. 1995). Instead of involving abstract changes, modeling should link population viability to landscape dynamic models associated with mechanisms of habitat conversion, vegetation dynamics, disturbances, species introductions, pollution, or overexploitation. Researchers also should develop new methods for subdividing landscapes into patches that can be related to population units or individuals (e.g., territories, home ranges) for applying demography and dispersal rates. These techniques should be compatible with widely used GIS.

The techniques (e.g., Gould and Nichols 1998) for partitioning the variance of demographic measurements into density dependence, sampling error, and environmental variation need to be synthesized into guidelines for biologists who are not statisticians. Guidelines need to review techniques and their limitations, include examples, recommend software, and describe mistakes to avoid while doing PVA.

The effects of uncertainty in landscape change need to be integrated into decision making because natural and socioeconomic factors influence landscape change. Approaches for incorporating uncertainty need synthesis into a general set of recommendations that include new approaches, such as fuzzy arithmetic and Bayesian approaches (Lee 2000). Fuzzy numbers can be employed to handle uncertainty associated with measurement error, natural variation, and vagueness, which are poorly handled by Monte Carlo analyses (Ferson et al. 1998); applications of fuzzy numbers can improve conservation planning substantially (Burgman et al. 2001).

To present results in the context of their uncertainty, sensitivity analyses should explore the full range of plausible values of the model, including deterministic and stochastic elements (McCarthy et al. 1996; Lindenmayer et al. 2000). Sensitivity analyses should identify which time horizons require the most accurate estimation (Akçakaya and Raphael 1998). Longer time frames for analyses need to be considered for species with long generation times (Armbruster et al. 1999). Sensitivity analyses should focus on how uncertainty influences the relative rank of management decisions.

Model validation studies should be synthesized into recommendations for reporting quantitative model predictions in the context of decisions. Despite skepticism regarding the usefulness of PVA (Ludwig 1998; Fieberg and Ellner 2000), predicted risks of population decline can sometimes be determined reliably (Brook et al. 2000b).

Theoretical outcomes that drive empirical research and management priorities should produce recommendations that are feasible and justified. The assumptions and structure of theoretical applications should not need to be deciphered to determine the most important parameters for particular questions. Data prioritization needs to consider the sensitivity of parameters and their influence on management alternatives (Heppell et al. 2000). Modeling and sensitivity analyses should explore whether improvements in model structure (e.g., Beissinger 1995) or parameterization can reduce uncertainty to levels acceptable to decision making. One approach may consider sensitivity analyses that vary several parameters at a time (Pulliam et al. 1992) using data that can be collected easily.

Information derived from metapopulation modeling is needed to enhance mapping of species' occurrences. Modeling results are generally reported as distributions for an entire population with little information on patch occupancy or importance. Useful information about the occupancy or the importance of a patch can be recorded during simulations of the entire population. This information could be tested against simulations that exclude each patch at a time to investigate how various measures of patch importance correspond.

Modeling of metapopulation viability and economics cannot be independent because the results of each are not always comparable when alternative strategies must balance cost and viability. Ecologists should collaborate directly with natural resource managers, land acquisition agents, and economists to integrate economics into metapopulation models.

22.6.2 Approaches for Empirical Research

Studies that map species distribution and habitat suitability should state the type and quality of data on which the map is based, as well as samples of species abundances across time intervals. Field studies are needed in many subpopulations to test habitat fragmentation effects and spatial models. The demographic testing of habitat relationships from a single subpopulation can be misleading if there are immigration, emigration, and density-dependent effects. Studies of metapopulation dynamics should consider the importance of dynamics of subpopulations in space (Kean and Barlow 2000).

Empirical data are needed to test metapopulation theory in which simulations represent a null model of population dynamics (Harrison and Taylor 1997). Testing should include alternative models (e.g., Doncaster et al. 1997; Nieminen and Hanski 1998; Boughton 1999). Model validations need to be conducted under different circumstances because models do well under some conditions but not others (Conroy et al. 1995). McCarthy et al. (In press) and McCarthy and Broome (In press) provide examples of validation strategies. Studies that mark and track all individuals in entire metapopulations or many coinciding subpopulations are needed (e.g., Doncaster et al. 1997; Breininger 1999; Sweanor et al. 2000). Alternative approaches using unmarked animals and occupancy data can sometimes be used in validations (Hanski 1999).

Demography and dispersal of animals inside reserve buffers and outside of reserves need to be investigated to determine whether they increase or decrease population viability. Cantrell and Cosner (1999) reference theoretical inferences that provide important hypotheses for the empirical testing of buffers.

National databases of demographic and dispersal data are needed to assess fragmentation impacts on representative taxa. Empirical studies need to be reviewed to summarize gaps in demographic data needed for metapopulation models. For example, there are many avian studies of how reproductive success is influenced by edge, habitat type, and fragmentation, but these studies fail to measure associated survival and dispersal differences that are important at the population level (Matthysen 1999). Studies must evaluate whether reproductive success and survival vary among habitat types, whether they are the same along edges and in the center of patches, and whether different types of edges influence demography (Mumme et al. 2000). Plant population studies must quantify the roles of seed banks, self-compatibility, and vegetative reproduction in PVA. Field studies should be designed in collaboration with modelers from the outset to prevent data gaps and to collect data in ways that can be used to test and parameterize models.

Long-term studies are needed to identify extinction and recolonization events and to partition demographic variance into sampling error, sources of environmental variability, and density dependence. Examples of approaches for partitioning demographic variance can be found in Gould and Nichols (1998), Akçakaya et al. (1999), and White (2000). Demographic data collection should evaluate whether parameter estimates need to be improved to estimate extinction risk or rank alternative land-use scenarios (e.g., Goldwasser et al. 2000; Meir and Fagan 2000).

Dispersal studies need to be designed according to the model appropriate for the ecology of the species, the question addressed, and the available demographic data. For occupancy or structured models, mark-recapture studies (White and Garrott 1990; Murray and Fuller 2000) may be appropriate. Individual-based applications sometimes require data collected from radio-telemetry studies or frequent censuses of individually marked animals to parameterize mechanistic models of animals moving through landscapes (Turchin 1998). Movements need to be translated into full statistical distributions in formats useful to modeling. Depending on the model, such distributions may involve the probability of immigration to different patches, the population fraction that disperses from each patch per unit time, mean and maximum dispersal distances, or the distance at which dispersers detect new patches. Dispersal studies should test for differences in dispersal propensities between sexes and determine how sex differences influence population dynamics in fragmented systems. It is important to characterize the habitat suitability and population density of both source and recipient patches of dispersing individuals.

22.7 Conclusions

Metapopulation models constitute a diverse toolbox for landscape planning. Models are increasingly being applied to planning because general principles do not always apply to the diversity of management actions and species that managers must address. Often, the improved understanding that comes with the development of a metapopulation model can be sufficient to solve a problem. In other circumstances, resolving the rank order of the consequences of management alternatives is enough. Increasingly, managers desire quantitative population goals, so modeling must prioritize additional data collection and research.

Many managers lack the experience to decide which kinds of models and scales of resolution are best to solve a given problem. As skills improve and modeling cultures extend through agencies and industry, the benefits of building metapopulation models are likely to be embraced further. Knowledge about how landscape changes and arrangements influence species is poor because long-term empirical data on population dynamics is lacking across diverse ranges of landscape patterns and changes. Even under these circumstances, metapopulation models can provide a heuristic platform to structure knowledge, to predict the future before the fates of populations are sealed, and to develop monitoring programs that become the basis of adaptive management.

Acknowledgments

We thank NASA and Applied Biomathematics for funding, and Brad Stith, Brean Duncan, Ross Hinkle, and Vickie Larson for reviewing the manuscript. Space limitations prevent us from thanking 60 scientists whose discussions contributed to our review and from citing a number of references. A complete reference list is available from the first author.

References

Akçakaya, H.R. 2000. Population viability analyses with demographically and spatially structured models. *Ecol. Bull.* 48:23–38.

Akçakaya, H.R., and Atwood, J.L. 1997. A habitat-based metapopulation model of the California Gnatcatcher. *Conserv. Biol.* 11:422–434.

Akçakaya, H.R., and Baur, B. 1996. Effects of population subdivision and catastrophes on the persistence of a land snail metapopulation. *Oecologia* 105:475–483.

Akçakaya, H.R., Burgman, M.A., and Ginzburg, L.R. 1999. *Applied Population Ecology.* Second Edition. Sunderland, Massachusetts: Sinauer Associates.

Akçakaya, H.R., McCarthy, M.A., and Pearce, J. 1995. Linking landscape data with population viability analysis: management options for the Helmeted Honeyeater. *Biol. Conserv.* 73:169–176.

Akçakaya, H.R., and Raphael, M.G. 1998. Assessing human impact despite uncertainty: viability of the Northern Spotted Owl metapopulation in the northwestern USA. *Biodiver. Conserv.* 7:875–894.

Akçakaya, H.R., and Sjogren-Gulve, P. 2000. Population viability analysis in conservation planning: an overview. *Ecol. Bull.* 48:9–21.

Antonovics, J., Thrall, P.H., and Jarosz, A.M. 1997. Genetics and the spatial ecology of species interactions: the Silene-Ustilago system. In *Spatial Ecology: The Role of Space in Population Dynamics and Interspecific Interactions,* eds. D. Tilman and P. Kareiva, pp. 158–184. Princeton: Princeton University Press.

Armbruster, P., Fernando, P., and Lande, R. 1999. Time frames for population viability analysis of species with long generations: an example with Asian elephants. *Anim. Conserv.* 2:69–73.

Beier, P. 1993. Determining minimum habitat areas and habitat corridors for cougars. *Conserv. Biol.* 7:94–108.

Beissinger, S.R. 1995. Modeling extinction in periodic environments: Everglades water levels and Snail Kite population viability. *Ecol. Appl.* 5:618–631.

Beissinger, S.R., and Westphal, M.I. 1998. On the use of demographic models of population viability analysis in endangered species management. *J. Wildl. Manage.* 62:821–841.

Boughton, D.A. 1999. Empirical evidence for complex source-sink dynamics with alternative states in a butterfly metapopulation. *Ecology* 80:2727–2739.

Boyce, M.S. 1997. Population viability analysis: adaptive management for threatened and endangered species. In *Ecosystem Management: Applications for Sustainable Forest and Wildlife Resources,* eds. M.S. Boyce and A. Haney, pp. 226–236. New Haven, Connecticut: Yale University Press.

Bradstock, R.A., Bedward, M., Kenny, B.J., and Scott, J. 1998. Spatially explicit simulation of the effect of prescribed burning on fire regimes and plant extinctions in shrublands typical of southeastern Australia. *Biol. Conserv.* 86:83–95.

Breininger, D.R. 1999. Florida Scrub-Jay demography and dispersal in a fragmented landscape. *Auk* 116:520–527.

Breininger, D.R., Burgman, M.A., and Stith, B.M. 1999. Influence of habitat, catastrophes, and population size on extinction risk in Florida Scrub-Jay populations. *Wildl. Soc. Bull.* 27:810–822.

Breininger, D.R., Larson, V.L., Duncan, B.W., and Smith, R.B. 1998. Linking habitat suitability to demographic success in Florida Scrub-Jays. *Wildl. Soc. Bull.* 26:118–128.

Brook, B.W. 2000. Pessimistic and optimistic bias in population viability analysis. *Conserv. Biol.* 14:564–566.

Brook, B.W., Burgman, M.A., and Frankham, R. 2000a. Differences and congruencies between PVA packages: the importance of sex ratio for predictions of extinction risk. *Conserv. Ecol.* [online] 4:6. Available from the Internet: www.consecol.org/vol4/iss1/art6.

Brook, B.W., O'Grady, J.J., Chapman, A.P., Burgman, M.A., Akçakaya, H.R., and Frankham, R. 2000b. Predictive accuracy of population viability analysis in conservation biology. *Nature* 404:385–387.

Burgman, M.A., Breininger, D.R., Duncan, B.W., and Ferson, S. 2001. Setting reliability bounds on habitat suitability indices. *Ecol. Appl.* 11:70–78.

Burgman, M.A., Ferson, S., and Akçakaya, H.R. 1993. *Risk Assessment in Conservation Biology.* London: Chapman and Hall.

Bushing, W.W. 1997. GIS-based gap analysis of an existing marine reserve network around Santa Catalina Island. *Intl. J. Marine Geodesy* 20:205–234.

Cantrell, R.S., and Cosner, C. 1999. Diffusion models for population dynamics incorporating individual behavior at boundaries: applications to reserve design. *Theor. Pop. Biol.* 55:189–207.

Conroy, M.J., Cohen, Y., James, F.C., Matsinos, Y.G., and Maurer, B.A. 1995. Parameter estimation, reliability, and model improvement for spatially explicit models of animal populations. *Ecol. Appl.* 5:17–19.

Cooper, A.B., and Mangel, M. 1998. The dangers of ignoring metapopulation structure for the conservation of salmonids. *Fisheries Bull.* 97:213–226.

Crooks, K.R., and Soulé, M.E. 1999. Mesopredator release and avifaunal extinctions in a fragmented system. *Nature* 400:563–566.

DeAngelis, D.L., and Gross, L.J. 1992. *Individual-Based Models and Approaches in Ecology: Populations, Communities and Ecosystems.* New York: Chapman and Hall.

DeAngelis, D.L., Gross, L.J., Huston, M.A., Wolff, W.F., Fleming, D.M., Comiskey, E.J., and Sylvester, S.M. 1998. Landscape modeling for Everglades ecosystem restoration. *Ecosystems* 1:64–75.

Diffendorfer, J.E. 1998. Testing models of source-sink dynamics and balanced dispersal. *Oikos* 81:417–433.

Doak, D.F. 1995. Source-sink models and the problem of habitat degradation: general models and applications to the Yellowstone grizzly. *Conserv. Biol.* 9:1370–1379.

Doak, D.F., and Mills, L.S. 1994. A useful role for theory in conservation. *Ecology* 75:615–626.

Doncaster, C.P., Clobert, J., Doligez, B., Gustafsson, L., and Danchin, E. 1997. Balanced dispersal between spatially varying local populations: an alternative to the source-sink model. *Am. Nat.* 150:425–445.

Donovan, T.M., Lamberson, R.H., Kimber, A., Thompson, F.R., III, and Faaborg, J. 1995. Modeling the effects of habitat fragmentation on source and sink demography of Neotropical migrant birds. *Conserv. Biol.* 9:1396–1407.

Drechsler, M., Lamont, B.B., Burgman, M.A., Akçakaya, H.R., Witkowski, E.T.F., and Supriyadi. 1998. Modeling the persistence of an apparently immortal *Banksia* species after fire and land clearing. *Biol. Conserv.* 88:249–259.

Drechsler, M., and Wissel, C. 1998. Trade-offs between local and regional scale management of metapopulations. *Biol. Conserv.* 83:31–41.

Driscoll, D.A. 1998. Genetic structure, metapopulation processes and evolution influence the conservation strategies for two endangered frog species. *Biol. Conserv.* 83:43–54.

Dunning, J.B., Jr., Stewart, D.J., Danielson, B.J., Noon, B.R., Root, T.L., Lamberson, R.H., and Stevens, E.E. 1995. Spatially explicit population models: current forms and future uses. *Ecol. Appl.* 5:3–11.

Ferson, S. 1996. What Monte Carlo methods cannot do. *Human Ecol. Risk Assess.* 2:990–1007.

Ferson, S., Root, W., and Kuhn, R. 1998. *RAMAS/RiskCalc: Risk Assessment with Uncertain Numbers.* Setauket, New York: Applied Biomathematics.

Fieberg, J., and Ellner, S.P. 2000. When is it meaningful to estimate an extinction probability? *Ecology* 81:2040–2047.

Garshelis, D.L. 2000. Delusions in habitat evaluation: measuring use, selection, and importance. In *Research Techniques in Animal Ecology: Controversies and Consequences,* eds. L. Boitani and T.K. Fuller, pp. 111–164. New York: Columbia University Press.

Goldwasser, L., Ferson, S., and Ginzburg, L. 2000. Variability and measurement error in extinction risk analysis: the Northern Spotted Owl on the Olympic Peninsula. In *Quantitative Methods for Conservation Biology,* eds. S. Ferson and M. Burgman, pp. 169–187. New York: Springer-Verlag.

Gould, W.R., and Nichols, J.D. 1998. Estimation of temporal variability of survival in animal populations. *Ecology* 79:2531–2538.

Haight, R.G., Mladenoff, D.J., and Wydeven, A.P. 1998. Modeling disjunct gray wolf populations in semi-wild landscapes. *Conserv. Biol.* 12:879–888.

Hanski, I. 1997. Habitat destruction and metapopulation dynamics. In *The Ecological Basis for Conservation: Heterogeneity, Ecosystems, and Biodiversity,* eds. S.T.A. Pickett, R.S. Ostfeld, M. Shachak, and G.E. Likens, pp. 217–228. New York: Chapman and Hall.

Hanski, I. 1999. *Metapopulation Ecology.* Oxford, United Kingdom: Oxford University Press.

Hanski, I., Moilanen, A., Pakkala, T., and Kuussaari, M. 1996. The quantitative incidence function model and persistence of an endangered butterfly metapopulation. *Conserv. Biol.* 10:578–590.

Harrison, S., and Taylor, A.D. 1997. Empirical evidence for metapopulation dynamics. In *Metapopulation Biology: Ecology, Genetics, and Evolution,* eds. I.A. Hanski and M.E. Gilpin, pp. 27–42. San Diego, California: Academic Press.

Heppell, S., Crouse, D.T., and Crowder, L.B. 2000. Using matrix models to focus research and management efforts in conservation. In *Quantitative Methods for Conservation Biology,* eds. S. Ferson and M. Burgman, pp. 148–168. New York: Springer-Verlag.

Hess, G. 1996. Disease in metapopulation models: implications for conservation. *Ecology* 77:1617–1632.

Hokit, D.G., Stith, B.M., and Branch, L.C. 1999. Effects of landscape structure in Florida scrub: a population perspective. *Ecol. Appl.* 9:124–135.

Holsinger, K.E., and Vitt, P. 1997. The future of conservation biology: what is a geneticist to do? In *The Ecological Basis for Conservation: Heterogeneity, Ecosystems, and Biodiversity,* eds. S.T.A. Pickett, R.S. Ostfeld, M. Shachak, and G.E. Likens, pp. 202–216. New York: Chapman and Hall.

Holthausen, R.S., Raphael, M.G., McKelvey, K.S., Forsman, E.D., Starkey, E.E., and Seamen, D.E. 1995. *The Contribution of Federal and Non-Federal Habitat to Persistence of the Northern Spotted Owl on the Olympic Peninsula, Washington: Report of the Reanalysis Team.* General Technical Report PNW-GTR 352. Portland, Oregon: USDA Forest Service, Pacific Northwest Research Station.

Howe, R.W., Davis, G.J., and Mosca, V. 1991. The demographic significance of "sink" populations. *Biol. Conserv.* 57:239–255.

James, M.K., Mason, L.B., and Bode, L. 1997. Larval transport modeling in the Great Barrier Reef. In *The Great Barrier Reef: Science, Use and Management: A National Conference,* eds. N. Turia and C. Dalliston, Volume 6, Session 6. Townsville, Australia: Great Barrier Reef Marine Park Authority.

Kareiva, P. 1990. Population dynamics in spatially complex environments: theory and data. *Phil. Trans. R. Soc. London, Ser. B.* 330:175–190.

Kareiva, P., Skelly, D., and Ruckelhaus, M. 1997. Reevaluating the use of models to predict the consequences of habitat loss and fragmentation. In *The Ecological Basis for Conservation: Heterogeneity, Ecosystems, and Biodiversity,* eds. S.T.A. Pickett, R.S. Ostfeld, M. Shachak, and G.E. Likens, pp. 156–165. New York: Chapman and Hall.

Kean, J.M., and Barlow, N.D. 2000. The effects of density-dependence and local dispersal in individual-based stochastic populations. *Oikos* 88:282–290.

Kindvall, O. 1999. Dispersal in a metapopulation of the Bush Cricket, *Metrioptera bicolor* (Orthoptera: Tettigoniidae). *J. Anim. Ecol.* 68:172–185.

Lacy, R.C. 1993. VORTEX: a computer simulation for use in population viability analysis. *Wildl. Res.* 20:45–65.

LaHaye, W.S., Gutierrez, R.J., and Akçakaya, H.R. 1994. Spotted Owl meta-population dynamics in southern California. *J. Anim. Ecol.* 63:775–785.

Lamberson, R.H., McKelvey, R., Noon, B.R., and Voss, C. 1992. A dynamic analysis of Northern Spotted Owl viability in a fragmented forest landscape. *Conserv. Biol.* 6:505–512.

Lande, R. 1988. Genetics and demography in biological conservation. *Science* 241:1455–1460.

Lawes, M.J., Mealin, P.E., and Piper, S.E. 2000. Patch occupancy and potential metapopulation dynamics of three forest mammals in fragmented Afromontane forest in South Africa. *Conserv. Biol.* 14:1088–1098.

Lee, D.C. 2000. Assessing land-use impacts on bull trout using Bayesian belief networks. In *Quantitative Methods for Conservation Biology,* eds. S. Ferson and M. Burgman, pp. 127–147. New York: Springer-Verlag.

Letcher, B.H., Priddy, J.A., Walters, J.R., and Crowder, L.B. 1998. An individual-based, spatially explicit simulation model of the population dynamics of the endangered Red-cockaded Woodpecker, *Picoides borealis. Biol. Conserv.* 86:1–14.

Lindenmayer, D.B., Lacy, R.C., and Pope, M.L. 2000. Testing a simulation model for population viability analysis. *Ecol. Appl.* 10:580–597.

Lindenmayer, D.B., and Possingham, H.P. 1995. Modeling the impacts of wildfire on the Australian arboreal marsupial, Leadbeater's possum, *Gymnobelideus leadbeateri*. *For. Ecol. Manage.* 74:197–222.

Lindenmayer, D.B., and Possingham, H.P. 1996. Ranking conservation and timber management options for Leadbeater's possum in southeastern Australia using population viability analysis. *Conserv. Biol.* 10:235–251.

Litvaitis, A.J., and Villafuerte, R. 1996. Factors affecting the persistence of New England cottontail metapopulations: the role of habitat management. *Wildl. Soc. Bull.* 24:686–693.

Liu, J. 1993. ECOLECON: An ECOLogical-ECONomic model for species conservation in complex forest landscapes. *Ecol. Model.* 70:63–87.

Ludwig, D. 1998. Is it meaningful to estimate a probability of extinction? *Ecology* 80:298–310.

MacDonald, D.W., Mace, G.M., and Barretto, G.R. 1999. The effects of predators on fragmented prey populations: a case study for the conservation of endangered prey. *J. Zool. Soc. Lond.* 247:486–506.

Matsinos, Y.G., Wolff, W.F., and DeAngelis, D.L. 2000. Can individual-based models yield a better assessment of population viability? In *Quantitative Methods for Conservation Biology,* eds. S. Ferson and M. Burgman, pp. 188–198. New York: Springer-Verlag.

Matthysen, E. 1999. Nuthatches (*Sitta europaea:* Aves) in forest fragments: demography of a patchy population. *Oecologia* 119:501–509.

McCarthy, M.A. 1996. Modeling extinction dynamics of the Helmeted Honeyeater: effects of demography, stochasticity, inbreeding and spatial structure. *Ecol. Model.* 85:151–163.

McCarthy, M.A., and Broome, L.S. In press. A method for validating stochastic models of population viability: a case study of the mountain pygmy-possum (*Burramys parvus*). *J. Anim. Ecol.*

McCarthy, M.A., Burgman, M.A., and Ferson, S. 1996. Logistic sensitivity and bounds for extinction risks. *Ecol. Model.* 86:297–303.

McCarthy, M.A., and Lindenmayer, D.B. 1999. Incorporating metapopulation dynamics of greater gliders into reserve design in disturbed landscapes. *Ecology* 80:651–667.

McCarthy, M.A., Lindenmayer, D.B., and Possingham, H.P. In press. Testing spatial PVA models of Australian Treecreepers (AVES: Climacteridae) in fragmented forest. *Ecol. Appl.*

McKelvey, K.B., Noon, B.R., and Lamberson, R.H. 1992. Conservation planning for species occupying fragmented landscapes: the case of the Northern Spotted Owl. In *Biotic Interactions and Global Changes,* eds. P.M. Kareiva, J.G. Kingsolver, and R.B. Huey, pp. 424–450. Sunderland, Massachusetts: Sinauer Associates.

McPeek, M.A., and Holt, R.D. 1992. The evolution of dispersal in spatially and temporally varying environments. *Am. Nat.* 140:1010–1027.

Meir, E., and Fagan, W.F. 2000. Will observation error and biases ruin the use of simple extinction models? *Conserv. Biol.* 14:148–154.

Menges, E.S. 1990. Population viability analysis for an endangered plant. *Conserv. Biol.* 4:52–62.

Menges, E.S., and Dolan, R.W. 1998. Demographic viability of populations of *Silene regia* in midwestern prairies: relationships with fire management, genetic variation, geographic location, population size and isolation. *J. Ecol.* 86:63–78.

Moilanen, A., Smith, A.T., and Hanski, I. 1998. Long-term dynamics in a metapopulation of the American pika. *Am. Nat.* 152:530–542.

Mooij, W.M., and DeAngelis, D.L. 1999. Error propagation in spatially explicit population models: a reassessment. *Conserv. Biol.* 13:930–933.

Mumme, R.L., Schoech, S.J., Woolfenden, G.E., and Fitzpatrick, J.W. 2000. Life and death in the fast lane: demographic consequences of road mortality in the Florida Scrub-Jay. *Conserv. Biol.* 14:501–512.

Murray, D.L., and Fuller, M.R. 2000. A critical review of the effects of marking on the biology of vertebrates. In *Research Techniques in Animal Ecology,* eds. L. Boitani and T.K. Fuller, pp. 15–64. New York: Columbia University Press.

Nieminen, M., and Hanski, I. 1998. Metapopulations of moths on islands: a test of two contrasting models. *J. Anim. Ecol.* 67:149–160.

Noon, B.R., McKelvey, K.B., and Murphy, D.D. 1997. Developing an analytical context for multispecies conservation planning. In *The Ecological Basis for Conservation: Heterogeneity, Ecosystems, and Biodiversity,* eds. S.T.A. Pickett, R.S. Ostfeld, M. Shachak, and G.E. Likens, pp. 43–59. New York: Chapman and Hall.

Noss, R.F., O'Connell, M.A., and Murphy, D.D. 1997. *The Science of Conservation Planning: Habitat Conservation Under the Endangered Species Act.* Washington, DC: Island Press.

Possingham, H.P. 1996. Decision theory and biodiversity management: how to manage a metapopulation. In *Frontiers in Population Ecology,* eds. R.B. Floyd, A.W. Sheppard, and P.J. DeBarro, pp. 391–398. Canberra, Australia: The Commonwealth Scientific and Industrial Research Organization.

Possingham, H.P. 1997. State-dependent decision analysis for conservation biology. In *The Ecological Basis of Conservation: Heterogeneity, Ecosystems, and Biodiversity,* eds. S.T.A. Pickett, R.S. Ostfeld, M. Shachak, and G.E. Likens, pp. 298–304. New York: Chapman and Hall.

Pulliam, H.R. 1996. Sources and sinks: empirical evidence and population consequences. In *Population Dynamics in Ecological Space and Time,* eds. O.E. Rhodes, R.K. Chesser, and M.H. Smith, pp. 45–70. Chicago: University of Chicago Press.

Pulliam, H.R., Dunning, J.B., and Liu, J. 1992. Population dynamics in complex landscapes: a case study. *Ecol. Appl.* 2:165–177.

Quintana-Ascencio, P.F., Dolans, R.W., and Menges, E.S. 1998. *Hypericum cumulicola* demography in unoccupied and occupied Florida scrub patches with different time-since-fire. *J. Ecol.* 86:640–651.

Ruckelshaus, M., Hartway, C., and Kareiva, P. 1997. Assessing the data requirements of spatially explicit dispersal models. *Conserv. Biol.* 11:1298–1306.

Rushton, S.P., Lurz, P.W., South, A.B., and Mitchell-Jones, A. 1999. Modelling the distribution of red squirrels (*Sciurus vulgaris*) on the Isle of Wight. *Anim. Conserv.* 2:111–120.

Schumaker, N.H. 1996. Using landscape indices to predict habitat connectivity. *Ecology* 77:1210–1225.

Sork, V.L., Campbell, D., Dyer, R., Fernandez, J., Nason, J. Petit, R., Smouse, P., and Steinberg, E. 1998. *Proceedings From a Workshop on Gene Flow in Fragmented, Managed, and Continuous Populations.* Research Paper Number 3. Santa Barbara, California: National Center for Ecological Analysis and Synthesis.

Starfield, A.M. 1997. A pragmatic approach to modeling for wildlife management. *J. Wildl. Manage.* 61:261–270.

Stith, B.M., Fitzpatrick, J.W., Woolfenden, G.E., and Pranty, B. 1996. Classification and conservation of metapopulations: a case study of the Florida Scrub-Jay. In *Metapopulations and Wildlife Conservation,* ed. D.R. McCullough, pp. 187–216. Washington, DC: Island Press.

Sweanor, L.L., Logan, K.A., and Hornocker, M.G. 2000. Cougar dispersal patterns, metapopulation dynamics, and conservation. *Conserv. Biol.* 14:798–808.

Taylor, B.L., Wade, P.R., Stehn, R.A., and Cochrane, J.F. 1996. A Bayesian approach to classification criteria for Spectacled Eiders. *Ecol. Appl.* 6:1077–1089.

Turchin, P. 1998. *Quantitative Analysis of Movement: Measuring and Modeling Population Redistribution in Animals and Plants.* Sunderland, Massachusetts: Sinauer Associates.

Turner, M.G., Arthaud, G.J., Engstrom, R.T., Hejl, S.J., Liu, J., Loeb, S., and McKelvey, K. 1995. Usefulness of spatially explicit population models in land management. *Ecol. Appl.* 5:12–16.

Van Horne, B. 1983. Density as a misleading indicator of habitat quality. *J. Wildl. Manage.* 47:813–901.

Verboom, J. 1996. *Modeling Fragmented Populations: Between Theory and Application in Conservation Planning.* Wageningen, The Netherlands: Institute for Forestry and Nature Research.

Watkinson, A.R., and Sutherland, W.J. 1995. Sources, sinks, and pseudo-sinks. *J. Anim. Ecol.* 64:126–130.

White, G.C. 2000. Population viability analysis: data requirements and essential analyses. In *Research Techniques in Animal Ecology: Controversies and Consequences,* eds. L. Boitani and T.K. Fuller, pp. 288–331. New York: Columbia University Press.

White, G.C., and Garrott, R.A. 1990. *Analysis of Radio-Tracking Data.* New York: Academic Press.

Wiens, J.A. 1997. Metapopulation dynamics and landscape ecology. In *Metapopulation Biology: Ecology, Genetics, and Evolution,* eds. I.A. Hanski and M.E. Gilpin, pp. 43–68. San Diego, California: Academic Press.

Wiens, J.A., and Rotenberry, J.T. 1981. Censusing and the evaluation of avian habitat occupancy. *Stud. Avian Biol.* 6:522–532.

Wooton, J.T., and Bell, D.A. 1992. A metapopulation model of the Peregrine Falcon in California: viability and management strategies. *Ecol. Appl.* 2:307–321.

23

Prescribing Habitat Layouts: Analysis of Optimal Placement for Landscape Planning

CURTIS H. FLATHER, MICHAEL BEVERS, AND JOHN HOF

23.1 Introduction

Physical restructuring of landscapes by humans is a prominent stress on ecological systems (Rapport et al. 1985). Landscape restructuring occurs primarily through land-use conversions or alteration of native habitats through natural resource management. A common faunal response to such land-use intensification is an increased dominance of opportunistic species leading to an overall erosion of biological diversity (Urban et al. 1987). Slowing the loss of biodiversity in managed systems will require interdisciplinary planning efforts that meld analysis approaches from several fields, including landscape ecology, conservation biology, and management science.

The objective of this chapter is to review emerging methods from this set of disciplines that enable analysts to make explicit recommendations (prescriptions) concerning the placement of wildlife habitat in managed landscapes. We first define what we mean by habitat prescriptions and compare two general habitat-placement planning strategies. We follow this with a discussion of critical thresholds and their influence on landscape planning analyses. This conceptual summary is followed by a review of two theoretical and two management application examples of these planning strategies. Our examples focus on conservation problems involving the response of a single species to varying amounts and arrangements of habitat. The last three sections of the chapter use the findings from these case examples to derive general principles related to spatially explicit habitat prescriptions; to define theoretical and empirical knowledge gaps that appear to be impeding the implementation of habitat prescriptions in landscape planning; and to propose specific research approaches that may help close those knowledge gaps and extend our capability to incorporate spatial complexity into resource planning.

23.2 Concepts, Principles, and Emerging Ideas

Wildlife are rarely distributed uniformly in space. Species exhibit a spatial structure in their occurrence and abundance that is caused by habitat heterogeneity, natural disturbance patterns, land use, and resource management activities. Accounting for the interplay among vegetation, animal abundance, and human activity poses increasingly complex problems for landscape planners. Questions concerning the likely persistence of populations within a network of habitat patches, or questions about the most appropriate layout and schedule for land management actions are now commonplace.

Despite a shared landscape perspective, these questions have typically been addressed using different analysis protocols. Population persistence questions are most often addressed by *spatially explicit simulations* that combine species population models with geographic depictions of habitat resources (Dunning et al. 1995). Such simulation approaches permit a detailed mechanistic representation of population dynamics and are used to address planning questions *predictively* (i.e., what "will" the animal response be to a given habitat arrangement). Conversely, questions pertaining to the placement and scheduling of resource management actions are being addressed by *spatial optimization methods* that enable analysts to select the "best" habitat arrangement from an essentially infinite number of spatial landscape choices (Hof and Bevers 1998). Because the outcome of spatial optimization analysis is a habitat design, management recommendations are *prescriptive* (i.e., what "should" the habitat arrangement be to meet specified population objectives most efficiently). Our focus in this chapter is on *prescriptive planning,* by which we mean planning that results in spatially explicit habitat placement recommendations. Although we have linked optimization with prescription, we do not mean to imply that simulation cannot be used to prescribe habitat layouts. However, the use of simulation in prescriptive planning has limitations because a great many potential habitat arrangements must be analyzed before planners can infer a specific habitat design.

Regardless of which strategy is chosen, it is important to realize that these planning approaches are decision tools that carry with them a set of strengths and weaknesses. Neither approach is inherently better than the other, and both involve simplifying assumptions about complex ecological systems. Whether these approaches capture sufficient ecological detail or not will have to be evaluated on a case-by-case basis, which serves to remind us that no one planning tool can be regarded as a panacea to resolve all conservation problems.

23.2.1 Habitat Prescription From Simulation

Resource managers concerned with maintaining target-species populations are typically presented with the task of predicting population response to management activities that alter the amount and geometry of habitat. A number of modeling approaches are available to address this type of planning problem, including *empirical models* derived from statistical estimation (e.g., Morrison et al. 1987), *analytical models* that have exact solutions derived from a few fundamental mathematical relationships (e.g., Lande 1987), and *simulation models* derived from mechanistic mathematical relationships whose number and complexity requires that solutions be explored numerically with digital computers (e.g., Fahrig 1997). Because empirical models tend to be correlative in nature and have not performed well predictively (Block et al. 1994), and because analytical models tend to become intractable in all but the simplest cases of spatial heterogeneity (Fahrig 1991), simulation modeling has become the method of choice for managers (Simberloff 1988).

Simulation modeling is often used to evaluate population response to alternative landscapes *a posteriori*. Management choices are thus made by ranking the various landscape alternatives by some criterion (e.g., organism abundance, persistence) and selecting the strategy that ranks highest. This use of simulated predictions in conservation planning has been termed *relative ranking* by Turner et al. (1995:13) and is a form of prescriptive planning because the result specifies a habitat arrangement that will "best" address some set of management objectives. The primary shortcoming of this approach is that the outcome is only "the best" alternative from among the limited number of landscape layouts investigated (Conroy 1993).

23.2.2 Habitat Prescription From Optimization

When managers must choose from a large set of possible habitat layouts, spatial optimization methods become useful as an alternative to running numerous simulations. With simulation, we can estimate effects in great detail one habitat layout at a time, whereas optimization models forego some ecological detail to search the entire set of available layouts by simultaneously solving a large system of equations. This equation set formalizes the relationship between various ecosystem goods and services that often involve tradeoffs. A *tradeoff* occurs when one ecosystem output cannot be increased without reducing another. Although simulation can generate different scenarios with different output mixes, only optimization can reduce the search to those tradeoffs that are efficient (e.g., provide the greatest abundance of a particular species for a given amount of land devoted to timber harvesting). From this reduced set of efficient output mixes, optimization can then select a solution that best meets the planner's objectives.

23.2.3 Critical Thresholds and Species Movement

Within the landscape-ecology literature, there is accumulating evidence that landscapes exhibit critical thresholds (Turner and Gardner 1991). *Critical thresholds* refer to changes in some attribute of landscape structure (usually habitat amount) that result in an abrupt shift in some property of landscapes or some ecological phenomenon of interest (Green 1994). One case receiving increased attention is the relation between habitat amount and species extinction (Lande 1987; With and King 1999). Although a relatively recent topic in landscape ecology, the notion of critical thresholds associated with species extinction was actually introduced in the early 1950s when Skellam (1951) mathematically demonstrated a threshold relationship between the size of a single habitat patch and species persistence using a reaction-diffusion model. *Reaction-diffusion* models represent a general class of spatial models that account for reproduction as well as movement. They are closely allied to biochemical reaction-diffusion models (Turing 1952), with reproduction and dispersal as the reaction and diffusion components, respectively (Kareiva 1990). The simplicity of reaction-diffusion models has made them particularly useful in situations in which information on the exact movement paths of individuals is lacking (Turchin 1998).

One implication of critical thresholds that we want to emphasize is the existence of regions in the parameter space (as defined by landscape attributes or population vital rates) where model behavior exhibits fundamental shifts. These regions not only define thresholds, but they also can be used to delimit *domains of applicability* for particular habitat arrangement recommendations, serving as a framework for defining habitat prescriptions for different ecological circumstances.

23.3 Recent Applications

It has been noted that the number of successful collaborations between research scientists and managers in a landscape planning situation is limited (Hobbs et al. 1993; Schemske et al. 1994). We found this to be true in our search for on-the-ground field applications of habitat-prescription concepts. We conducted a review of the literature, sent requests to colleagues, and sent a survey to the Wildlife, Fish, and Rare Plants staff of the U.S. Forest Service requesting information on case examples in which habitat layouts had been prescribed in a resource planning context. This survey was sent to Wildlife Program Managers, Threatened and Endangered Species Program Managers, and Landscape Ecologists within each of the nine National Forest System Regions of the U.S. Forest Service. This review produced examples that were classified into one of two groups. One group was research-driven and focused on examining the response of modeled populations to various habitat layouts on hypothetical landscapes (e.g., Frank and Wissel 1998; Etienne and Heesterbeek 2000).

The second group consisted of conservation planning examples where habitat layouts in actual managed landscapes were proposed for a target species, or set of target species (e.g., Suring et al. 1993; Mellen et al. 1995). Although landscape design concepts were used in these conservation planning examples, conflicts among alternative resource uses, the public review process, and negotiated plan modifications often superseded the proposed habitat prescription. This made it difficult to evaluate the success of these various applications. Furthermore, insufficient monitoring data to date prevented any quantitative evaluation of a particular habitat layout in terms of target species response.

In this section, we review a sample of case examples while acknowledging that none of them represents the situation in which the habitat layout is prescribed, the plan is implemented, and the success of the plan is evaluated through monitoring. We describe four applications of habitat-prescription concepts that have occurred in either a research context (two theoretical examples) or a resource-planning context (two management-oriented applications).

23.3.1 Diffusion and Critical Thresholds—A Theoretical Investigation Using Simulation

When a landscape undergoes fragmentation, habitat is lost, patch sizes decrease, and the distance separating habitat fragments increases (Andrén 1994). The erosive influence of habitat loss on species is widely recognized (Simberloff 1988), but the effects of habitat arrangement on populations is disputed (Fahrig 1997). The contention stems from the multivariate nature of fragmentation and the difficulty in distinguishing the relative influences of habitat amount and habitat arrangement. In this example, we use a spatially explicit reaction-diffusion model (Bevers and Flather 1999a) to simulate long-term equilibrium populations of a hypothetical species with particular attention given to the possible existence of critical thresholds. By varying habitat amount and arrangement systematically, we illustrate the value of simulation experiments in developing general management rules related to prescribing habitat layouts across a landscape.

Many model formulations are available to represent animal movement (Turchin 1998). For species that establish and defend distinct breeding territories, territory size defines a convenient scale for modeling populations across a landscape (Bevers and Flather 1999a). In this example, population abundance over a landscape is estimated using the following reaction-diffusion model:

$$v_{it} = \min(b_{it}, \sum_{j=1}^{N} [1 + f_j(v_{j,t-1})] v_{j,t-1} g_{ji}) \qquad \text{for all } i,\ t,$$

where i and j each index all cells (i.e., potential breeding territories) in the landscape; v_{it} is the population in cell i at time t; b_{it} represents adult carrying capacity in each cell for each time period; $f_j(v_{j,t-1})$ is the net per capita rate of reproduction within a cell (not accounting for mortality associated with dispersal); and g_{ji} reflects the probability of an individual emigrating from breeding territory j to territory i a full breeding season later by any number of possible routes. In this example, we have assumed that the per capita net rate of reproduction is constant across all habitat cells, as is carrying capacity (i.e., we do not let habitat quality vary). We also assume that individuals disperse identically from the center of each habitat cell in uniform random directions with the probability of successful dispersal between habitat cells decaying with distance. Mortality occurs when individuals disperse into nonhabitat, into saturated territories, or outside of the boundaries of the landscape.

We systematically examined the influence of habitat amount and arrangement (Flather and Bevers 2002) by simulating landscape structure using RULE, a computer program that permits the generation of random landscape replicates with user-specified characteristics (Gardner 1999). We generated a total of 2,430 binary (habitat, nonhabitat) landscapes, each composed of 1,024 cells (32 rows × 32 columns). Habitat amount was varied from 10% to 90% of the total area in 10% increments. A fragmentation parameter also was systematically varied over nine levels to cover a range from highly fragmented to highly aggregated spatial configurations. This resulted in a 9 × 9 factorial simulation experiment with each treatment replicated 30 times (Flather and Bevers 2002). We parameterized the reaction-diffusion model to represent a generic bird species that breeds in forest-interior habitats and defends a 1-ha territory (the area of a landscape cell). The adult carrying capacity within any habitat cell was one breeding pair.

If total bird abundance within each landscape was determined solely by the amount of habitat, we would expect the population response to be proportional to habitat amount. A plot of mean population against habitat amount and fragmentation level appears consistent with a habitat amount effect (Figure 23.1a). Analysis of variance (ANOVA) confirmed this visual inspection—habitat amount accounted for nearly 97% of the variation in abundance (Table 23.1). Fragmentation and its interaction with habitat amount accounted for an additional 1.3% of the variation.

These results suggest that habitat arrangement effects, at least for the hypothetical species being modeled, are inconsequential. However, some distortion of the population response surface away from a pure habitat effect can be observed as habitat was reduced below 40% of our total landscape (Figure 23.1a). That distortion appears to coincide with a persistence threshold as indicated by a rapid decline in the probability of landscapes supporting viable populations (Figure 23.1b). If fragmentation effects become important as landscape structure approaches a persistence threshold, then we would expect to see evidence for a strong difference in the fragmentation effect as our landscapes cross that persistence threshold.

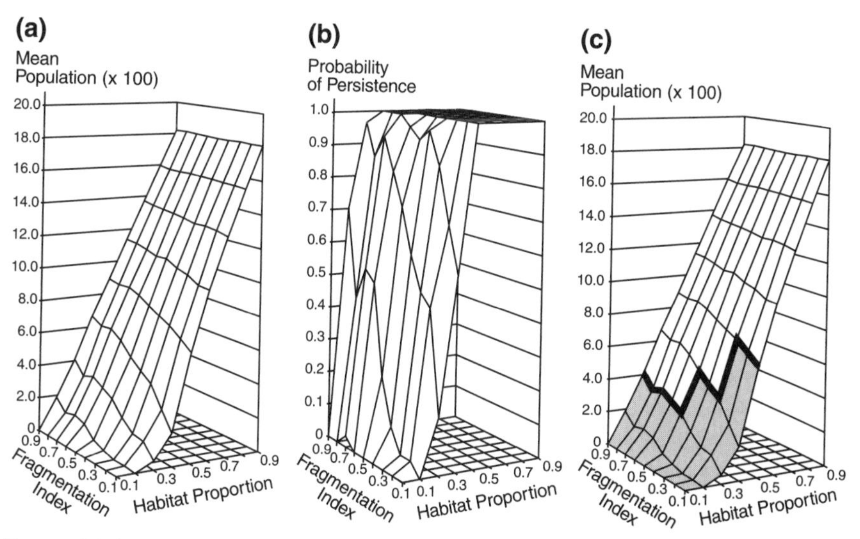

FIGURE 23.1. Population response of a hypothetical species to simulated landscapes with specified amounts of habitat and levels of fragmentation: (a) mean abundance across 30 replicates per treatment pair, (b) probability of replicates supporting viable populations, and (c) identification of the persistence threshold (bold line). Adapted from Flather and Bevers (2002).

TABLE 23.1. Summary ANOVA findings for simulated landscapes for the full 9 × 9 factorial experiment, above a persistence threshold, and below a persistence threshold (*DF* = degrees of freedom, *SS* = partial [Type III] sums of squares). Persistence threshold is as defined in Figure 23.1c.

Source of variation	*DF*	% of Total *SS*	*F*	*P*
Full experiment				
Habitat amount	8	96.8	15871.5	0.0001
Fragmentation	8	0.7	124.4	0.0001
Habitat amount × Fragmentation	64	0.6	13.3	0.0001
Error	2349	1.8	—	—
Above threshold				
Habitat amount	6	96.3	10497.2	0.0001
Fragmentation	8	0.7	55.8	0.0001
Habitat amount × Fragmentation	42	0.5	8.3	0.0001
Error	1653	2.5	—	—
Below threshold				
Habitat amount	3	30.3	122.3	0.0001
Fragmentation	8	6.2	9.4	0.0001
Habitat amount × Fragmentation	12	6.1	6.2	0.0001
Error	696	57.4	—	—

We used Classification and Regression Trees (CART; Breiman et al. 1984) to define a persistence threshold across our set of landscapes. CART defined a simple set of rules based on habitat amount and fragmentation level that maximizes the accuracy of a classification into species-persistent and species-nonpersistent landscapes. The set of rules estimated by CART can then be mapped directly onto the population response surface (Figure 23.1c). Defining the persistence threshold in this manner enabled us to partition landscapes objectively into sets above and below the threshold. Rerunning the ANOVA on these two sets of landscapes supports the hypothesis that fragmentation effects become more important as one approaches a population persistence threshold (Table 23.1). Above the persistence threshold, habitat amount accounts for 96% of the variance in population abundance across treatments. Below the persistence threshold, habitat amount only accounts for 30% of the abundance variation. Note, however, that the error term in the below-threshold ANOVA indicates that 57% of the variation in abundance is still unexplained. This observation is consistent with Green's (1994) finding that variation is greatest close to the critical threshold.

To reduce the uncertainty associated with explaining variation in abundance below the persistence threshold, we characterized the structure of each landscape according to the following simple measures: fragmentation level (as specified during landscape generation), number of habitat patches, mean size of habitat patches, mean distance to nearest patch, total length of edge, size of largest habitat patch, shape (perimeter-to-area ratio) of largest patch, and edge length of

TABLE 23.2. Summary stepwise regression results using landscape structure variables within simulated landscapes to predict species abundance. This analysis was run on those landscape scenes that fell below the persistence threshold defined in Figure 23.1c.

Variable	R^2	F	P
Size of largest patch	0.572	956.06	0.0001
Edge length of largest patch	0.692	279.78	0.0001
Fragmentation	0.727	91.64	0.0001
Average patch size	0.728	2.43	0.1193
Shape of largest patch	0.729	2.74	0.0985

largest patch. Stepwise regression resulted in a five-variable model that accounted for 73% of the variability in population abundance (Table 23.2). Two variables, the size and edge length associated with the largest patch, accounted for nearly 70% of the variation in abundance. This suggests that population persistence in these landscapes was critically dependent on having a single patch that was large and "blocky" enough (i.e., had relatively little edge) to sustain the population within the landscape.

This example has demonstrated, at least theoretically, that manipulation of habitat arrangement can be an important management consideration in maintaining persistent populations in patchy landscapes. However, simulation approaches often cannot feasibly examine all possible habitat layout choices. As a result, habitat arrangement designs prescribed from simulations may be suboptimal (Van Deusen 1996).

23.3.2 Optimal Placement—A Theoretical Investigation Using Spatial Optimization

Given that resources devoted to conservation are limited, managers should seek a habitat layout that is in some sense optimal. In this section, we review an analysis (Hof and Flather 1996) that demonstrates the utility of spatial optimization in making general habitat layout recommendations. We assume, for this particular management problem, that resource extraction and conservation of some target species are simultaneous objectives for a given landscape. Furthermore, resource extraction is assumed to destroy the species' habitat. The problem presented to the resource manager is to determine how to arrange the habitat remaining after resource extraction in a way that will most benefit the target species.

When habitat is going to be lost, it is common to see planning recommendations to maintain well-connected habitats. These recommendations stem from observations that populations found in fragmented habitats may be more prone to local extinction events (Pimm and Gilpin 1989). But wildlife populations also are affected by environmental disturbances (e.g., extreme weather events, fire, epizootics) that have different degrees of spatial covariation (Gilpin 1987). Consideration of these environmental disturbances raises the possibility that population persistence times may actually be longer in fragmented landscapes because the risk of local extinction is spread among independent populations (Palmqvist and Lundberg 1998). These various population pressures define an ecological tradeoff: minimizing demographic extinction pressures would lead managers to recommend clumping remaining habitat, whereas spatially correlated catastrophic events suggest that some degree of habitat spreading would be advantageous (Hof and Flather 1996).

We can define a statistically based formulation that enables us to capture the tradeoff between habitat connectivity and spatial covariation by treating this as a problem of estimating the total population (T) with a certain level of confidence:

$$T = E(P) + \delta\ V(P)^{1/2}, \tag{1}$$

where $E(P)$ is the expected value of population size across all habitat patches, $V(P)$ is the variance of population size in space and time, and δ is the standard normal deviate for some given confidence level (e.g., $\delta = 1.96$ for 95% confidence).

The expected value of the population among all patches in a landscape is given by

$$E(P) = \sum_{i=1}^{N} d_i A_i PR_i,$$

where d_i is the density of individuals in patch i, A_i is the area of patch i, and PR_i is the joint probability that the patch i is "connected" to any of the N patches in the complex. The joint probability (PR_i) of each patch being connected is given by

$$PR_i = 1 - \left[\prod_{i=1}^{N}(1 - pr_{ij})\right] \qquad \text{for all } i \text{ patches},$$

where pr_{ij} is the probability that patch i is connected to patch j. Ecologically, one expects pr_{ij} to decrease with interpatch distance, but conditioned on the dispersal capability of the species and the resistance of the interpatch matrix to species movement. A continuous function decaying with distance that captures these conditions is given by

$$pr_{ij} = 1 - \left(1 - \theta^{D_{ij}}\right)^{\beta}, \tag{2}$$

where D_{ij} is the distance between patch i and patch j, β reflects the dispersal capability of the target species, and θ reflects the resistance of the matrix to species movement.

Populations of species inhabiting separate patches show differing levels of covariation in their population dynamics (Gilpin 1987). The definition of covariation between any two patches is given by

$$\sigma_{ij}^{2} = \rho_{ij}\sigma_i\sigma_j,$$

where σ_{ij}^2 is the covariance between the population in patch i and the population in patch j, ρ_{ij} is the correlation between population sizes in patch i and patch j, and

σ_i and σ_j are the standard deviations of population sizes in patches *i* and *j*, respectively. The variance in population size, *V(P)* from Eq. 1, is then given by

$$V(P) = \sum_{i=1}^{N} \sum_{j=1}^{N} \rho_{ij} \sigma_i \sigma_j .$$

We assume that ρ_{ij} is negatively related to interpatch distance and that its decay can be represented by a function similar to Eq. 2:

$$\rho_{ij} = 1 - \left(1 - \omega^{D_{ij}}\right)^{\tau},$$

where τ reflects the threshold distance beyond which the degree of correlation among patch population sizes declines rapidly, and ω is the rate at which patch covariation declines with distance and reflects the resistance of the interpatch matrix to the spread of a particular disturbance agent. Both τ and ω are specific to a particular kind of disturbance agent.

For purposes of illustration, consider a 10,000-ha management area of homogeneous habitat inhabited by a hypothetical target species that defends a 1-ha territory (occupied by a single breeding pair). We assume that 20% of the habitat will be retained following resource extraction. We define the retained patches to have a circular shape and arbitrarily set the maximum number of patches to be four. The optimization problem is thus to choose a size and location for each habitat patch in such a way as to maximize *T* in Eq. 1. For most conservation problems, the manager is interested in guarding against catastrophically low populations. Consequently, our focus is on the lower bound of the confidence interval specified in Eq. 1. For this example, we want to maximize the population *T* with 80% certainty, and set $\delta = -0.84$ accordingly. We fixed β and τ to 50 and systematically varied θ and ω over the values of 0.75, 0.85, and 0.95 to explore how optimal habitat placement varied among species with low, moderate, and high dispersal rates and disturbance agents causing low, moderate, and high degrees of spatial covariation. Disturbances with low degrees of spatial covariation affect patch populations locally, and correlation among patches decays rapidly with distance. Disturbances with high degrees of spatial covariation have broad spatial effects, and correlation among patches decays slowly with distance.

The resulting optimal habitat layouts vary greatly among ecological circumstances (Figure 23.2). Species with low dispersal rates in environments with high degrees of spatial covariation reach maximum populations in a clumped arrangement (Figure 23.2a)—disturbances that affect landscapes broadly decrease the risk-spreading advantage of more isolated patches. Species with high dispersal rates inhabiting landscapes with low (Figure 23.2b) to moderate (Figure 23.2c) spatial covariation reach higher equilibrium populations if at least some of the habitats are separated. Finally, species with low-to-moderate dispersal rates in environments with moderate-to-low degrees of spatial covariation attain maximum populations in habitat layouts that provide "corridors" (Figure 23.2d) or "stepping stones" (Figure 23.2e) to facilitate exchange of individuals.

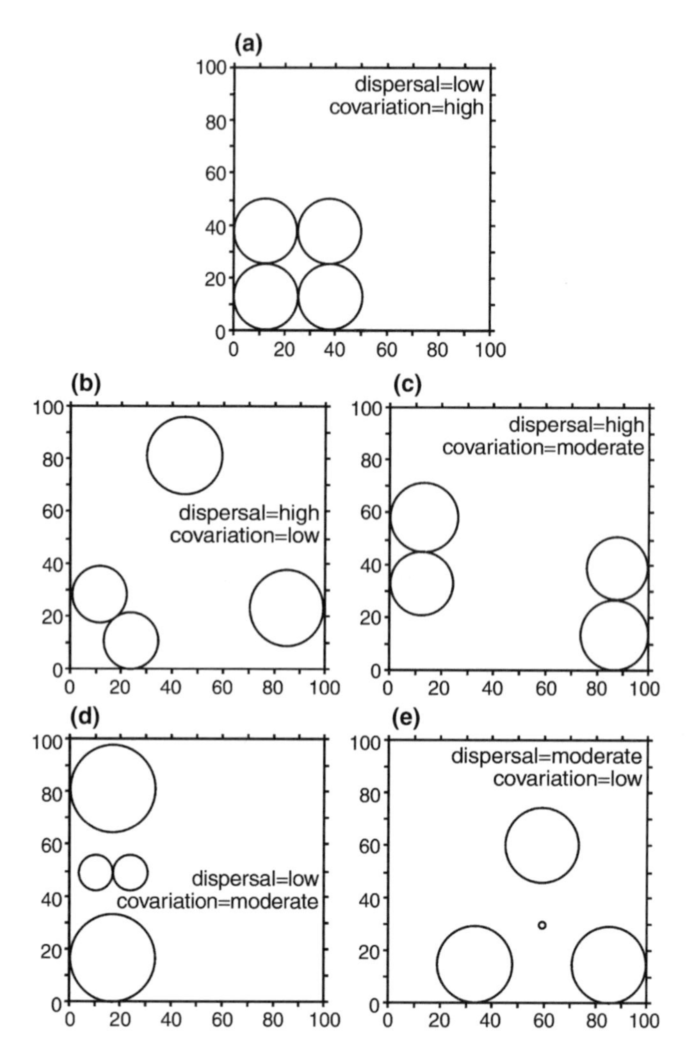

FIGURE 23.2. Spatially optimized habitat layouts for species with different dispersal rates inhabiting landscapes with different degrees of spatial covariation. All layouts are based on the assumption that 20% of the original habitat would be retained in four patches on the landscape after resource extraction. Reprinted from Ecological Modelling, 88, J. Hof and C.H. Flather, Accounting for connectivity and spatial correlation in the optimal placement of wildlife habitat, 143–155, Copyright 1996, with permission from Elsevier Science.

We acknowledge that these two theoretical analyses (in Sections 23.3.1 and 23.3.2) involve conditions that are simple compared with the ecological complexities encountered in an actual landscape planning problem. To balance the simplicity characterizing analyses on hypothetical landscapes, we now review two case examples of habitat placement prescription in actual landscapes.

23.3.3 Reserve Placement for Northern Spotted Owls—A Simulation Application

The Northern Spotted Owl (*Strix occidentalis caurina*) occurs in the coastal mountain region of the northwestern United States and southwestern Canada (Forsman et al. 1984). Throughout its geographic range, the species is associated with old (>150 years), dense, large-diameter coniferous forest stands (Thomas et al. 1990). Timber harvesting, agricultural development, and urban expansion have reduced the owl's preferred habitat to <10% of its original area (Murphy and Noon 1992). Even with such substantial reductions in habitat, the vulnerability of the owl to extinction was greatly debated. In an attempt to resolve the issue, an Interagency Scientific Committee (ISC), including both researchers and managers, was charged with developing a long-term, scientifically defensible conservation plan for the owl (Thomas et al. 1990). The reserve design task was to determine a configuration of habitat areas that would maintain stable local populations of owls throughout the bird's historic range (Noon and McKelvey 1996).

Addressing the question of how to "best" configure owl habitat reserves relied predominately on simulation modeling (McKelvey et al. 1993; Lamberson et al. 1994). Of particular importance to these simulation experiments was the demonstration of sharp extinction thresholds (Lande 1988; Lamberson et al. 1992). As old-growth habitat was reduced in a series of simulation experiments, a point was reached (~20% habitat) when the ability of juveniles to locate suitable territories became compromised, leading eventually to population extinction (Noon and Murphy 1994). Extensive exploration of model parameter sensitivity resulted in the following habitat reserve design criteria: individual habitat areas should be large enough to support at least 20 breeding pairs; habitat areas should be no farther than 19 km apart; and the matrix between habitat areas should be 50% forested with tree diameters >28 cm and with canopy closure >40% to facilitate juvenile dispersal (Noon and McKelvey 1996). Two findings related to these design criteria warrant remark. First, the amount of habitat prescribed in the reserve design was close to a theoretical persistence threshold, raising some concern about owl viability in the face of parameter uncertainty and catastrophic events (Harrison et al. 1993). Second, the ISC noted that the prescribed habitat layout was not a unique solution to the landscape design question, implying that equally good or better layouts may exist.

In a study following the work of the ISC, Hof and Raphael (1997) formulated a spatial optimization model that located areas for old-growth habitat retention such that owl numbers would be maximized on a small portion of the owl's range, the Olympic Peninsula in northwestern Washington. The value of this study was that habitat layout prescriptions derived from both simulation and optimization approaches could be compared, permitting a verification of the more ecologically simple optimization formulation against the more ecologically detailed simulation model. For the optimization model to be beneficial, the number of owl pairs simulated in the habitat layout generated under optimization should be greater than or equal to the number of owl pairs estimated from the simulation-prescribed habitat layout. Two parameter sets (one with optimistic and one with pessimistic demographic parameters) from Holthausen et al. (1995) were explored. In both cases, populations simulated on the optimization-derived habitat layout produced a marginally greater number of owl pairs than did the populations simulated on the habitat layout derived from simulation modeling (0.01% for the pessimistic parameter set and 5.2% for the optimistic parameter set). The minimal opportunity for finding better habitat arrangements given a pessimistic parameter set near threshold conditions is consistent with reaction-diffusion theory (Bevers and Flather 1999b). As the parameter set moved away from threshold conditions, the optimization-based layout became more beneficial, even with the simplifications.

A unique contribution of the Hof and Raphael (1997) study is that it demonstrates that simulation and optimization need not be mutually exclusive approaches to conservation planning. The interaction has the potential to offer new planning insights and to suggest which aspects of the respective formulations are key in understanding and predicting species response to landscape structure.

A decision to retain late-successional habitats is a decision that is essentially permanent (Hof and Bevers 1998). The "lifetimes" of these habitats are long relative to those of the target species. Under these circumstances, a static perspective on the habitat layout problem was appropriate. However, if habitats are dynamic over a period that is short relative to the lifetimes of target species, then planning analyses must account for habitat change in the allocation of management activities over time. The applied planning case study described in the next section addresses the problem of prescribing habitat configurations dynamically.

23.3.4 Black-Footed Ferret Reintroduction—A Spatial Optimization Application

Historically, the black-footed ferret (*Mustela nigripes*) ranged sympatrically with prairie dogs (*Cynomys* spp.) across most of the North American grasslands (Anderson et al. 1986). Ferrets live principally in prairie dog burrows and depend primarily on prairie dogs for prey (Clark 1989). The black-footed ferret has become one of the world's most endangered mammals, and in 1987, the last known free-ranging members of the species were taken into captivity (Thorne and Belitsky 1989). Demise of the species in the wild was attributed to loss and fragmentation of habitat due to extensive prairie dog eradication programs, sylvatic plague, and changes in land use (U.S. Fish and Wildlife Service, National Park Service, and USDA Forest Service 1994). A successful captive breeding program was established by the Wyoming Game and Fish Department, setting the stage for a national recovery program of releasing captive-bred ferrets back into the wild (Clark 1989).

In 1994, an area centered around Badlands National Park, South Dakota, was established as a release site for captive-bred ferrets. The Badlands reintroduction site includes a region approximately 1575 km^2 in size, comprising portions of the Park, Buffalo Gap National Grassland, and other ownerships. Federally managed lands with active or readily recoverable prairie dog colonies (i.e., suitable habitat for black-footed ferret) are fragmented, occupying less than one-tenth of the area. Ferrets disperse widely and randomly, particularly as juveniles (Richardson et al.

1987), and the spatial arrangement of prairie dog colonies is expected to affect the number of ferrets that can be supported (Minta and Clark 1989). Because rodenticides are used to control prairie dog populations (Roemer and Forrest 1996), ferret recovery also will be affected by the location and timing of rodenticide treatments in the area.

Bevers et al. (1997) built an optimization model to locate prairie dog colonies in the reintroduction area. For recovery planning purposes, it was assumed that prairie dog control would preclude the establishment of ferret populations on nonfederal ownerships. The expected adult black-footed ferret population on federal lands was maximized over a 25-year planning horizon subject to six different levels of prairie dog population control ranging (in equal 20% increments) from no additional ferret habitat (beyond 1994 conditions) to ceasing rodenticide treatments and allowing maximum ferret habitat. Spatial optimization methods were used to select the timing and locations for rodenticide treatments required to meet each prairie dog population policy constraint. Black-footed ferret populations were estimated using a cell-based reaction-diffusion model similar to that described in Section 23.3.1. Carrying capacity for ferrets was based on expected prairie dog population size, modeled as a function of the location and timing of rodenticide treatments within the recovery area.

The projected populations for the six alternatives (from lowest to highest habitat allocation) were approximately 51, 240, 369, 483, 590, and 647 adult black-footed ferrets by the end of the planning horizon (Figure 23.3a). The population increases for each 20% increment in habitat amount (adding potential capacity for about 104 adult ferrets per increment) were 189, 129, 114, 107, and 57 adult ferrets, respectively, demonstrating that those habitat locations with the greatest potential to increase total abundance were added to the complex first. By strategically locating new habitats, population increases greater than the potential capacity of each increment (104 adult ferrets) were possible because the initial complex consisted of widely scattered prairie dog colonies that would otherwise be underused.

Preferred habitat arrangements and the resulting ferret populations (Figure 23.3b–d) changed over time. Early in the planning horizon, the model tended to arrange the available habitat thinly across the entire landscape (Figure 23.3c) to support small expanding population "waves" that spread outward from the selected ferret release locations (Figure 23.3b). Once the prairie dog population capacity constraints became binding, however, the model shifted preferred habitat arrangements into dense clusters around the Badlands National Park boundary (Figure 23.3d), consistent with theoretical findings reviewed in Bevers and Flather (1999b). This sort of dynamic habitat layout strategy may be particularly difficult to discern using simulation analyses alone.

23.3.5 Success of Conservation Planning Applications

The two specific planning applications (Sections 23.3.3 and 23.3.4) we have reviewed were based on rigorous, repeatable protocols that resulted in tangible

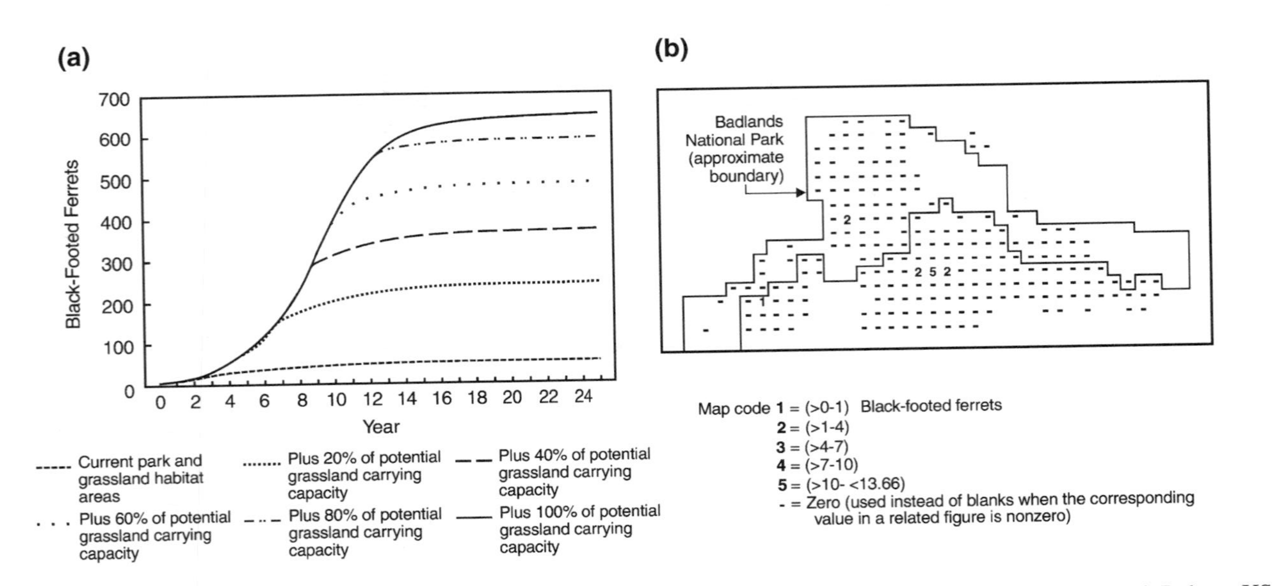

FIGURE 23.3. Black-footed ferret distribution and abundance responses in and around Badlands National Park (South Dakota, USA): (a) population response to 1994 habitat management plans and five alternatives strategies, (b) recommended number of released ferrets and their locations under the +20% alternative, (c) the projected number of black-footed ferrets and their locations under the +20% alternative seven years after release, and (d) the projected number of black-footed ferrets and their locations under the +20% alternative 25 years after release (near equilibrium). Reprinted by permission, M. Bevers et al., Spatial optimization of prairie dog colonies for black-footed ferret recovery, Operations Research, volume 45, number 4, July–August, 1997. Copyright 1997, the Institute for Operations Research and the Management Sciences (INFORMS), 901 Elkridge Landing Road, Suite 400, Linthicum, MD 21090 USA.

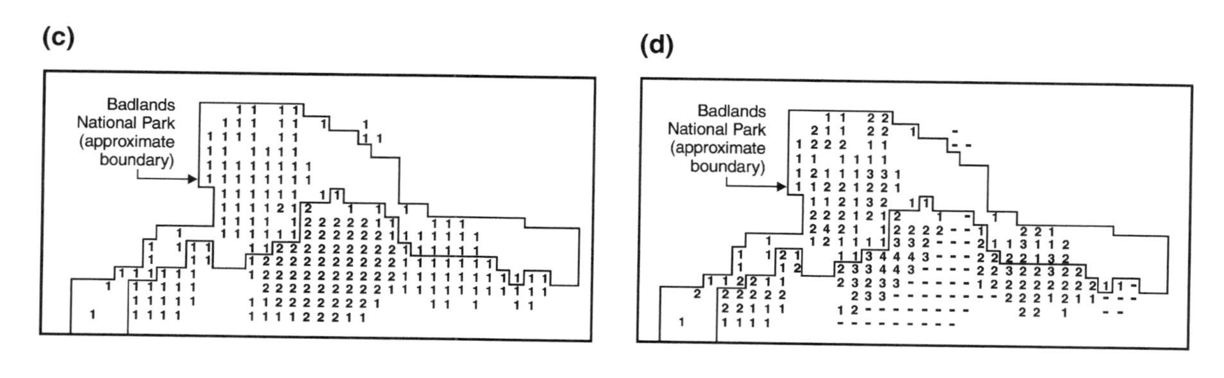

FIGURE 23.3. *Continued.*

management recommendations. Although habitat designs derived from simulation or optimization have yet to be fully implemented, the framework and analyses used make them successful from the perspective of the *process* of conservation planning. From the perspective of *outcome,* the success of these planning efforts must ultimately be judged against the long-term persistence of the species in its natural environment. Unfortunately, evaluating the success of these planning recommendations in terms of population response is premature. The effectiveness of a given habitat prescription may not be detectable for long periods of time because of a species' longevity or for reasons associated with extensive time lags in population responses (Chapter 18). Monitoring in the short term may only document transient behavior not indicative of the prospects for a species' long-term viability (Bevers and Flather 1999a).

23.4 Principles for Applying Landscape Ecology

At present, habitat-placement principles derived from simulation and optimization studies such as those described in this chapter come from theoretical analyses rather than from a distillation of real-world trial-and-error experiences. This situation occurs for two reasons. First, applied spatial habitat prescription, especially spatial optimization, is a relatively new field of study. Second, landscape-level planning is driven as much by political compromise as by analysis. In neither of our planning case studies, involving the Northern Spotted Owl and the black-footed ferret, were the layouts prescribed by the models actually implemented. For these reasons, there are few cases where habitat-placement prescriptions have been tested to the extent that definitive planning guidelines can be formulated (Kareiva and Wennergren 1995). From general theoretical models and planning analyses, however, we can derive several application principles for planners who are involved in using landscape ecology and habitat prescription concepts in biological conservation.

23.4.1 Simulation Versus Optimization

When should planners use simulation or spatial optimization? The answer, to a large degree, depends on the questions being asked and the circumstances surrounding the planning problem. For example, habitat placement choices may be limited because the pattern of land development has reduced the configuration possibilities (Saunders et al. 1991). When habitat-placement options are restricted to a small set, simulation modeling can offer a useful approach for ranking alternative configurations. If, however, placement choices are numerous, then spatial optimization may be useful in determining a layout that is "the best" given the objectives and constraints of the planning problem. Joint use of both strategies, as described in Section 23.3.3, offers planners the opportunity to take advantage of the ecological detail captured by simulation models and the analytical power of spatial optimization to select the best solution.

23.4.2 Placement Rules

An important principle arising from our review is that habitat placement is conditioned on ecological context as defined by species and disturbance. As we have shown theoretically, species movement capability and the degree of spatial covariation caused by different disturbance agents leads to fundamentally different habitat arrangements that will maximize population size within a landscape (see Section 23.3.2). These results suggest that it may be possible to define domains of applicability for different habitat placement recommendations. That is, there may be particular sets of ecological conditions for which different habitat-placement recommendations would offer the best arrangement for a target species (Figure 23.4). However, broadly applicable habitat placement generalizations may be difficult to define because of differences among species' life histories and differences in landscape context. If detailed planning recommendations are necessary, then detailed studies of the target species in a particular setting will be required (Lawton 1999). But incorporation of ecological realism may obscure the detection of general management recommendations that can emerge when studying simplified systems (Doak and Mills 1994). The interplay between ecological realism and generality of recommendations tends to place landscape planning in a dilemma—quick answers based on general rules of thumb will tend to be imprecise, whereas answers specific to a given situation will be expensive and require patience.

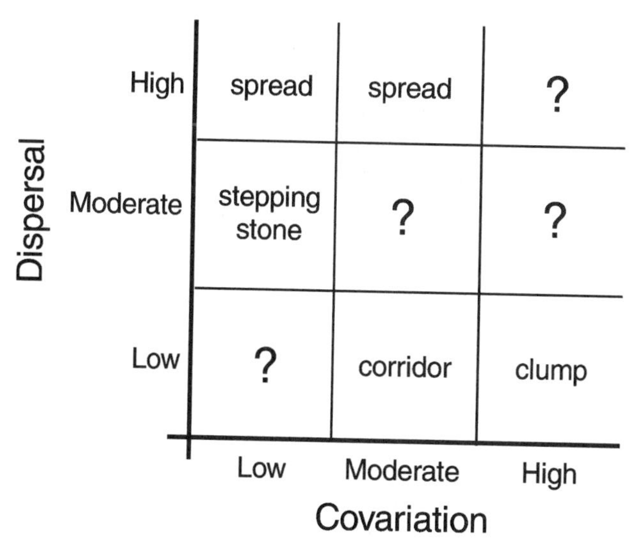

FIGURE 23.4. General habitat placement rules in an ecological parameter space defined by species dispersal rate and degree of spatial covariation based on the results shown in Figure 23.2. The "?" indicates those parameter combinations for which an optimal layout was not estimated.

23.4.3 Critical Thresholds

Do all resource plans need to be spatially explicit? A landscape planning principle that has emerged from our review suggests that spatially explicit habitat prescriptions are not necessarily required under all circumstances. We observed evidence that habitat arrangement effects were unimportant in explaining variation in species abundance as long as the amount of habitat in a landscape remained well above a critical threshold associated with species extinction (see Section 23.3.1). A systematic search for threshold responses would enable planners to tease out conditions for which arrangement effects are important and to identify landscape structure attributes that help assure species persistence in patchy landscapes. An important caveat, however, is that planners should not interpret this principle to mean that habitat arrangement can be ignored prior to reaching a critical threshold. As habitat is lost, future habitat arrangements are constrained by the spatial pattern of habitat destruction—habitat lost randomly offers little opportunity to cluster remnant habitats once the persistence threshold is reached.

23.5 Knowledge Gaps

23.5.1 Theoretical Voids

The recognition that environmental heterogeneity and organism movement interact to affect population dynamics has been key to what Turchin (1998:2) called a "paradigmatic shift from the aspatial equilibrium view . . . to a spatially explicit view." Although theoretical developments in spatial ecology are relatively recent, theoretical research on the concepts reviewed in this chapter has far out paced empirical research. The disparity between theory and empirical research has resulted in a baffling array of models and results that can appear contradictory to conservation planners (de Roos and Sabelis 1995). Consequently, the lack of theoretical synthesis is an important knowledge gap hindering the application of these concepts in conservation planning. There is a need to summarize how different modeling approaches and parameterizations affect habitat prescriptions so that domains of applicability can be recognized by those who develop and implement conservation plans. Our attempt to do this was incomplete (see Figure 23.4). In the absence of a more comprehensive synthesis, landscape planners will continue to be reluctant to implement habitat prescriptions derived from purely theoretical analyses.

Another important theoretical gap concerns how one prescribes habitat layouts that are relevant to the broader species assemblage occupying the landscape. We defined the scope of our chapter to be the response of individual populations to varying habitat arrangements, avoiding the issue of conflicting habitat requirements among multiple species. Even if planning models could be developed for some suite of species inhabiting the planning area, the feasibility of deriving habitat prescriptions that are relevant to the group is questionable due to species interactions. The extension to biodiversity conservation as a whole is even less clear (Turner et al. 1995).

23.5.2 Empirical Voids

Quantifying the movement of individuals among habitat patches is seen as critical to understanding the consequences of varying habitat arrangements across the landscape (Law and Dickman 1998). Unfortunately, organism movement is poorly understood and is an extremely difficult area of empirical research (Turchin 1998), particularly at the landscape scale (Harrison and Bruna 1999). It may seem trivial to highlight species movement as a knowledge gap given its oft-cited importance, but we would be remiss not to emphasize the need for more empirical data regarding how (e.g., random or biased diffusion), when (e.g., what ecological conditions promote dispersal), and to what degree (e.g., how often and how far) species move in landscapes.

Even if landscape planners had access to more complete information on species movement, a more fundamental empirical gap is the paucity of tests of spatially explicit conservation plans. The fact that our review of recent applications focused on theoretical examples and conservation planning case studies in which the prescribed habitat design was not implemented is a symptom of this empirical gap. An especially critical need is for well-designed and active monitoring programs that are linked to plan implementation. Monitoring information is needed to assess the success or failure of a given plan, to test basic ecological concepts on which the plan is based, and to identify the mechanisms underlying population response to habitat prescriptions (Hansson and Angelstam 1991). For example, critical thresholds have been demonstrated theoretically, but there have been very few attempts to establish the existence of critical thresholds in field studies of species distribution and abundance (but see With and Crist 1995; Trzcinski et al. 1999).

23.6 Research Approaches

23.6.1 Approaches for Theoretical Research

Although synthesis may lack the intellectual excitement associated with new theoretical developments, it is nonetheless important. We identify two broad approaches to theoretical synthesis. The first is through structured reviews of the literature. This will not be an easy task. Numerous nuances to spatial planning models can quickly overwhelm attempts to distill rules for application (de Roos and Sabelis 1995). It is not uncommon for models of similar ecological phenomena to reach divergent conclusions (e.g., compare the conclusions reached by Hill and Caswell [1999] and Fahrig [1998] regarding the importance of habitat arrangement in facilitating population persistence). Part of the problem is that both model structure and parameterization differ among models, making it difficult to determine what is causing the variability in model results. It is this complexity that may render traditional narrative reviews for theoretical synthesis flawed, suggesting that the more structured analyses formalized in meta-analysis may be better suited for identifying common patterns among published models (Arnqvist and Wooster 1995).

A second approach to theoretical synthesis involves the systematic manipulation of specific models using sensitivity analysis (Conroy et al. 1995). This would involve the quantification of model response following purposeful and systematic alteration of model parameters (singly and in combination). Comprehensive exploration of model behavior under a wide range of ecologically relevant parameter settings is important for at least two reasons. First, it can help identify those aspects of species life history or habitat arrangement that are critically important for understanding population response to landscape changes. Second, it can determine whether there are fundamental shifts in habitat arrangements in response to relatively small changes in the parameter space (see Figure 23.4).

Moving from single-species to multispecies conservation planning is certainly difficult, and we know of no definitive strategy. One approach that has emerged repeatedly is the use of indicator species (related incarnations include keystone, umbrella, or focal species) to reflect the status of the overall species assemblage. Unfortunately, indicators are often chosen opportunistically (e.g., well-studied and well-surveyed taxa). Tests of this strategy using broad taxonomic groups (e.g., birds, mammals, butterflies) do not support its general applicability (Flather et al. 1997). However, rejection at broad taxonomic levels should not preclude the search for indicators among narrower sets of species. A more refined search to identify life-history attributes that could serve to group species based on critical vital rates (e.g., reproductive potential, dispersal ability) may offer an alternative approach to defining indicators because species that share similar life histories may respond to habitat geometry in a similar fashion (see Noon et al. 1997). In a slight variation of this approach, mathematical taxonomy and ordination techniques (Gauch 1982) could be used to define clusters of species based on measured life-history attributes such that the species pool for a given planning area is partitioned into sets that may be similarly sensitive to habitat arrangement effects.

23.6.2 Approaches for Empirical Research

Techniques for studying animal movement are less-developed than are methods for estimating demographic parameters (Turchin 1998). Choice of methods will depend to a large degree on whether one needs to quantify the movement paths or simply the population-level pattern of redistribution. The kind of movement data required will be dictated by the information needed to address the planning problem. For example, quantifying species-specific search rules used to locate vacant territories would require information on the actual movement paths taken by individual organisms. Alternatively, estimating the distribution of dispersal distances or the resistance of the interpatch environment to movement may only require information on the rate and directionality of population spread. Turchin (1998) provides a comprehensive review of data collection and analysis methods that are appropriate for both types of questions.

Although the use of new approaches to study species movement will extend our abilities to prescribe landscape configurations, the spatial and temporal extent of conservation plans often make detection of population response difficult, replication of conservation treatments infeasible, and identification of mechanisms underlying population change enigmatic. These problems pervade empirical testing of conservation plans, and they have been addressed with several atypical research protocols. Carpenter (1990) reviews pre-treatment and post-treatment time series analyses to infer nonrandom change in system response. Similarly, Hargrove and Pickering (1992) outline the use of "quasi-experiments" to take advantage of natural disturbances that alter habitat configuration. An approach directed specifically at increasing replication is to repeat unreplicated studies in different systems (Carpenter 1990). Such an approach can be approximated by accumulating the efforts of others and analyzing independent research efforts with meta-analysis procedures (Arnqvist and Wooster 1995). Finally, when replication and experimental controls within the planning problem are feasible, active adaptive planning (Walters and Holling 1990) could provide opportunities for stronger inferences from management experiments.

Regardless of the approach used, it is critical that a more concerted effort be directed at designing and implementing monitoring strategies that allow conservation plans to be tested (Havens and Aumen 2000). Because long time periods are required to quantify species response to modified landscapes, researching the relative merits of alternative monitoring designs will be difficult. Wennergren et al. (1995) offer an intriguing and perhaps imperative suggestion for researchers—use spatially explicit population models to test the ability of alternative monitoring designs to detect the effects of habitat layouts prior to implementing those monitoring designs in the field. Failure to devote more effort toward monitoring will perpetuate our current reliance on untested concepts and heighten the contention and uncertainty that surround the use of planning models to derive spatially explicit habitat prescriptions.

Acknowledgments

We thank Debbie Pressman, National Wildlife Program Leader, U.S. Forest Service, for coordinating our request to agency biologists for examples of habitat layout prescriptions. Mike Knowles is gratefully acknowledged for his assistance in several of the analyses, as is Joyce VanDeWater for her graphical support. A special thanks is extended to Carolyn Hull Sieg and Rollie Lamberson for their thoughtful comments on an earlier draft of this chapter.

References

Anderson, E., Forrest, S.C., Clark, T.W., and Richardson, L. 1986. Paleobiology, biogeography, and systematics of the black-footed ferret, *Mustela nigripes* (Audubon and Bachman) 1851. *Great Basin Nat. Mem.* 8:11–62.

Andrén, H. 1994. Effects of habitat fragmentation on birds and mammals in landscapes with different proportions of suitable habitat: a review. *Oikos* 71:355–366.

Arnqvist, G., and Wooster, D. 1995. Meta-analysis: synthesizing research findings in ecology and evolution. *Trends Ecol. Evol.* 10:236–240.

Bevers, M., and Flather, C.H. 1999a. Numerically exploring habitat fragmentation effects on populations using cell-based coupled map lattices. *Theor. Pop. Biol.* 55:61–76.

Bevers, M., and Flather, C.H. 1999b. The distribution and abundance of populations limited at multiple spatial scales. *J. Anim. Ecol.* 68:976–987.

Bevers, M., Hof, J., Uresk, D.W., and Schenbeck, G.L. 1997. Spatial optimization of prairie dog colonies for black-footed ferret recovery. *Oper. Res.* 45:495–507.

Block, W.M., Morrison, M.L., Verner, J., and Manley, P.N. 1994. Assessing wildlife-habitat-relationships models: a case study with California oak woodlands. *Wildl. Soc. Bull.* 22:549–561.

Breiman, L., Friedman, J.H., Olshen, R.A., and Stone, C.J. 1984. *Classification and Regression Trees.* Belmont, California: Wadsworth.

Carpenter, S.R. 1990. Large-scale perturbations: opportunities for innovation. *Ecology* 71:2038–2043.

Clark, T.W. 1989. *Conservation Biology of the Black-Footed Ferret, Mustela nigripes.* Wildlife Preservation Trust Special Scientific Report Number 3. Philadelphia, Pennsylvania: Wildlife Preservation Trust International.

Conroy, M.J. 1993. The use of models in natural resource management: prediction, not prescription. *Trans. N. Am. Wildl. Nat. Resour. Conf.* 58:509–519.

Conroy, M.J., Cohen, Y., James, F.C., Matsinos, Y.G., and Maurer, B.A. 1995. Parameter estimation, reliability, and model improvement for spatially explicit models of animal populations. *Ecol. Appl.* 5:17–19.

de Roos, A.M., and Sabelis, M.W. 1995. Why does space matter? In a spatial world it is hard to see the forest before the trees. *Oikos* 74:347–348.

Doak, D.F., and Mills, L.S. 1994. A useful role for theory in conservation. *Ecology* 75:615–626.

Dunning, J.B., Jr., Stewart, D.J., Danielson, B.J., Noon, B.R., Root, T.L., Lamberson, R.H., and Stevens, E.E. 1995. Spatially explicit population models: current forms and future uses. *Ecol. Appl.* 5:3–11.

Etienne, R.S., and Heesterbeek, J.A.P. 2000. On optimal size and number of reserves for metapopulation persistence. *J. Theor. Biol.* 203:33–50.

Fahrig, L. 1991. Simulation methods for developing general landscape-level hypotheses of single-species dynamics. In *Quantitative Methods in Landscape Ecology,* eds. M.G. Turner and R.H. Gardner, pp. 416–442. New York: Springer-Verlag.

Fahrig, L. 1997. Relative effects of habitat loss and fragmentation on population extinction. *J. Wildl. Manage.* 61:603–610.

Fahrig, L. 1998. When does fragmentation of breeding habitat affect population survival? *Ecol. Model.* 105:273–292.

Flather, C.H., and Bevers, M. 2002. Patchy reaction-diffusion and population abundance: the relative importance of habitat amount and arrangement. *Am. Nat.* 159: In press.

Flather, C.H., Wilson, K.R., Dean, D.J., and McComb, W.C. 1997. Identifying gaps in conservation networks: of indicators and uncertainty in geographic-based analyses. *Ecol. Appl.* 7:531–542.

Forsman, E.D., Meslow, E.C., and Wight, H.M. 1984. Distribution and biology of the Spotted Owl in Oregon. *Wildl. Monogr.* 87:1–64.

Frank, K., and Wissel, C. 1998. Spatial aspects of metapopulation survival—from model results to rules of thumb for landscape management. *Landsc. Ecol.* 13:363–379.

Gardner, R.H. 1999. RULE: map generation and a spatial analysis program. In *Landscape Ecological Analysis: Issues and Application,* eds. J.M. Klopatek and R.H. Gardner, pp. 280–303. New York: Springer-Verlag.

Gauch, H.G. 1982. *Multivariate Analysis in Community Ecology.* Cambridge, United Kingdom: Cambridge University Press.

Gilpin, M.E. 1987. Spatial structure and population vulnerability. In *Viable Populations for Conservation,* ed. M.E. Soulé, pp. 125–139. Cambridge, United Kingdom: Cambridge University Press.

Green, D.G. 1994. Connectivity and complexity in landscapes and ecosystems. *Pac. Conserv. Biol.* 1:194–200.

Hansson, L., and Angelstam, P. 1991. Landscape ecology as a theoretical basis for nature conservation. *Landsc. Ecol.* 5:191–201.

Hargrove, W.W., and Pickering, J. 1992. Pseudoreplication: a *sine qua non* for regional ecology. *Landsc. Ecol.* 6:251–258.

Harrison, S., and Bruna, E. 1999. Habitat fragmentation and large-scale conservation: what do we know for sure? *Ecography* 22:225–232.

Harrison, S., Stahl, A., and Doak, D. 1993. Spatial models and Spotted Owls: exploring some biological issues behind recent events. *Conserv. Biol.* 7:950–953.

Havens, K.E., and Aumen, N.G. 2000. Hypothesis-driven experimental research is necessary for natural resource management. *Environ. Manage.* 25:1–7.

Hill, M.F., and Caswell, H. 1999. Habitat fragmentation and extinction thresholds on fractal landscapes. *Ecol. Letters* 2:121–127.

Hobbs, R.J., Saunders, D.A., and Arnold, G.W. 1993. Integrated landscape ecology: a western Australian perspective. *Biol. Conserv.* 64:231–238.

Hof, J., and Bevers, M. 1998. *Spatial Optimization for Managed Ecosystems.* New York: Columbia University Press.

Hof, J., and Flather, C.H. 1996. Accounting for connectivity and spatial correlation in the optimal placement of wildlife habitat. *Ecol. Model.* 88:143–155.

Hof, J., and Raphael, M.G. 1997. Optimization of habitat placement: a case study of the Northern Spotted Owl in the Olympic Peninsula. *Ecol. Appl.* 7:1160–1169.

Holthausen, R.S., Raphael, M.G., McKelvey, K.S., Forsman, E.D., Starkey, E.E., and Seaman, D.E. 1995. *The Contribution of Federal and Nonfederal Habitat to Persistence of the Northern Spotted Owl on the Olympic Peninsula, Washington.* General Technical Report PNW-GTR-352. Portland, Oregon: USDA, Forest Service, Pacific Northwest Research Station.

Kareiva, P. 1990. Population dynamics in spatially complex environments: theory and data. *Phil. Trans. R. Soc. London, Ser. B* 330:175–190.

Kareiva, P., and Wennergren, U. 1995. Connecting landscape patterns to ecosystem and population processes. *Nature* 373:299–302.

Lamberson, R.H., McKelvey, R., Noon, B.R., and Voss, C. 1992. A dynamic analysis of Northern Spotted Owl viability in a fragmented landscape. *Conserv. Biol.* 6:505–512.

Lamberson, R.H., Noon, B.R., Voss, C., and McKelvey, K.S. 1994. Reserve design for territorial species: the effect of patch size and spacing on the viability of the Northern Spotted Owl. *Conserv. Biol.* 8:185–195.

Lande, R. 1987. Extinction thresholds in demographic models of territorial populations. *Am. Nat.* 130:624–635.

Lande, R. 1988. Demographic models of the Northern Spotted Owl (*Strix occidentalis caurina*). *Oecologia* 75:601–607.
Law, B.S., and Dickman, C.R. 1998. The use of habitat mosaics by terrestrial vertebrate fauna: implications for conservation and management. *Biodiver. Conserv.* 7:323–333.
Lawton, J.H. 1999. Are there general laws in ecology? *Oikos* 84:177–192.
McKelvey, K., Noon, B.R., and Lamberson, R.H. 1993. Conservation planning for species occupying fragmented landscapes: the case of the Northern Spotted Owl. In *Biotic Interactions and Global Change,* eds. P.M. Kareiva, J.G. Kingsolver, and R.B. Huey, pp. 424–450. Sunderland, Massachusetts: Sinauer Associates.
Mellen, K., Huff, M., and Hagestedt, R. 1995. *HABSCAPES: Reference Manual and User's Guide.* Portland, Oregon: USDA, Forest Service, Pacific Northwest Region.
Minta, S., and Clark, T.W. 1989. Habitat suitability analysis of potential translocation sites for black-footed ferrets in north-central Montana. In *The Prairie Dog Ecosystem: Managing for Biological Diversity,* eds. T.W. Clark, D. Hinckley, and T. Rich, pp. 29–46. Billings, Montana: USDI, Bureau of Land Management.
Morrison, M.L., Timossi, I.C., and With, K.A. 1987. Development and testing of linear regression models predicting bird-habitat relationships. *J. Wildl. Manage.* 51:247–253.
Murphy, D.D., and Noon, B.R. 1992. Integrating scientific methods with habitat conservation planning: reserve design for Northern Spotted Owls. *Ecol. Appl.* 2:3–17.
Noon, B.R., and McKelvey, K.S. 1996. Management of the Spotted Owl: a case history in conservation biology. *Ann. Rev. Ecol. Syst.* 27:135–162.
Noon, B., McKelvey, K., and Murphy, D. 1997. Developing an analytical context for multispecies conservation planning. In *The Ecological Basis of Conservation: Heterogeneity, Ecosystems, and Biodiversity,* eds. S.T.A. Pickett, R.S. Ostfeld, M. Shachak, and G.E. Likens, pp. 43–59. New York: Chapman and Hall.
Noon, B.R., and Murphy, D.D. 1994. Management of the Spotted Owl: the interaction of science, policy, politics, and litigation. In *Principles of Conservation Biology,* eds. G.K. Meffe and C.R. Carroll, pp. 380–388. Sunderland, Massachusetts: Sinauer Associates.
Palmqvist, E., and Lundberg, P. 1998. Population extinctions in correlated environments. *Oikos* 83:359–367.
Pimm, S.L., and Gilpin, M.E. 1989. Theoretical issues in conservation biology. In *Perspectives in Ecological Theory,* eds. J. Roughgarden, R.M. May, and S.A. Levin, pp. 287–305. Princeton: Princeton University Press.
Rapport, D.J., Regier, H.A., and Hutchinson, T.C. 1985. Ecosystem behavior under stress. *Am. Nat.* 125:617–640.
Richardson, L., Clark, T.W., Forrest, S.C., and Campbell, T.M., III. 1987. Winter ecology of black-footed ferrets (*Mustela nigripes*) at Meeteetse, Wyoming. *Am. Midl. Nat.* 117:225–239.
Roemer, D.M., and Forrest, S.C. 1996. Prairie dog poisoning in the Northern Great Plains: an analysis of programs and policies. *Environ. Manage.* 20:349–359.
Saunders, D.A., Hobbs, R.J., and Margules, C.R. 1991. Biological consequences of ecosystem fragmentation: a review. *Conserv. Biol.* 5:18–32.
Schemske, D.W., Husband, B.C., Ruckelshaus, M.H., Goodwillie, C., Parker, I.M., and Bishop, J.G. 1994. Evaluating approaches to the conservation of rare and endangered plants. *Ecology* 75:584–606.
Simberloff, D. 1988. The contribution of population and community biology to conservation science. *Ann. Rev. Ecol. Syst.* 19:473–511.
Skellam, J.G. 1951. Random dispersal in theoretical populations. *Biometrika* 38:196–218.

Suring, L.H., Crocker-Bedford, D.C., Flynn, R.W., Hale, C.S., Iverson, G.C., Kirchhoff, M.D., Schenck, T.E., Shea, L.C., and Titus, K. 1993. *A Proposed Strategy for Maintaining Well-Distributed, Viable Populations of Wildlife Associated with Old-Growth Forests in Southeast Alaska.* Report to an Interagency Committee. Juneau, Alaska: USDA, Forest Service, Alaska Region.

Thomas, J.W., Forsman, E.D., Lint, J.B., Meslow, E.C., Noon, B.R., and Verner, J. 1990. *A Conservation Strategy for the Northern Spotted Owl.* Report of the Interagency Scientific Committee to Address the Conservation of the Northern Spotted Owl. Portland, Oregon: USDA, Forest Service.

Thorne, E.T., and Belitsky, D.W. 1989. Captive propagation and the current status of free-ranging black-footed ferrets in Wyoming. In *Conservation Biology and the Black-Footed Ferret,* eds. U.S. Seal, E.T. Thorne, M.A. Bogan, and S.H. Anderson, pp. 223–234. New Haven, Connecticut: Yale University Press.

Trzcinski, M.K., Fahrig, L., and Merriam, G. 1999. Independent effects of forest cover and fragmentation on the distribution of forest breeding birds. *Ecol. Appl.* 9:586–593.

Turchin, P. 1998. *Quantitative Analysis of Movement: Measuring and Modeling Population Redistribution in Animals and Plants.* Sunderland, Massachusetts: Sinauer Associates.

Turing, A.M. 1952. The chemical basis of morphogenesis. *Phil. Trans. R. Soc. London, Ser. B* 237:37–72.

Turner, M.G., Arthaud, G.J., Engstrom, R.T., Hejl, S.J., Liu, J., Loeb, S., and McKelvey, K. 1995. Usefulness of spatially explicit population models in land management. *Ecol. Appl.* 5:12–16.

Turner, M.G., and Gardner, R.H. 1991. Quantitative methods in landscape ecology: an introduction. In *Quantitative Methods in Landscape Ecology,* eds. M.G. Turner and R.H. Gardner, pp. 3–14. New York: Springer-Verlag.

Urban, D.L., O'Neill, R.V., and Shugart, H.H. 1987. Landscape ecology. *BioScience* 37:119–127.

U.S. Fish and Wildlife Service, National Park Service, and USDA Forest Service. 1994. *Black-Footed Ferret Reintroduction, Conata Basin/Badlands, South Dakota, Final Environmental Impact Statement.* Pierre, South Dakota: USDI, Fish and Wildlife Service.

Van Deusen, P.C. 1996. Habitat and harvest scheduling using Bayesian statistical concepts. *Can. J. For. Res.* 26:1375–1383.

Walters, C.J., and Holling, C.S. 1990. Large-scale management experiments and learning by doing. *Ecology* 71:2060–2068.

Wennergren, U., Ruckelshaus, M., and Kareiva, P. 1995. The promise and limitations of spatial models in conservation biology. *Oikos* 74:349–356.

With, K.A., and Crist, T.O. 1995. Critical thresholds in species' responses to landscape structure. *Ecology* 76:2446–2459.

With, K.A., and King, A.W. 1999. Extinction thresholds for species in fractal landscapes. *Conserv. Biol.* 13:314–326.

24

Aquatic Conservation Planning: Using Landscape Maps to Predict Ecological Reference Conditions for Specific Waters

PAUL W. SEELBACH, MICHAEL J. WILEY, PATRICIA A. SORANNO, AND MARY T. BREMIGAN

24.1 Introduction

Regional planning for conservation and management of aquatic ecosystems is an extremely challenging task, involving integration of information across the breadth of disciplines involved in water resources management. Consideration must be given to hydrologic, sediment, and water-quality regimes; local geomorphic processes and habitat structures; network connectivity of water bodies; and maintenance of source populations of both characteristic and rare biota.

In this chapter, we highlight a promising new approach to estimating specific expected or reference conditions for various ecological parameters that is based on ideas from the field of landscape ecology. In Section 24.2, "Concepts, Principles, and Emerging Ideas," we provide some definition, historical background, and description of this emerging approach. In Section 24.3, we review "Recent Applications" of this approach across a variety of aquatic ecosystem types and in several areas of the world. In Section 24.4, we provide "Principles for Applying Landscape Ecology" derived from reviewing recent applications. In Section 24.5, we identify primary theoretical and empirical "Knowledge Gaps" that hinder further development. Finally, in Section 24.6, "Research Approaches," we lay out a series of steps to provide guidance for development of new applications.

24.2 Concepts, Principles, and Emerging Ideas

24.2.1 Information Required for Aquatic Conservation Planning

Resource inventory and assessment are commonly the first steps in regional conservation planning. *Inventory* involves enumerating the distribution and status of waters, and it is a prerequisite to strategic prioritization of management opportunities within a region. *Assessment* involves normalizing observed ecological conditions for a water body through the use of some potential or reference condition (Gallant et al. 1989; Claessen et al. 1994). The *reference condition* is typically estimated from characteristics of a regional set of least-disturbed (reference) waters. The *assessed status* of a water body can be expressed either as a deviation from the reference condition (e.g., IBI-type scores; Karr et al. 1986) or as a ratio of observed and reference conditions (Hakanson 1996).

24.2.2 Historical Approaches to Modeling Reference Conditions Across a Large Region

To simplify their task, aquatic resource managers responsible for large geographic areas have often turned to classifying water bodies (Davis and Henderson 1978; Zonneveld 1994). Classifications allow extrapolation of attributes from sampled to unsampled water bodies, to create comprehensive regional coverage. Traditional aquatic classifications have generally followed either a site-based or a regionalization approach (these are sometimes termed bottom-up and top-down approaches, respectively; Figure 24.1; Zonneveld 1994).

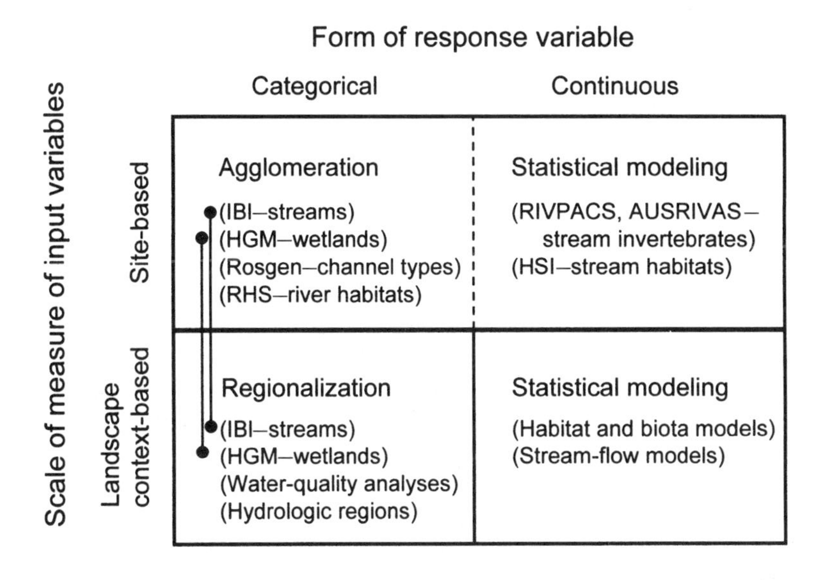

FIGURE 24.1. Classification of common approaches (shown in parentheses) to developing reference characteristics for aquatic systems. The dashed line indicates a weaker differentiation between classes. The lines connecting dots indicate that some approaches bridge multiple classes. Abbreviations are IBI—Index of Biotic Integrity; HGM—Hydro-Geomorphic Approach; RHS—River Habitat Survey; RIVPACS and AUSRIVAS refer to stream invertebrate assessment protocols used in Great Britain and Australia, respectively; HSI—Habitat Suitability Indices.

Site-Based Approaches (Agglomeration and Statistical Modeling)

Classification by agglomeration groups sampled, similar water bodies and asks, "What are water bodies in this group like?" The primary goal is simplification, achieved by categorizing many individual water bodies into groups that share similar attributes. Both diagnostic (easily measured indices) and response (e.g., biota) attributes are first measured for a subset of water bodies (learning set). Attributes are measured on-site with a fair degree of accuracy and precision. Water bodies with similar diagnostics are then clustered into groups whose attributes are summarized (e.g., mean and range). Group membership for additional waters is based on shared diagnostics. The reference condition for a water body is modeled from the summarized response attributes of a "least-impacted" subset within each group. Regional assessment is only achieved when an adequate sample of all system types is collected.

The strength of agglomeration lies in the accuracy and precision of local diagnostic measures. Confidence in these initial data is especially valued when managers are responsible for estimating difficult-to-measure attributes or complex processes, or when response attribute predictions are controversial (such as in regulatory programs). Examples of agglomeration-based assessment programs include the U.S. Hydrogeomorphic (HGM) Wetland Classification program (Hauer and Smith 1998); many lake and stream classifications used for prediction of water quality, habitat quality, or fishery potential (Schupp 1992; Rosgen 1996; Thorn and Anderson 1999; Emmons et al. In press); and the European River Habitat Survey (RHS) program (Raven et al. 1997).

A continuous-data modeling variation on the site-based approach also is used to predict aquatic reference conditions. Examples include: the U.S. Habitat Suitability Index (HSI) models (Terrell et al. 1982) that are widely used to predict potential distributions of fishes in streams; and the closely related River InVertebrate Prediction And Classification System (RIVPACS; Wright et al. 1997) and AUStralian RIVers Assessment Scheme (AUSRIVAS; Simpson and Norris In press) models that are being used in Great Britain and Australia (respectively) to predict potential occurrence of stream invertebrates.

Despite wide use, site-based approaches have several weaknesses. First, it is prohibitively expensive to measure on-site attributes everywhere within a region (Meixler and Bain 1998). Although fairly comprehensive data on selected attributes will eventually accumulate (e.g., Raven et al. 1997; Wright et al. 1997), this will not always be a viable regional approach. Second, much of the site-specific accuracy inherent in this process is actually lost in the classification process, because each water body assumes the "average" attributes of the group. Third, this process generally includes no information on the positional context of a water body in the landscape, so one cannot draw inferences about the landscape-scale processes that largely shape the character of each aquatic ecosystem (Zonneveld 1994). This emphasis on accuracy of local measures rather than on landscape context sets up the risk of sometimes knowing "what" but not "why" (sensu Holling 1998; Davies 1999) and can provide a false sense of confidence in the final estimates.

Regionalization (Landscape Context-Based Classification)

An alternative classification approach is regionalization, in which we group water bodies that lie together within a relatively similar geographic subregion and ask the question, "What are waters like that share the coarse-scale, ecological processes characteristic of this region?" Goals include simplification for planning and communication, and comprehensive regional coverage through use of landscape maps. Regionalization begins with the study of a series of overlay maps showing spatial concurrence of selected landscape characteristics thought to be diagnostic of ecosystem processes (e.g., climate, geology, soils, and topography). The larger landscape is divided into relatively similar subregions (i.e., ecoregions), each containing a number of water bodies (Davis and Henderson 1978; Gallant et al. 1989). These subregions are typically large; for example, a midwestern U.S. state may contain three to five ecoregions. Attributes of these regions are typically measured for a subset of waters (learning set), and then summarized (as means and ranges). Group membership is based on a shared location within the subregion such that all waters within the subregion would have the same predicted potential condition (Gallant et al. 1989). The model for predicting potential condition of any given water body is the summarized condition for the learning set (a selected reference set).

Regionalization's strength is comprehensive geographic coverage. Its generalized descriptions of both landscape-scale attributes and processes, and selected water body (site-scale) attributes, also provide valuable insights into the hierarchical processes controlling regional ecosystems, as well as a useful, albeit coarse stratification for sampling design. It is widely used as the basis for determining ecological potentials for water quality and aquatic biota (Gallant et al. 1989; Klijn 1994; Davis and Simon 1995; Davis et al. 1996). However, this top-down approach also has limitations because ecoregions are quite heterogeneous at the scale pertinent to aquatic ecosystems (Bryce and Clarke 1996). Thus, regional generalizations often do not provide accurate estimates for specific unsampled water bodies, and it becomes impossible to differentiate between a deviation from the reference condition caused by human impacts and one caused by geographic variation. Furthermore, because river catchments often are not nested cleanly within subregions, analyses across multiple contiguous regions may be required.

Two common variations on basic regionalization help to overcome these weaknesses. Regionalization can be combined with agglomeration by developing detailed classifications of measured water body characteristics within specific subregions. This brings some generalized landscape setting to the agglomeration process but still demands heavy investment in on-site measures. In the HGM wetlands classification (Hauer and Smith 1998) and various biological (Davis et al. 1996; Yoder and Smith 1998) and water quality monitoring programs (Gallant et al. 1989), agglomeration classes have been developed within subregions. This effectively stratifies subregions by water body type (determined from site-level diagnostics), and produces fairly useful predictive models. Alternatively, subregions can be further divided into smaller, more homogeneous units such as land-type associations (Corner et al. 1997) or ecotopes (Claessen et al. 1994). Relating aquatic systems to a mosaic of these smaller land units represents movement toward the modeling approach described in Section 24.2.3.

24.2.3 Statistical Modeling From a Landscape Context

Many of the weaknesses of traditional classification and site-based modeling approaches are addressed by the integrative discipline known as landscape ecology. Through landscape ecology, we explore the patterns, dynamics, and ecological consequences of spatial heterogeneity in the environment (Risser et al. 1984; Turner 1998). We emphasize the importance of a site's unique placement in the larger landscape in explaining local ecological characteristics (Turner 1998). Landscape ecology encourages study of hierarchical relationships between coarse-scale variables descriptive of landscape character and local ecological attributes, and the identification of system-level patterns and processes that only emerge when viewed at coarse scales (Levin 1992; Wessman 1990). It explicitly recognizes the importance of human effects in the landscape.

Evaluating a water body's position in the landscape enables us to incorporate information on landscape-scale processes that shape the character of rivers (Lotspeich 1980; Schlosser 1991), lakes (Hakanson 1996; Kratz et al. 1997; Soranno et al. 1999), and wetlands (Hauer and Smith 1998). We can consider potential movements of water, sediments, and nutrients across landscape units and into the water body (Turner 1998). We can characterize movements and storage of these materials within water networks (connected bodies of lakes, streams, and wetlands), and we can consider movements of organisms among critical habitats within the system (Schlosser 1991; Kratz et al. 1997). We can consider the influence of landscape geomorphology on water body morphology. To assist planners, who examine mostly managed landscapes, we also can incorporate human effects on riverine processes (Risser et al. 1984; Wessman 1990).

A landscape-ecology approach suggests that coarse-scale information can be useful for characterizing and understanding potential characteristics of individual water bodies (Klijn 1994; Rabeni and Sowa 1996; Higgins et al. 1998; Davies 1999). Hierarchical systems theory suggests that higher-level, coarse-scale variables constrain and shape variables at lower levels and finer scales (O'Neill et al. 1986; Bourgeron and Jensen 1994). In addition, because coarse-scale variables are typically mapped for entire regions, it is possible to describe the unique geographical position of each water body within a region. This idea challenges the traditional view that site-scale measures are needed to address site-management issues, whereas coarse-scale measures are useful only for planning regional policy.

This ability of the landscape perspective to characterize hierarchical relationships provides both a conceptual and an empirical basis for development of statistical models that relate patterns in coarse-scale, contextual variables to site-scale, ecological response variables (Risser et al. 1984). Only modeling allows for exploration of relationships across large ranges in scale (Levin 1992; Holling 1998). Coarse-scale descriptions of the climatic, physiographic, and land-cover settings of a water body, and its position and connectivity within the hydrologic network, are readily derived from available maps. By matching location-specific, coarse-scale contextual data with site-level ecological data, statistical models can be developed that use maps to predict the likelihood of occurrence of a habitat or species within specific waters. Ultimately, the potential distribution of an ecological characteristic can be predicted for waters across an entire region (Meffe and Carroll 1997; Higgins et al. 1998; Turner 1998).

We argue that ecological classification and statistical modeling are similar in intent and form. Perhaps less clear is the relationship between empirical modeling from landscape data and what is often called ecosystem process modeling. Process models attempt to infer structure from function (i.e., by integrating state variable derivatives) and typically focus on temporal intrasite variation via dynamic mathematical models. Process models can be driven by landscape-scale input variables, as in lumped-parameter water-quality models (e.g., Cosby et al. 1985). However, an unfortunate (and we believe false) dichotomy is often made between linear regression-based landscape models and process models (e.g., Thomann 1987; Thierfelder 1998), suggesting that process models are mechanistic and empirically parameterized regression models are not. In both cases, the modeling approaches are actually mechanism-neutral. In fact, both methods can be used to parameterize models based on explicit mechanistic hypotheses. Building regression models from explicit causal hypotheses (*causal modeling* sensu Retherford and Choe 1993) has a long and productive history in ecology (Asher 1983; Wooton 1994).

Classification, statistical, and dynamic-process models can all be used (singly or in combination) to infer site characteristics from landscape context (Wessman 1990; Haber 1994). Statistical models may be particularly appropriate for problems of intermediate complexity in which intersite variance is important and predictable, and when time-averaged characteristics are useful (Haber 1994). Conservation planning and regional management typically involve assessing potential value at an array of sites and making strategic decisions about the investment of limited management dollars. Detailed temporal behaviors are less important in this context than are long-term average conditions. On the other hand, detailed restoration strategies for particular sites may require additional detailed, time-dependent process modeling.

24.2.4 Application of Landscape-Context Statistical Modeling to Aquatic Ecosystems

The emerging practice of landscape-context statistical modeling has a key role to play in the assessment of the reference condition and status of aquatic ecosystems. Such modeling is very similar to traditional methods of ecological classification (agglomeration or regionalization) in both philosophy and intent. All are modeling exercises in the sense that they abstract complex realities into simpler representations of essential features (Kerr 1976). Ecological classification, for example, generates inferences about a particular site based on decision rules for class membership generated from analyses of data from a subsample of sites. Statistical models likewise generate inferences about individual sites based on parameterization of models from large subsamples of site-referenced data. Both statistical modeling and classification focus on the intersite variation in ecological properties and use methods to partition and explain the observed variance.

Landscape-context modeling builds on the strengths of agglomeration and regionalization. We can identify any specific water body and ask, "What are the expected properties of this water body, given its unique position in the landscape?" The goal is to provide maximum descriptive information about potential properties of a water body.

Landscape-context models are developed using learning sets of map-scale and local diagnostic data compiled for subsets of waters (Figure 24.2). Statistical models are developed that predict potential water body properties from position-specific landscape data. For example, the composition and proximity of upstream landscape units might be used to predict the discharge regime of a particular stream (Wiley and Seelbach In press). Quantitative models are typically applied using position-specific geographic information system (GIS) measurements as inputs (Hakanson 1996), but interpretive models also have been applied by expe-

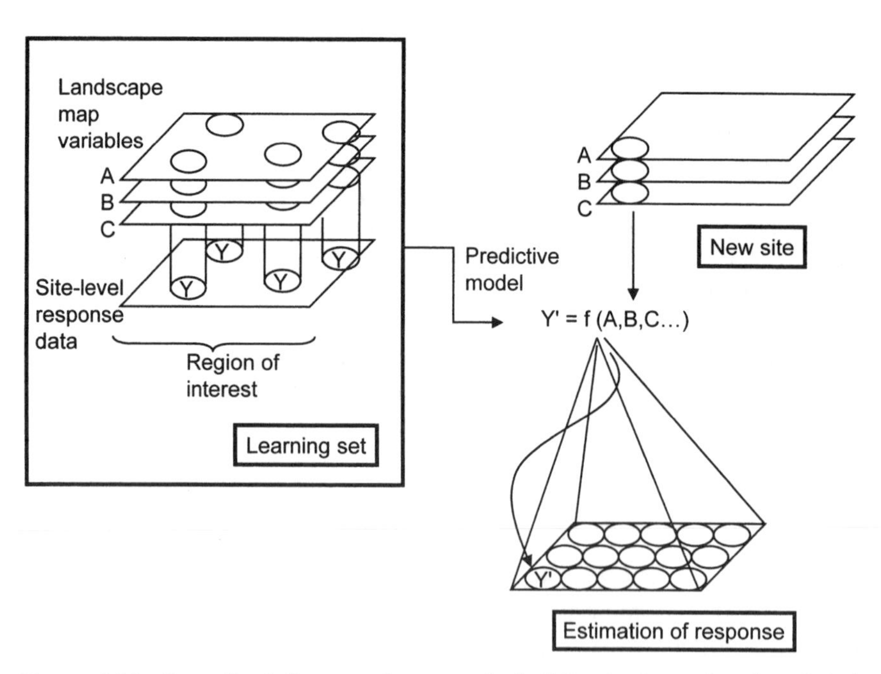

FIGURE 24.2. Generalized diagram of process for building landscape-based statistical models and applying them to estimate response characteristics of sites not in the learning set (indicated by Y′).

rienced ecologists to draw qualitative inferences. For example, Seelbach et al. (1997) and Higgins et al. (1998) developed estimates of "first-cut" attribute types for all river valley segments across a region based on interpretation of landscape map overlays.

The landscape-based modeling approach has a number of strengths. Like regionalization, it provides comprehensive regional coverage, so any and all water bodies within the modeled region can be addressed. Like agglomeration, it often provides accurate characterization of individual water bodies. It provides an explicit description of the specific landscape context of each water body, so a sense of the hierarchical processes driving and constraining each system is obtained (Bourgeron and Jensen 1994). In addition, the use of continuous data often provides more accuracy than does the use of categorical or class data (Latour et al. 1994; Zonneveld 1994), and statistical modeling can provide error bounds for predicted conditions (Hakanson 1996). Finally, such models enable planners to explore costs and benefits of alternative management scenarios (Claessen et al. 1994; Hakanson 1996).

An often-overlooked value of a modeling approach is that it provides not only specific outputs, but also opportunities to study and learn about the structure and function of ecological systems. The initial steps of describing and mapping data characteristics and patterns are invaluable (Seelbach et al. 1997; Emmons et al. In press). Hypotheses and assumptions about how system components interact can be stated, tested, and revised. This heuristic value of modeling is important in shaping the thinking and judgments of resource managers and planners, and in providing a common conceptual base to a diverse set of users. Even as our information systems grow, human interpretation and judgment will remain a critical part of the management process.

The primary weakness of this approach is similar to that of regionalization: predictions do not necessarily provide accurate estimates of the local attributes of a specific unsampled water body. This uncertainty should, however, be less than that associated with regionalization because of the added positional and landscape information included for each unsampled water body. In addition, prediction uncertainty can be quantified as a confidence range on the estimate of a response variable. Integration of map-based modeling with agglomeration or regionalization can greatly reduce this prediction uncertainty. Predictive power can be strengthened by adding some site-level diagnostic measures to models, or by developing a hierarchical two-step approach (e.g., Claessen et al. 1994; Higgins et al. 1999).

24.3 Recent Applications

We did an extensive search of the recent literature, and we personally contacted key experts, to build a rough inventory of the extent to which landscape-context statistical modeling is actually being applied in aquatic assessment efforts. Our

TABLE 24.1. Recent applications of map-based modeling to aquatic ecosystems. Examples covered in the text are highlighted in bold type.

Applications	Dependent variable(s)	Locations	References
Comprehensive regional planning			
Streams and rivers	**Macrohabitat and faunal classes**	**USA, Great Lakes Basin, Illinois River Basin**	**Higgins et al. 1998; S. Miller et al., unpublished report[a]**
	Macrohabitat and faunal classes	USA, Missouri	Sowa et al. 1999
	Macrohabitat and fish distribution	USA, Michigan	Seelbach et al. 1997
	Fish distribution	France	T. Oberdorff et al., unpublished report[b]
	Stream flow	USA, Michigan	Wiley and Seelbach, In press
Lakes	**Water quality and yields of biota**	**Sweden**	**Hakanson and Peters 1995; Hakanson 1996; Thierfelder 1998**
	Sensitivity to acidification	USA, Northeast	Young and Stoddard 1996
Wetlands	**Floral community distribution**	**The Netherlands**	**Claessen et al. 1994; Latour et al. 1994; R. van Ek, personal communication.**
	Riparian habitat and floral community classification	USA, Michigan	M. Baker, unpublished manuscript[c]
Subregional model development			
Streams and rivers	Fish community structure	USA, New York	Meixler and Bain 1998
	Macrohabitat and faunal classes	USA, Colorado	A. Reed et al., unpublished report[d]
	Habitat and fish rehabilitation targets	**USA, Michigan**	**Wiley et al. 1998**
	Fish distribution	USA, Rocky Mountains	Nelson et al. 1992; Rahel and Nibbelink 1999
	Invertebrate indices of ecological integrity	Australia	Davies 1999

Lakes	**Water chemistry and biota**	**USA, Northern Wisconsin**	**Riera et al. In press**
	Water chemistry	USA and Canada, scattered	Soranno et al. 1999
	Fish distribution	USA, Minnesota	Cross and McInerny 1995
Wetlands	**Habitat and floral distribution**	**USA, Michigan**	**D. Merkey, unpublished manuscript[e]**

[a] S. Miller, J. Higgins, and J. Perot. 1998. The Classification of Aquatic Communities in the Illinois River Watershed and Their Use in Conservation Planning. Peoria: The Nature Conservancy of Illinois.

[b] T. Oberdorff, D.Chessel, B. Hugueny, D. Pont, P. Boet, and J. P. Porcher. A Statistical Model Characterizing Riverine Fish Assemblages of French Rivers: A Framework for the Adaptation of a Fish-Based Index. Contact D. Pont, Laboratory Ecologie des Hydrosystemes Fluviaux, Universite Lyon 1, France.

[c] M. Baker. 1998. Doctoral Dissertation Proposal. School of Natural Resources and Environment. Ann Arbor: University of Michigan.

[d] A. Reed, J. Higgins, and R. Wiginton. 1998. Aquatic Community Classification Pilot for the San Miguel Watershed. Boulder, Colorado: The Nature Conservancy.

[e] D. Merkey. 1999. Doctoral Dissertation Proposal. School of Natural Resources and Environment. Ann Arbor: University of Michigan.

search naturally focused on North America, but we did locate parallel activities, in several cases quite intensive, around the globe. It appeared that recent development and use of such models to assess and manage aquatic ecosystems is fairly widespread, apparently following on the heels of rapid growth in the availability of GIS technology (Table 24.1). Variations on the common approach diagrammed in Figure 24.2 are being applied across riverine, lake, and wetland ecosystems, and at two different planning scales (Tables 24.2–24.7). Although water body

TABLE 24.2. Ecological classification of river valley segments for U.S. regional and national conservation planning.

Water body:	River valley segment
Management region:	U.S. ecoregions
Management goal:	Prioritizing aquatic sites for biodiversity conservation
Developing agency:	The Nature Conservancy, Freshwater Initiative (Higgins et al. 1998; Higgins et al. 1999)
Map input variables:	Climate, landform
	Catchment size, network position, surficial geology, bedrock geology, topography
	Channel, valley, and lake morphology
	Connectivity to other aquatic ecosystems
Model form:	Developed by U.S. Ecoregions (regionalization)
	Delineation of river valley segment and lake units
	Assignment of attribute types
	Assignment to potential hydrologic regime type using visual interpretation of GIS map overlays and decision rules (Figure 24.3)
	Assignment to potential macrohabitat types using cluster analysis
	Macrohabitats, combined with zoogeographic patterns, used as a surrogate for potential biodiversity
Ecological outputs:	GIS database with river segment and lake attribute classes, including size, network position, valley slope, connectivity, and estimated hydrologic regime
	Potential macrohabitat type membership
Implementation status:	Initial draft completed for entire U.S. Great Lakes Basin, Idaho batholith and prairie forest border ecoregions, Illinois River Basin; evaluation and revision underway
	Work ongoing in lower New England and superior mixed forest ecoregions
	Validated assumption that macrohabitats are predictive of biological communities in Michigan (Higgins et al. 1998; M. Wiley et al., unpublished report[a]) and Illinois River Basin (S. Miller et al., unpublished report[b])
	Used with terrestrial classifications in ecoregional conservation prioritization process (Higgins et al. 1999)

[a] M. Wiley, M. Baker, and P. Seelbach. 1998. Summer Field Sampling and Preliminary Analysis of Michigan Stream Assemblages. Chicago: The Nature Conservancy, Great Lakes Program Office.
[b] S. Miller, J. Higgins, and J. Perot. 1998. The Classification of Aquatic Communities in the Illinois River Watershed and Their Use in Conservation Planning. Peoria: The Nature Conservancy of Illinois.

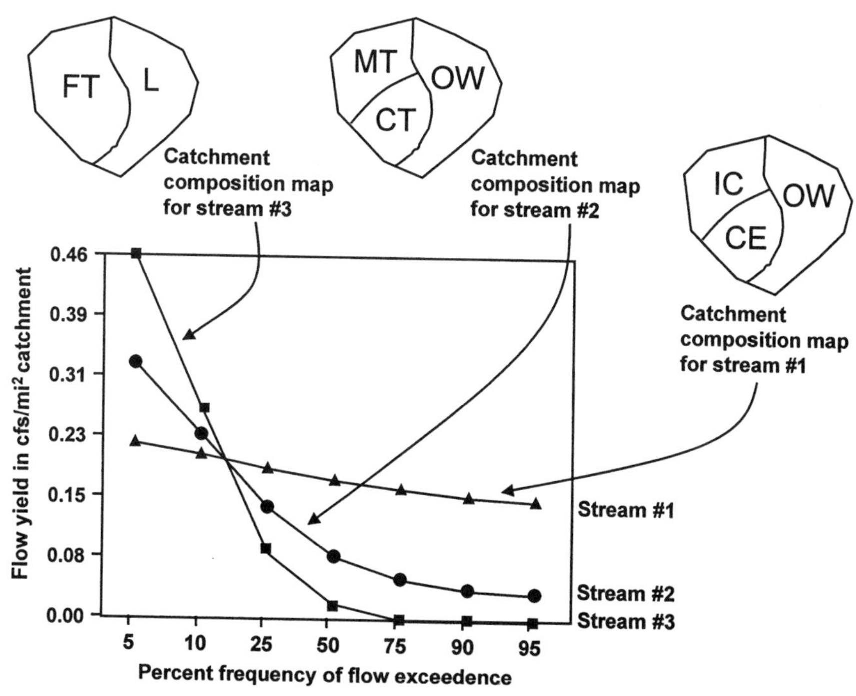

FIGURE 24.3. Examples of typical flow duration curves for streams in Michigan's Lower Peninsula. Assignment rules using catchment surficial geology data are: Stream #1—groundwater dominated; catchment composed of coarse outwash (OW) plain in stream valley, downslope of sizable coarse-textured ice-contact (IC) or end moraine (CE) ridges. Stream #2—runoff dominated with some groundwater; catchment with gentle topography composed of diverse mixture of coarse- (CT) and medium-textured (MT) till plains, and coarse outwash. Stream #3—runoff dominated; catchment with flat topography composed of fine-textured tills (FT) and lacustrine silts, clays, and sands (L). Stream flow yield at a specific exceedence frequency represents the stream flow generated per unit area of the upstream catchment that is exceeded for that specified percentage of each year. Stream flow yield is represented in cubic feet per second (cfs) per square mile (1 cubic foot = 0.0283 cubic meters, and 1 square mile = 2.59 square kilometers).

types and management regions and goals are unique in each example, common steps included: using map-based catchment and local landscape attributes as input variables; developing empirical models relating such variables to site-scale ecological response attributes; predicting reference conditions for additional waters across a large region based on unique landscape position data for each; and in some cases, integrating either an agglomeration or a regionalization approach.

Work on rivers appeared to be most popular, perhaps reflecting a gradient in the relative importance of landscape position versus local geomorphic and biological factors in determining ecological character. Rivers, as predominantly

TABLE 24.3. Modeling potential water quality for lakes across Sweden.

Water body:	Small glacial lakes
Management region:	Swedish subregions; potentially all of Sweden (81,000 lakes!)
Management goal:	Assessment of ecological status of lakes
Developing agency:	Uppsala University, Institute of Earth Sciences (Hakanson 1996)
Map input variables:	Catchment size and relief, land covers and till depth
	Lake morphometry
Model form:	Multiple linear regression
	Correction factors accounting for widespread temporal changes
Ecological outputs:	Attributes, including potential lake pH, total phosphorus, water color
	Indices of potential abundance of fishes, phytoplankton, benthic invertebrates; contaminants in fishes
	Actual condition/potential condition = index of status
	Combined indices into overall index of lake ecosystem status
Implementation status:	Summarized for lakes within selected subregions; illustrated use of indices to track lake attributes and overall status through time
	Estimated response of water quality and biological communities to chemical management actions, and developed a cost/benefit analysis to compare alternative actions
	Evaluated and supported use of spatially distributed catchment characteristics as input variables (Theirfelder 1998)
	Apparently not widely used to date

rapid flow-through systems, are strongly driven by catchment deliveries of water and sediment. In contrast, basin morphology is a known key in structuring physical habitat in lakes, and scientists are only recently beginning to examine the relative importance of catchment inputs (Eilers et al. 1983; Rochelle et al. 1989; Webster et al. In press). Wetlands range from systems clearly driven by position in the hydrologic landscape (e.g., northern white cedar [*Thuja occidentalis*] swamps or river floodplains), to those depressions where specific basin morphologies help define their hydrologic character, to bogs that (through time) become largely divorced from landscape influence. The form of models linking landscape characteristics to aquatic ecosystems also varies. Multiple linear regression models (Hakanson 1996), statistical summaries (Wiley et al. 1998; Riera et al. In press), empirically based decision rules (Seelbach et al. 1997; Higgins et al. 1998), and literature-based decision rules (Meixler and Bain 1998; Sowa et al. 1999) have all been used successfully.

24.4 Principles for Applying Landscape Ecology

Our literature review indicated that landscape-based statistical modeling is emerging as a viable tool for planners to determine ecological potential of aquatic ecosystems. Model-based information systems allow not only among-site comparison and prioritization of management actions, but also comparisons of costs

TABLE 24.4. Raster-based modeling of potential wetland values across The Netherlands.

Water body:	Wetlands
Management region:	The Netherlands
Management goal:	To estimate responses of wetland ecosystems to alternative regional and local water management scenarios
Developing agency:	Institute for Inland Water Management, Ministry of Transport, Public Works and Water Management (Claessen et al. 1994)
Map input variables:	1-km raster map of ecoseries and ecotope landscape units; these denote areas of relatively uniform water table elevation and soil types
Model form:	Develop by ecoseries units (regionalization)
	DEMNAT has empirically derived dose-response functions that translate changes in water table elevation into changes in soil moisture, nutrient levels, and acidity
	Additional functions that translate physical and chemical changes into changes in vegetative association
	Calculation of nationally normed, potential nature value through summing indices of vegetative community structure (rarity, completeness, and percent coverage for specific associations)
Ecological outputs:	Potential vegetation associations
	Potential nature value; useful for examining regional resource patterns under alternative water management scenarios, and as a reference condition for status assessment
Implementation status:	Recent improvements in the model include sensitivity analyses, added range coverage for dose-response models, links to site-scale hydrology and conservation value models, improved ecoseries classifications (founded on improved national vegetative survey database), and an improved computer interface (R. van Ek, personal communication)
	DEMNAT 2.1 has been operational since 1996
	Has been fundamental to a nation-wide study of future desiccation problems and management scenarios, climate change, and land subsidence—the "Dutch Aquatic Outlook"
	Has been used to develop ecologically based water management policy for subregions

and benefits of alternative management strategies at each site (Hakanson 1996; Wiley et al. 1998). Based on research conducted to date, we identified several principles that planners can apply to the conservation and management of aquatic ecosystems.

Successful model response variables usually have high among-system variability. Hakanson (1996) suggested that variables with high within-system variance may not be modeled easily; however, such high-frequency variation can be summarized (e.g., stream discharge exceedence frequencies; Wiley and Seelbach In press) to portray significant variation in pattern among water bodies. Landscape-context modeling also implicitly requires large sample sizes of waters in the learning set. Thus, subregions need to be extensive and heterogeneous enough to provide both a large sample of waters and some variation among them.

Conservation planning is often considered a "crisis discipline" (Meffe and Carroll 1997), meaning that management decisions and actions often cannot afford to wait for development of fully accurate scientific understanding and information. Thus, decision-support tools must be developed and iteratively updated using the best available science. In our review, applications seeking broad regional coverage often employed some form of analytical shortcuts: regionalization (Claesson et al. 1994; Latour et al. 1994), raster-classified landscape units (Claesson et al. 1994), interpretive estimation of landscape position and characteristics (Seelbach et al. 1997; Higgins et al. 1998), literature-based decision rules (Meixler and Bain 1998; Sowa et al. 1999), or estimation of current status from land-use maps (Higgins et

TABLE 24.5. Modeling ecological targets for rehabilitation of the Rouge River, Michigan.

Water body:	River valley segment
Management region:	Rouge River, Michigan
Management goal:	To develop a suite of ecological targets to guide stormwater rehabilitation efforts on this severely degraded urban river; specific focus on fish community targets to serve as integrated signal of future system recovery
Developing agencies:	University of Michigan, School of Natural Resources and Environment; and Institute for Fisheries Research, Michigan Department of Natural Resources (Wiley et al. 1998), for the Wayne County, Rouge Project Office
Map input variables:	For stream discharge: catchment area, precipitation, slope, surficial geology, soils, land covers
	For fishes: catchment area, estimated baseflow yield
Model form:	Regression models of stream discharge following standard hydraulic geometry relations (Wiley and Seelbach In press)
	Statistical summary of fish abundance—coarse-scale habitat affinities for Michigan streams (Zorn et al. 1997); standard deviations from the mean were used to gauge likelihood of occurrence given a particular attribute value
	For selected rehabilitation target fishes: the large regional, relational database was queried in reverse and typical summer thermal and stormflow regimes calculated (agglomeration of sites by fishes)
Ecological outputs:	Potential fish community structure
	Acceptable summer thermal regime to support selected fishes
	Acceptable summer and stormflow regimes to support selected fishes (Figure 24.4)
Implementation status	Enabled feasibility assessment of alternative stormwater mitigation measures for specific segments by Rouge Project Office; allows determination of where rehabilitation is not feasible
	Illustrated overall ecological structure of the river system, and highlighted priority problems and opportunities
	Provided realistic fishery rehabilitation goals for specific segments; for example, it became clear that some tributaries should not be expected to support sport fisheries. Alternatively, new attention has been drawn to recreational potentials in other segments

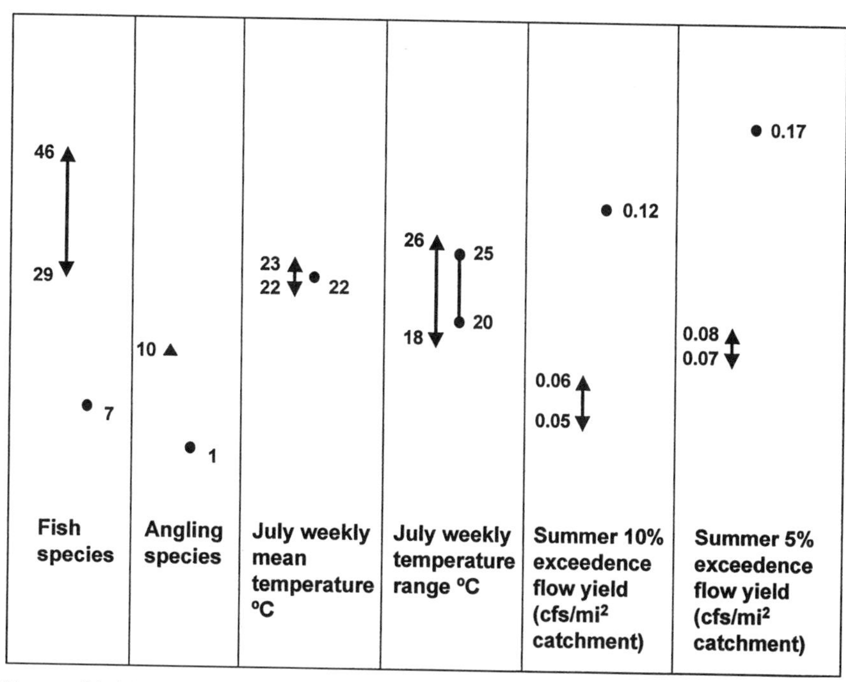

FIGURE 24.4. Predicted reference conditions (triangle, or lines bounded by triangles) and observed conditions (circle, or lines bounded by circles) for a suite of ecological parameters on the Rouge River, Michigan. July temperatures fell within expected ranges; however, measures of fish community structure and summer stormflows were clearly different from predicted conditions.

TABLE 24.6. Examining importance of landscape context to characteristics of Wisconsin lakes.

Water body:	Lakes
Management region:	Northern Wisconsin
Management goal:	Explanation of landscape-driven heterogeneity in lake ecological attributes
Developing agency:	University of Wisconsin, Center for Limnology (Riera et al. In press)
Map input variables:	Lake order (position in surface and subsurface flow networks)
Model form:	One-way ANOVA to test for differences among lakes in different lake-order classes
Ecological outputs:	Potential lake and catchment size
	Major ions, nutrients, fish species richness, chlorophyll
	Distribution and abundance of aquatic biota and humans
Implementation status:	Has not yet been applied to management; in early stages of exploring relationships and building models

TABLE 24.7. Initial modeling of hydrologic context for depressional wetlands in southern Michigan.

Water body:	Depressional wetlands
Management region:	Subregions in southern Michigan
Management goal:	To estimate potential wetland types and associated functions across subregions in Michigan for use as the reference condition in assessment and planning
Developing agency:	University of Michigan, School of Natural Resources and Environment; and Institute for Fisheries Research, Michigan Department of Natural Resources (D. Merkey, unpublished manuscript[a])
Map input variables:	Wetland size Catchment summaries of climate, topography, surficial geology, and soils Potential groundwater deliveries derived from theoretical interactions between local topography and surficial geology textures (Wiley and Seelbach In press) Position in the surficial flow network
Model form:	Develop by subregions (regionalization) Multiple linear regression to predict hydrologic, chemical, and vegetative attributes Ecological classification based on statistical analysis and literature guidelines (Hauer and Smith 1998) Ecological functions assigned according to literature guidelines
Ecological outputs:	Potential hydrologic source and hydroperiod Potential water chemistry Potential vegetative associations Potential wetland hydrogeomorphic type (Hauer and Smith 1998) Additional ecological context related to a wetland's landscape position, including relationships with interconnected water bodies and nearby uplands
Implementation status:	Has not yet been applied to management; new study

[a] D. Merkey. 1999. Doctoral Dissertation Proposal. School of Natural Resources and Environment. Ann Arbor: University of Michigan.

al. 1999). Despite their generalities, these applications have been immediately and eagerly employed in management planning processes (R. van Ek, personal communication; Higgins et al. 1999).

Many examples also emphasize that creative use of several approaches will be needed to simplify and describe complex ecosystems effectively. Most studies incorporated some form of agglomeration or regionalization with landscape-context modeling to control for coarse-scale climatic and geologic variables. For example, Higgins et al. (1999) used a two-step procedure, with a regionalization accounting for coarse-scale patterns in climate and historical zoogeography, and landscape-context interpretive modeling used to estimate site-specific ecological potentials.

It is clear that landscape-context modeling cannot account for all variables important to aquatic ecosystems (Claessen et al. 1994). In fact, models ideally should contain only a small number of independent landscape variables to highlight the fundamental system processes (Thierfelder 1998). Thus, there will be variance (noise) associated with predictions, and this should be explicitly communicated as model output. Some of this noise is due to the following factors: the inability to account for important local, modifying processes; noise in the landscape input variables (maps); the limits of modeling from finite, subsample databases; and current limits in our conceptual understanding of ecosystem processes (e.g., processes driving the distribution of biota are especially complex).

24.5 Knowledge Gaps

Landscape-context modeling is a nascent discipline, and many challenges lie ahead. Three general areas need development: conceptual models of hierarchical processes that drive aquatic systems, assembly of standardized regional data sets, and integration of site-scale data into landscape-based models.

Models will improve as our conceptual understanding and empirical bases improve. Currently, significant gaps in our conceptual understanding of aquatic ecosystems limit development of formal models (hypotheses) of landscape-driven, hierarchical aquatic processes. Such models have only recently begun to be developed and tested. The examples reviewed above are among the first to explore such processes quantitatively. These initial hypothetical models will highlight cross-scale relationships and expose conceptual holes, leading to new rounds of hypotheses, data collection, and improved model development.

Landscape-context modeling requires extensive map and site-level data. Extensive digital map coverages exist, but these can suffer from coarse scales of resolution or classification schemes inappropriate for analyses of local aquatic ecosystems. In addition, modeling requires that numerous catchment or local summaries be developed for waters of interest (and for waters to be extrapolated to). Large sets of site-scale assessment data exist for many areas, typically for certain water quantity, water quality, and biological variables. However, these data vary greatly in their variable selection, representative site coverage, and sampling methodologies. We feel that such shortcomings will be overcome, in time, through the iterative, adaptive management process. Only by assembling existing data sets, and analysis of the performance and sensitivities of formal models, will we be able to design more appropriate sampling schemes.

With their conscious focus on hierarchy, efficiency, and simplicity, initial landscape-based models have stressed map-scale physical input variables. It is likely that the addition of certain easily measured site-scale habitat variables could greatly improve predictive power of the models. For example, the current AUSRIVAS models of stream invertebrate communities (Simpson and Norris In press) are based on site-level habitat variables that provide useful predictions.

The addition of key landscape-scale drivers would provide such models with landscape context, map efficiencies, and potentially greater statistical power, ideas that are beginning to be explored (Davies 1999). The idea of requiring some limited site data for models would not make the assessment process less efficient, as some on-site measurement of observed condition is needed for complete assessment anyway. In some systems, especially lakes and wetlands (perhaps less controlled by abiotic variables), the addition of key biological variables may be critical. For example, in lakes, food-web structure accounts for roughly half of the variance observed in pelagic primary production (Carpenter et al. 1991). Thus, food-web structure would appear to be a desirable variable in water-quality models. We do not yet understand whether landscape forces act directly on primary production or indirectly through influences on food-web structure.

Initial models have generally targeted response variables that characterize the average condition over several years. Such low-frequency information is appropriate for coarse-scale modeling and often useful for providing ecosystem reference conditions; however, many management questions will undoubtedly require estimation of responses over shorter time periods. These questions will require employing a different class of models that operate on shorter time steps and that perhaps have more detailed input variables and mechanistic analyses—referred to earlier as ecosystem process models. We see these process models as complementary to, not opposing, landscape-context models. Map-based models could provide an initial rough cut, or stratification for a site, leading to the appropriate application of various site- or type-specific process models. We need to identify the respective strengths and limitations of the different modeling scales, and use all scales and approaches to full advantage.

24.6 Research Approaches

Implementation of a landscape-context modeling approach to aquatic conservation planning requires commitment to a long-term, iterative development program, a form of adaptive management (Meixler and Bain 1998). The overall goal is to link regional landscape (map) data and detailed site databases already in existence (or to be collected in the future) through ecologically realistic models.

Step 1 is to use process-level theory and the local experience of managers to develop initial conceptual hypotheses and models regarding relationships between landscape-level data and aquatic ecosystems of interest.

Step 2 is to compile map and site data from existing assessment programs into a GIS; this information system must be designed to handle a variety of queries about a suite of response variables, and to provide opportunities for building and testing the models. Often, existing assessment programs house plenty of site and map data, but these are not being used in the proper context or in combination with each other. Both map and local site data have their weaknesses and associated errors related to data resolution, prior goals, and criteria for classification and sampling. These errors must be acknowledged and incorporated into error assessments throughout the process. However, we argue strongly that model development should proceed using available data, at available resolutions and data quality. We can learn much from incomplete or semiquantitative data, and the iterative process will produce continuous improvements.

Step 3 is to develop and test analytical models that predict ecological response from landscape context. Modeling approaches may include graphing, statistical summary, multiple linear regression, logistic regression, or path analysis. Analytical techniques such as path analysis (Retherford and Choe 1993) were specifically designed to explore hierarchical relationships (e.g., Hinz and Wiley 1999; Wehrly 1999). Model calibration and validation are used to identify the utility and limits of current models, and additional sampling needs (Haber 1994; Walters 1997). Understanding limitations and weaknesses of the models is critical. In particular, sensitivity analyses are used to highlight the variables and scales (Hakanson 1996; Rabeni and Sowa 1996) that are most influential and that should therefore be focused on in future sampling designs.

Step 4a is to use the current models in management planning, with explicit recognition of prediction error (perhaps as confidence limits). These would be the best available, regionally comprehensive estimates of ecological potential for each water body.

Step 4b occurs simultaneously with Step 4a and involves a return back to Step 1: development of new conceptual models of how the hierarchical system works, thus beginning a second cycle of modeling in which strategic data are gathered and an improved set of models is constructed.

Modeling in The Netherlands (Claessen et al. 1994; R. van Ek, personal communication) of wetland responses to water table fluctuations provides an excellent example of the suggested iterative approach. They initially used existing science to formulate conceptual models of how water table levels should influence soil moisture and chemistry and, therefore, vegetation communities. They compiled existing data on water table, landscape, and vegetation characteristics and then built and tested initial predictive models of potential vegetation. These models were used to estimate wetland responses to proposed alternative water management scenarios throughout The Netherlands. At the same time, the investigators initiated a second cycle of model testing and development. They incorporated improved and expanded landscape and vegetation data sets; performed sensitivity analyses; determined probabilities of occurrence of site-level features (e.g., clay lenses) within landscape types; established links to site-level hydrology and conservation-value models for increased management resolution; made the models more user-friendly; and were able to use model predictions to assist in national water management planning. Ongoing work (R. van Ek, personal communication) identified weaknesses in the second-version model, setting the stage for continued development.

Acknowledgments

Thanks to the many colleagues who contributed materials and ideas during our review of current work. M. E. Baker, T. P. Simon, J. Higgins, and K. J. Gutzwiller edited the manuscript.

References

Asher, H. 1983. *Causal Modeling.* Newbury Park, New Jersey: Sage Publications.

Bourgeron, P.S., and Jensen, M.E. 1994. An overview of ecological principles for ecosystem management. In *Volume II: Ecosystem Management: Principles and Applications,* eds. M. Jensen and P. Bourgeron, pp. 45–57. General Technical Report PNW-GTR-318. Portland, Oregon: USDA Forest Service, Pacific Northwest Research Station.

Bryce, S.A., and Clark, S.E. 1996. Landscape-level ecological regions: linking state-level ecoregion frameworks with stream habitat classifications. *Environ. Manage.* 20:297–311.

Carpenter, S.R., Frost, T.M., Kitchell, J.F., Kratz, T.K., Schindler, D.W., Shearer, J., Sprules, W.G., Vanni, M.J., and Zimmerman, A.P. 1991. Patterns of primary production and herbivory in 25 North American lake ecosystems. In *Comparative Analyses of Ecosystems: Patterns, Mechanisms, and Theories,* eds. J. Cole, G. Lovett, and S. Findlay, pp. 67–96. New York: Springer-Verlag.

Claessen, F.A.M., Klijn, F., Witte, J.P.M., and Nienhuis, J.G. 1994. Ecosystem classification and hydro-ecological modelling for national water management. In *Ecosystem Classification for Environmental Management,* ed. F. Klijn, pp. 199–222. Dordrecht, The Netherlands: Kluwer Academic Publishers.

Corner, R.A., Albert, D.A., and Delain, C.J. 1997. *Landtype Associations of the Leelanau and Grand Traverse Peninsulas: Subsection VII.5.* Technical Report and Map. Lansing: Michigan Natural Features Inventory.

Cosby, B.J., Hornberger, G.M., Galloway, J.N., and Wright, R.F. 1985. Modeling the effects of acid deposition: assessment of a lumped-parameter model of soil water and streamwater chemistry. *Water Resour. Res.* 21:51–63.

Cross, T.K., and McInerny, M.C. 1995. *Influences of Watershed Parameters on Fish Populations in Selected Minnesota Lakes of the Central Hardwood Forest Ecoregion.* Fisheries Investigational Report 441. St. Paul: Minnesota Department of Natural Resources.

Davies, N. 1999. *Prediction and Assessment of Local Stream Habitat Features Using Large-Scale Catchment Characteristics.* Honours Thesis in Applied Science. Belconnen, Australia: University of Canberra.

Davis, L.S., and Henderson, J.A. 1978. Many uses and many users: some desirable characteristics of a common land and water classification system. In *Classification, Inventory, and Analysis of Fish and Wildlife Habitat,* general chairman A. Marmelstein, pp. 13–34. Report FWS/OBS-78/76. Washington, DC: U.S. Fish and Wildlife Service.

Davis, W.S., and Simon, T.P. 1995. *Biological Assessment and Criteria: Tools for Water Resource Planning and Decision Making.* Boca Raton, Florida: Lewis Press.

Davis, W.S., Snyder, B.D., Stribling, J.B., and Stoughton, C. 1996. *Summary of State Biological Assessment Programs for Streams and Wadable Rivers.* Report EPA-230-R-96-007. Washington DC: U.S. Environmental Protection Agency.

Eilers, J.M., Glass, G.E., Webster, K.E., and Rogalla, J.A. 1983. Hydrologic control of lake susceptibility to acidification. *Can. J. Fish. Aquat. Sci.* 40:1896–1904.

Emmons, E.E., Jennings, M.J., and Edwards, C. In press. An alternative classification method for northern Wisconsin lakes. *Can. J. Fish. Aquat. Sci.*

Gallant, A.L., Whittier, T.R., Larsen, D.P., Omernik, J.M., and Hughes, R.M. 1989. *Regionalization as a Tool for Managing Environmental Resources.* Report EPA/600/3-89/060. Corvallis, Oregon: U.S. Environmental Protection Agency.

Haber, W. 1994. Systems ecological concepts for environmental planning. In *Ecosystem Classification for Environmental Management,* ed. F. Klijn, pp. 49–68. Dordrecht, The Netherlands: Kluwer Academic Publishers.

Hakanson, L. 1996. Predicting important lake habitat variables from maps using modern modelling tools. *Can. J. Fish. Aquat. Sci.* 53(Suppl. 1):364–382.

Hakanson, L., and Peters, R.H. 1995. *Predictive Limnology—Methods for Predictive Modelling.* Amsterdam, The Netherlands: SPB Academic Publishing.

Hauer, F.R., and Smith, R.D. 1998. The hydrogeomorphic approach to functional assessment of riparian wetlands: evaluating impacts and mitigation on river floodplains in the U.S.A. *Freshwater Biol.* 40:517–530.

Higgins, J., Lammert, M., and Bryer, M. 1999. *Including Aquatic Targets in Ecoregional Portfolios: Guidance for Ecoregional Planning Teams.* Designing a Geography of Hope Update Number 6. Arlington, Virginia: The Nature Conservancy.

Higgins, J., Lammert, M., Bryer, M., DePhilip, M., and Grossman, D. 1998. *Freshwater Conservation in the Great Lakes Basin: Development and Application of an Aquatic Community Framework.* Final Project Report. Chicago: The Nature Conservancy, Great Lakes Program Office.

Hinz, L.C., and Wiley, M.J. 1998. *Growth and Production of Juvenile Trout in Michigan Streams: Influence of Potential Ration and Temperature.* Fisheries Research Report 2042. Ann Arbor: Michigan Department of Natural Resources.

Holling, C.S. 1998. Two cultures of ecology. *Conserv. Ecol.* [online] 2:4. Available from the Internet: www.consecol.org/vol2/iss2/art4.

Karr, J.R., Fausch, K.D., Angermeier, P.L., Yant, P.R., and Schlosser, I.J. 1986. *Assessing Biological Integrity in Running Waters: A Method and Its Rationale.* Special Publication 5. Champaign: Illinois Natural History Survey.

Kerr, S.R. 1976. Ecological analysis and the Fry paradigm. *J. Fish. Res. Board Can.* 33:329–332.

Klijn, F. 1994. Spatially nested ecosystems: guidelines for classification from a hierarchical perspective. In *Ecosystem Classification for Environmental Management,* ed. F. Klijn, pp. 85–116. Dordrecht, The Netherlands: Kluwer Academic Publishers.

Kratz, T.K., Webster, K.E., Bowser, C.J., Magnuson, J.J., and Benson, B.J. 1997. The influence of landscape position on lakes in northern Wisconsin. *Freshwater Biol.* 37:209–217.

Latour, J.B., Reiling, R., and Wiertz, J. 1994. A flexible multiple stress model: who needs a priori classification? In *Ecosystem Classification for Environmental Management,* ed. F. Klijn, pp. 183–198. Dordrecht, The Netherlands: Kluwer Academic Publishers.

Levin, S.A. 1992. The problem of pattern and scale in ecology. *Ecology* 73:1943–1967.

Lotspeich, F.B. 1980. Watersheds as the basic ecosystem: this conceptual framework provides a basis for a natural classification system. *Water Resour. Bull.* 16:581–586.

Meffe, G.K., and Carroll, C.R., eds. 1997. *Principles of Conservation Biology.* Second Edition. Sunderland, Massachusetts: Sinauer Associates.

Meixler, M.S., and Bain, M.B. 1998. *Aquatic Gap Analysis: Demonstration of a Geographic Approach to Aquatic Biodiversity Conservation.* Final Project Report. Reston, Virginia: U.S. Geological Survey, Biological Resources Division.

Nelson, R.L., Platts, W.S., Larsen, D.P., and Jensen, S.E. 1992. Trout distribution and habitat in relation to geology and geomorphology in the North Fork Humboldt River drainage, northeastern Nevada. *Trans. Am. Fish. Soc.* 121:405–426.

O'Neill, R.B., DeAngelis, D.L., Waide, J.B., and Allen, T.F.H. 1986. *A Hierarchical Concept of Ecosystems.* Princeton: Princeton University Press.

Rabeni, C.F., and Sowa, S.P. 1996. Integrating biological realism into habitat restoration and conservation strategies for small streams. *Can. J. Fish. Aquat. Sci.* 53(Suppl. 1): 252–259.

Rahel, F.J., and Nibbelink, N.P. 1999. Spatial patterns in relations among brown trout (*Salmo trutta*) distribution, summer air temperature, and stream size in Rocky Mountain streams. *Can. J. Fish. Aquat. Sci.* 56(Suppl. 1):43–51.

Raven, P.J., Fox, P., Everard, M., Holmes, N.T.H., and Dawson, F.H. 1997. River Habitat Survey: a new system for classifying rivers according to their habitat quality. In *Freshwater Quality: Defining the Indefinable?,* eds. P.J. Boon and D.L. Howell, pp. 215–234. Edinburgh: Scottish Natural Heritage Press.

Retherford, R.D., and Choe, M.K. 1993. *Statistical Models for Causal Analysis.* New York: John Wiley and Sons.

Riera, J.L., Magnuson, J.J., Kratz, T.K., and Webster, K.E. In press. A geomorphic template for the analysis of lake districts applied to the Northern Highland Lake District, Wisconsin, U.S.A. *Freshwater Biol.*

Risser, P.G., Karr, J.R., and Forman, R.T.T. 1984. *Landscape Ecology—Directions and Approaches.* Special Publication 2. Champaign: Illinois Natural History Survey.

Rochelle, B.P., Liff, C.I., Campbell, W.G., Cassell, D.L., Church, M.R., and Nusz, R.A. 1989. Regional relationships between geomorphic/hydrologic parameters and surface water chemistry relative to acidic deposition. *J. Hydrol.* 112:103–120.

Rosgen, D.L. 1996. *Applied River Morphology.* Pagosa Springs, Colorado: Wildland Hydrology.

Schlosser, I.J. 1991. Stream fish ecology: a landscape perspective. *BioScience* 41:704–712.

Schupp, D.H. 1992. *An Ecological Classification of Minnesota Lakes with Associated Fish Communities.* Section of Fisheries Special Publication 417. St. Paul: Minnesota Department of Natural Resources.

Seelbach, P.W., Wiley, M.J., Kotanchik, J.C., and Baker, M.E. 1997. *A Landscape-based Ecological Classification System for River Valley Segments in Lower Michigan.* Fisheries Research Report 2036. Ann Arbor: Michigan Department of Natural Resources.

Simpson, J., and Norris, R.H. In press. Biological assessment of water quality: development of AUSRIVAS models and outputs. In *RIVPACS and Similar Techniques for Assessing the Biological Quality of Freshwaters,* eds. J.F. Wright, D.W. Sutcliffe, and M.T. Furse. Ambleside, United Kingdom: Freshwater Biological Association and Environmental Agency.

Soranno, P.A., Webster, K.E., Riera, J.L., Kratz, T.K., Baron, J.S., Bukaveckas, P.A., Kling, G.W., White, D.S., Caine, N., Lathrop, R.C., and Leavitt, P.R. 1999. Spatial variation among lakes within landscapes: ecological organization along lake chains. *Ecosystems* 2:395–410.

Sowa, S.P., Morey, M.E., Sorensen, G.R., and Annis, G. 1999. Predicting the biological potential of each valley segment. In *Implementing the Aquatic Component of Gap*

Analysis in Riverine Environments: A Training Workbook, ed. S.P. Sowa, Section 5. Columbia: Missouri Resource Assessment Partnership.

Terrell, J.W., McMahon, T.E., Inskip, P.D., Raleigh, R.F., and Williamson, K.L. 1982. *Habitat Suitability Index Models: Appendix A. Guidelines for Riverine and Lacustrine Applications of Fish HSI Models with the Habitat Evaluation Procedures.* Report FWS/OBS-82/10.A. Washington, DC: U.S. Fish and Wildlife Service.

Thierfelder, T.K.E. 1998. *An Inductive Approach to the Modeling of Lake Water Quality in Dimictic, Glacial/Boreal Lakes.* Comprehensive summaries of Uppsala Dissertations from the Faculty of Science and Technology 399. Uppsala, Sweden: Uppsala University.

Thomann, R.V. 1987. *Principles of Surface Water Quality Modeling and Control.* New York: Harper and Row.

Thorn, W.C., and Anderson, C.S. 1999. *A Provisional Classification of Minnesota Rivers with Associated Fish Communities.* Special Publication 153. St. Paul: Minnesota Department of Natural Resources.

Turner, M.G. 1998. Landscape ecology. In *Ecology,* ed. S.I. Dodson, pp. 77–122. New York: Oxford University Press.

Walters, C. 1997. Challenges in adaptive management of riparian and coastal ecosystems. *Conserv. Ecol.* [online] 1:1. Available from the Internet: www.consecol.org/vol1/iss2/art1.

Webster, K.E., Soranno, P.A., Baines, S.B., Kratz, T.K., Bowser, C.J., Dillon, P.J., Campbell, P., Fee, E.J., and Hecky, R.E. In press. Structuring features of lake districts: geomorphic and landscape controls on lake chemical responses to drought. *Freshwater Biol.*

Wehrly, K.E. 1999. *The Influence of Thermal Regime on the Distribution and Abundance of Stream Fishes in Michigan.* Ph.D. dissertation. Ann Arbor: University of Michigan.

Wessman, C.A. 1990. Landscape ecology: analytical approaches to pattern and process. In *An Ecosystem Approach to the Integrity of the Great Lakes in Turbulent Times,* eds. C.J. Edwards and H.A. Regier, pp. 285–299. Special Publication 90-4. Ann Arbor, Michigan: Great Lakes Fishery Commission.

Wiley, M.J., and Seelbach, P.W. In press. *Hydrology of Rivers in Michigan's Lower Peninsula.* Fisheries Research Report 2039. Ann Arbor: Michigan Department of Natural Resources.

Wiley, M.J., Seelbach, P.W., and Bowler, S.P. 1998. *Ecological Targets for Rehabilitation of the Rouge River.* Final Report RPO-PI-SR21.00. Detroit, Michigan: Wayne County Rouge Project Office.

Wooton, J.T. 1994. The nature and consequences of indirect effects in ecological communities. *Ann. Rev. Ecol. Syst.* 25:443–466.

Wright, J.F., Moss, D., Clarke, R.T., and Furse, M.T. 1997. Biological assessment of river quality using the new version of RIVPACS (RIVPACS III). In *Freshwater Quality: Defining the Indefinable?,* eds. P.J. Boon and D.L. Howell, pp. 102–107. Edinburgh: Scottish Natural Heritage Press.

Yoder, C.O., and Smith, M.A. 1998. Using fish assemblages in a state biological assessment and criteria program: essential concepts and considerations. In *Assessing the Sustainability and Biological Integrity of Water Resources Using Fish Communities,* ed. T.P. Simon, pp. 17–56. Boca Raton, Florida: CRC Press.

Young, T.C., and Stoddard, J.L. 1996. The TIME project design: I. Classification of Northeast lakes using a combination of geographic, hydrogeochemical, and multivariate techniques. *Water Resour. Res.* 32:2517–2528.

Zonneveld, I.S. 1994. Basic principles of classification. In *Ecosystem Classification for Environmental Management,* ed. F. Klijn, pp. 23–47. Dordrecht, The Netherlands: Kluwer Academic Publishers.

Zorn, T.G., Seelbach, P.W., and Wiley, M.J. 1997. *Patterns in the Distributions of Stream Fishes in Michigan's Lower Peninsula.* Fisheries Research Report 2035. Ann Arbor: Michigan Department of Natural Resources.

Section V

Synthesis and Conclusions

25

Applying Landscape Ecology in Biological Conservation: Principles, Constraints, and Prospects

KEVIN J. GUTZWILLER

25.1 Introduction

When detailed biological information about a conservation problem is unavailable and there is insufficient time to acquire it, conservation design, management, and policy decisions may need to be based on working principles. For situations in which landscape ecology is to be applied, these principles can be derived from patterns or relations that are common among studies within a given subject area. For example, if various studies indicate that indigenous species are better supported by large rather than small habitat patches, it would be reasonable, in the absence of specific data for a given situation, to apply this principle. This basic approach, supplemented whenever possible with knowledge about general ecological principles and actual management experience, was used to formulate the application principles presented in previous chapters of this book.

Another way to derive working principles for applying landscape ecology in conservation is to identify principles with common relevance to many different conservation subject areas. The value of this approach is that such principles are likely to be applicable as well in conservation situations for which the relevance of landscape ecology has not yet been fully explored or determined. For instance, habitat restoration holds promise for conserving various organisms, but we have little experience in restoring habitat at broad spatial scales. Principles for applying landscape ecology that are simultaneously relevant to colonization, local extinction, and long-term population viability, for example, would be valuable in guiding the initial steps of broad-scale habitat restoration. If studies from various conservation subject areas indicated that native species tend to repopulate sooner and persist longer in naturally connected landscapes, then the overall principle of maximizing natural connectivity of landscapes could be applied to restore habitat at broad scales. The first objective of this chapter is to identify such overarching principles for applying landscape ecology.

Neither the number nor the effectiveness of conservation applications of landscape ecology is likely to increase substantially until major constraints to applications are identified. Thus, the second objective of this chapter is to expose scientific and nonscientific barriers that limit applications; I address constraints that pervade many conservation subject areas. My third objective is to discuss some basic approaches in science, education, personnel management, and public activism that can be used to reduce these constraints. I conclude this chapter by considering the prospects for expanding conservation applications of landscape ecology. The perspectives that follow are derived from information from preceding sections of this book, other literature, and my personal experience.

25.2 Overarching Principles for Applying Landscape Ecology

The principles I suggest in this section are interrelated, are applicable to many of the subject areas considered in this volume, and in many cases represent a synthesis of the more specific application principles presented in previous chapters. I offer these principles as initial, overall rules of thumb for increasing effective applications of landscape ecology in biological conservation.

25.2.1 Scientific Basis for Applications

- Applications of landscape ecology must be founded on science, and many forms of research (e.g., experimental, observational, basic, applied) can contribute usefully to this science.
- Because of multiple perspectives that can be gained, the use of several integrated scientific approaches (e.g., interconnected field experiments, simulations, and empirical modeling) is more likely to lead to a sound basis for applying landscape ecology than is the use of only one scientific approach.
- Characterization of possible nonlinear and threshold effects (not just linear effects) of landscape conditions on organisms is important for successful applications of landscape ecology.
- To minimize unforeseen and undesirable effects of landscape-ecology applications on target and nontarget species, possible cascading, synergistic, and antagonistic effects of applications should be considered whenever feasible.
- Organism responses to full ranges of natural and anthropogenic variability in landscape conditions should be studied.
- Although scientists should continue to try to uncover general principles that relate organism responses to landscape structure and function, for many conservation issues, such principles may remain elusive because of varied, context-related responses. Under these circumstances, species-specific research in different environmental settings will be necessary to underpin landscape-ecology applications.
- Theories and concepts from economics, geography, and physics may be useful for discovering what drives landscape change and organism-landscape relations (see O'Neill 1999); understanding reality related to these issues is crucial for devising effective applications.
- Spatial and temporal uncertainties about organism responses to landscape conditions (e.g., Gutzwiller and Barrow 2001) must be quantified.

- Knowledge about the history of a landscape and its organisms (e.g., natural and anthropogenic disturbances, organism distributions, local extinctions and colonizations) is indispensable for deciding where, when, and how to implement landscape-ecology concepts and principles.

25.2.2 Implementing Field Applications

- Landscape management actions should reflect a holistic approach that treats aquatic and terrestrial systems as inseparable.
- Applications of landscape ecology should reflect information about the various spatial scales of environmental variation that organisms perceive and to which they respond.
- Landscape-ecology principles and concepts should be applied at all venues (city, county, state, and national levels) where land-use decisions are made.
- All landscape elements (patches, corridors, matrix) can influence organisms, so the ramifications of landscape-ecology applications for each of these elements must be considered.
- Management actions should preempt alteration of intact portions of landscapes, minimize or mitigate degradation of declining areas, and rehabilitate or restore destroyed portions.
- Logistical complexities, limits on management resources, and species' biological idiosyncrasies make it virtually impossible to conserve multispecies assemblages through single-species approaches. Field applications of landscape ecology will therefore have to target subsets of species or species' characteristics that are surrogates for the full complement of local endemism, taxonomic breadth and depth, biological uniqueness, genetic variation, evolutionary potential, and aesthetic, economic and cultural values that we hope to conserve.
- To be maximally effective and economical, applications of landscape ecology should be implemented before species and landscapes have undergone serious decline.
- Widespread, continual use of landscape ecology will require support from all stakeholders. For a given project, mutual trust among stakeholders (developed through honesty and integrity) and knowledge of potential benefits for all will be necessary to garner this support in the beginning and to maintain it indefinitely thereafter.
- For long-term effectiveness, applications of landscape ecology must reflect not just conservation goals, but also economic, social, and political factors.
- Interest in applying landscape ecology to solve conservation problems will depend in part on how well information about successes and failures is publicized and shared. A site on the World Wide Web should be established and continually updated so that those interested in applying landscape ecology to a given problem can quickly access information about possible approaches, their cost and conservation effectiveness, associated literature, and experienced contacts.

25.3 Major Constraints on Applications of Landscape Ecology

In this section, I summarize the nature and extent of applications of landscape ecology for the subjects considered in previous chapters. I then discuss key scientific and nonscientific barriers that limit field applications of landscape ecology in those subject areas.

25.3.1 General Nature and Extent of Landscape-Ecology Applications

The authors of previous chapters addressed 18 subject areas that I have categorized into three groups: (1) Scale, Connectivity, and Movement (Chapters 5–10); (2) Landscape Change (Chapters 12–18); and (3) Conservation Planning (Chapters 20–24). Applications of landscape-ecology concepts have occurred in research settings for all of the subject areas (Figure 25.1). Planning applications are evident for most of the subject areas (Figure 25.1); the single exception was that there have not yet been any documented planning applications of the concepts associated with time lags in metapopulation responses to landscape change. Field

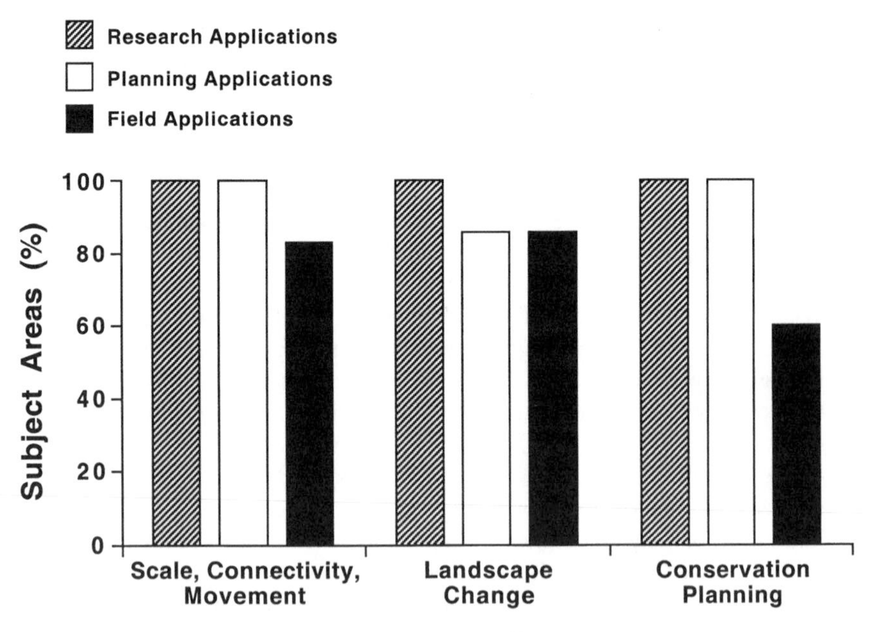

FIGURE 25.1. Percentages of subject areas whose concepts have been applied in research, planning, and field contexts. Eighteen subject areas were categorized into three general groups: (1) Scale, Connectivity, and Movement; (2) Landscape Change; and (3) Conservation Planning. See text for subject areas (chapters) in each general group.

implementation of landscape-ecology concepts has been most common for subject areas related to scale, connectivity, and movement, and for subject areas related to landscape change; field application of concepts has been least common for subject areas associated with conservation planning (Figure 25.1). Concepts associated with percolation theory, time lags in metapopulation responses to landscape change, landscape cohesion assessment, and spatial optimization methods have not been applied on the ground.

Considering the 18 subject areas collectively, landscape-ecology concepts from 100% of the subject areas have been applied in a research context; concepts from 94% of the areas have been applied in conservation planning; and concepts from 78% of the subject areas have been applied in the field. These summary percentages involve documented applications as well as unpublished cases communicated to authors through personal contacts; they include subject areas whose concepts have been applied often or rarely; and they involve situations in which many or only a few of the concepts in a given subject area have been applied. Consequently, these percentages are only rough indices of the relative degrees of application.

With these caveats in mind, the percentages seem to reflect the relative ease of, and investment necessary for, applying landscape-ecology concepts. That is, it is relatively easy and inexpensive to involve the concepts in conservation research; it is not always a simple matter to use them or to secure their acceptance in formulations of conservation plans; and it can be difficult and costly to put landscape ecology into practice in the field. Applications in all three contexts (research, planning, field) are crucial for effective application of landscape ecology in biological conservation. The lower percentage of on-the-ground applications observed here is consistent with my belief that science-based field applications need to be increased.

25.3.2 Scientific Constraints That Limit Field Applications

For the 18 subject areas considered, several pervasive scientific constraints block or retard on-the-ground applications (Figure 25.2). I believe that these barriers prevent field applications by generating significant uncertainties about the probable success of applications, the relevance of landscape ecology, or both. As a consequence, many governments, land-management agencies, corporations, and individuals have not committed resources to field applications.

Perhaps the most limiting of these constraints is our rudimentary understanding of the biology that underlies relations between landscape conditions and the biotic patterns and processes that are important in conservation. Because of this limited knowledge, some efforts to preserve and manage living systems through applications of landscape ecology may not amount to much more than ecological intuition or educated guesswork.

The relative paucity of field applications is a problem because it limits information about application effectiveness. Managers cannot gain confidence or develop objective information about the effectiveness of applying landscape-ecology concepts without witnessing the outcome of actual field applications.

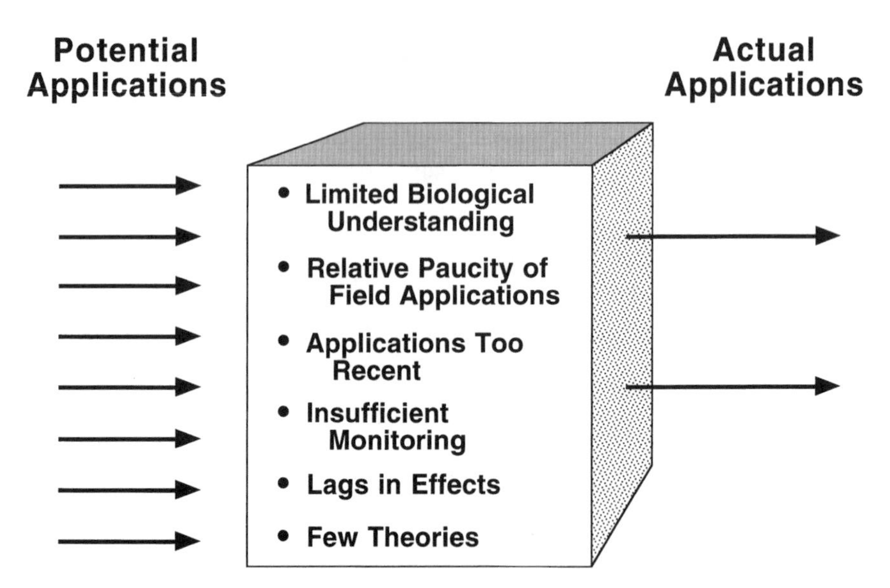

FIGURE 25.2. Pervasive scientific barriers that limit field applications of landscape ecology in biological conservation.

Conservationists will not be convinced that landscape ecology offers realistic solutions to conservation problems without supporting field data. The most convincing evidence will come from successful field applications, but these are currently in short supply for many subject areas.

Unfortunately, long time periods may have to elapse before success can be adequately assessed. In a conservation context, the pertinent time frames for population viability, persistence time, and maintenance of genetic diversity, for example, may be on the order of many years or decades. Considering these time scales, most field applications have been implemented far too recently for a clear assessment of their long-term effects. Offering advice about how to apply landscape ecology based on short-term data is tantamount to asking a manager to gamble with limited resources (personnel, time, funding) and the fates of organisms.

Systematic monitoring efforts are uncommon. Without data from monitoring programs, managers do not have information about whether field applications are influential, or how such influences develop over space and time. This lack of information prevents conservationists from assessing application effectiveness and from determining the relevance of landscape ecology to conservation. Moreover, it is difficult to gain support for additional field applications when previous applications have not been monitored because, without monitoring data, neither success nor reasons to modify previous application approaches can be objectively substantiated.

Lags in effects of field applications, if confounded with other biotic and abiotic factors, can complicate efforts to determine application effectiveness. Lags also delay our understanding of the influence of applications. At present, there are very few situations for which we know the duration of such lags. In effect, the return time on investment of precious management resources remains a mystery. Decision makers are not usually willing to commit resources to a project without knowing how long they will have to wait before effects accrue.

Few theories link landscape ecology to various conservation issues (With 1999). Indeed, the theoretical framework of landscape ecology itself is poorly developed (Wiens 1999a,b). The availability of developed theory is important for field applications of landscape ecology because, when landscape-management decisions cannot wait for the results of specific empirical studies, theory can be used to help decide how to implement landscape ecology. Without well-developed theory, the relevance of landscape ecology to a given conservation issue may not be apparent. And lacking at least theoretical support, managers may not implement landscape ecology in the field even under dire circumstances.

25.3.3 Nonscientific Barriers That Limit Field Applications

For the subjects we studied, several major nonscientific constraints also block or retard field applications (Figure 25.3). Not all conservationists have adequate scientific training to decide whether or how to apply concepts of landscape ecology, and some conservationists may not be convinced that landscape ecology is especially relevant (Hobbs 1997; With 1999). We cannot expect people to apply ideas that they do not understand or support.

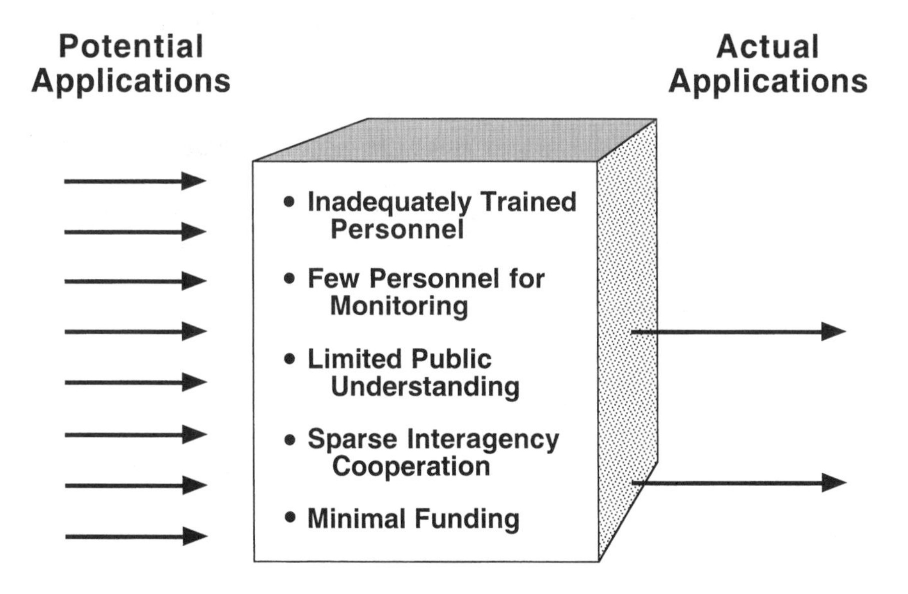

FIGURE 25.3. Major nonscientific factors that block or retard field applications of landscape ecology in biological conservation.

Too few personnel are available for regular monitoring of the outcomes of field applications. Useful monitoring is expensive and time-consuming, and personnel engaged in monitoring cannot simultaneously work on other important conservation tasks. But as indicated above, without data from monitoring and hence information about the effectiveness of applications, many unknowns about how to proceed will persist.

The landscape perspective seems to be appreciated primarily by lay and professional members of the environmental and natural resource communities. Those in disciplines other than ecology, conservation, and environmental planning typically do not have training in landscape ecology. Rarely have I encountered in the general public someone who is knowledgeable about the basic connections between landscape ecology and biological conservation. The limited understanding of landscape ecology that exists among members of Society as a whole makes it difficult for conservationists to garner public support for the broad-scale approaches (field applications of landscape ecology) that are frequently necessary to solve conservation problems.

Although the situation is improving (see references cited in Dramstad et al. 1996), there has been relatively sparse interagency cooperation in field applications of landscape ecology. This has retarded applications across natural land units such as watersheds, migration routes, and species' geographic ranges, which often span multiple jurisdictions. Disparities among agency missions, mandates and legal authorities, and perhaps even an unwillingness to share costs, successes, failures, and responsibilities, seem to be at the heart of the problem.

Today, developed nations probably have more overall funding for conservation than they have ever had in the past, but they still have minimal funding when one considers the amount that is needed to apply landscape ecology in activities such as habitat acquisition, management, and restoration. In developing countries, and in countries facing political unrest, economic woes, or burgeoning human populations, money for such work may be extremely limited, if available at all. For effective on-the-ground applications of landscape ecology, there must also be money to support the research and planning necessary to underpin and define appropriate field action. Even in the United States, with its stable government and economy, there is no adequate, regular source of funding for the research, planning, and implementation stages of applying landscape ecology in biological conservation.

25.4 Reducing Scientific Constraints

Experimental and observational field studies can be used to improve understanding of the biology that underlies relations between organisms and landscape conditions. Only properly designed experiments are capable of reliably determining causal relations. Observational studies do not enable one to draw inferences about causation, but they can be used to infer associations (e.g., organism-landscape relations), which can be put to valuable use in conservation. Ideally, broad-scale ecological experiments (e.g., Lovejoy et al. 1986; Margules 1992) should be used to improve our understanding of what governs organism-landscape relations, but the spatial scale of such studies makes them expensive and logistically difficult to conduct.

Fortunately, the space-for-time substitution technique, which uses presumed chronosequences, and quasi-experiments (e.g., before/after studies or impact studies) that take advantage of the effects of broad-scale natural phenomena, can be used inductively (Hargrove and Pickering 1992). In addition, some important broad-scale questions can be addressed through classic experiments at finer, more-tractable scales (Ims 1999). Experimental studies at the scale of patch boundaries, for example, can be used to discern mechanisms of organism movement and material transfer—processes that are believed to be responsible for many landscape ecological phenomena (Ims 1999).

Management experiments (Macnab 1983; Walters and Holling 1990) involving landscape manipulations can be used to increase the number of field applications of landscape ecology, which would provide more opportunities to learn from application successes and failures. A series of well-designed management experiments would simultaneously enable conservationists to implement landscape ecology in the field and distill practical lessons about how to apply landscape ecology. With the new knowledge gained from each experiment, principles for implementing landscape ecology could be refined incrementally.

Management experiments involving field applications of landscape ecology need to be initiated as soon as possible for two major reasons. First, because sufficient time (perhaps years or decades, depending on the issue of concern) must elapse before we can obtain a clear assessment of the results. And second, because experiment, assessment, and application phases must transpire while there is still time left to conserve a substantial portion of biodiversity. With each system that is degraded, with each habitat that is lost, our ability to protect biological resources is dwindling.

Careful monitoring across space and through time can be used to determine the effectiveness of landscape-ecology applications. Monitoring also can be used to determine the length of time it takes applications to have an effect (lag times). Recent discussions about monitoring aquatic systems (e.g., Ringold 2000), although not about monitoring applications of landscape ecology per se, provide valuable insights about how to monitor application effects in both terrestrial and aquatic systems. For example, monitoring should address clear questions; a limited number of key variables will probably be sufficient; thoughtful decisions must be made about how to use limited resources to measure variables that are subject to numerous sources of variation; "baseline" conditions must be defined carefully for valid interpretations of monitoring data; monitoring should produce data for various temporal and spatial scales; and these efforts should be implemented collaboratively across institutional boundaries (Ringold 2000).

To provide reliable assessments, monitoring protocols must incorporate statistical sampling standards. For instance, monitoring should include samples from temporal and spatial controls, enough samples for reasonable statistical power, and samples over sufficiently wide ranges of space and time to enable change detection. Useful information about these and associated issues of data analysis can be found in Green (1979), Cohen (1988), Hairston (1989), and Zar (1999). Monitoring should be conducted in terrestrial and aquatic systems because data on changes in both are needed to develop a holistic understanding of the effects of applications. Specific recommendations for monitoring aquatic systems can be found in Karr and Chu (1999).

Experimental model systems (EMS)—involving experiments at microcosm scales—offer a means by which we may test basic assumptions of landscape ecological theory (Wiens et al. 1997). Examples of studies in which EMS have been used to study relations between organisms and landscapes are cited in Ims (1999) and described in Gonzalez et al. (1998) and McIntyre and Wiens (1999). Because of their value for assessing causal relations, EMS may be useful in developing and refining theories that relate landscape ecology to conservation.

By considering a range of multiscale biotic and abiotic conditions, and responses of organisms to them, simulation modeling can be used to derive testable hypotheses and help formulate theories about organism-landscape relations. This approach will be especially valuable for situations in which we cannot conduct broad-scale experiments on the ground, or when we cannot implement them with enough replication for valid inferences. Simulation modeling can provide us with information that is relevant to the long temporal scales and broad spatial scales that are often at issue in conservation. Verboom and Wamelink (1999) outlined the strengths and weaknesses of various modeling approaches pertinent to linking landscape ecology and conservation.

A theoretical framework that integrates landscape ecology with various aspects of biological conservation would provide a basis for making urgent management decisions when specific empirical data for a given problem are not available. This framework also would help stimulate and guide research and thereby promote science-based field applications of landscape ecology.

25.5 Reducing Nonscientific Constraints

To ensure that personnel understand landscape-ecology concepts and how they can be used in conservation, employers can require college training in landscape ecology and management. Under the guidance of a seasoned practitioner, experiences involving field applications of landscape ecology, or use of landscape ecology in planning, decision-making, and policy formation, would also be valuable. For those already in the workforce, a series of well-designed, practical short courses may be effective. I have frequently seen advertisements for short courses on geographic information systems, modeling, and sampling techniques, but I have not seen advertisements for short courses that cover landscape ecology and its field application in conservation. Filling this void would provide additional chances to increase the number of personnel who are knowledgeable about landscape ecology and its use.

Additional monitoring personnel should be hired, or current personnel should be trained, to assess the effects of landscape-ecology applications. Such positions, even if only seasonal, will have to be incorporated into budget plans well in advance of their need to ensure that funding is in place and personnel are continuously available for long time periods or multiple seasons. Monitoring during brief periods or just one or two seasons may not sufficiently characterize the effects of applications.

Additional effort should be made to inform the general public about the need for a landscape perspective in conservation. Such information could be part of what is presented by wildlife- and land-management agencies at various outreach events. The media, especially prominent newspapers and magazines with interests in environmental issues, could address this issue. Scientific meetings that involve landscape ecology and conservation could be focal points for some of this coverage. During the last two or three decades, the extent to which school curricula have exposed children to ideas about ecology has increased. Fundamental information about how broad-scale conditions and processes can affect biodiversity could be integrated into such environmental educational efforts.

More interagency cooperation must be encouraged so that agencies jointly apply landscape ecology across legal boundaries. Expectation of this can be expressed during public comment periods for environmental impact assessments, or at public hearings on zoning, land-use planning, or conservation policy, for example. Citizens also can voice their desire for agency collaboration by calling and writing to officials at various levels of government. Professional societies can develop science-based position statements that advocate multijurisdictional collaboration. Conservation, economic, and public-relations incentives for cooperation should be identified and presented to agencies. Legal barriers to cooperation must be dissolved, and formal collaborative agreements should be established.

Noss (2000) pointed out that, largely because the applied aspects of conservation biology tend to exclude it from certain dependable sources of support, funding for conservation biology research is not regularly available. Landscape research is integral to much of conservation biology research. Accordingly, Noss's recommendations for improving funding for conservation biology research are relevant to securing money for research needed to underpin (hence ultimately increase) field applications of landscape ecology. Adapting Noss's recommendations to the context of funding for landscape-ecology research leads to a strategy of education about landscape-ecology science. Specifically, additional effort must be made to convince those with money that landscape-based conservation science is clearly needed; research involving various blends of basic and applied landscape science is capable of providing more information and insight than is either basic or applied research alone; landscape research useful to conservation will involve a wide gamut of basic and applied approaches; and human health is connected to landscape health (cf., Noss 2000).

We should vigorously support funding initiatives pertinent to acquiring, managing, and restoring habitat at broader spatial scales than current funding permits. Although some money for these purposes (e.g., Federal Aid in Wildlife Restoration Act money in the United States) has been available for a number of years, much more is needed to realize a substantial increase in field applications of landscape ecology. People can support local, state, and national funding initiatives through direct contacts with elected officials, sign-on letters from organizations (e.g., coalitions of conservation, recreation, tourism, and business groups), and outreach events aimed at informing and organizing the general public. Because individuals and private foundations that fund conservation typically prefer to support conservation action (not science) (Noss 2000), they may be good sources of money for actual field applications of landscape ecology. Recent efforts in the United States to increase funding for the Land and Water Conservation Fund and to pass the Conservation and Reinvestment Act are examples of the widespread, grass-roots efforts that are often necessary.

25.6 Prospects for Expanding Applications of Landscape Ecology

Frequently, an organism within a local area is affected not just by factors in its immediate environment, but also by broader-scale conditions and processes. We can use knowledge of such landscape influences to improve the realism of the concepts, principles, and models we use to make conservation decisions. In turn, the better our data, ecological understanding, and decision tools reflect reality, the more effective we will be as conservationists. In other words, through applications of landscape ecology, we should be able to increase our conservation effectiveness. If we expand science-based field applications of landscape ecology to include more taxa, more scales in time and space, and more levels of biological organization, we should be able to improve our conservation effectiveness with additional taxa, additional scales, and additional levels of organization.

In recent years, a number of field applications have occurred in reserve and corridor establishment, timber extraction, and in efforts to reduce the dissecting effects of roads through construction of road overpasses and underpasses (e.g., Langton 1989; Mansergh and Scotts 1989; Bennett 1990; Saunders and Hobbs 1991; Smith 1993; Meffe and Carroll 1997; Noon and Murphy 1997). But field applications have not been extensive for a wide array of other conservation issues to which landscape ecology is highly pertinent. For example, landscape ecology is relevant to rehabilitation and restoration of large land areas (Risser 1992; Hobbs and Saunders 1993; Saunders et al. 1993), but very few field applications for this purpose and scale have occurred (see Hobbs 1999). Judging from the relative absence of documented information, on-the-ground applications of landscape ecology also are uncommon or nonexistent in species-reintroduction programs, regulation of rural development, control of grazing impacts, management of recreational disturbance in wildlands, and efforts to minimize anticipated climate-change effects. Furthermore, landscape ecology has not yet been applied extensively on city, county, state, and provincial lands, or on corporate and other privately owned lands. Most field applications to date seem to have occurred on federal or national lands, or to have been initiated most frequently by federal or national agencies. In short, a vast potential for field implementation of landscape ecology remains untapped.

I believe that the possibility and effectiveness of applying landscape ecology in biological conservation will continue to improve. My optimism originates from several sources. I am optimistic because, as noted above, many field applications have already occurred. I am optimistic because of increasing recognition of the conservation benefits of applying landscape ecology; this recognition is growing not just among conservationists, but also among landscape architects and land-use planners (e.g., see Dramstad et al. 1996). I am optimistic because the number of interagency efforts to address broad-scale conservation issues seems to be increasing. I am optimistic because I have noticed that during the last decade the number of universities offering training in landscape ecology, and the number of universities and agencies hiring landscape ecologists, have increased. Finally, I am optimistic because we can use established approaches in science, education, personnel management, and public activism to reduce some of the barriers that currently limit applications. By increasing science-based applications of landscape ecology in biological conservation, we have a chance to improve the fate of natural systems—if we act now.

Acknowledgments

I thank the contributors to this book for the information they provided in their chapters; I used this information to help develop parts of the present chapter. John A. Bissonette, Frank B. Golley, and Monica G. Turner supplied valuable advice for improving this chapter, and I am grateful for their recommendations.

References

Bennett, A.F. 1990. *Habitat Corridors: Their Role in Wildlife Management and Conservation.* Melbourne, Australia: Department of Conservation and Environment.

Cohen, J. 1988. *Statistical Power Analysis for the Behavioral Sciences.* Second Edition. Hillsdale, New Jersey: Lawrence Erlbaum.

Dramstad, W.E., Olson, J.D., and Forman, R.T.T. 1996. *Landscape Ecology Principles in Landscape Architecture and Land-Use Planning.* Washington, DC: Island Press.

Gonzalez, A., Lawton, J.H., Gilbert, F.S., Blackburn, T.M., and Evans-Freke, I. 1998. Metapopulation dynamics, abundance, and distribution in a microecosystem. *Science* 281:2045–2047.

Green, R.H. 1979. *Sampling Design and Statistical Methods for Environmental Biologists.* New York: John Wiley and Sons.

Gutzwiller, K.J., and Barrow, W.C., Jr. 2001. Bird-landscape relations in the Chihuahuan Desert: coping with uncertainties about predictive models. *Ecol. Appl.* 11:1517–1532.

Hairston, N.G. 1989. *Ecological Experiments: Purpose, Design, and Execution.* New York: Cambridge University Press.

Hargrove, W.W., and Pickering, J. 1992. Pseudoreplication: a *sine qua non* for regional ecology. *Landsc. Ecol.* 6:251–258.

Hobbs, R. 1997. Future landscapes and the future of landscape ecology. *Landsc. Urban Plan.* 37:1–9.

Hobbs, R.J. 1999. Restoration ecology and landscape ecology. In *Issues in Landscape Ecology,* eds. J.A. Wiens and M.R. Moss, pp. 70–77. Guelph, Ontario, Canada: International Association for Landscape Ecology.

Hobbs, R.J., and Saunders, D.A., eds. 1993. *Reintegrating Fragmented Landscapes: Towards Sustainable Production and Nature Conservation.* New York: Springer-Verlag.

Ims, R.A. 1999. Experimental landscape ecology. In *Issues in Landscape Ecology,* eds. J.A. Wiens and M.R. Moss, pp. 45–50. Guelph, Ontario, Canada: International Association for Landscape Ecology.

Karr, J.R., and Chu, E.W. 1999. *Restoring Life in Running Waters: Better Biological Monitoring.* Washington, DC: Island Press.

Langton, T.E.S., ed. 1989. *Amphibians and Roads.* Bedfordshire, United Kingdom: ACO Polymer Products.

Lovejoy, T.E., Bierregaard, R.O. Jr., Rylands, A.B., Malcolm, J.R., Quintela, C.E., Harper, L.H., Brown, K.S. Jr., Powell, A.H., Powell, G.V.N., Schubart, H.O.R., and Hays, M.B. 1986. Edge and other effects of isolation on Amazon forest fragments. In *Conservation Biology: The Science of Scarcity and Diversity,* ed. M.E. Soulé, pp. 257–285. Sunderland, Massachusetts: Sinauer Associates.

Macnab, J. 1983. Wildlife management as scientific experimentation. *Wildl. Soc. Bull.* 11:397–401.

Mansergh, I.M., and Scotts, D.J. 1989. Habitat continuity and social organization of the mountain pigmy-possum restored by tunnel. *J. Wildl. Manage.* 53:701–707.

Margules, C.R. 1992. The Wog Wog habitat fragmentation experiment. *Environ. Conserv.* 19:316–325.

McIntyre, N.E., and Wiens, J.A. 1999. Interactions between habitat abundance and configuration: experimental validation of some predictions from percolation theory. *Oikos* 86:129–137.

Meffe, G.K., and Carroll, C.R., eds. 1997. *Principles of Conservation Biology.* Second Edition. Sunderland, Massachusetts: Sinauer Associates.

Noon, B.R., and Murphy, D.D. 1997. Management of the Spotted Owl: the interaction of science, policy, politics, and litigation. In *Principles of Conservation Biology,* Second Edition, eds. G.K. Meffe and C.R. Carroll, pp. 432–441. Sunderland, Massachusetts: Sinauer Associates.

Noss, R.F. 2000. Science on the bridge. *Conserv. Biol.* 14:333–335.

O'Neill, R.V. 1999. Theory in landscape ecology. In *Issues in Landscape Ecology,* eds. J.A. Wiens and M.R. Moss, pp. 1–5. Guelph, Ontario, Canada: International Association for Landscape Ecology.

Ringold, P. 2000. What limits regional stream monitoring design? *Bull. Ecol. Soc. Am.* 81:143–145.

Risser, P.G. 1992. Landscape ecology approach to ecosystem rehabilitation. In *Ecosystem Rehabilitation, Volume 1: Policy Issues,* ed. M.K. Wali, pp. 37–46. The Hague, The Netherlands: SPB Academic Publishing.

Saunders, D.A., and Hobbs, R.J., eds. 1991. *Nature Conservation 2: The Role of Corridors.* Chipping Norton, Australia: Surrey Beatty and Sons.

Saunders, D.A., Hobbs, R.J., and Ehrlich, P.R., eds. 1993. *Nature Conservation 3: Reconstruction of Fragmented Ecosystems—Global and Regional Perspectives.* Chipping Norton, Australia: Surrey Beatty and Sons.

Smith, D.S. 1993. Greenway case studies. In *Ecology of Greenways: Design and Function of Linear Conservation Areas,* eds. D.S. Smith and P.C. Hellmund, pp. 161–208. Minneapolis: University of Minnesota Press.

Verboom, J., and Wamelink, W. 1999. Spatial modeling in landscape ecology. In *Issues in Landscape Ecology,* eds. J.A. Wiens and M.R. Moss, pp. 38–44. Guelph, Ontario, Canada: International Association for Landscape Ecology.

Walters, C.J., and Holling, C.S. 1990. Large-scale management experiments and learning by doing. *Ecology* 71:2060–2068.

Wiens, J.A. 1999a. The science and practice of landscape ecology. In *Landscape Ecological Analysis: Issues and Applications,* eds. J.M. Klopatek and R.H. Gardner, pp. 371–383. New York: Springer.

Wiens, J.A. 1999b. Toward a unified landscape ecology. In *Issues in Landscape Ecology,* eds. J.A. Wiens and M.R. Moss, pp. 148–151. Guelph, Ontario, Canada: International Association for Landscape Ecology.

Wiens, J.A., Schooley, R.L., and Weeks, R.D., Jr. 1997. Patchy landscapes and animal movements: do beetles percolate? *Oikos* 78:257–264.

With, K.A. 1999. Landscape conservation: a new paradigm for the conservation of biodiversity. In *Issues in Landscape Ecology,* eds. J.A. Wiens and M.R. Moss, pp. 78–82. Guelph, Ontario, Canada: International Association for Landscape Ecology.

Zar, J.H. 1999. *Biostatistical Analysis.* Fourth Edition. Upper Saddle River, New Jersey: Prentice-Hall.

Index

Italicized page numbers indicate pages with a figure or table.

B

C

F

G

H

I

J

K

M

N

O

Q

R

S